Structural Renovation of Buildings

Structural Renovation of Buildings

Methods, Details, and Design Examples

Alexander Newman

McGraw-Hill

New York   Chicago   San Francisco   Lisbon   London
Madrid   Mexico City   Milan   New Delhi   San Juan
Seoul   Singapore   Sydney   Toronto

Cataloging-in-Publication Data is on file with the Library of Congress

McGraw-Hill

A Division of The McGraw·Hill Companies

Copyright © 2001 by The McGraw-Hill Companies, Inc. Printed in the
United States of America. Except as permitted under the United States
Copyright Act of 1976, no part of this publication may be reproduced
or distributed in any form or by any means, or stored in a data base or
retrieval system, without the prior written permission of the publisher.

9 0 BKM/BKM 0 9 8

ISBN-13: 978-0-07-047162-7

ISBN-10: 0-07-047162-2

*The sponsoring editor of this book was Larry Hager, the editing
supervisor was Steven Melvin, and the production supervisor was Sherri
Souffrance. It was set in New Century Schoolbook by Kim Sheran and
Paul Scozzari of McGraw-Hill's Professional Book Group Hightstown
composition unit.*

Printed and bound by Book-Mart Press, Inc.

McGraw-Hill books are available at special quantity discounts to use as
premiums and sales promotions, or for use in corporate training programs.
For more information, please write to the Director of Special Sales,
Professional Publishing, McGraw-Hill, Two Penn Plaza, New York, NY
10121-2298. Or contact your local bookstore.

To Margarita, Laura, and David.

Contents

Preface

Structural engineers often fret about renovation projects. They complain about owners and architects who are proposing to convert old buildings to other uses—without having the original construction drawings or an adequate budget for comprehensive testing and investigation. This sentiment probably has been repeated countless times in design offices around the country.

Whenever an existing building is modified, strengthened, or repaired, it presents an enigma. What kind of structure lies behind its wall and floor finishes? What condition is it in? Is it safe to modify—or even to disturb? Even after extensive field investigation, the answers to some of these questions may remain uncertain. And, when conditions are uncertain, additional services by designers and change orders by contractors are likely to follow. By contrast, with new construction, such headaches are largely sidestepped.

In the past, outmoded and functionally obsolete buildings were routinely demolished; today, they are often remodeled and restored. Because owners and the public often share an ethos of conservation and adaptive reuse, the volume of renovation work is bound to increase. In fact, as zoning and environmental regulations make it ever more difficult to construct new buildings, especially in cities, renovation may be the most practical course of action. Also, as the concept of recycling now extends beyond collecting glass bottles and old newspapers, recycling of buildings can be viewed as a way to conserve resources and reduce landfill demand.

Surprisingly, the reference literature on structural renovation of buildings is quite scarce, and design schools and textbooks tend to ignore the topic. Construction professionals must therefore rely on information presented in technical journals, such as those on concrete repair, masonry maintenance, and historic preservation; attend specialty conferences; and learn from one another. To the extent broader-circulation construction periodicals cover renovation, they tend to focus on prestigious projects and unique challenges affecting landmark structures. What is missing is a comprehensive textbook that addresses the issues that arise in the structural renovation of ordinary buildings.

This book was written to fill the vacuum. It is intended to give structural engineers, architects, facility managers, and building officials a single condensed source of information about all aspects of structural renovation of buildings—all buildings. Because it avoids impenetrable technical terminology, the discussion should be easily followed by architects, graduate students, and knowledgeable contractors who are seeking to broaden their knowledge of building renovation methods. Much of the material applies to buildings of any size—and to structures that would not even be characterized as buildings.

The author understands that a hundred books of this size could not exhaustively treat this immense subject. This book is a necessary compromise between a rudimentary primer and a technical monograph. It addresses the structural issues that are most commonly encountered in building renovation. While it provides only general guidance on subjects for which there are adequate referenced sources available, it focuses on issues that are less likely to be fully explored and on those that invite controversy. Rehabilitation of historic buildings, for example, is well covered by others; although relevant information is dispersed throughout this book, its thrust is broader. Conversely, the chapter on renovation of pre-engineered buildings contains material that is not found in any other book.

Structural Renovation of Buildings was written by a practicing structural engineer and was not intended as a theoretical exercise—differential equations and esoteric jargon are conspicuously absent. Instead, the book attempts to clearly identify problems that are often encountered during renovations, explain why they occur, and suggest how they can be solved. It offers practical tips, design details, examples, and reference data to help design professionals accomplish tasks ranging from feasibility studies to preparation of construction documents. This information has been collected by the author—and by several contributors—over many years from hundreds of sources. The material is arranged by chapters dealing with various building materials. Throughout the book, every attempt has been made to keep the presentation clear, readable, and enjoyable.

This wealth of material could not have been assembled without a wealth of assistance. The author is indebted to Kevin Chan for expert illustrations. He appreciates the thoughtful review comments by LeMessurier Consultants, Inc., and Prof. Howard I. Epstein of The University of Connecticut. He is also grateful to the many participants, whether acknowledged by name in the text or not, who contributed the material and helped make this book a reality.

Alexander Newman

The Challenge of Renovation

1.1 Terminology

Philosophers have long recognized that a fruitful discourse requires agreement on the terms of discussion. Various "R words" are used in this book to describe building renovation activities; these words sound similar but refer to slightly different concepts. Since there is no universal agreement on the meaning of these terms, the following common definitions are used here:

Renovation is a process of substantial repair or alteration that extends a building's useful life.

Rehabilitation is an upgrade required to meet the present needs; it implies sensitivity to building features and a sympathetic matching of original construction.

Restoration is a more restrictive term than rehabilitation; it suggests replicating the structure as originally built. This term is most commonly applied to buildings of historical value.

Repair is a process of reconstruction and renewal of the existing buildings, either wholly or in part.

Retrofit is an upgrading of certain building systems, such as mechanical, electrical, or structural, to improve performance, function, or appearance.

Remodeling is essentially the same as renovation; the term is commonly applied to residential construction.

Sometimes the word *renovation* is used to refer to inappropriate alterations to historic properties, as opposed to *rehabilitation,* which implies a deferential approach to the historic fabric. *Renovation,* as used here, is intended to encompass any measures needed to make a building stronger and more appealing.

1.2 When to Renovate

A confluence of several factors, whose relative importance varies from project to project, usually establishes the need for building renovations. Some of the most common ones are described here. Whatever the reason for renovation, however, structural work is often needed to accomplish the objectives of the project.

1.2.1 Change in use

Functionally inadequate and underutilized buildings are common candidates for rehabilitation. Numerous old mill buildings throughout the country have been turned into office and research-and-development space, and numerous aged downtown office buildings have been converted into apartments and condos.

The historic Board of Trade Building in Boston, originally constructed in 1906, is a case in point. It was completely gutted and converted into apartments (Fig. 1.1). The project was privately financed and without tax credits, and it was completed within a year of acquisition. What made it a success?

First, the building's location in the heart of Boston's financial district made it attractive to executives visiting the large companies nearby. Second, its impressive and heavily decorated façade provided a certain cachet that newer buildings simply could not match. Last, but not least, its plan and shape were readily adaptable to the intended use. According to the project architect, the 11-story H-shaped building provided ample natural light and ventilation to all dwelling units.[1]

Another famous Boston building that has undergone a drastic change of use is the Custom House. Constructed between 1837 and 1847—the foundation alone took 3 years to build—this structure was the focus of mercantile activity in the port of Boston. A tower was added to it in 1915, making it the tallest building in the city (Fig. 1.2). The Custom House was originally located on the water, where duties were paid, but as the harbor was gradually filled in, the building lost its proximity to the water. Eventually, the original use was abandoned, and the property was sold to private developers. It was converted to offices in the late 1990s. Despite the fact that the Custom House was not very suitable for these occupancies because its floor plates were

Figure 1.1 The historic Board of Trade Building in Boston was converted into apartments.

Figure 1.2 The Custom House, one of Boston's landmarks, was converted into offices

very small, its outstanding location and rich history have helped overcome this limitation, and the building is now revitalized by its new uses.

Changes in use are commonplace and sometimes rather frequent. A historic building in Golden, Colorado, shown in Fig. 1.3, is a good example. It was constructed in the mid-1860s by a famous Colorado pioneer, William A. H. Loveland, and is still known as the Loveland Building. It housed the first Colorado legislature (at the time, Golden was competing with Denver to be the new territory's capital). Later on, the structure was converted to classrooms for the Colorado School of Mines, housed railway offices, and was transformed into a dry-goods store. Now, it is a popular restaurant. It has been renovated numerous times, most recently in 1993.

Vacant buildings are prime candidates for rehabilitation. After the 1991 Los Angeles riots, local government vowed to improve living conditions in the area by converting several vacant county buildings into youth centers. In one such conversion, a former jail was reorganized into a mix of educational and recreational facilities, including a library, cafeteria, gym, and computer room. Cells were transformed into tutoring rooms, with the bars retained in one of them to suggest the consequences of bad behavior. The $400,000 project included installation of bulletproof paneling and graffiti-resistant exterior metal finishes.[2]

Figure 1.3 The Loveland Building, Golden, Colorado, has seen many changes in use and has undergone numerous renovations.

Even when a building's use stays essentially the same, its interior may benefit from circulation improvements, new stairways and escalators, and the development of new lobby and atrium spaces. Such work requires rearrangement of structural members, sometimes extensive. For example, during renovation of the Chicago Board of Trade Building, new steel transfer trusses were installed to support the columns above, thereby creating an uninterrupted trading floor.[3]

Sometimes, structural improvements are necessitated by *added* uses. Many church steeples now contain telecommunications equipment. Cellular-service providers have realized that many churches, in addition to having steeples, are ideally situated on hills—a perfect place for the transmission antennas, which otherwise are difficult to locate because of local opposition. Another factor that makes the steeples attractive for housing antennas is that they are typically made of wood, which does not interfere with wireless signals. These old structures require engineering evaluation, and quite often strengthening, to qualify as structural supports for the expensive gear. The cash-strapped churches are happy to embrace the new substantial source of revenue.[4]

1.2.2 Upgrading mechanical and electrical systems

Improvements to heating, ventilation, and air-conditioning (HVAC) systems are among the most common reasons for building renovations. The requirements for climate control contained in today's codes are quite different from those used even two or three decades ago. Building owners are increasingly recognizing that improvements in indoor air quality bring substantial benefits in terms of increased worker productivity—and that is in addition to any reductions in operating costs. One 1994 study documented eight cases in which improvements in energy-efficient lighting, heating, and cooling increased worker productivity, decreased absenteeism, and improved the quality of work.[5]

A typical HVAC retrofit involves installation of cooling towers, chillers, fans, air shafts, and ductwork, all of which typically require at least some structural work. Unfortunately, the upgrading of mechanical systems often entails substantial changes to the building envelope and appearance that are not always considered appropriate in historic structures. Sometimes they are not allowed at all, and the occupants of many historic buildings are left to suffer the consequences. A good example is the Statue of Liberty, one of the most important landmarks in this country. As reported in *The Wall Street Journal,* the historic status of the copper statue prevents it from being air-conditioned, despite its stifling summer heat and some 400 to 500 resulting medical problems annually, ranging from heat exhaustion to heart attacks.[6]

Electrical systems in vintage buildings also frequently require upgrading. To stay competitive, many owners elect to upgrade communication lines in their structures. This work is normally done in conjunction with general interior renovation programs and rarely affects structure.

1.2.3 Deterioration of building envelope

Along with HVAC upgrades, leaking roofs and walls are among the most common reasons for building renovations. Other building envelope problems include buckled or deteriorated wall structure, failing sealants, and cracking. As discussed in Chap. 13, age, prolonged exposure to the elements, and deferred maintenance all contribute to gradual disintegration of building façades and roofing.

Masonry façades are susceptible to mortar deterioration, cracking caused by settlement or lack of movement joints, efflorescence, and other troubles. The intricate stone and masonry details of many century-old buildings often suffer from the effects of pollution. If their soft mortar is not repointed in a timely manner, erosion of it allows the intrusion of water, which may freeze within the wall, expand in the process, and cause further crumbling and deterioration. Yet, properly designed, constructed, and maintained, solid masonry walls can last practically forever, and many such structures have survived for millennia. As tourist guides explain, it is not natural degradation that caused the damage to the Roman Colosseum but poaching of its stonework by succeeding generations for other uses.

In our time, the prevalent method of masonry construction is brick veneer, which requires periodic maintenance and replacement of joint sealants. Without this, the joints become conduits for water infiltration and structural deterioration.

Similarly, many building roofs have been ruined by a lack of proper maintenance. An unnoticed small leak can lead to roof insulation's becoming saturated with water. If the water cannot readily escape, it may attack and corrode the structural substrate. By the time rust stains on the ceiling are noted and investigated, a minor problem may have mushroomed into a major roof removal and replacement project involving structural work.

Sometimes cladding rehabilitation must be followed by other structural renovations. The high-rise building shown in Fig. 1.4 was clad in dry-joint granite panels that did not provide effective moisture protection. The panels were eventually removed, trimmed, and reinstalled to incorporate joint sealants.

1.2.4 Structural damage or failure

Damage to structural elements can be attributed to various causes, of which corrosion and weather-related deterioration are the most

Figure 1.4 This high-rise building not only had its exterior wall panels redone, but was also equipped with fluid inertial dampers to help it resist earthquakes.

common. A typical example is corrosion of reinforcing bars in concrete, which leads to the concrete's cracking and spalling. This type of corrosion frequently occurs in parking garages, where the use of deicing salts exposes concrete decks to chloride-ion infiltration. Concrete structures can also be damaged by persistent leakage (Fig. 1.5). Other materials are even more vulnerable to deterioration: Steel structures can suffer from the effects of oxidation (rusting), while damp wood is sensitive to moisture and can be attacked by fungi and insects.

Structural damage may also be caused by overloading. For example, floors designed for light loading may fail when heavy materials and

Figure 1.5 This concrete column has been damaged by persistent leakage through a broken drain pipe.

equipment are placed on them. An enlightened owner will consult a structural engineer to evaluate the load-carrying capacity of the floor before placing any unusual loads on it; a shortsighted one will engage a structural engineer to evaluate the resulting damage. (The oblivious owner does not even heed the signs of structural distress and becomes alarmed only when a collapse is imminent.)

Other sources of structural distress include hurricanes and earthquakes, which leave cracked, crumbled, and distorted buildings.

Television screens invariably show teams of engineers and building officials evaluating damaged structures and making decisions as to whether to condemn or renovate them. The violent shaking of an earthquake may also result in buildings that appear undamaged but experience a host of serviceability problems arising from distortion, such as inoperable windows, binding doors, and buckled floors and partitions. Dealing with distortion is more difficult than fixing the obvious signs of failure, as centuries of efforts to straighten the famous Leaning Tower of Pisa, which started to lean after its third story was constructed, have demonstrated.

Occasionally, structural damage can be traced to the original construction, either because a building's design was deficient in some respect or because the execution was wanting. Frank Lloyd Wright's world-famous Fallingwater now has its dramatic cantilevered floors shored up (until a more elegant solution is in place) because of excessive sagging attributable to inadequate support and long spans. Despite having been warned by structural engineers back in 1936 about potential problems with his design, the architect proceeded anyway, and now the caretakers of the landmark have the task of fixing the structure.[7]

The most unfortunate cause of structural damage or failure is sabotage. Urban terrorists who think that blowing up buildings will help them accomplish their goals can leave little to restore. On the other hand, the effects of a less powerful attack might be able to be overcome by structural rehabilitation. To safeguard against the threat of sabotage, a renovation program may involve steps to improve the structural resistance of the building to blast loading, although a more common and cost-effective method of preventing bombing attacks is to limit access to the critical areas by erecting traffic barriers and redirecting the traffic flow.

1.2.5 Upgrading buildings for lateral loads

To prevent damage from hurricanes and earthquakes, many local building codes require renovations to be accompanied by improved resistance to lateral loads. In some areas of high seismicity, such as Los Angeles, local ordinances require *all* buildings of certain types to be seismically strengthened, even in the absence of renovations. (In Los Angeles, this ordinance was later declared advisory in nature and compliance with it voluntary.)

Lateral-load upgrading involves more than installation of new shear walls or braced frames. Quite often, seismic upgrading involves bracing and anchoring nonstructural elements, such as partitions, lights, HVAC equipment, exterior parapets and ornamentation, and even file cabinets and bookshelves. A complete retrofit of a building for lateral loads may become quite complex and involve specialized products and expertise. For example, Fig. 1.6 shows the historic Hotel Woodward, in Woodland, California. Constructed in 1928, the building is on the National Register of Historic Places. It was one of the first buildings in North America to be seismically retrofitted with fluid viscous dampers, a solution that allowed preservation of its exterior, while enabling it to resist the code-mandated levels of earthquake loading.

A spectacular example of renovation for lateral loads occurred in 1978, when LeMessurier Consultants, one of the country's premier structural engineering firms, decided to strengthen the newly constructed Citicorp Center in New York City, which it had designed. An innocent question posed by an engineering student started an internal investigation into the strength of the structure, during which the design firm's chairman became convinced that the connections in the lateral-load-resisting braces did not have sufficient capacity. This startling conclusion culminated in a dramatic sequence of events profiled in *The New Yorker*.[8] The designers brought the bad news to the owner of their own accord and suggested a program of strengthening, which was carried out at night in order to minimize the disruption. Doing the right thing in this difficult situation not only did not diminish, but actually enhanced, the reputation of William LeMessurier and his firm.

Figure 1.6 The historic Hotel Woodward was one of the first buildings in North America to be seismically retrofitted with viscous dampers. (*Taylor Devices Inc.*)

1.2.6 Reducing serviceability problems

A building suffering from serviceability problems may require renovation even though its strength is adequate. As discussed previously, inoperable windows, binding doors, and buckled floors and partitions can be the result of earthquake damage; however, these problems are more commonly caused by an overly flexible structure. A perfectly safe building may be perceived as dangerous if the beams sag and windows break. In addition to consequences of such excessive vertical framing deflections, serviceability concerns can also arise when the building is too flexible in the horizontal direction, that is, when its lateral drift under strong winds is excessive. A building swaying in the wind is rarely considered safe by the occupants.

Another common serviceability problem involves annoying floor vibrations. What used to be a frequent complaint in office buildings tenanted by dancing or aerobic studios has become increasingly common in other structural steel buildings as well. What are the reasons for these increased vibration problems? Sophisticated structural design methods, the use of high-strength steels, and composite-design construction all contribute to more slender and lighter framing. This lightweight framing tends to have very low natural frequencies, which happen to coincide with the harmonic frequencies induced by the building's occupants.[9] Correcting the natural frequency of framing is very difficult, especially in an occupied building.

1.2.7 Government mandates

Sometimes building renovations are driven by government regulations. For example, all buildings with public access must conform to the guidelines of the Americans with Disabilities Act (ADA). Installation of handicapped-accessible ramps, removal of asbestos, and other modifications are common sights in schools and other public buildings throughout the country (Fig. 1.7).

Multistory public buildings, such as courthouses, may be required to have interior elevators installed. Local building ordinances—Local Law 5 in New York City is one—require installation of certain fire-protection systems, such as sprinklers, fire alarms, and smoke-containment devices. Mandated asbestos abatement and lead-paint removal often serves as a catalyst for comprehensive general renovation work.

1.2.8 Undoing the effects of botched renovations

Unfortunately, poorly conceived and poorly executed renovations frequently necessitate remedial work later. The 95-year-old Haupt

Figure 1.7 Like many other schools throughout the country, this one is undergoing renovations.

Conservatory in New York City, the largest public conservatory in the country, underwent five renovation efforts to correct maintenance and environmental problems, including glass that slipped away from its seals, jammed windows, leakage, and structural corrosion. The sixth renovation effort, completed in 1997, consisted of a comprehensive rehabilitation program which may finally have solved most of the building's problems.[10]

Similarly, when the 200-year-old Octagon in Washington, D.C., was "renovated" in 1954, the original second-floor wood joists were replaced with steel framing and concrete deck. The original joists were framed directly into the brick walls, but were also interconnected by handmade mortise-and-tenon joints which allowed for their free expansion and contraction. The newer rigid concrete diaphragm restricted this movement of the building and was blamed for the ensuing brick cracking.[11] A later restoration project undertaken in the 1990s changed the rigid concrete diaphragm to a system that replicated the flexibility of the original structure.

Incidentally, wood framing in general has a large share of the problems stemming from thoughtless renovation activities. How many times have we seen cuts and notches in wood beams and joists to allow for passage of plumbing pipes or electrical conduits? (Some methods of fixing this damage are explored in Chap. 8.)

Figure 1.8 illustrates an unfortunately all too common example of renovations that compromise the building's original character. The massive stone archways of the historic Fort Trumbull in New London, Connecticut, constructed around 1850, were filled with brick to "convert" the majestic fort into laboratory space. Decades later, a new renovation program was commissioned to restore the fort to its original condition and reconvert it into a tourist attraction.

A more tragic example is the partial roof collapse in the famed Basilica di San Francesco in Assisi, Italy. On September 26, 1997, a strong earthquake struck the region and cracked the roof vaults of this thirteenth-century masonry structure, which had withstood thousands of tremors before. During the next round of ground shaking, which happened to occur when a delegation of friars and technical staff was evaluating the damage, a part of the roof collapsed and killed four. Some professionals suggested that the tragedy was precipitated by a roof restoration project undertaken in the 1950s, in which the

Figure 1.8 These "renovations" compromised the character of the historic Fort Trumbull in New London, Connecticut, and had to be undone later. (*Maguire Group Inc.*)

original timber roof beams were replaced with a brittle concrete structure less capable of absorbing seismic shocks.[12]

Sometimes renovations are needed in order to undo the effects of previously abandoned *partial* renovation efforts. Such a situation arose when a telephone company purchased a former factory building. The structure had been constructed in the 1930s and partly renovated in the 1980s by a design-build firm. The abandoned renovation work included construction of a mezzanine—without, as was later discovered, any column footings. Also, in an effort to install strip windows and another floor, much of the building's lateral-load bracing had simply been removed. So, the first design effort of a later renovation project was to provide new foundations for the mezzanine and lateral bracing for the building.[13]

1.2.9   Taking advantage of credit for historic rehabilitation

Tax credits for historic rehabilitation, to promote urban renewal and preservation of historic properties, are offered by the federal government and by some state governments. Without the credits, some rehabilitation projects would not be feasible. Many states have enacted programs intended to facilitate rehabilitation of owner-occupied historic houses.

1.3   Beginning a Renovation Project

Each renovation project is unique in one respect or another, but there are many common issues that have to be tackled at the beginning of most renovations.

1.3.1   Assessing economic feasibility and political factors

The project must make economic sense. Renovations that would bring the building's value well above its neighbors' might be ill advised, at least in residential construction, although for well-situated properties higher spending limits may be justified. For building restoration projects, adequate funds must be available. If the project relies on rental income or involves a condo conversion, pro forma financial statements are prepared prior to any brick-and-mortar work.

The budgets for renovations of historic structures should include the cost of complying with requirements for historic preservation. Even if the building is just a familiar landmark in the neighborhood, any change to its appearance is likely to be fought by local conservation groups, who might prefer restoration instead.[14]

Whether or not the building exterior stays the same, some public opposition to any change in occupancy, based on changes in traffic patters, sewage usage, and similar factors, may be encountered. The project budgets and schedules must reflect these realities. The owners and designers of renovation projects are well advised to keep community representatives informed of the renovation plans and to form working relationships with them. The societal, economic, and political implications of building renovation are further examined by Chandler.[15]

1.3.2 Condition assessment

Before a lot of effort is invested in design, the condition of the building must be assessed. A survey of building condition is typically conducted by a multidisciplinary team that includes architects and structural, mechanical, and electrical engineers. Asbestos and lead-paint surveys may be performed by specialized consultants or testing agencies. During this survey, the participants verify whether the building is generally constructed in conformance with the original construction documents, if these are available, or sketch out some of the observed building elements and note their condition.

An architectural survey might include verifying or generating a building layout and checking the overall condition of the roof, windows, doors, elevators, and finishes. The inspecting architect evaluates the existing building hardware for functionality and for compliance with ADA and other regulations, checks the egress routes for compliance with the present codes, and examines the condition and weathertightness of doors and windows.

Mechanical and electrical surveys include general verification or investigation of these systems. Whether the existing mechanical system can be reused depends on many factors. Even a relatively inefficient system is often retained if it works well and if its removal would entail asbestos-abatement work. (Asbestos in existing pipe insulation may be relatively inexpensively encapsulated in situ rather than being laboriously removed.[16]) The condition of an existing mechanical system is of some interest to structural engineers, because a new system generally requires new equipment supports and a fair number of floor openings.

The structural survey, of main importance to the readers of this book, is intended evaluate, the general condition of the building structure and identify any areas of deficiency. This work requires a certain level of curiosity, knowledge of old structural systems, and mental alertness. All too often condition assessment is delegated to junior engineers, when what is needed is the trained eye of an experienced practitioner. A structural survey is generally conducted starting from the roof and proceeding down to the foundations; it includes all visible roof, floor, and wall framing. Chapter 2 describes various aspects and techniques of this process.

1.3.3 The multidisciplinary design effort

A program of code compliance includes determination of code require- ments for structural work. As discussed in Sec. 1.5, these requirements can be far from straightforward. Some discussions and negotiations with building officials might be involved, sometimes quite prolonged and requiring variances from local building and zoning regulations. A preliminary design, at least, has to be completed to assist in these efforts and to illustrate the proposed renovation solutions. To save time, some owners may elect to proceed with the final design, hoping that the building permit will be eventually granted.

In a successful building renovation project, the contributions of vari- ous team members are coordinated by bringing the team together for periodic project meetings. Some examples of common coordination problems include the HVAC designer's neglecting to inform the struc- tural engineer of the need for floor openings for mechanical ducts; the structural engineer's failure to inform the architect of the need for new wall cross-bracing; and the architect's failure to alert the structural engineer of planned openings in shear walls. The earlier these and many other potential coordination conflicts are resolved, the better the chances of bringing in the project within budget and on schedule.

1.4 Typical Structural Challenges

Certain structural design issues are repeated in one renovation project after another, and it is worthwhile to review some of them. In this sec- tion, we take a broad overview of some common structural challenges; additional information on many of them is provided in other chapters.

1.4.1 Dealing with archaic and proprietary structural systems

Many old buildings, especially those constructed before World War II, uti- lize proprietary structural systems that are no longer in use. Design data for many of these systems are difficult to find. Even if it is available, the information may be of little use, because some of the systems were empir- ically designed and therefore do not readily lend themselves to analysis. This should come as no surprise, as most structures in human history were constructed on the basis of field experience. How can one evaluate the load-carrying capacity of archaic framing? Undoubtedly, countless structural engineers involved in building renovations have struggled with this question at one time or another. This is a difficult issue indeed.

When a proprietary system involving structural steel, open-web joists, or engineered wood is discovered in the field, the sizes and spacing of the framing members can be measured and documented. Based on field measurements of the framing, it might be possible to determine its design designation and load-carrying capacity. The capacities of open-web joists can be found if their sizes and series are established and relevant design load tables are available. Structural steel beams of unknown sizes can be readily measured and an approximate designation determined from the AISC publications discussed in Chap. 3. Properties of engineered wood systems can be estimated based on information contained in manufacturers' literature.

For concrete framing, the situation is more complicated. In some cases it may be possible to determine the bar sizes, slab thicknesses, and concrete strengths using some of the methods described in Chap. 2, and then to analyze the framing. Otherwise, making a load test—an expensive and somewhat dangerous procedure—may be the only solution.

1.4.2   Reinforcing floor and roof framing

If analysis or load tests indicate that the structure is inadequate for the proposed loading, the framing can be strengthened. Structural steel members can be reinforced by adding new members between the existing members, by field-welding new sections, by introducing composite action with a concrete floor, or by other methods discussed in Chap. 3. Reinforced-concrete members are more difficult to strengthen, but it is possible. Methods of upgrading concrete framing include adding steel members or reinforcing sections, cutting down the span, and external post-tensioning (see Chap. 4). Wood structures can be strengthened by adding new members or reinforcing, as explained in Chap. 8.

1.4.3   Finding analytical ways to increase design loading on structure

Strengthening is the most direct way of upgrading an existing structure for a new design loading. Before starting on this road, however, efforts should be made to determine whether the existing framing has some additional load-carrying capacity and can support an increased load without reinforcing. Quite often, existing floors indeed have some extra capacity—it is roofs that tend to present problems. Design provisions for snow loading have undergone significant changes over the years, including snow-drift requirements for roofs adjacent to elevation changes.

There are many methods of "unlocking" the reserve capacity of building framing, some of which are discussed in the chapters dealing with specific materials. The investigation may be difficult and time-consuming, requiring a lot of design and testing efforts, but it is still usually less expensive than actually strengthening the members. Unfortunately, sometimes the owners resist paying for extensive analysis, and the design fees are insufficient to permit detailed investigation.

The problem is particularly acute when renovations of small but complex structures are contemplated by public-sector owners. The proposed work could have a rather small construction cost but require extensive design efforts. If the owners adhere to artificially set upper limits on the design services—say, equal to 6 percent of the construction cost—they will not allow the value of the portion of the design services involving analysis and investigation to exceed 30 percent of the construction cost. And yet, in some small but demanding projects (for example, the project mentioned in Sec. 1.7.1), these kinds of design efforts have been required. Ironically, if the engineers are diligent and actually succeed in unlocking the hidden capacity of the structure—justifying the status quo without a need for any strengthening—there will be no construction work relating to structure at all, and the ratio of structural design fee to construction cost will be infinity!

1.4.4 Repairing deteriorated structural members

Repairing damaged framing is among the most common renovation tasks. Some effects of deterioration have already been noted: corrosion of reinforcing bars in concrete structures, rusted steel members, damage to wood by rot and insects. Some specific methods of repairing various construction materials are discussed in the appropriate chapters of this book. For example, Chap. 5 is devoted to repair of deteriorated concrete, while Chap. 8 deals with repairing wood members.

1.4.5 Upgrading lateral-load-resisting systems

Many old buildings lack rationally designed structural systems to resist lateral loads. Even if such systems are present, the levels of loading for which they were designed are likely to be considered inadequate by today's standards. Building codes often require that a general renovation program exceeding a certain monetary threshold include an upgrade of the lateral-load-resisting system. This effort is frequently the most difficult and expensive part of structural renovations.

A typical lateral-load upgrading project could include installation of new shear walls or braced frames, strengthening of floor and roof diaphragms, and provision of new connections between the building floors and walls. Complex buildings could require more sophisticated solutions, such as base isolation and installation of tuned-mass dampers. A program of seismic retrofit can also include seismic bracing of nonstructural elements—partitions, parapets, lights, bookshelves, and the like. Chapter 11 addresses the issue of upgrading lateral-load-resisting systems.

1.4.6   Making floor, roof, and wall openings and filling existing openings

Most ·building renovation projects include upgrading of HVAC systems. More often than not, this work entails making new floor, roof, and wall openings for ducts, conduits, and pipes. Making floor openings between the beams in one-way concrete slabs or in metal deck is relatively straightforward; the situation becomes more complicated in two-way concrete slabs, as discussed in Chap. 4.

Running a large heating duct through a major floor beam is not a course of action most structural engineers would favor, but sometimes there is no other choice. The existing building might offer insufficient headroom to install new mechanical pipes or ducts under the beams, and making web openings in existing beams might be the only alternative. Depending on the relative sizes of beams and openings, the solution could be easy or tortured. Chapter 3 discusses the challenge of making web openings in steel members.

Making openings of moderate size in loadbearing walls is relatively easy if the walls are in good structural condition. Normally, a lintel can be installed over the new opening; however, making a very large opening sometimes requires rebuilding the whole wall. Making openings in existing masonry walls is addressed in Chap. 9.

Conversely, existing floor and wall openings may need to be filled. There are challenges involved even in this seemingly straightforward task. (For one, the existing framing at the sides of the opening might not be strong enough to support the infill structure and the gravity or lateral loads on it.) Floor openings framed with steel or concrete beams of adequate strength and rigidity can be filled with composite steel floor deck carrying concrete topping (Fig. 1.9).

1.4.7   Adding framing for rooftop HVAC equipment

Whenever new HVAC equipment is installed on the roof of an existing building, structural engineering efforts are required. These efforts include making floor and roof openings, providing equipment supports, and checking the capacity of existing framing to take the load.

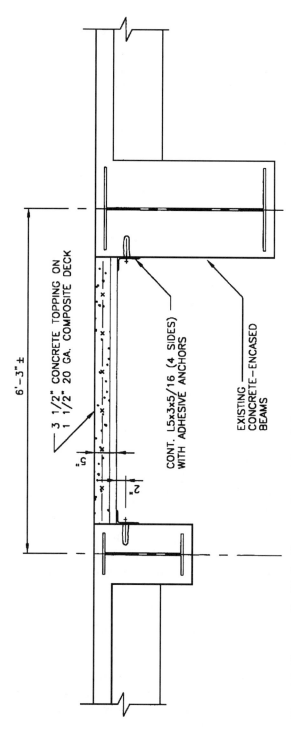

6'-3" ±

3 1/2" CONCRETE TOPPING ON
1 1/2" 20 GA. COMPOSITE DECK

5"

2"

CONT. L5x3x5/16 (4 SIDES)
WITH ADHESIVE ANCHORS

EXISTING
CONCRETE-ENCASED
BEAMS

Figure 1.9 Floor opening filled with composite steel floor deck carrying concrete topping.

Depending on the type and weight of the unit and on the recommendations of its manufacturer, the unit can be supported either on a curb or on a frame on legs.

In some cases, either the existing roof structure is not strong enough to support the weight of the new equipment (including the snow drift accumulating around it), or the old framing is too deteriorated to carry the load. In these cases, a common solution is to bypass the roof deck and beams altogether, erecting a structural-steel framework on top of the existing columns and supported only by them. A detail of one such support is shown in Fig. 1.10.

1.4.8 Installing mezzanines and additional floors

Mezzanines typically occupy only a part of the building's floor plate. They are often installed in existing buildings to provide additional storage or office space. In single-story structures, mezzanines can have their own supports, and their installation may not require any strengthening of building columns. However, there are pitfalls even in this seemingly simple task. One common problem with laying out mezzanine supports is aligning them with the building columns. While this design looks nice on paper, it becomes much less elegant when the time comes to place the new foundations and it is discovered that the existing ones are in the way.

A better tactic may be to stagger the new columns with respect to the existing ones. A solution involving staggered columns usually allows installation of new footings without incident but requires close coordination among the team members and the owner, because it makes circulation more difficult. Another possibility is to align the two sets of columns, separating them by a distance adequate for footing installation, and cantilever mezzanine beams over the gap.

Installation of full additional floors is a much more difficult task involving a good deal of structural investigation and upgrade, unless the building was originally designed for this purpose. Typically, both columns and their foundations will require strengthening. The original lateral-load-resisting system will also need reinforcing.

1.4.9 Removing loadbearing walls

Structural engineers routinely deal with removal of loadbearing walls. This work is part of almost every renovation project that requires rearrangement of interior partitions. Also, structural engineers are frequently called to help in "opening up" building space.

When only a small part of a wall needs to be removed (for a wide door opening, for example), the solution is relatively straightforward.

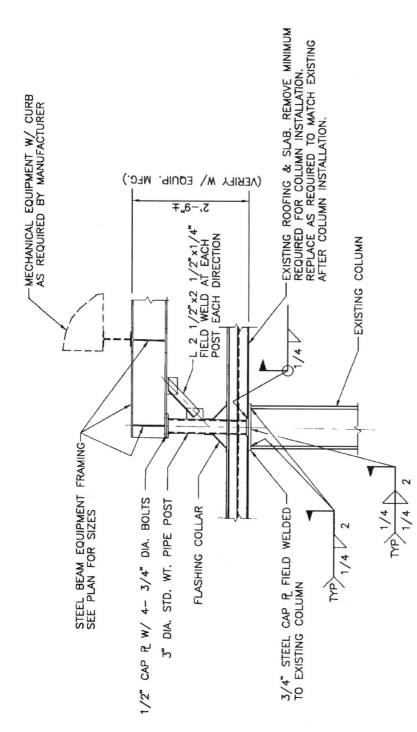

MECHANICAL EQUIPMENT W/ CURB
AS REQUIRED BY MANUFACTURER

2'-9"±

(VERIFY W/ EQUIP. MFG.)

L 2 1/2"x2 1/2"x1/4"
FIELD WELD AT EACH
POST EACH DIRECTION

EXISTING ROOFING & SLAB. REMOVE MINIMUM
REQUIRED FOR COLUMN INSTALLATION.
REPLACE AS REQUIRED TO MATCH EXISTING
AFTER COLUMN INSTALLATION.

EXISTING COLUMN

STEEL BEAM EQUIPMENT FRAMING
SEE PLAN FOR SIZES

1/2" CAP ℞ W/ 4— 3/4" DIA. BOLTS

3" DIA. STD. WT. PIPE POST

FLASHING COLLAR

3/4" STEEL CAP ℞ FIELD WELDED
TO EXISTING COLUMN

TYP 1/4 2

1/4

TYP 1/4 2

1/4 2

Figure 1.10 Structural-steel framework for rooftop mechanical equipment arches over the roof, supported only by the existing columns. (*Maguire Group Inc.*)

It involves the obvious: installation of a lintel bearing on the remaining parts of the wall or on new bearing posts installed to support the lintel. The frequently overlooked task of continuing the load path downward includes checking, and reinforcing if necessary, wall elements in the floors below that have to support concentrated loads from the lintel supports.

Total removal of walls brings other challenges. It normally requires installation of new support framing, such as steel beams supported by steel columns, in place of the walls. A detail that has been used in wood-framed buildings where headroom was tight is illustrated in Fig. 1.11. Here, wood floor joists are shored on both sides of the original wall and cut. After the wall is removed, joist hangers are attached to the joists. Then, steel framing with bolted-on wood blocking is installed, and the hangers are connected to the blocking. (The sequence of construction might be different with different contractors.) Sometimes, a new plywood overlay is required in order to restore the cut floor diaphragm to its original capacity, or to improve it.

1.4.10 Installing elevators

Government mandates relating to accessibility often require installation of elevators in renovated buildings. This invites all sorts of structural complications, from making holes in the floors to adding new elevator pits and protecting them from water intrusion. The pits can be installed without incident if the elevators are deliberately kept away from the existing foundations (Fig. 1.12). Otherwise, some of the

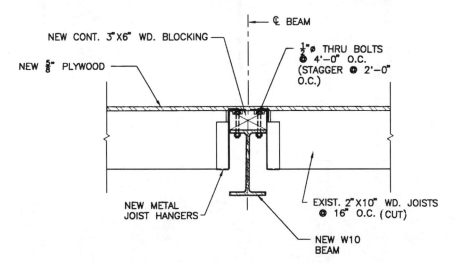

Figure 1.11 Steel beam replacing a removed bearing wall.

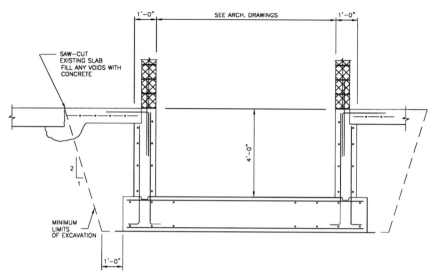

Figure 1.12 New elevator pit installed in an existing building.

existing footings (assuming that those foundations are present) may have to be partly removed and lowered, a complicated and potentially dangerous operation. In any case, soil excavation in the middle of an existing building requires experience to protect the adjoining construction from caving in.

A knowledgeable contractor should be responsible for the design and construction of the soil excavation support system, for protection of existing construction, and for determination of the exact limits of excavation. If circumstances make excavation of the pit too difficult, it is possible to specify a so-called holeless elevator that requires only a minimum depression rather than a deep pit.

Perhaps the worst-case scenario occurs when elevators must be placed in the middle of pile-supported structural floor slabs. Then, not only will the slab continuity be violated by the new openings, but also new piles will in all probability be required to support the pit. Minipiles are typically used for this purpose.

1.4.11   Dealing with foundation problems

Addition of new floors and mezzanines automatically requires checking the existing building foundations. Similarly, attempts to increase the design live load on the structure should include verification of the foundation's load-carrying capacity. Foundations designed "to the limit"—or, worse, underdesigned even for the existing loading—are common obstacles to the most optimistic plans.

Foundations may require attention even if no change of occupancy is contemplated, if there are signs of foundation distress, such as diagonal wall cracking in the characteristic "stair" fashion, building settlement or heaving, and water leakage. Foundation problems are difficult to rectify because they require removal of large areas of floor slabs and some disruption of operations. The remedial actions required to undo the effects of settlement might include jacking up the building, which can further damage the finishes and expand the scope of repairs. Foundation problems can also include tilting, cracking, settling, and otherwise failing retaining walls.

1.5   Role of Building Codes in Renovation

1.5.1   Why do we need the codes?

The local building code may have a lot to say on the subject of building renovations, and can answer many questions. Does it allow the existing structure to be kept in service, even though it was designed to a set of standards different from those used today? Can any sort of alterations be made to the structure—floor openings for HVAC ducts and stairs, for example—without necessitating the upgrading of the whole building to fit the code requirements for new construction? The answers to these questions may spell the difference between undertaking a renovation program or demolishing the building and starting anew.

In the words of the late Prof. Hardy Cross, the codes offer "standardization as a check on fools and rascals."[17] Indeed, as long as fools and rascals are with us, we need some set of minimal standards to prevent gross errors of judgment from harming the public. But these standards constantly change, and what was common wisdom only two or three decades ago may no longer be accepted. The building code requirements for snow-drift, wind, and earthquake loading have all been drastically revised in the last few decades and are continuing to evolve even now.

Provisions of various building codes dealing with structural renovations are slightly different, reflecting differing local approaches to this issue. Since most local jurisdictions in the United States have historically adopted one of three model codes, with some modifications, the relevant provisions of these model codes, and a few others, are worth examining.

1.5.2   The origins of the first codes

Until the twentieth century, each U.S. city or jurisdiction developed its own individual building code. The very first code meant to be applied nationwide was developed in 1905 by the National Board of Fire

Underwriters, whose *National Building Code* survived until 1980, when the rights to its name were eventually acquired by Building Officials and Code Administrators International, Inc. (BOCA).[18] The code was primarily concerned with prevention of fire losses and is not examined here in further detail.

The true origins of all contemporary model building codes can be traced elsewhere—to Herbert Hoover's efforts of the 1920s. In 1921, Hoover, who was then secretary of commerce, launched a campaign to replace a multitude of local building ordinances with a few standardized codes.[19] Each of the new codes was intended to apply to a large part of the country, and the codes were to reflect regional differences, as well as geographic and climatic conditions. Hoover's Department of Commerce established the Building Code Committee, which developed the suggested code formats and published a number of bulletins dealing with various technical code requirements.

The committee made an in-depth study of structural loads acting on buildings. In 1924, it prepared a report entitled "Minimum Live Loads Allowable for Use in Design of Buildings" and provided for its wide distribution through the Government Printing Office. This report, the granddaddy of all model codes, was continually revised and became known as the authoritative national standard ANSI 58. Its present designation is ASCE 7.[20] A brief review of some of its provisions may help establish the original design criteria for many structures built in the 1930s and 1940s.

The minimum live load for office buildings (and "for rooms with fixed seats, as in churches, school classrooms, reading rooms, museums, art galleries, and theaters") was 50 pounds per square foot (psf). The office floors were required to support a 2000-lb concentrated load acting on a square with sides 2.5 feet long. A larger load, 75 psf, was assigned to light manufacturing facilities, retail salesrooms with light merchandise, and stables.

A minimum live load of 100 psf was specified for "aisles, corridors, lobbies, public spaces in hotels and public buildings, banquet rooms, assembly halls without fixed seats, grandstands, theater stages, gymnasiums, stairways, fire escapes or exit passageways, and other spaces where crowds of people are likely to assemble." The same minimum load was required for special storage uses, printing plants, wholesale stores with light merchandise, and garages housing all types of vehicles. The largest live load, 250 psf, was to be used in the design of general storage areas.

The report acknowledged the need to provide an allowance for partitions that could be moved during the life of the building, although it did not specify the actual design numbers for this partition allowance. A live-load reduction was allowed only for "columns, piers or walls, foundations, trusses, and girders," except for storage buildings. The allowed reduction was 10 percent for supports carrying loads from two floors, 20 percent for three floors, 30 percent for four floors, 40 percent for five floors, 45 percent for six floors, and 50 percent for seven or more floors. A further reduction of one-half the live load was allowed for foundations in buildings with human occupancy.

The report's provisions dealing with roof loads varied markedly from those of today's codes. For roofs with pitches of 4 to 12 or less, a vertically acting live load of 30 psf was specified, while for roofs with pitches over 4 to 12 but less than 12 to 12, a vertically acting live load of 20 psf was to be used. For roofs with pitches more than 12 to 12, no vertical load needed to be considered, but the roof had to be designed for "a wind force acting normal to the roof surface (one slope at a time) of 20 pounds per square foot of each surface."

1.5.3   Development of model codes

Herbert Hoover's efforts toward the establishment of nationwide and regional standards culminated in the development of three model codes, all based to some degree on the 1924 Department of Commerce report. The first comprehensive building code was produced in the west by the Pacific Coast Building Officials Conference (now the International Conference of Building Officials, or ICBO). The first edition of the *Uniform Building Code* (UBC) was published in 1927, three years after the Building Code Committee report appeared.

In the southeast, the *Standard Building Code* (SBC) appeared in 1945–1946. In the east, the *Basic Building Code* was first published in 1950 by Building Officials and Code Administrators International, Inc.; the code's name was changed to *National Building Code* in the 1980s, as already stated. The UBC has been adopted primarily in the western and central states, the SBC in the southeast, and the BOCA code mainly in the northeastern United States.

Because the three codes continued to develop separately, a potential for confusion and a need for uniformity gradually became acknowledged. The pressure was on to produce a single model building code. The three model codes continued their separate lives until in the 1970s the efforts to unify them culminated in the establishment of the Board for Coordination of Model Codes (BCMC). The board consisted of two representatives from each model code and two members of the National Fire Protection Association. One of the BCMC's first tasks was to produce a uniform format for the three model building codes, even if their provisions remained different. As a result, the subject of existing buildings is found in Chapter 34 of the 1993 and later editions of all three codes.

In 1994, a new organization called the International Code Council (ICC) was formed to develop a unified document, the *International Building Code*. Another major task of the ICC was to develop performance-based code provisions to replace the prescriptive requirements found in the three model codes. With performance-based requirements, designers and contractors could develop innovative methods for meeting various functional requirements, instead of simply following the sometimes stifling letter of the code. Design of renovations, with its characteristic variety of field conditions that do not always fit neatly in the prescriptive code provisions, could be the primary beneficiary of performance-based requirements.

1.6   Renovation Provisions of Model Building Codes

1.6.1   The *Uniform Building Code* (UBC)

The *Uniform Building Code** has been generally accepted in the western half of the country, from New Mexico to Minnesota. In the 1994[21] and later editions, Chapter 34 and Appendix Chapter 34 deal with existing structures.

The code states that, generally, those buildings and structures not undergoing any changes in occupancy, but undergoing additions, alterations, or repairs, must comply with all the code provisions for new construction. This statement is clarified by another paragraph that prohibits making any additions or alterations that would cause the existing building to become unsafe or noncompliant with any of the code's provisions. An unsafe condition, as defined by the UBC, includes not only structural overload, but also inadequate egress, fire-protection hazards, and situations where the building's fire resistance is reduced below an acceptable level. Moreover, the code does not allow making additions or alterations to existing noncomplying buildings, except when the resulting structures are no more hazardous than before.

However, the UBC specifically allows alterations of existing structural members and adding new structural members "for the purpose of increasing the lateral-force-resisting strength or stiffness of an existing structure." The code further states that this work need not be designed for the forces required by the code for new construction if an engineering analysis satisfies several criteria. Those include verification that the affected existing structural members are not weakened by the proposed work and do not become overloaded beyond their

---

*Uniform Building Code is a registered trademark of International Conference of Building Officials.

capacities; that the detailing of new structural *and* nonstructural members and their connections to the existing building conform to the code provisions for new construction; and that unsafe conditions, as defined in the previous paragraph, are not created.

The UBC allows the local building official to authorize renovations of historic buildings without full conformance to the current code, provided that the building does not become more hazardous than before and that all unsafe conditions are corrected.

What about alterations to buildings that do not conform to the code provisions for new construction? The specific life-safety requirements that must be met in this case can be found in Appendix Chapter 34. Alternatively, the *Uniform Building Code* allows adherence to a separate standard entitled the *Uniform Code for Building Conservation* (UCBC), described separately in Sec. 1.6.7.

1.6.2 The *BOCA National Building Code*

The *BOCA National Building Code**[22] is generally followed in the northeastern quarter of the country and in the Midwest. Chapter 34, "Existing Structures," describes the requirements for rehabilitation of buildings.

In general, unless a change of occupancy is involved, the code allows alterations to buildings, an alteration being defined as any construction or renovation other than repair or addition. The altered work—but not the portions of the building unaffected by the alteration—must conform to the requirements for new construction. An important nuance of these provisions allows the design live load at the time of the building's erection to be used (Section 1614.4), "provided that the public safety is not endangered thereby." A change of occupancy requiring a higher floor live load than the existing building was designed for is not necessarily prohibited; it can be accommodated by posting of the original live load. A change of occupancy requires the approval of the building official. The official has broad discretion in approving renovations of historic structures.

Additions are required to conform to the code for new construction. The code further stipulates that no addition should increase the forces in any existing structural member by more than 5 percent, unless the member still complies with the code after the increase. Also, the strength of any member is not allowed to be decreased to less than that required by the code for new construction.

At the same time, the code offers compliance alternatives intended to permit renovation of existing buildings without requiring full compli-

---

*National Building Code is a registered trademark of Building Officials & Code Administrators International, Inc.

ance with the code for new construction. The compliance alternatives do not apply to buildings with high-hazard and institutional uses. Alterations to noncomplying buildings are not allowed if they reduce the existing level of safety. Any alteration that reduces the building's safety must essentially conform to the code for new construction.

In order for the compliance alternatives to be approved by the building official, some specific steps must be taken. The first of these is an investigation by a structural engineer, who has to demonstrate that the building can safely support the live load stipulated in the current code for the required occupancy. The results of this investigation, together with any proposed compliance alternatives, are to be submitted to the building official. The official, in turn, is to undertake a detailed evaluation process to determine if the building, as altered, complies with the life-safety provisions of the code.

1.6.3   The *Standard Building Code*

The *Standard Building Code** was first published in 1946. Chapter 34, "Existing Buildings," of the 1994 and later editions[23] deals with alteration, repair, and rehabilitation. The provisions of the SBC are relatively laconic in comparison with those of the previous two model codes. Essentially, alteration, repair, and rehabilitation of existing buildings are allowed without the whole structure's having to comply with all the requirements of the code for new construction, provided that the renovation work itself complies with them. To what degree the existing structure has to comply with the code for new construction is left to the building official to determine.

According to the SBC, if a change of occupancy is involved, the building systems must "conform to the intent of the technical codes" for new construction as determined by the building official. Historic buildings can be altered, restored, or renovated without full conformance with the code for new construction if this is allowed by the building official. When a change of occupancy entails reclassification of the building to a higher Seismic Hazard Exposure Group, the code requires that the building comply with the seismic requirements for new buildings.

Whenever an attached addition is constructed, the whole building must conform to the seismic provisions of the code for new construction. The only exception is made when the addition itself conforms to the seismic provisions of the code for new construction, the addition does not increase the seismic forces in any of the existing members by more than 5 percent, *and* the addition does not lessen the

---

*The *Standard Building Code* is copyrighted by Southern Building Code Congress International, Inc.

earthquake-resisting capacity of any member below that required for new buildings.

1.6.4 The *International Building Code**

This new model code, introduced around the year 2000, is intended to eventually replace the three codes previously discussed. This task may seem rather straightforward, at least in the area of building rehabilitation. It remains to be seen how quickly the *International Building Code* (IBC) will be accepted by the states and how widespread the acceptance will be. The proponents of a single code include many design professionals who prefer to deal with a single set of rules rather than with a patchwork of code and state regulations—a "regulatory nightmare," in the words of the American Institute of Architects, which passed a resolution supporting a single code in 1991.[24]

Chapter 34 of the IBC, "Existing Structures," simply states: "Additions, alterations or repairs to any building or structure shall conform with the requirements of the code for new construction."[25] An addition or alteration shall not increase the force in any structural element by more than 5 percent, unless the member will still be in compliance with the code for new construction, and shall not decrease its strength below the new construction code level. Strengthening of deficient members also has to conform to the code for new construction.

1.6.5 State building codes dealing with renovations

While the building codes of most states generally follow one of the three model codes, several state codes have their own provisions dealing with renovations. An example of the approach to renovations that used to be commonplace is contained in the *Rhode Island State Building Code*.[26] That code stipulates that if all the alterations and repairs made to a building within a one-year period exceed 50 percent of its physical value, the code requirements for new construction apply. Similar provisions are made for buildings damaged by fire or another disaster, when damage exceeds 50 percent. If the cost of alterations is between 25 and 50 percent of the building's value, the extent of compliance with the current code is to be determined by the building official. For alterations costing less than 25 percent of the building's value, the building may be restored with materials of the original qual-

---

ity, although the code does not seem to expressly extend this approach to the altered parts.

Some other local codes have unique features; among these is the sixth edition of the Massachusetts Building Code,[27] enacted in 1997. This code has its own version of Chapter 34 entitled "Repair, Alteration, Addition, and Change of Use of Existing Buildings." This chapter allows renovation work to be undertaken, under certain circumstances, without full compliance with the code requirements for new construction. The provisions of this chapter are much more complex than those of the model codes. The degree of compliance required depends on a hazard index for the building, a number between 1 and 8 assigned to various use groups.

If the proposed renovation work keeps the building in the existing use group or changes its hazard index by only one point, the code allows alterations and repairs to the building with materials similar to the original ones. The individual structural members may be repaired or replaced in kind, without having to comply with the code for new construction, unless the whole building system is replaced and unless such compliance is required by other parts of the chapter. Proposed work that changes the existing hazard index by two or more points activates the code requirements for new construction. The code allows some compliance alternatives to be accepted by the building official, when meeting all the provisions of the code for new construction is impractical or creates regulatory conflicts.

The foregoing provisions are augmented by a complex and seemingly independent Section 3408, "Structural Requirements for Existing Buildings." This section appears to require that any existing building in need of a building permit undergo a structural evaluation. The section then establishes a level of structural performance that these buildings must meet.

1.6.6 Other design documents dealing with renovations

In addition to model building codes and state codes, other documents dealing with building renovations are promulgated by a variety of U.S. government agencies. For example, the General Services Administration (GSA) has its own program of evaluation, maintenance, repair, and improvement for the sizable building stock it controls. This program includes a schedule, according to which each building receives a comprehensive multidisciplinary evaluation, summarized in a Building Evaluation Report (BER) prepared by an independent design firm. A BER provides information about building conditions, recommendations for a program of improvements, and a cost estimate for these improvements. The information is listed by "work items," and for each of those, feasible alternatives and courses of corrective action are proposed. Depending on the severity and urgency of the uncovered problem, the proposed work is placed in one of three priority time slots. The Immediate Action Plan (IAP) includes urgent work affecting the life and safety of the building's occupants that must be completed within a year. The Short-Term Improvement Plan (STIP) includes work to be completed within 5 years. The Long-Term Improvement Plan (LTIP) includes major capital expenditures required within 20 years.[28]

The U.S. Army Corps of Engineers publishes "The REMR Notebook,"[29] where REMR stands for repair, evaluation, maintenance, and rehabilitation activities. Geared mostly toward civil-works projects undertaken by the corps, this document is nevertheless of interest to those involved in evaluation and monitoring of steel and concrete structures, determination of causes of deterioration, nondestructive testing techniques, instrumentation, and data acquisition.

ASCE 7, "Minimum Design Loads for Buildings and Other Structures,"[20] has been increasingly integrated into the provisions of model and state building codes. This document does not offer detailed prescriptive requirements dealing with alterations and additions, but its Article 1.6 requires that in any building "enlarged or otherwise altered, structural members affected shall be strengthened" so that the stresses in these members are kept within the allowable limits, or so that their design strength is sufficient to resist the factored loads. Presumably, this means that the requirements of the code for new construction apply only to the "affected" members and not to the whole existing building.

1.6.7   The *Uniform Code for Building Conservation*

The *Uniform Code for Building Conservation** (UCBC) is published by the International Conference of Building Officials, the same organization that publishes the Uniform Building Code.[30] The UCBC offers alternative methods of achieving safety in existing buildings in lieu of complying with all the provisions of the code for new construction.

Like other building codes, the UCBC allows a continuation of the existing use of buildings designed in accordance with the previous codes if such use was legal at the time of enactment of the current code, unless a dangerous condition or a conflict with another applicable regulation exists. Building alterations and repairs are generally permitted without a requirement that the whole structure be upgraded

---

*Uniform Code for Building Conservation is a registered trademark of International Conference of Building Officials.

to the code for new construction, provided that such work does not make the existing building unsafe or overloaded. Additions must comply with the code for new construction, including the building-size limitations. A change of use is permitted in accordance with the UCBC's provisions. The UCBC grants the building official the discretion of accepting compliance alternatives when following the letter of the code is impractical.

The code establishes minimum standards for existing buildings. These standards mostly concern life safety, but also include some structural requirements, such as strengthening unreinforced masonry (URM) buildings located in seismic zones 3 and 4. The detailed upgrade methodology for the URM buildings is contained in Appendix Chapter 1, which is arguably the most important part of the UCBC for structural engineers.

Also very important is UCBC Chapter 5, which contains requirements for changes of occupancy. It states that for gravity loads, the building has to conform with the requirements of the code for new construction, except that the analysis methods of the original code may be used and the existing roof structures may be retained. (The latter exception does not apply to unsafe roofs or those that have received any additional dead load from reroofing or equipment.) The UCBC Chapter 5 also requires that all buildings undergoing changes of occupancy conform to the earthquake provisions of the code for new construction, unless the hazard group of the new occupancy does not exceed that of the existing use or the building is of a fire-resistive construction.

The bulk of the code's book is devoted to *UCBC Guidelines,* prepared for the U.S. Department of Housing and Urban Development (HUD) by a variety of participants. The guidelines concern exit systems, fire ratings of archaic materials (archaic construction being defined as that used generally prior to 1950), allowable stresses for archaic materials, and electrical and plumbing work.

1.6.8 Summary

At least two salient points are quite clear from the foregoing. First, in most codes, the degree of compliance with the code for new construction that is required is no longer tied to the relative dollar amount of the renovations. In the past, an upgrade to the code for new construction was typically required if the cost of the renovations exceeded 50 percent of the building's assessed value.

Second, the question of whether the renovated building must comply with the code for new construction is far from clear, despite the seemingly unambiguous wording of the codes. (See additional discussion on this topic in Chap. 11.) As a result, the local building official has a substantial power to establish the degree of compliance with the current code. Some codes give building officials more power than others, but in any jurisdiction it is a wise designer who confers with building officials early in the planning process.

1.7   Renovate or Rebuild?

The final issue we consider in this chapter is when it is more advantageous to renovate an old building and when it is better to demolish it and start anew.

1.7.1   Dealing with obsolete historic buildings

Renovations of historic buildings are likely to be more expensive than new construction, in the opinion of some experienced architects.[31] Still, when the structure enjoys historic status, it is rather difficult to totally demolish it, even if the building is totally obsolete and unsuitable for the proposed use. In some cases, the owner may be forced to retain the *appearance* of the building while actually demolishing it, a fact that has spawned a mini-industry of façade-keeping—retaining the historic exterior walls facing the streets and erecting a new, usually taller structure behind them. Without passing judgment on the merits of this approach, we simply acknowledge it as a challenging project that requires careful engineering.

Quite commonly, the news of a pending demolition galvanizes the neighborhood to find a way to save the familiar landmark. The historic but dilapidated Zinzer Building in Birmingham, Alabama, was saved from the wrecking ball by a consortium of local citizens and businesses that asked the owners for a month's delay to find a buyer—and found one.[31] The building was eventually converted into law offices. Another example is the fate of the venerable Lit Bros. Building, a former department store in downtown Philadelphia. The building, scheduled to be demolished, received a last-minute reprieve and was eventually converted into a modern office and retail center.[32]

Apart from such heroic initiatives, the decision to renovate may be sweetened by tax credits for historic rehabilitation offered by the federal government and by some state governments to promote urban renewal and preservation of historic properties, as mentioned previously. When blighted urban areas become eligible for these tax credits, developers' interest is more likely to follow, especially if abandoned or inexpensive properties are involved.

Corporate clients are rarely interested in renovated old structures, because these buildings typically do not meet the requirements for

Class A office space. Moreover, corporations demand the availability of high-technology communication cables, maximum planning flexibility, and easy access—traits more common in new suburban office space than in renovated old landmarks. The adaptive reuse, therefore, usually targets architectural and law offices, design studios, discount retail, start-ups, and similar prospective owners who might appreciate the historic charm, low rent for Class B office space, and central location typical of many old buildings.

For truly unique historic structures, demolition is obviously out of the question; these are usually rehabilitated at any cost. A case in point is fixing another failing Frank Lloyd Wright landmark (recall the discussion in Sec. 1.2.4), the Great Hall of his "Wingspread" in Wisconsin. The pinwheel-shaped room had clerestories that sagged as much as 2 in overnight under heavy snow. According to the *ENR,* designing the fix for the problem in the 3000-ft^2 room were the project architect, two preservation architects, a structural engineer, and a number of specialist consultants. The professional services alone consumed about 30 percent of the budget.[33]

1.7.2   Dealing with buildings that are not historic

The situation is different for structures that cannot be characterized as historic. Here, the cost-effectiveness of the solution is often the primary motivator. Major corporations are even less likely to get involved in adaptive reuse, often preferring demolition to renovation.

There are some notable exceptions to these observations, and some highly visible downtown office buildings have been successfully renovated for corporate clients. Among them is the building at 320 Park Avenue in New York City. This structure attracted Mutual of America Life Insurance because it had a leasable floor area larger than could be built in a new structure, because it was affordably priced, and because the city offered a tax break. The last was the deciding factor; without the incentive, it would have been cheaper and easier for the company to demolish and rebuild on the same site.[34] The project's architects decided to remove and replace all the elevators, the HVAC system, and even all the exterior curtain walls. Moreover, in this remarkable project, the small and awkward upper floors were expanded by "borrowing" and redistributing space from the larger lower floors, keeping the total building area essentially the same.

For industrial and commercial facilities, the renovate-or-rebuild dilemma is sometimes resolved in unconventional ways. When foundation conditions permit, the building can be expanded upward by raising the roof. This solution was used in a shopping mall in Ross

Township, Pennsylvania, where the entire steel-framed existing roof was lifted 8 ft.[35]

An even bolder solution was found for a Jefferson Smurfit Corp. plant in Highland, Illinois: The aging plant structure was encapsulated in a new 632-ft by 154-ft metal building system. After the new shell was completed, the existing building inside was demolished. Plant operations continued through the construction. This scheme was estimated to save the owner $5 million compared to the cost of building a new plant elsewhere.[36]

Still, these examples are the exceptions rather than the rule. In most instances, the required changes to the building plan and appearance are so extensive, and the cost of complying with government mandates so extreme, that it is more economical to demolish and rebuild. And this observation does not apply only to small, decrepit structures—many high-rise buildings in relatively good condition have been demolished to make space for other construction.

A great illustration of this trend is a recently announced plan by the City of Chicago to demolish almost all of the 51 high-rise public housing buildings it owns, among them the infamous Stateway Gardens, Cabrini-Green, and Robert Taylor Homes. What prompted this decision? When these skyscrapers were built in the 1950s, they were promoted as a means of ending urban blight, of doing away with ghettos. And yet, these futuristic high-rises have become, in the words of *The Wall Street Journal,* "the worst kind of ghettos—monstrosities of modern life."[37] The planned replacement? Low-rise townhouses, today considered to better relate to human scale. Will they be there in another 40 years?

A case study that recalls the real-life drama of a building that was scheduled to be renovated but was eventually demolished because the price of the proposed upgrade was too high can be found in Case 2 of Chap. 12.

References

1. Hue Cook, "Second Career for Obsolete Office Buildings," Building Design and Construction, May 1997, p. 46.
2. "Recycled Government Buildings Perform New Civic Duty," *Building Design and Construction,* December 1997, p. 11.
3. Daniel A. Cuoco et al. "New Life for Old Skyscrapers," Proceedings of Structural Congress XIII, Boston, Mass., April 1995.
4. Jon G. Auerbach, "Holy Toll Calls: Telecom Companies Now Turn to Heaven," *The Wall Street Journal,* December 23, 1997.
5. William D. Browning, "Boosting Productivity with IEQ Improvements," *Building Design and Construction,* April 1997, p. 50.
6. Danielle Reed, "Where the Huddled Masses Wait All Day," *The Wall Street Journal,* September 19, 1997.

7. Timothy Aeppel, "Famed Fallingwater House Is Slowly Falling Down," *The Wall Street Journal,* October 24, 1997.
8. Joe Morgenstern, "The Fifty-Nine-Story Crisis," *The New Yorker,* May 29, 1995, p. 45.
9. Nadine M. Post, "Annoying Floors," *ENR,* May 19, 1997, p. 33.
10. John Gregerson, "In the Clear," *Building Design and Construction,* May 1997, p. 37.
11. Raul A. Barreneche, "Octagon's Progress," *Architecture,* November 1993, p. 108.
12. Frederica Randall, "The Basilica After the Quake," *The Wall Street Journal,* October 17, 1997.
13. Thomas Fisher, "Redoing a Rehab," *Building Renovation,* Fall 1994, p. 36.
14. Eric C. Freund and Gary L. Olsen, "Renovating Commercial Structures: A Primer," *The Construction Specifier,* July 1985, p. 36.
15. Ian E. Chandler, *Repair and Renovation of Modern Buildings,* McGraw-Hill, New York, 1991.
16. Derek H. Trelstad, "Toys in the Attic," *Building Renovation,* Summer 1994, p. 7.
17. Robert C. Goodpasture, "Hardy Cross, Engineers and Ivory Towers," Books for Libraries Press, McGraw-Hill, Freeport, 1969, quoted in Gene W. Corley, "Protecting the Public from Fools and Rascals: Building Codes for the Millennium," *Concrete International,* September 1992, p. 57.
18. Robert M. Dillon, "Development of Seismic Safety Codes," in "Societal Implications: Selected Readings, Earthquake Hazard Reduction Series 14," Federal Emergency Management Agency, Washington, D.C., 1985.
19. Almon H. Fuller and Frank Kerekes, "Analysis and Design of Steel Structures," Van Nostrand, New York, 1936.
20. "Minimum Design Loads for Buildings and Other Structures," ASCE Standard 7-95, American Society of Civil Engineers, New York, 1995.
21. *Uniform Building Code,* International Conference of Building Officials, Whittier, Calif., 1994.
22. *The BOCA National Building Code,* Building Officials and Code Administrators International, Inc., Country Club Hills, Ill., 1996.
23. *Standard Building Code,* Southern Building Code Congress International, Inc., Birmingham, Ala., 1994.
24. "Development Work Continues on a Single Model Code," *Building Design and Construction,* December 1997, p. 14.
25. International Building Code, International Code Council, Final Draft, July 1998.
26. *Rhode Island State Building Code,* 7th ed., Providence, R.I., 1997.
27. *The Commonwealth of Massachusetts Building Code,* 6th ed., Boston, Mass., 1997.
28. "Repair and Alteration Building Engineering Report Training Manual," General Services Administration, Washington, D.C., 1993.
29. "The REMR Notebook," U.S. Army Engineer Waterways Experiment Station, Vicksburg, Miss., 1985.
30. *Uniform Code for Building Conservation,* International Conference of Building Officials, Whittier, Calif., 1991.
31. John Gregerson, "Preservation's Role in Reviving Main Street," *Building Design and Construction,* October 1988, p. 89.
32. Margaret Doyle, "Venerable Retail Complex Restored to Former Glory," *Building Design and Construction,* May 1988, p. 66.
33. Nadine M. Post, "To Cure Roof Spread...," *ENR,* August 18, 1997.
34. Derek H. Trelstad, "Cladding Revisited," *Building Renovation,* Fall 1994, p. 32.
35. "Mall Raises Its Roof to Avoid New Construction," *Building Design and Construction,* June 1987, p. 145.
36. "New Plant Constructed Atop Its Predecessor," *Building Design and Construction,* September 1988, p. 40.
37. Alan Ehrenhalt, "If You Build It, They Will Yawn," *The Wall Street Journal,* October 13, 1999.

Investigating Existing Conditions

2.1 Why Investigate?

The beginning of a typical renovation project resembles a medical checkup of a first-time patient. Engineers, like doctors, can see the outside of their subjects but can't tell for certain what is happening inside. Their first guess is necessarily based on the appearance; they know from experience that buildings, like people, deteriorate with age and that it is possible to extend the life span of both with proper care. The condition of the exterior can often tell whether the subject of their study has been pampered or abused, although this information may be obscured by facelifts. It is the task of the professionals to determine whether their patient is a good-looking wreck or a dirty workhorse still full of energy.

An investigation of existing conditions is intended to determine the state of the building's health, to establish a diagnosis, and to arrive at a prognosis. As with people, such an investigation usually includes a visual checkup, a few tests, and an examination of the patient's history. However, unlike people, buildings cannot tell what ails them, and it is the task of the engineer to piece together the answer from a variety of scattered symptoms.

The most fundamental question the investigation is intended to answer is whether the structure can safely support the proposed loading. Alas, simply checking the original drawings for the loading information is not enough. Even if the original design loads are adequate for the proposed use, they probably contain no allowances for such factors as accidental overload, member deterioration, and improper alterations—and all of these can occur during the life span of the structure.

Worse, the building may not have been constructed as its designers intended.

An unsophisticated owner may assume that a quick walk-through by a structural engineer is sufficient to determine the building's condition, but in reality such a determination often involves a lot of time and effort. It may take some persuading to get such an owner to agree on a formal program of assessment that is likely to involve additional funds and some disruption to operations.

It is wise to make a written agreement with the client to define the scope of the assessment services, including provisions for additional investigative efforts, should those be required by the observed field conditions. The agreement should state that the engineer (or the engineer's agents) has the authority to perform the investigation and any required testing and has rights of entry. It should define the responsibility for removal and repair of finishes and building members, as might be needed during the investigation. The agreement may state the extent of liability accepted by the engineer.

In this chapter, we first examine general methods of assessing building structures and then move on to discussing specific evaluation methods for various types of framing materials.

2.2   Assessing Building Condition

A building can be investigated in a variety of ways, depending on the type of structure, its apparent condition, and whether the original design drawings are available. To standardize the assessment process, the American Society of Civil Engineers has published ASCE 11-90, *Guideline for Structural Condition Assessment of Existing Buildings.*[1] This seminal document, which we cite throughout this chapter, should be on the bookshelf of every engineer involved in building renovations. The standard recommends a multilevel approach to structural assessment of buildings. The first level is a preliminary assessment that includes review of existing construction documents, site inspection, preliminary analysis of the structure, and arrival at the preliminary conclusions and recommendations. Depending on the results of this stage, a second level, involving a more detailed assessment that deals with the same items in much more detail, may or may not be required. This multilevel approach helps avoid unnecessarily detailed assessment efforts where they are not justified by the circumstances.

2.2.1   Reviewing existing construction documents

If the existing construction documents are available, the project is off to a good start. Structural drawings may indicate the design loads, material properties, member layout, and construction details. Knowing this information is half the battle, since the theoretical load capacity of the framing can then be analytically determined. Even if only the architectural drawings can be obtained, they may at least tell the age of the building and the general type of structure.

Actually, architectural drawings for old buildings often include much more structural information than is typical for today's projects. It is common for them to include the overall designations of structural members, size of lintels, and large-scale sections through exterior walls. These drawings might also include the names of proprietary framing systems, enough to point the engineers in the right direction. (Later chapters deal with various strategies to determine the load-carrying capacities of some proprietary building systems.)

Of particular help to those doing these investigations are as-built drawings, which in theory incorporate the changes and substitutions made to the framing during construction. Unfortunately, it is a rare case when all the modifications are actually documented. If a steel-framed building is involved, one may find the erection drawings instead of the original construction drawings. As is the practice today, steel-erection drawings contain dimensions and piece marks, but no member sizes, and are of little use to the investigator besides indicating the general layout of the structural members.

Where can the existing drawings be found? Obviously, the older the building, the lower the chances of any drawings surviving intact. Still, there are many potential sources, some not so obvious. The first place to start is the client's files. Many large owners maintain libraries of existing drawings, including those of alterations performed over the years. If your contact is unsure about where to look, ask him or her to check with the building superintendent, the custodian, or a plant engineer.

Other sources of drawings include the city's building department and the original architect's office. Some drawings may be retained by contractors who have done major work in the building, by one of their subcontractors, or even by equipment installers. Insurance companies sometimes maintain drawings of the buildings they insure. Drawings of some historic buildings are preserved by local or national historical societies.

If the building is of some importance or has utilized innovative framing methods, it could have been written about in engineering publications or in the general press. Both these sources could contain pictures of construction in progress that are helpful in understanding the workings of the existing structure. This method came in handy during renovations of historic Union Station in Worcester, Massachusetts, for which the design drawings were not available. The engineers were unsure whether some intermediate columns under the main girder were loadbearing and whether they could be removed as the proposed plan dictated. Review of an old photograph taken during construction and published in the local paper helped: The picture clearly showed the girder in place without any interior columns. Apparently, those were added later.

Frequently, the search for the existing drawings involves a bit of detective work. In one project, an engineer (the author) was told by the project architect that no original drawings could be found and that all search avenues had been exhausted. The only pieces of information available were the schematic-design sketches for the unfinished renovations to the property. The sketches bore the name of a design-build firm. The architects attempted to contact that firm, but it had long since gone out of business.

Was this a dead end? No! The engineer, taking the initiative in his own hands, searched the telephone directories of several outlying towns and eventually found an establishment with a similar-sounding name—a factory. As it happened, the factory was a successor to the design-build firm. After a few telephone calls, the engineer located an employee who claimed that he had the design drawings for the aborted renovations. An appointment was made to review the drawings, which, alas, turned out to be the same sketches that the architects already had. However, there was one more sheet—a partial copy of an original design drawing that gave the name and address of the out-of-state architectural firm that had originally designed the building.

Armed with this information, the engineer attempted to call the firm, only to find that no such firm existed...it had probably gone out of business, too. The end? No! The telephone operator was asked to check whether a firm containing the names of any of the partners in the defunct firm was listed. One such firm was indeed found; it happened to be the successor to the designer of the building, and it happened to have not only the pertinent drawings but even structural calculations! After mailing a check to cover the copying costs, the engineer soon had a full set of drawings and calculations at his disposal. The drawings indicated a very complicated structural scheme and proved truly invaluable for assessing the building's condition. Needless to say, many tens of thousands of dollars in potential field-investigation costs were thus saved.

2.2.2   Field investigation

Field investigation follows the review of the existing drawings. The extent of the field work depends on the conclusions reached during the office review. In the best-case scenario when the existing drawings contain the information about the design loading and show all the member sizes and material properties—only a visual inspection may be needed. At the start of the preliminary walk-through, it is wise to talk to the building's superintendent or, if the superintendent is new on the job, to some of the "old-timers," who may be able not only to advise the investigators of current problems, but also to remember some prior renovations and the ways the previous problems were fixed.

Visual inspection consists of a walk-through to verify that the building is indeed framed in the way the drawings indicate, that the actual member sizes correspond to those shown, that no major modifications have taken place, and that there are no signs of overload or major deterioration. If the answers to all those questions are satisfactory, the building may be presumed to be constructed in accordance with the original drawings. Observed deviations can be mentioned in the existing-conditions drawings if these are prepared as part of the investigation. These drawings will serve as a basis for construction documents for the renovations.

The actual sequence of the walk-through differs among various practitioners. Most investigators inspect the interior of the building first and then move to the exterior. It is good practice to start at the top floor and move downward, covering all rooms in a clockwise direction.[2] Any damage to structural elements and to finishes is noted during the interior inspection and verified during an examination of the exterior and the roof.

Visual inspection may be supplemented by field measurements and some physical probing and testing. A lot depends on whether the structure is readily accessible or covered with finishes and on the age and condition of the building. Here are some typical situations:

- The existing drawings are silent about member sizes and material properties, but the structure can be readily measured and analyzed—it is made of sawn wood or structural steel, for example—and seems to be in good shape. Then, visual inspection can include the laborious task of measuring the spans, spacing, and sizes of structural members. Essentially, the missing structural drawings are reconstructed piece by piece. This mapping is accompanied by testing of material properties, such as the bending strength of wood or the yield point of structural steel.

- The existing drawings are incomplete and include only a part of the structure and not the modifications and additions made later. Assuming that a search for additional drawings proves fruitless, the program of field measurement and material testing described in the previous paragraph is undertaken for the missing pieces only, while the sizes of the members shown on the existing drawings are spot-checked.

- The existing drawings contain adequate structural data, but the members are visibly deteriorating. Then, the purpose of field investigation is to determine the extent of the damage and to assess the actual load-carrying capacities of the members, by analytical methods or by load tests. This task may require total removal of all the ceiling finishes.

- The drawings contain no loading data at all, *and* the structural components cannot be readily identified or analyzed—the worst possible case. This situation is common when unidentified concrete framing or a proprietary structural system of any sort is involved. Then, the main purpose of the field work is to determine the load-carrying capacities of the members by load testing or other investigative techniques, which usually involve destructive methods.

- The drawings indicate member sizes but not the connections, which are crucial to understanding the system's performance. This situation frequently arises in assessing the seismic resistance of concrete framing, where it is important to know whether there are dowels extending from columns into beams. Similar questions may arise about the ability of structural steel framing connections to resist bending moments. In this case, field investigation is likely to include both destructive and nondestructive investigation methods.

2.2.3 Probing and exploratory demolition

Frequently, field conditions do not favor visual inspection, and framing is obscured by fireproofing, ceilings, and wall finishes. To expose structural members, exploratory demolition is needed—someone has to partially remove the finishes and obstructions by making openings so that an engineer can review and measure the members (Fig. 2.1).

This work typically has to be done prior to the start of construction, often while the building is occupied. Whether exploratory demolition is performed before the design for renovations is completed or at the beginning of construction, when some demolition usually takes place anyway, depends on such factors as the age of the structure, its condition, the availability of the drawings, and the ease of visual observations. Obviously, a recently constructed steel building for which a full set of design drawings is available warrants less scrutiny at the preconstruction stage than an undocumented old concrete structure. Some probing and testing prior to the start of construction are a must if the design drawings are not found.

Apart from causing the occupants consternation, the exploratory demolition work is messy, it leaves unsightly gaping holes that are not likely to be patched until the start of construction—and it is expensive. The expense is rooted in the fact that the engineer is usually prohibit-

Figure 2.1 An opening made in a plastered ceiling reveals wood framing. (*Briggs Engineering and Testing Co.*)

ed by the insurer from engaging in this type of work, and an independent testing laboratory or a contractor has to be hired instead.

Still, exploratory demolition is a necessity. Starting building alterations without knowing what the framing is sounds foolish but, unfortunately, is all too common. Predictably, this shortsighted approach tends to result in large change orders and project delays during construction as problems are discovered that need to be fixed. Donald Friedman, an authority on historic building rehabilitation, recalls a renovation case where the existing drawings for an apartment building indicated concrete framing. The visual inspection confirmed that the framing looked like concrete—but did not quite match the drawings. A further search for information uncovered another set of drawings in which the building was redesigned in steel—the actual framing material. Had the first set of drawings been uncritically accepted without verification, design revisions—and delays—during construction would have been inevitable.[3]

Apart from making openings in ceilings, exploratory demolition may include verification of material properties and details of construction. In concrete buildings, for example, such probing includes measuring slab thicknesses and sizes and spacing of reinforcing bars. A typical effort involves taking floor cores to determine the slab thickness and composition (Fig. 2.2). For exterior masonry walls, the thickness and composition can also be determined by coring (Fig. 2.3) or simply by removing a part of the wall (Fig. 2.4). Partial wall removal allows the engineers not only to measure the thicknesses of the various wall layers, as shown in

Fig. 2.4, but also to inspect the condition of the interior layers that are hidden from view and to verify the presence of wall ties.

How extensive should the exploratory demolition be? Depending on the age, size, framing type, and condition of the building, the extent of the investigation program can vary from making a few ceiling openings to major efforts involving exposing—and testing where required—every element of the structure. (The major-efforts scenario is illustrated in Case 2 of Chap. 12.) If the structure is obviously deteriorating to a degree where replacement is needed, the extent of exploratory demolition may be minimal—just enough to confirm the poor condition of the structure.

Exploratory demolition is a relatively dangerous exercise, and the dangers extend beyond being hit on the head with a piece of plaster. Any removal of framing or finishes, however seemingly innocuous, may result in unintentionally removing a vital piece of the structure. Friedman[3] documents two cases in which exploratory demolition could have led to disaster if it had not been halted in time. In the first case, a new stair opening had been made in an existing wood-framed floor and a header installed to support the cut floor joists. Incredibly, the cut joists, which were supposed to be attached to the header by joist hangers, had never been fastened to it and were simply resting on top of the dropped ceiling. The ceiling was acting as a structural

Figure 2.2 A floor core is taken to determine slab thickness and composition. (*Briggs Engineering and Testing Co.*)

Figure 2.3 A core is taken in an exterior masonry wall to determine its thickness and composition. (*Briggs Engineering and Testing Co.*)

Figure 2.4 Removing part of the wall to measure its thicknesses and inspect the condition of interior layers.

support—and it was slated to be partly removed for exploratory demolition! Fortunately, the problem was recognized in time and the impending collapse averted.

In the second case, a steel floor beam, instead of being attached to the column, was actually supported by nothing more than the column's concrete fireproofing. Since the plan was to partially remove the fireproofing below the beam in order to examine the connection, the consequences for the inspectors could have been severe indeed.

These two cases demonstrate why it is very important to have a structural engineer present during the exploratory demolition: Only a person with a good understanding of structural behavior could have averted these two near-disasters. Also, in some situations the additional safety measure of installing temporary shoring under the members being probed makes a lot of sense.

2.2.4 Testing materials

As already mentioned, it is important to know not only the sizes, but also the actual material properties of the existing structural members. For example, depending on the actual yield strength of the reinforcing bars in a slab, the slab may be considered either adequate or grossly overstressed for the given loading. The best way to find out is to cut out a piece of rebar and have it tested in a laboratory.

Like exploratory demolition, removing pieces of structure is a dangerous exercise that can weaken the structure if it was done the wrong way and in a wrong place. It is quite obvious that the best place to remove a steel coupon is from a filler beam rather than from the main girder, and from a lightly loaded column rather than from the potentially overloaded one, although the engineer is actually more interested in the latter. Most structural engineers know that in a simple-span steel beam, the best place to remove a coupon for testing is the tip of the flange near one of the supports; in a one-way concrete slab, the best place to remove a core is near a support as well. However, as discussed in more detail in Chap. 4, the safest place to cut an opening in a two-way slab is in the middle of the bay, where two middle strips intersect. But what if one does not know whether the slab spans in a one-way or two-way fashion? Taking a core near one of the supports in a two-way slab would mean taking it from the column strip, and therefore from the wrong place. In such cases, engineers are left to rely on their judgment and intuition as the only guides.

2.2.5 Analyzing existing framing

Once the sizes of the existing structural members have been determined by reviewing existing drawings and making field observations, it is wise to analytically check at least some members for the proposed loading. The checking should be much more comprehensive when the existing drawings lack information about structural design loads or do not even show the member sizes.

As noted in Chap. 1, some building codes allow the use of the design methods and material properties common at the time of the original construction. However, more often than not, present-day analysis methods are more liberal than those of the old codes and yield larger theoretical member capacities than those assumed by the original designers. The ultimate-design methods used with the actual material properties usually result in larger allowable member capacities than those predicted by the old working-stress methods in combination with the lowest possible material strengths. This efficiency has to be tempered by some inevitable deterioration of the existing members. For deteriorated structural elements—or even for cases where the detailing is different from today's practices—the analysis may include capacity reduction factors in excess of those used in the present-day codes.

The author's own preference is to start with the simplest approximate methods of analysis, and to move to the more sophisticated methods only if those do not yield the desired results. It could well be, for example, that an existing continuous concrete floor slab has adequate load-carrying capacity even based on the single-span assumption; there is no need then to conduct a finite-element analysis of the floor structure—or any computer analysis, for that matter.

In addition to simply checking the isolated members, the engineers should also examine how well the structure functions overall and how well its components fit together. Is there a load path for wind and earthquake loading? Are there discontinuities in any of the load-resisting systems? Are the exterior walls properly tied to the floor structure? Are nonbearing partitions isolated from the floors but still laterally supported at the tops? The stability of nonstructural building elements and their interaction with the structure is worth examining. For instance, the most immediate seismic hazard may be posed not by the insufficiently reinforced masonry walls but by the poorly anchored roof parapets, which are likely to fall during an earthquake.

While the structural analysis is going on, the other members of the design team perform their evaluations. The problems they observe may dwarf any structural deficiencies and may cut short or at least curtail the scope of structural evaluation efforts. For this reason, structural engineers should avoid working in isolation from the other team members. The client does not need a detailed dissertation about structural problems if the building is found to be clearly unsuitable for the proposed use by the other design disciplines—too small or inaccessible, for example.

2.2.6   Making an evaluation

After the available documents have been reviewed and the building structure has been visually examined in the field, tested if necessary, and analyzed, the engineers can at last formulate their conclusions. The most basic question that must be answered here is whether the existing structure is capable of serving in the new capacity. There can be only three answers: yes; no; and yes, but certain elements must be strengthened or replaced first. The last answer is, of course, the most common. In the evaluation stage, all the structural deficiencies are tallied and integrated, and the program of renovations is proposed.

The preliminary evaluation is typically followed by a cost-impact study—essentially a cost estimate for the proposed renovation program. Estimating renovation costs is a thankless occupation, but the client needs this information in order to determine the fate of the building and the extent to which the acceptable renovations are economically feasible. From the client's standpoint, the actual renovation costs are only a part of the total project costs, which also includes the monetary impact of temporary relocation and disruption to the operations, if this is applicable, the cost of financing, and the efforts required to obtain permission to build.

It may well be that the structural condition of the building is on the border of being unacceptable and no firm conclusion can be reached in the preliminary assessment stage. In this case, the evaluation moves to a new level—the detailed assessment. As the name implies, during the detailed assessment, a more thorough investigation is undertaken, a more invasive and extensive program is implemented, and a more refined analysis is made. All these efforts obviously cost more than the preliminary assessment, but they result in more reliable recommendations. The general procedures for detailed assessment parallel those already discussed.

2.2.7   Preparing a report of condition assessment

The information gathered during the previous steps is now organized into a report that is submitted to the client. We suggest this format, which is largely based on the ASCE 11 00 recommendations:

A. An optional executive summary

B. Introduction

C. Description of the existing building structure

D. The desired design loading and other performance criteria

E. Description of the assessment and evaluation processes

F. Conclusions and findings

G. Recommendations

The recommendations may include the proposed renovation steps or a need for a detailed assessment if that is required. An appendix to ASCE 11-90 contains a more detailed format for the report.

2.2.8 A word of caution

In the interest of fairness, we should note that, despite all the investigation, testing, and analysis efforts, some major structural problems can elude detection and become obvious only during construction. Such incidents happen with the best of firms; they should not be viewed as errors on the investigator's part. The owners should realize that there can be no guarantees in the assessment business and that some unanticipated field surprises will probably occur and should be budgeted for.

Baumert[4] tells a story of a prominent building that was slated for reroofing. The structure, which his firm was retained to investigate, suffered from a long-term problem of excessive deflections of the wood trusses. The design firm proceeded with a substantial program of condition assessment involving documents review, field investigation, testing, and computer analysis, none of which could provide a clue to the truss-sagging problem. The firm then prepared a comprehensive report following the accepted procedures and issued the bid documents for renovation work.

During construction, when the pole gutter and deteriorated sheathing was removed, some decay was found in the end of the bottom chord of a newly exposed truss. Further investigations, which involved a nationally recognized wood expert, found that there was major deterioration in the bottom chords of most trusses. The decay was blamed on the leaking pole gutter, which allowed water to penetrate the end grain of the bottom-chord timbers. The water could then travel along the interior wood fibers and continue hollowing out the wood while its exterior stayed intact. The plans for construction were changed to incorporate the new findings. The building rehabilitation, proceeding on an accelerated schedule, was completed on time.

2.3 Material Properties in Steel Systems

In the remainder of this chapter, we discuss various destructive and nondestructive methods of testing materials. Material testing methods that involve removal and destruction of a portion of the member to determine its properties are called *destructive testing. Nondestructive testing* does not alter the members' properties or affect the service of the structure.

2.3.1 Introduction

In this section, we review specific tests and investigative techniques that apply to steel-framed buildings. We continue the topic of renovating steel buildings in Chap. 3.

Steel-framed buildings may contain only structural steel framing, such as familiar wide-flange members, or may also include open-web steel joists. Open-web joists, also known as bar joists, may be supported either on structural steel framing or on masonry walls. Buildings framed with structural steel should not be confused with metal building systems, also known as pre-engineered buildings—metal structures designed and fabricated by their manufacturers. Metal building systems typically include cold-formed framing and custom-made rigid frames with tapered profiles. The specific challenges of renovating metal building systems are described in Chap. 10.

Some archaic structures built in the nineteenth and early twentieth centuries include columns and other framing made of cast iron and wrought iron, all discussed in Chap. 3. Aluminum structures, on the other hand, may be encountered in newer buildings. Field investigation of these structures can be performed based on the same principles as for structural steel.

Quite often it is important to assess the structural condition of steel connections made with welds, rivets, or bolts. Unfortunately, the steel industry has historically promoted the view that connections should be designed for the calculated reactions of the members they join, such as end reactions of beams, rather than for the maximum capacities of the members or some percentage of those. The author has questioned before[5] whether connections should be the weakest points of the structure. Indeed, in renovation practice, a situation occasionally arises in which the existing beams could support larger loads than they were designed for, but the connections are inadequate and need expensive retrofitting.

Additional sources of information for the material tests listed below are contained in ASCE 11-90.

2.3.2 Nondestructive testing of structural steel

Existing steel can be tested by the following nondestructive methods[1,6,7] to determine the condition of steel members and their connections.

Visual

How it works: The inspector examines the object visually, sometimes aided by a magnifying glass ($\times$10 or less), flashlight, borescope, or weld gages.

Typical uses: Rapid evaluation of surface conditions to detect member distortion, excessive sagging or buckling, lack of bracing, rust, cracking, sloppy workmanship, and missing or loose fasteners, and to judge the apparent quality of welds. Visual inspection is still the most common method of evaluating steel structures.

Strong points: The method is simple and inexpensive; problem areas can be located rapidly; very effective in combination with other methods.

Drawbacks: Obviously, this method cannot detect subsurface problems or provide quantitative data.

Ultrasonic testing (ASTM E164)

How it works: Electrically timed ultrasonic waves in the range of 0.1 to 25 MHz are applied to the steel surface and are reflected by interior defects, voids, and changes in density. The results are transmitted to a screen or a meter and are compared with control data.

Typical uses: Detecting hidden cracks, voids, laminar tearing, porosity, changes in composition, and inclusions that are too small to be detected by other testing methods.

Strong points: This method is very efficient, fast, can detect minute flaws, has a wide range of applications, and can be used to check pieces of large thicknesses (up to 60 ft); the equipment is small and portable; the results can be recorded automatically and reproduced electronically; the technique can be applied only to one surface.

Drawbacks: Ultrasonic testing may not produce good results with objects having complex shapes or rough surfaces, is highly dependent on user expertise, does not readily provide a hard copy, and requires careful calibration of equipment.

Radiography (ASTM E94)

How it works: The x-rays or gamma rays are applied to the object. The radiation identifies internal voids, changes in structure, and other defects that offer less resistance to penetration than sound metal. As in medical x-rays, the defects can be seen as dark spots on a film. X-rays can penetrate up to 30 in, gamma rays up to 10 in.

Typical uses: The same as ultrasonic; can also be used to detect undercutting and incomplete penetration in welds.

Strong points: The equipment is portable. The test is reliable; it can detect both external and internal flaws; it produces permanent records on film. Gamma-ray equipment is smaller and somewhat lower in cost than x-ray equipment.

Drawbacks: This method is very expensive, is potentially dangerous and requires shielding, and has to conform to health and government standards. Film typically has to be developed separately, although some laboratories can develop it in on-site trailers. Large test installations may require sources of cooling water and electric power. The test cannot detect defects less than 2 percent of the object's thickness and those oriented in certain ways. Difficult to use with complex shapes.

Magnetic particle (ASTM E709)

How it works: The object is magnetized and covered with magnetic powder. Any flaws in the metal create leakages in the magnetic field, affecting the orientation of the particles above those areas. A variation of this test involves using wet fluorescent particles visible in black light through a borescope.

Typical uses: Locating surface cracks, laps, voids, seams, and other irregularities. Some subsurface defects can also be detected to a depth of about 0.25 in.

Strong points: The test is relatively fast, simple to administer, and inexpensive; it can detect cracks filled with nonferrous debris.

Drawbacks: There is a limited depth penetration. The object must be clean and originally demagnetized. The test requires a source of high-current electric power to magnetize the object. After the test, the object may need to be demagnetized again. Careful surface preparation is required, and the operation is relatively messy. Needs good operators' skills to interpret results. Often requires applications along two orthogonal axes. May not detect defects parallel to the magnetic field.

Infrared

How it works: A source of infrared radiation is applied to one side of the object, and the flow of infrared energy is monitored and analyzed. As with the other test methods, voids and discontinuities alter the flow of the emission and are therefore detected by the device.

Typical uses: Detecting hidden cracks, voids, porosity, changes in composition, and inclusions.

Strong points: The test is very sensitive and can handle complex shapes, results can be recorded on a computer.

Drawbacks: Results can be distorted by coatings and colors. The test is rather slow.

Liquid penetrant (ASTM E165)

How it works: Liquid penetrant is applied to the surface of the object and gets into any defects by capillary action. When coated with a developing solution, the penetrant becomes visible and identifies the defects. There are two types of liquid penetrants: a dye that works with a developer and is visually evaluated and a fluorescent penetrant that glows in darkness and when seen in ultraviolet light.

Typical uses: Locating surface cracks, areas of porosity, material seams, incomplete fusion in welds, and similar defects. Sometimes used to complement the magnetic particle test.

Strong points: The test is simple and inexpensive, can be made at the site, and can accommodate complex shapes.

Drawbacks: The test requires that the checked parts be clean, can detect only small surface defects, and requires careful preparation and operator expertise.

Eddy current (ASTM E566)

How it works: A probe coil is moved along the object's surface and sends eddy currents into the material being tested. Any interior defects affect the flow of electricity. These changes in impedance are detected by a meter or a screen graph and analyzed.

Typical uses: Locating hidden cracks, voids, porosity, changes in composition, and inclusions. This test is commonly used to monitor thicknesses of produced metal shapes and wire, it can also detect the location of repair welds in machined or ground surfaces.

Strong points: The test can be performed continuously and can be easily automated, the cost is moderate, and a hard copy of records can be produced if required.

Drawbacks: A variety of factors can affect the flow of electricity and skew the results. The test does not produce absolute measurements, only comparative ones; it may not detect cracks in some directions. Operators with good skills are needed to conduct the test and to interpret results. Only relatively shallow objects can be tested.

Hardness (ASTM E10, E18)

How it works: A hardened object or a steel ball is forced into the object, the resulting mark is measured, and a hardness number—such as the commonly used Brinell or Rockwell—is established.

Typical uses: Measuring steel's hardness (resistance to deformation), from which its tensile strength and the effects of cold working can be roughly determined.

Strong points: The test is simple and inexpensive, and can be made at the site if portable units are used.

Drawbacks: The test may require surface preparation and is subject to some testing constraints.

Another tool of nondestructive evaluation that was not specifically mentioned in this list is the strain gage. Strain gages can be used to measure changes in relative deformations, and therefore in levels of stress, in steel structures. One example of their use is offered by Steficek,[8] who describes a project in which two steel columns had to be moved in a 25-story building that supported a 150,000-lb rooftop water tank. The drawings for the building were not available, and the distribution of the tank load among the various columns was not known. To help determine how much of this loading each column actually carried, the engineers placed strain gages on the columns. By varying the water level in the tank and taking readings on the gages, it was possible to confirm the assumed distribution of the water loading. The strain gages were also used to verify the total actual load on the new columns.

Some additional information about nondestructive tests of steel is contained in Ref. 9.

2.3.3   Common destructive tests of structural steel

If the existing drawings do not provide adequate information about structural steel properties, one can make an assumption about the material properties from the information contained in Chap. 3 and in other relevant publications. Alternatively, existing steel can be tested by the following methods, adapted from Ref. 1.

For all the test methods listed in this section, a specimen of the predetermined size has to be removed from the steel member being evaluated. In some cases this procedure can weaken the member, a common limitation of all these tests. The size of the removed coupon is typically less than the sizes required for standardized mill testing of new steel that are used in ASTM specifications. These so-called subsize specimens may still be used in the testing laboratory, but some correction factors may have to be applied to the results.[10]

Chemical (ASTM E30)

How it works: The specimen is analyzed to determine the chemical composition of the metal, including carbon content.

Typical uses: Determination of weldability, ductility, and corrosion resistance of unknown steel.

Strong points: This is the most common laboratory test method of identifying unknown metals and finding their properties.

Drawbacks: The test requires a specialized chemical-testing lab with experienced personnel.

Bend (ASTM E190, E290)

How it works: A specimen of a standard size is removed from the tested member and bent into an U shape with a certain inside radius; its outside surface is then checked for cracks.

Typical uses: Measuring ductility of metal and welds.

Strong points: The test is easy, fast, and inexpensive.

Drawbacks: The test lacks quantifiable results.

Tension (ASTM E8)

How it works: The test specimen is removed by flame cutting, machined into a testable sample, and subjected to axial tension. Its properties such as strength, elongation, and reduction of area are measured; a load-elongation curve may be constructed.

Typical uses: Determination of yield and ultimate tensile strengths and modulus of elasticity.

Strong points: This is the most direct and common method of finding the mechanical properties of steel.

Drawbacks: Removal of a test specimen may weaken the structure; the test is slow and expensive.

Compression (ASTM E9)

How it works: The specimen undergoes axial compression, its mechanical properties are measured, and a stress-strain curve is constructed.

Typical uses: Determination of yield strength, compressive strength, and the modulus of elasticity.

Strong points: This is a fast and direct method of measuring the mechanical properties of steel.

Drawbacks: None are significant, other than that removal of a test specimen is required.

Charpy, Izod, and drop-weight impact (ASTM E23, E208, A673)

How it works: A specimen notched in a standard way is fractured by the impact of a dropped weight.

Typical uses: Assessing the brittle-fracture potential of steel, and measuring its toughness and resistance to shock loading.

Strong points: It is direct and relatively inexpensive.

Drawbacks: The test could be affected by heat treatment and the composition of the metal.

Fatigue (ASTM E466 and E606)

How it works: A standard specimen is stretched, twisted, or bent in a repetitive predetermined pattern of loading.

Typical uses: Assessing the fatigue resistance of steel.

Strong points: This is a direct and well-established procedure.

Drawbacks: The test is expensive and requires experienced personnel.

2.3.4 Test to determine stress level in structural steel members

When the existing steel members are in need of strengthening or modification, information about the level of stress they carry may be needed. The following test method can help (after Ref. 1).

Measuring residual stress by hole drilling (ASTM E837)

How it works: A hole of predetermined size is drilled in the existing member (typically in the beam or column flange), and three strain gages are attached around the hole. The stresses around the hole are relaxed and are measured by the gages. From these, the principal surface stresses are determined.

Typical uses: Determination of existing principal stresses at the steel surface.

Strong points: This is a relatively simple and direct test method that does not significantly damage the existing structure and does not require removal of a coupon.

Drawbacks: The member must be accessible and capable of having the gages attached. The test requires experienced personnel for proper interpretation of results.

2.4 Concrete Framing

2.4.1 Introduction

As O'Connor et al.[11] put it, concrete is a durable but not eternal construction material. On the one hand, it gains strength with age; on the other, it tends to deteriorate as time progresses. When an existing concrete building is being renovated, its load-carrying capacity is uncertain. How strong is this concrete? Is it suffering from corrosion? Are there steel reinforcing bars? What is their size and spacing? Is that big crack running through the floor a sign of danger? Anyone who has ever walked across an unfamiliar concrete building can continue the list.

The structural condition of concrete is very difficult to evaluate visually. As explained in Chap. 5, concrete can be damaged by corrosion of reinforcing steel bars, because the products of rusting dramatically expand in volume and cause delamination and cracking. Visual inspection can detect the outward signs of advanced stages of corrosion, such as cracking and spalling of concrete, but only tests can provide a comprehensive picture of the overall condition of the structure and the extent of its deterioration.

The general guidelines and procedures for evaluating concrete structures prior to their renovation are contained in ACI 364.1R-93, *Guide for Evaluation of Concrete Structures Prior to Rehabilitation.*[12] The guide deals with such issues as specifics of preliminary and detailed investigation, documentation, field observations, testing, evaluation, and reporting.

Like steel, concrete can be tested on-site in a nondestructive manner and in a laboratory destructively. Some of the most common tests are described here. Additional sources of information are contained in ASCE 11-90[1] and in the *Handbook of Nondestructive Testing of Concrete.*[13]

2.4.2 Nondestructive testing of concrete

Existing concrete can be tested by the following nondestructive methods (after Ref. 1 and others).

Visual

How it works: The inspector examines concrete visually, sometimes aided by a magnifying glass, flashlight, ruler, or crackscope. The crackscope is a microscope with a built-in scale; it allows the inspector to quantify the sizes of visible surface cracks (Fig. 2.5).

Typical uses: Rapid evaluation of surface conditions to detect cracking, spalling, efflorescence, and other surface defects; assessing condition of joints; detecting excessive sag and creep of framing. Visual inspection is the most common method of evaluating concrete.

Strong points: It is simple and inexpensive, problem areas can be located rapidly; it is very effective in combination with other test methods.

Drawbacks: This method cannot detect subsurface problems, except in cases of deep cracking and spalling; it can provide only subjective data; it requires experienced investigators.

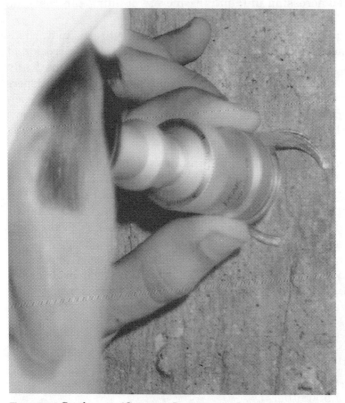

Figure 2.5 Crackscope. (*Germann Instruments, Inc.*)

Windsor probe (ASTM C803)

How it works: A special gun drives the probes into concrete, and the depth of penetration can be approximately related to concrete strength. An advanced electronic version of this system is shown in Fig. 2.6.

Typical uses: Assessing the compressive strength of concrete and finding areas of unsound concrete.

Strong points: It is among the least expensive, simplest to operate, and most commonly used test methods.

Drawbacks: It is best for approximate evaluations.

Rebound hammer (ASTM C805)

How it works: The hammer strikes a concrete surface, and the rebound distance is converted into the approximate concrete strength based on calibration curves. Electronic devices utilizing the same concept are now available (Fig. 2.7).

Typical uses: Assessing the compressive strength and quality of concrete.

Strong points: It is among the least expensive, simplest to operate, fastest, and most commonly used test methods. It can be used to locate areas of poor-quality concrete.

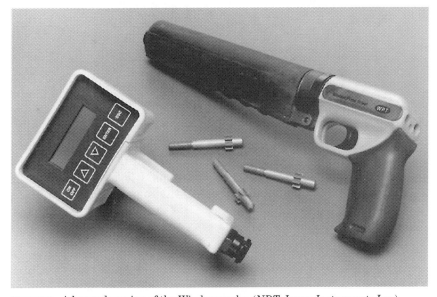

Figure 2.6 Advanced version of the Windsor probe. (*NDT James Instruments Inc.*)

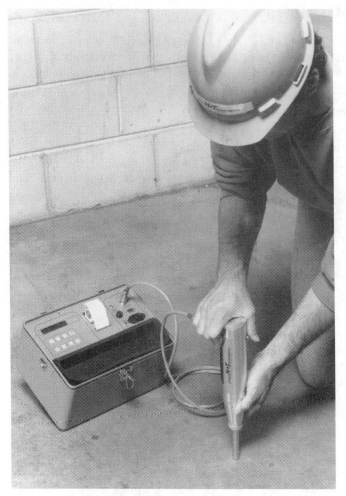

Figure 2.7   Digital rebound hammer with printer. (*NDT James Instruments Inc.*)

Drawbacks: The test requires frequent calibrations of the hammer; it is good for approximate evaluations; the results may be skewed by surface conditions and concrete proportions. It is best used when its results can be correlated with those of the compression test.

Chain drag

How it works: A heavy chain is dragged on the surface of concrete. A change in sound indicates a hollow, and therefore delaminated, area of concrete.

Typical uses: Quickly finding badly delaminated areas of slabs.

Strong points: It is fast and inexpensive.

Drawbacks: This test is very inexact and requires a trained ear. It cannot be used if overlays are present. This test has been largely superseded by the more advanced testing methods.

Hammer strike

How it works: Concrete is struck with a hammer, and a change in the reflected sound tone indicates internal voids or delaminations. The principle of operation is essentially the same as that of the chain-drag method.

Typical uses: Finding delaminated areas and internal voids.

Strong points: This test is traditionally performed during a visual investigation because of its simplicity and ease of use.

Drawbacks: It is very inexact and requires a trained ear.

Impact-echo

How it works: A mechanical impact is generated by tapping a steel ball against concrete. The resulting low-frequency stress waves are reflected by internal flaws and external surfaces and are recorded by a receiver placed near the point of impact. The frequency content provides information about the condition of the concrete, working on essentially the same principle as the hammer-strike test. The history of the impact-echo test is documented by Sansalone.[14] A computerized impact-echo device is shown in Fig. 2.8.

Typical uses: Detecting delaminated areas and internal voids.

Strong points: The equipment is portable and has a wide range of applications. The impact-echo test is more precise than the hammer strike.

Drawbacks: The geometry and mass of the object may influence the results.

Ground-penetrating radar

How it works: A portable radar unit is moved on top of the slab while sending electromagnetic impulses, which are reflected back to the transmitter. An internal void reflects the signal. The reflection depends on the electrical properties of the material, which are influenced by moisture content and degree of reinforcement.

Figure 2.8 Computerized impact-echo device. (*Germann Instruments, Inc.*)

Typical uses: Detecting areas of delamination and voids.

Strong points: This test provides faster and more accurate results than the chain drag and acoustic-impact methods.

Drawbacks: It is expensive ($3000 to $4000 per day). The results may be skewed by moisture content and steel reinforcement, are provided in strips trailing the radar's movement instead of providing a complete picture (as, for example, is possible with x-rays), are qualitative in nature, and require on-site calibration, and their interpretation depends on the experience of the personnel conducting the test.

Pulse velocity (ultrasonic) (ASTM C803)

How it works: Ultrasonic pulse waves are generated by a portable soniscope. Internal defects and voids reflect the waves in a different fashion from sound concrete. This makes them stand out and be detected. Equipment utilizing digital signal processing is now available.

Typical uses: Very popular not only for finding internal defects, cracks, and voids, but also for estimating the compressive strength, modulus of elasticity, and quality of concrete.

Strong points: The equipment is inexpensive and easy to operate, and gives relatively accurate results.

Drawbacks: Trained personnel are required in order to interpret results and conduct the test properly; density and variations in aggregate can affect results.

X-ray and gamma-ray radiography

How it works: The procedure is same as for medical x-rays—the emissions penetrate the object and are recorded on film.

Typical uses: X-rays are used to locate reinforcing bars and to examine concrete composition and density. Gamma rays can also locate voids in concrete and determine bar sizes and, to some degree, condition.

Strong points: The equipment is portable. The test is reliable, and it can detect both external and internal flaws and produce permanent records on film. Gamma-ray equipment is smaller and somewhat lower in cost than x-ray equipment.

Drawbacks: The test is ineffective for concrete more than 12 in thick, so beams generally cannot be tested. It is expensive, is potentially dangerous, requires shielding, and has to conform to health and government standards. Film has to be developed separately. Access to both sides of the object is required. X-ray equipment is heavy, and large test installations may require sources of cooling water and electric power. Gamma-ray equipment requires licensed operators.

Electric potential

How it works: Field measuring of the electric potential of reinforcing bars by portable equipment is performed to determine the probability of active corrosion. Some tests use the half-cell method of measurement given in ASTM C876, in which a copper/copper sulfate half-cell is connected to a voltmeter and to reinforcing steel. The electric potential is read; if the values are more negative than -0.35 V, the probability of corrosion is in excess of 90 percent. A series of measurements can be taken at grid points to map the probable corrosion activity. One corrosion-mapping system is shown in Fig. 2.9.

Typical uses: To detect signs of rebar corrosion, an electrochemical process (see Chap. 5).

Strong points: Measurements are readily made in the field.

Drawbacks: Trained operators are required. The actual rate of corrosion (such as percent loss of section) is not provided by this test.

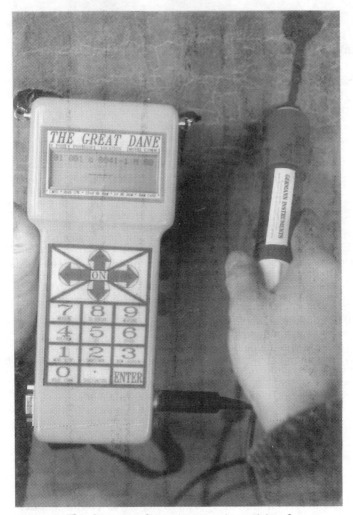

Figure 2.9 This device can determine corrosion activity of reinforcing bars by measuring electric potential gradients. (*Germann Instruments, Inc.*)

Infrared thermography

How it works: A source of infrared radiation is applied to one side of the object, and the flow of infrared energy is monitored and analyzed. Voids and discontinuities alter the flow of the emission and can be detected.

Typical uses: Detecting delaminations, hidden cracks, voids, porosity, and changes in composition. Commonly used for surveying bridge decks by mounting scanner heads and video cameras on vehicles, to measure the uniformity of natural surface temperatures 2 to 3 h after sunrise or sunset, the times of rapid heat transfer.[13]

Strong points: The test is very sensitive and accurate.

Drawbacks: Qualified technicians and specialized equipment are required.

Nuclear moisture meter

How it works: A portable machine sends a flow of neutrons into the object. Since water decreases the speed of radiation, the amount of moisture can be determined by measuring the speed of the neutrons.

Typical uses: Determination of moisture content in concrete.

Strong points: It is the most reliable moisture-content test available.

Drawbacks: The test is very expensive, and the equipment requires a Nuclear Regulatory Commission license to operate.

Pull-off test

How it works: A circular steel plate is bonded to a concrete surface and pulled off by the machine until the concrete breaks. Using charts, the force at failure is converted into concrete compressive strength. One such device is shown in Fig. 2.10. When the adhesion of overlays is tested, a core drill is used to cut through an overlay and into concrete, to a depth of at least $3/_8$ in; the core is then removed and a circular plate bonded with epoxy. A minimum rupture strength of 250 psi is typically required for bridge-deck overlays.

Typical uses: Assessment of concrete compressive strength. It can also be used to measure the adhesive capacities of various coatings and the bonding strength of overlays. Failures within the base concrete at depths less than $1/_4$ in often indicate microcracking damage. Failure at the bond line might mean poor surface preparation or poor quality of the overlay concrete. Failures deeper than $1/_4$ in accompanied by rupture strengths under 150 psi probably mean that the parent concrete is too weak to support an overlay.[15]

Strong points: The test is inexpensive and easy to use.

Drawbacks: The test causes local damage to concrete; the test procedure is still being refined.

Pachometers, cover meters, and rebar locators

How it works: The pachometer includes a horseshoe-shaped pad pressed against the concrete. It generates a magnetic field, which is

Figure 2.10 Device for conducting pull-off bond tests. (*Germann Instruments, Inc.*)

affected by reinforcing steel, and the interference is detected and quantified. Cover meters and rebar locators operate on the same principle as pachometers.

Typical uses: A pachometer can determine the size of the bars if the cover is known, or vice versa, but not both at the same time. A cover meter measures the depth to reinforcing steel (Fig. 2.11). The purpose of a rebar locator is obvious.

Strong points: These are portable, inexpensive, and popular devices. Some pachometers are sold for under $100.

Figure 2.11 Locating rebars in concrete column with a rebar locator. (*NDT James Instruments Inc.*)

Drawbacks: These devices are best used in slabs with one layer of bars. They may not detect a bottom layer of bars located exactly below the top bars because of the shadowing effect. The devices may not be effective in slabs reinforced with welded wire fabric.

2.4.3   Destructive testing of concrete

All the tests listed in this section have an obvious common limitation of damaging the existing structure; this drawback is not listed again for each test type.

Taking cores (ASTM C42) and compression testing (ASTM C39)

How it works: A cylindrical core is removed from the object and tested. The tests normally include compressive strength and modulus of elasticity. The coring is commonly done using water for lubrication; the equipment for making dry cores is available but is expensive and prone to failures. A minimum of three cores is usually needed for compression testing. Generally, a core will go right through the reinforcement, which can later be extracted for measurements. However, cores containing rebars often cannot be tested for compressive strength, because the rebar introduces a weakness into the concrete. If the rebar is removed, the core may be too stubby to test because it must be at least as long as its diameter. A specialized system for removing cores is shown in Fig. 2.12. (For budgetary purposes, the daily rate of a two-man coring crew might be $1000, with an additional $25 to $30 for extraction of each core sample, and another $25 to $30 for compression testing of each sample.)

Typical uses: Determination of compressive and tensile strength and modulus of elasticity of concrete, as well as condition of the aggregate, chloride content, density, and soundness. Taking cores is the most direct way to determine floor composition and thickness.

Strong points: This is the most reliable and widely used method of testing the properties of existing concrete.

Drawbacks: The process is slow and intrusive. Making slab cores requires access from underneath to retain the cored concrete and to contain the water used during coring. (Normally, a person with a 5-gal pail is assigned for this purpose.) Otherwise, water damage to the finishes below could be significant.

Petrographic analysis (ASTM C295)

How it works: Sections are cut from concrete cores, polished, and examined under a powerful microscope to determine the chemical and physical properties of the concrete.

Figure 2.12   A system for removing concrete cores. (*Germann Instruments, Inc.*)

Typical uses: Determination of concrete proportions, air content, denseness of cement, and the presence of chemically reactive aggregates and deleterious admixtures. The test can also detect the depth of paste carbonation and the extent of cracking in the concrete.[16]

Strong points: This method is extremely powerful, reliable, and accurate.

Drawbacks: The method requires specialized equipment and a trained petrographer.

Rapid soluble-chloride test

How it works: After the cores are removed from concrete, their ends are normally trimmed for compression testing. The trimmed pieces are ground into dust, mixed with a chloride-extraction liquid, and shaken for 5 min. The amount of acid-soluble chlorides, expressed as percent weight of concrete, is determined by a submerged calibrated electrode.[17]

Typical uses: Determination of soluble chloride content.

Strong points: This method is quick and inexpensive.

Drawbacks: The procedure is relatively new.

Tension test of reinforcing bars (ASTM E8)

How it works: The existing concrete is removed to cut a piece of reinforcing 12 in long. The removed rebar undergoes axial tension, and its properties such as strength, elongation, and reduction of area are measured.

Typical uses: Determination of yield and ultimate tensile strengths of existing reinforcing bars.

Strong points: This is the most accurate method of finding mechanical properties of steel.

Drawbacks: Removal of reinforcing bars weakens the structure. The test is slow and relatively expensive.

2.5   Load Testing of Concrete Structures

2.5.1   When is load testing appropriate?

Load testing is frequently the last resort for evaluation of structures for which the original drawings are not available and whose composition is unclear. The structures that are most commonly load-tested are those of concrete, because structural steel and wood members can be field measured and analyzed, but concrete composition cannot be readily learned. Even to determine concrete thickness and rebar spacing, the concrete must be cored or evaluated by some of the methods described in the previous section. Worse, concrete framed with one of the early proprietary systems is difficult to analyze even with the original drawings at hand. (To be sure, proprietary wood and metal systems such as bar joists of unknown designation are also excellent candidates for load testing.) Load testing is commonly performed for deteriorated framing and for structures that are theoretically overstressed by the proposed loading.

2.5.2   Applying the test load

During the load test, the structure is subjected to predetermined loading and its behavior monitored. It is common to test flexural framing—slabs, beams, and girders; compression members, such as columns and footings, are more often evaluated by analytical methods. Since the area to be tested serves as a proxy for the whole structure, it is important to choose it carefully, perhaps picking the worst-looking segment of the floor. The magnitude and pattern of the test loading should correspond to the proposed design load.

For two-way construction, the deflections should be measured at the geometric center of the panel rather than at the midpoint of the column strips. Testing of two-way slabs may involve loading large areas—four adjacent bays, for example. As Klein[18] points out, testing in this fashion applies a disproportionally high load on the slab at the column in the middle of the loaded area. Still, the only alternative to this simplified approach is to more than double the loaded area by increasing it on all sides by two-thirds of the bay length.

If it is impossible to design a single test that results in the maximum deflection and stress at all the critical areas, several tests may have to be conducted. For example, the building code may specify that patterned loading be used in the design of continuous slabs, with some loaded and some unloaded spans. For economic reasons, slabs are seldom tested in this fashion, and more commonly a single test is conducted. The suggested test setup for patterned loading in two-way construction is illustrated in Fig. 2.13.

The first step is to attach measuring devices to all five points shown in Fig. 2.13. Then the test loading is applied to the middle panel, and the deflections produced by this load are measured at the centers of the four other bays. To arrive at the maximum deflection at the center panel, the sum of the four deflections can simply be added to the one measured directly at the center panel. One can also increase the test load to reproduce this increase in the deflection at the center panel.[19]

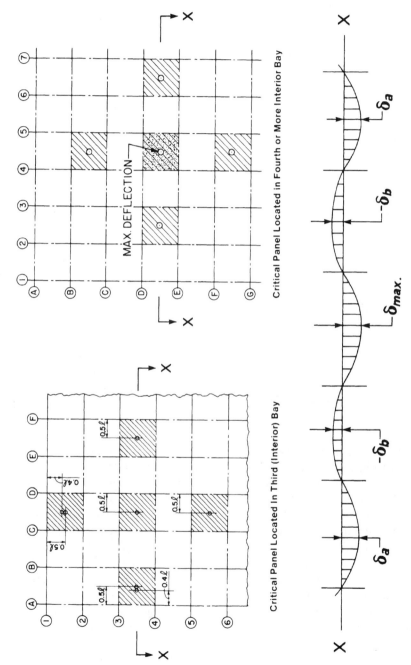

Critical Panel Located in Fourth or More Interior Bay

Critical Panel Located in Third (Interior) Bay

Figure 2.13 Load testing a two-way slab for deflection. (*CRSI.*)

Either way, the physical load testing of five full bays of a two-way slab is avoided.

For testing simply supported framing, it is best to load a complete framing bay. If a one-way slab is being tested, the loaded area may include a square with the sides equal to the slab's span. For single-span beams, the load should cover at least the area between the adjacent beams. The deflections should be measured at the center of the slab or beam.

To test multiple-span one-way framing, at least two bays should be loaded and at least two deflection measurements made. The first involves loading one span to measure the deflection of the slab or joists by subtracting the deflection at the ends from the deflection in the middle, to exclude the effects of the girder deflection. The second involves loading both spans to measure the deflection of the girder. The procedure is illustrated in Fig. 2.14.

The load can be applied by means of water contained within a wood-framed pool (Fig. 2.15), concrete blocks, or sandbags. All of these are inexpensive and readily available. The goal is to produce a truly uniform loading on the floor, if that is what the test is for, and to avoid any arching action of the ballast. This means, for example, that concrete blocks should not be laid on the floor in a running bond; if they were, they would span over the deflecting floor.

A disadvantage of using these materials is difficulty in removing the load quickly if the floor starts to fail during the test. Some other testing methods include hydraulic jacks, vacuum, and air pressure. All of these allow for quick load removal, but they are quite expensive and therefore are rarely used for in situ testing of floor framing. Hydraulic jacks are typically used for testing steel beams, wood trusses, and similar framing in the laboratory.

A load test resolves questions about the load-carrying capacity of the structure in a reliable and definitive manner. However, it is not cheap, especially in an occupied building, it is very slow, and it is potentially dangerous to the structure and to those performing the test—after all, the structure may fail the test! As a safeguard against failure, shoring is constructed in the floor below, and perhaps even in all the floors below that, all the way to the foundation level, if there are doubts about the lower structure's being able to take the weight of the falling concrete and ballast. So that the shoring does not interfere with the test, it is constructed with a gap under the structure being tested. The gap should be large enough to allow the tested structure to deflect under the load.

The expense of load testing in an occupied building is explained by, in addition to shoring, the need to partially remove the ceiling below the test area, so that test gages can be attached to the underside of the

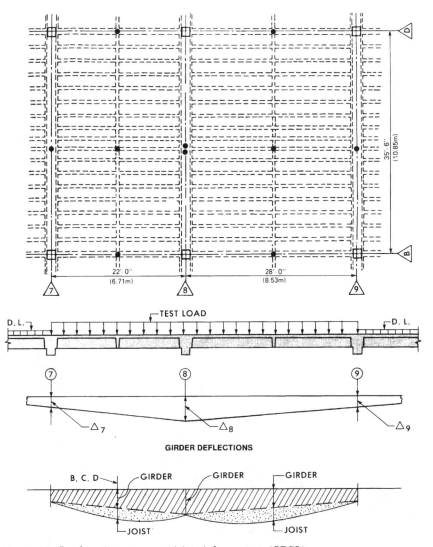

Figure 2.14 Load testing a one-way joist-girder system. (*CRSI.*)

structure. To measure member deflections, the gages may be hung from threaded rods attached to concrete (Fig. 2.16) and connected to a system of base plates and steel rods mounted on the floor below. A backup system for measuring member deflections, such as simple steel rulers, may be employed to verify the results. Alternatively, survey equipment may be used to record the deflections. An important aspect of testing is to attach the gages or other means of deflection measurement at the critical points, rather than where it is convenient.

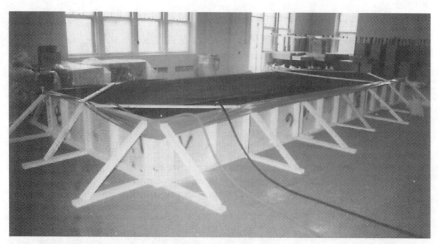

Figure 2.15 Load testing for uniform loading by means of a water pool. (*Thielsch Engineering, Inc.*)

Figure 2.16 Dial gages are suspended from the underside of the floor being tested. (*Thielsch Engineering, Inc.*)

The load test should be conducted under the supervision of a licensed engineer, who should be responsible for selecting the area to be tested, establishing the magnitude of the test load, and judging the acceptability of the results.

2.5.3 Load tests in accordance with ACI 318

The procedure for conducting load tests in concrete structures is described in Chap. 20 of ACI 318,[20] "Strength Evaluation of Existing Structures." This chapter also provides guidance on establishing the magnitude of the test load and evaluating of the results. It suggests that the test load be placed in stages, to eliminate impact and to allow the structure some time to deform under the load. After recording of the base readings (which should be taken not more than 1 h before loading), the load is typically applied in four or more equal increments approximately every half hour. The deflections are measured after each load increment and 24 h after the test. The load is then removed and the residual deflections measured 24 h after that.

According to ACI 318, the total test load should not be less than

$$0.85 \, (1.4D + 1.7L)$$

where D is the dead load and L is the live load on the structure, including the dead load already in place.

Load tests for both uniform and concentrated loading can be performed. The test for the 2000-lb concentrated load typically specified by the building codes for office occupancies can be conducted using piled-up sandbags or concrete blocks (Fig. 2.17). The magnitude of the concentrated test load can be determined by computing

$$0.85 \, (2000) \, (1.7) = 2890 \text{ lb}$$

ACI 318 deems the test satisfactory if the structure does not show any evidence of failure (spalling and crushing, for example) and if the measured deflection of the member does not exceed the value

$$l_t^2/20,000h$$

where l_t is the member's span and h is its overall thickness. Alternatively, the maximum *residual* deflection should not exceed 25 percent of the maximum deflection.

If the test fails to meet these two criteria, the code allows a repetition of the test, which can be performed not earlier than 72 h after the first test. The second test may be considered acceptable if the maximum residual deflection does not exceed 20 percent of the maximum deflection during the second test, measured from the level at the beginning of the second test.

Any cracking that occurs during load testing is a cause for concern and should be investigated. Cracks that indicate an imminent shear failure of the tested members are of special importance and should be viewed with alarm. The code specifically mentions long cracks (those

Figure 2.17 Load testing for concentrated loading. (*Thielsch Engineering, Inc.*)

with a horizontal projection longer than the depth of the member) occurring in regions where transverse reinforcement is absent. Similarly, horizontal or short inclined cracks in the regions of probable anchorage or lap splices of reinforcement should be investigated, because such cracks may be precursors of brittle failure.

2.5.4 Load tests in accordance with other codes

Model and local building codes may specify their own procedures for conducting load tests. The procedure contained in the *BOCA National Building Code*[21] is representative. This code requires applying a test load equal to twice the design loading, leaving it in place for 24 h, and measuring the deflections. If the maximum deflections do not exceed the limits established in the code for new construction, *and* the deflection recovery measured within 24 h after removal of the test load is at least 75 percent, *and* the structure does not show any evidence of failure, the structure is considered to have been successfully tested.

The procedure outlined in ACI 318 is intended for concrete structures but is sometimes used for testing of steel and wood joists supporting concrete floor and roof slabs. The procedure of the BOCA code may be used for any type of framing, but since it requires a larger test loading for acceptance than the ACI procedure, it is rarely used for concrete structures.

2.5.5   Case study: an attempted evaluation of an old reinforced-concrete structure

This short case study is included to illustrate the challenges of evaluating loadbearing conditions of old concrete buildings on a limited budget. A longer example illustrating the procedure for conducting a load test for both uniform and concentrated loads is included in Case 2 of Chap. 12.

A design firm was called by a prison administrator. The administrator wished to move the prison's library from the basement to the second floor of an old concrete building. She had already contacted the state department responsible for allocating the construction funds. An employee of that department, a structural engineer, had told the administrator (correctly) that the state building code required a minimum live-load capacity of 150 psf for library stack rooms but only 40 psf for prison cells. The administrator wanted to know whether the floor would take the load; she was looking for a formal fee proposal by the design firm to make this determination.

Since no existing structural drawings were available for the old concrete building, which was occupied under heavy security, the first reaction of the design firm's engineer assigned to the case was to advise the project manager against accepting the commission. He felt that the required investigative efforts would probably include a full-blown load test and would be so extensive that they could not possibly be justified for such a small area. Still, the client insisted on a written proposal, and the project manager wished to oblige.

A site meeting was arranged to review the field conditions. During a preliminary walk-through, the designers learned that only an overall small-scale plan of the whole facility was available. The structure appeared to have been constructed in the 1930s, a time when there were few concrete design standards and variable concrete practices.

The underside of the proposed second-floor library room was covered with ceiling, and the space below was occupied by the prison adminis-tration. Even a brief visit to those rooms required filling out a lot of paperwork and disturbed the occupants; it yielded little information other than the interior room dimensions.

Next, the team visited an area in the basement two levels below the proposed library location. There, the framing was exposed. It consist-ed of 4-ft-wide shallow concrete beams spaced 11 to 12 ft on center car-rying the slab, supported by concrete columns. The slab thickness was estimated as 7 to 8 in, based on measurements conducted in an adja-cent stairway, because no slab openings were present. The concrete seemed to be in good condition, with little observed damage.

The team then examined the exterior of the building to verify the validity of the assumed framing orientation and to check for any signs of settlement and other damage (none were found). Back in the office, a presumed framing plan of the room was sketched out to scale and the thinking began.

Looking for ideas and rough costs for the detailed investigation, the engineer contacted two reputable material-testing firms. As a result of the discussions, two approaches were considered theoretically feasible for this case—load testing and a combination of destructive and non-destructive testing.

Approach 1: Load testing. Both firms confirmed that conducting a load test was the best method of condition assessment in this situation. Both asked for the magnitude of the test uniform load, which was com-puted by the engineer as shown below.

The existing dead load was estimated as

7-in slab	85
Beams	31
Finishes	3
Total dead load	119 psf

Using a live load of 150 psf, the total required uniform test load was

$$0.85\,(1.4D + 1.7L) - 1.0D = 0.19D + 1.445L$$

$$0.19 \times 119 + 1.445 \times 150 = 239.4 \text{ psf} \qquad \text{say 240 psf}$$

It would probably be necessary to test for 240 psf twice. The first time, the load would be placed between the assumed beam centerlines, to test the slab. The second time, the load would be centered over the beam, to test the beam. An alternative was to apply the load to the whole area between two adjacent beams, as described in Sec. 2.5.2. To determine deflections under the load, special gages would have to be attached to the underside of the concrete and extend almost to the floor below. The loading medium could be concrete masonry units (CMU blocks), but most likely would be water pumped in specially designed heavy plastic bags. The required loading would equal 240/62.4=3.85 ft of water.

The load testing would be very expensive, advised both firms. One firm insisted on shoring two levels of framing during the test; the other did not. The first gave an estimate of about $20,000, mostly for protective shoring that would be built to within 1 in from the underside of the floor slab. Not surprisingly, the second firm, which did not feel that shoring was needed, thought this work to be worth only $10,000.

The cost estimates for conducting the tests did not include the cost of restoring the finishes in the rooms below. Also, what was to be done about the partitions below that were attached to the underside of the slab? Ideally, those should have been removed, or at least a wide slot cut at their tops to avoid damage from deflecting slabs. All this work could easily cost another $5000.

Approach 2: A combination of destructive and nondestructive testing. Nondestructive work could consist of pachometer testing to determine the location and size of the existing rebars on a sample section with a minimum size of 3 by 3 ft. Destructive work would include four to six slab cores to determine the slab depth at various parts of the room. As mentioned in Sec. 2.4.2, a pachometer can determine the size of the bars if the cover is known or vice versa, but not the cover and the size at the same time. A pachometer could also be used from the bottom, but then the ceiling would have to be removed in the same 3- by 3-ft section. Slab coring or chipping would reveal the bar locations, and this could be followed by pachometer use. Once the compressive test had been performed on the cores and the bar location was known, the load-carrying capacity of the floor would be analytically determined.

This method obviously would not be as direct as the first one, but promised to be less expensive. However, it would still be quite intrusive, since making slab cores would require access from below to catch the cored concrete before it fell down and to contain the water used during coring. Another problem was the possibility that the pachometer would not detect a bottom layer of bars if it was located exactly below the top bars, because of the shadowing effect.

Instead of a pachometer, one could try a ground-penetrating radar (GPR) to determine the location and size of the existing rebars, even though the GPR is best for finding voids in concrete, not rebars, because GPR also suffers from the shadowing effect. The GPR can determine the bar depth, but not necessarily the bar size. Some slab cores to determine the slab depth would still be needed. The work would have to be repeated several times to determine rebar sizes and locations in the beams—and even to determine where the beams actually were. The repair of the finishes and the efforts involved in moving the offices below would still be required.

This approach, while theoretically capable of doing the job less expensively than the load testing, was essentially open-ended, and its effectiveness was less than certain.

Eventually, the design firm submitted a preliminary proposal for some $30,000 to evaluate the structural capacity of the building, the vast majority of which was assigned to the load test and repairs of finishes. Needless to say, this was way above what the prison's budget allowed, and the actual investigation never took place.

The most ironic part of this case was not the result—that was expected from the beginning. The engineers involved in this study felt that the library relocation could probably be accomplished without incident—if the administration just did it without requiring a formal engineering evaluation.

2.6   Post-Tensioned Concrete Framing

Post-tensioned concrete buildings present special challenges to investigators. In a safe post-tensioned structure, not only the concrete, but also the tendons and their anchors must be reasonably free from deterioration. If the existing drawings for this type of framing are not available, the load-carrying capacity is often determined by load testing. Small differences in tendon sizes may not be reliably detected by nondestructive methods of evaluation but could be crucial for analysis.

Evaluation of post-tensioned concrete structures includes many of the above-mentioned methods of condition assessment. The first step is a review of the available contract and shop drawings for the post-tensioned structure. During this review, the engineer can determine the type of structure and the design locations of tendons and anchors. Next, a walk-through is conducted to review the general condition of the structure. The inspector should be on the lookout for the following telltale signs of problems, as recommended by Ref. 22.

- Cracks and joints that leak
- Slabs and beams that show excessive deflection
- Failed tendons that have erupted through the slab or anchorage
- Evidence of grease on the underside of a slab (this may signal broken sheathing)

- Corroded tendon anchors exposed to view, such as those located in the expansion joints or in the exterior edges of slabs; similarly, any rusted embedded steel members
- Tendon sheathing protruding through the slab

Posttensioned tendons and their connectors can be inspected using the assessment methods for structural steel and reinforcing bars. The cover and location of tendons can be found by ground-penetrating radar, pachometers or cover meters, gamma radiography, x-rays, and probe holes (these must be made with extra caution, for obvious reasons). Once a tendon is located, a small section of concrete can be carefully removed to expose it.

A tendon can then be inspected or a portion of it removed for testing. The inspection involves cutting the sheathing and examining the tendon steel and the protective grease that covers it. Grease that has dried out or contains moisture is a cause for concern. Reference 22 suggests attempting to insert a screwdriver between the wires; if this can be done, one of the wires may have lost tension or be broken.

Removing a tendon specimen is quite involved. Before a portion of the tendon can be safely taken out, lock-off devices usually must be installed to maintain tension in the rest of the tendon.[23] A tension test can be used to determine the breaking strength of the tendons; a chemical test, their composition.

The evaluation and repair of post-tensioned structures is described in more detail in Chap. 7.

2.7 Wood Framing

Being one of the oldest construction materials, wood framing can be found in many existing buildings. Among the available types of wood structures are sawn lumber, engineered trusses, glued-laminated lumber (glulam), laminated veneer lumber (LVL), composite wood and steel members such as flitch beams and tied timber arches, beams with plywood webs and timber chords, and stressed-skin panels.

Wood is susceptible to damage from rot, insects, and moisture. Heavy timbers are prone to horizontal cracking and splitting. Much of this damage can be detected during field investigation by visual observation and some probing. Naturally, if the framing is covered with finishes, exploratory demolition has to be done first. In cases where framing is proprietary or cannot be readily analyzed for other reasons, a load test can help in determining the load-carrying capacity. Additional discussion concerning various types of damage and methods of repair for wood structures can be found in Chap. 8.

2.7.1 Visual inspection

As usual, the first task facing the inspector conducting a field investigation is to determine the type of structure and check it against any available drawings. For sawn lumber, the species can be identified by checking the sides of the joists and rafters for stamped information, by visual identification, or by removing a piece of wood and sending it to a laboratory. Visual inspection can detect many signs of distress, such as excessive member deflections, delamination of arches and other glulam structures, inordinately large checks and splits in large timbers, and loose, missing, or corroded mechanical connectors in trusses. The inspection can also detect signs of water penetration or excessive condensation that, if not fixed in time, may invite damage from rot and insects. Any evidence of fire damage will also become apparent during the walk-through.

Visual inspection should cover the whole building, but there are a few specific places where damaged wood is most likely to be found. These include bathrooms and kitchen areas, where warm temperatures, moisture, and occasional water leaks are likely to promote both wood rot and attack by insects. Wood framing around bathrooms is also likely to have been notched by generations of careless plumbers laying and replacing pipes. In addition, these areas are likely to be covered with floor tile, sometimes mudset—installed over a thick level of mortar—which adds substantial dead load to the already weakened framing. Heavy cast-iron bathtubs and other appliances contribute to the load as well. As a result of all these factors, sagging floors under bathrooms are common.

Other areas that should not be missed include framing and sheathing around skylights, chimneys, and vents, and any wood in contact with exterior masonry. Sill plates, headers, and the ends of floor joists are especially susceptible to wood rot and insect attacks.

Throughout the building, the inspectors should watch out for the telltale signs of insect damage. Subterranean termites, for example, can be detected by their earthen tubes extending from the ground to the source of wood, by the presence of the shed wings, and by the swarming winged adults in the spring. Termites eat the middle of the wood while leaving the exterior intact, and wood suspected of being infested can be probed with a knife. The presence of carpenter ants and powder post beetles is evidenced by small piles of fine dust, known as frass, below the infested areas. To identify the insect species, the frass may be examined through a microscope.[2]

Wood rot is caused by microscopic fungi that grow within the wood and feed on the content of the wood cells. When the cells are destroyed, the wood disintegrates and decay becomes visible. In the early stages of decay, wood simply looks discolored, with some brown streaks and blotches. Later, it loses its luster and gets a stale, musty smell reminiscent of mushrooms. Still later, it may become brownish and crumbly and easily break into small pieces, a condition commonly called "dry rot," even though the decay actually occurred when the wood was wet. Alternatively, wood may look bleached white and spongy—"wet rot."

In either case, the damage can be detected by probing with a screwdriver, which will easily penetrate the decayed wood. Other signs of probable decay include moss and mushrooms growing on the surface of the wood. The areas vulnerable to rot include the same ones commonly attacked by insects. In addition, the end cuts of exterior wood framing, trim, and siding are especially prone to rot damage, because the exposed ends absorb water much more easily than other areas.[24]

2.7.2 Nondestructive testing of wood

Like steel and concrete, wood can be tested on-site in a nondestructive manner and in a laboratory destructively. Some of the most common tests are described here. Additional sources of information are contained in ASCE 11-90.[1]

Moisture meter

How it works: A moisture meter is a portable device with two needle-type prongs (Fig. 2.18). The prongs are driven into the wood, and the moisture content is indicated by the meter, which operates on either a resistance or a dielectric principle.

Typical uses: Determination of moisture content in wood, which is critical for evaluation of strength and susceptibility to rot.

Strong points: This is a small and easy-to-use device commonly carried by engineers during field investigation. The results are obtained on the spot, and many areas can be quickly checked.

Drawbacks: Correction charts supplied with the meter must be used. The meter may not properly evaluate pressure-treated and decayed wood. The results depend on the temperature and wood species.

Probing

How it works: A pointed tool is jabbed into wood; a decreased resistance may indicate rot. Early decay may be found by prying the inserted probe in order to split the wood. Sound wood breaks in long splinters, while partly rotted wood tends to break in shorter pieces with abrupt across-grain breaks.[25]

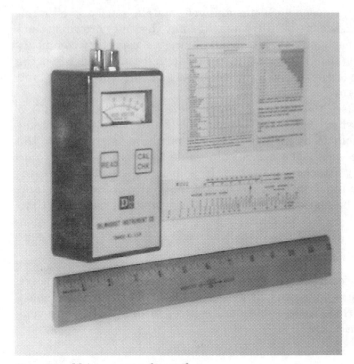

Figure 2.18   Moisture meter for wood.

Typical uses: Finding areas of decay.

Strong points: This method is very easy and inexpensive, and is commonly used during field investigation.

Drawbacks: The method may not detect deep decay in large and preservative-treated members.

Hammer sounding

How it works: Wood is rapped with a hammer. If the hammer does not rebound, or if it makes a hollow sound, there is a good probability of internal voids or decay.

Typical uses: Finding decayed areas. Also good for checking loose mechanical connectors, which may vibrate when struck, as opposed to making a clear-sounding ring.

Strong points: Because of its simplicity and ease of use, the test may be performed during visual investigation.

Drawbacks: The test is very inexact, requires a trained ear, and can detect only advanced stages of decay. Hammer sounding can be supplemented by coring or drilling.

Radiography

How it works: The x-rays are applied to the object. The radiation identifies internal voids, changes in grain direction, and other defects that offer less resistance to penetration than sound wood.

Typical uses: Finding internal defects, such as rot and insect damage.

Strong points: The equipment is portable. The test is reliable and can detect both external and internal flaws. Permanent records on film can be produced.

Drawbacks: The test is expensive and potentially dangerous, and requires shielding; it has to conform to health and government standards. Film typically has to be developed separately, although some laboratories can develop it on-site. Access to both surfaces of the tested wood member is required.

Impact-echo

How it works: A mechanical impact produces low-frequency stress waves, which are reflected by internal flaws and external surfaces and are recorded by a receiver placed near the point of impact.

Typical uses: Detecting areas of decay and voids. The test can also be used to determine the modulus of elasticity and strength of wood.

Strong points: The equipment is portable and can be used for a wide range of applications.

Drawbacks: The test is relatively expensive; the wood density must be obtained first.

2.7.3   Destructive testing of wood

All the tests listed have an obvious common limitation of damaging the existing structure, even if a small sample is taken or a minor area disturbed; this limitation is not stated again for each test type.

Identification of species (ASTM D245)

How it works: A specimen 2 in long and about 0.25 in thick is removed and examined under a microscope, through a 10× magnifying glass, or even by a trained naked eye to determine species.

Typical uses: Knowing the species permits an approximate determination of the allowable stresses in wood.

Strong points: The test is accurate and reliable.

Drawbacks: It requires submitting the sample to a trained wood specialist.

Drilling

How it works: A portable drill is used to find the areas where resistance to drilling decreases, a potential sign of rot. The drilled-out dust can be examined visually or sent for a laboratory analysis.

Typical uses: Finding areas of decay.

Strong points: The test is easy and inexpensive.

Drawbacks: The test may become time-consuming for large decay areas. The drilled holes must be treated with a preservative to prevent them from becoming avenues of further deterioration.

Coring

How it works: A core of the wood is taken, visually examined, and sent for a laboratory analysis.

Typical uses: Finding the depth of decay and fire damage. The lab test can determine density, type of decay, and other data.

Strong points: The test is direct and reliable. Some initial information is obtained on the spot.

Drawbacks: This is time-consuming for large decay areas. The drilled holes must be filled with plugs treated with preservative to prevent them from becoming avenues of further deterioration. Final results are not immediately available.

Laboratory analysis (ASTM D143)

How it works: The cores, dust, or other samples are analyzed in the laboratory to determine a variety of mechanical properties.

Typical uses: Determination of mechanical properties of wood members and steel connectors.

Strong points: The test is most definitive. A wide range of information can be obtained.

Drawbacks: The test is costly and takes time; the results apply only to the members tested.

2.8 Masonry

2.8.1 Introduction

Masonry, along with wood, is one of the oldest and most common construction materials. It is found in many framing types, although masonry walls are its most typical form. In this book, the discussion is confined to investigation and strengthening of masonry walls, of both solid and cavity types. The discussion of structural masonry can be found in Chap. 9, and that of exterior masonry walls in Chap. 13.

Buildings with loadbearing brick walls are ubiquitous. Traditionally, when renovations of such buildings were required, the investigation rarely went beyond visual inspection and perhaps taking a few cores to determine the wall composition. Increasingly, detailed information about material properties (strength, modulus of elasticity) is required, prompted in large part by seismic design considerations. A somewhat different challenge arises during restoration of historic buildings, where determination of some other material properties is needed in order to find matching replacements.

A typical masonry wall assembly consists of brick, block, or stone units bonded together by mortar. It can also include horizontal and vertical reinforcing, embedded anchors, plaster, and insulation. This section deals primarily with investigation of structural properties and those relating to weathertightness. A more detailed treatment of masonry testing can be found in ASCE 11-90.

2.8.2 Visual inspection

Visual inspection can uncover a lot of masonry problems. As usual, the first task facing an inspector is to determine the type of masonry structure and to check it against the available construction drawings. Old architectural drawings usually indicate the thickness and composition of the walls; the most reliable way to check this information is to take wall cores (see Fig. 2.3).

Visual inspection can determine whether masonry walls are in good physical condition or are bowing, bulging, settling, and cracking. On-site examination can also tell whether individual masonry units or the mortar that binds them is deteriorating. The areas of deterioration that can be detected include chipping, spalling, pitting, and crumbling of units and loose, crumbling, and missing mortar. Efflorescence can also be readily detected by observation. To check the condition of ties between various layers (wythes) of masonry, the investigator may require some exploratory demolition (see Fig. 2.4).

To document qualitative field observations, visual inspection can include, or be complemented by, a program of physical measurements by surveying instruments or other methods. The measurements might be needed to describe the effects of differential settlement, out-of-plumbness, bowing, and locations of wall defects such as cracks and missing bricks. Performing meaningful measurements of multistory structures may require exterior scaffolding, which drives up the cost of this program.

2.8.3   Nondestructive testing of masonry

If the information collected during the visual inspection stage needs amplification, masonry can be tested on-site in a nondestructive manner and in a laboratory destructively. Masonry can be tested in situ by some of the same methods described for other materials, especially concrete (Sec. 2.4). These include ultrasonic (pulse velocity), gamma radiography, acoustic impact (such as the hammer strike), and cover meters. Some other methods are described in Chap. 9 and in ASCE 11-90.[1] A good book devoted exclusively to this subject is *Nondestructive Evaluation and Testing of Masonry Structures* by Bruce A. Suprenant and Michael P. Schuller.[26] It describes, among other topics, shear testing, flatjack tests, and in-place bond wrench methods.

Investigation of leakage through masonry walls may involve some specific techniques, such as using a metal detector to find corroded embedded steel elements. Infrared thermography can be used to find the areas of mortar bridging in the insulated walls (on cold days, they will show as warm spots) and to pinpoint the exact areas of inward or outward air leakage. The borescope can be used to look inside a wall through a small opening. This thin periscope-like device contains a fiber-optic light source that illuminates the area being viewed. The borescope can help identify corroded steel ties and areas of mortar bridging.[27]

The procedure for in-place masonry shear testing, a popular new method for evaluation of mortar strength, is briefly described in Chap. 9 and illustrated in Case 2 of Chap. 12. Further information about this procedure can be found in the *Uniform Building Code*.[28]

2.8.4   Destructive testing of masonry

All the tests listed here have an obvious common limitation of damaging the existing structure by removing a sample of masonry; this limitation is not stated again for each test type. Unlike samples of wood or steel, a masonry sample is substantial in size, and the effect of its removal on wall strength is not insignificant.

Compressive (ASTM C67, C140, C170)

How it works: A single unit of masonry (a brick, block, or stone) or an assembly (a masonry prism) is removed from the wall and tested in compression.

Typical uses: Determination of allowable stresses in existing masonry.

Strong points: The procedure is standardized and reliable.

Drawbacks: There are none, other than those mentioned above.

Modulus of elasticity (ASTM C67, C140, C170)

How it works: An assembly is removed from the wall and loaded in a prescribed fashion.

Typical uses: Determination of modulus of elasticity of existing masonry. A similar test can determine modulus of rupture.

Strong points: The procedure is standardized and reliable.

Drawbacks: There are none, other than that mentioned above.

Petrographic analysis (ASTM C856)

How it works: A chemical and physical analysis of masonry is done.

Typical uses: Determination of mixture proportions and detection of evidence of alkali-silica and alkali-carbonate reactions.

Strong points: This is an extremely powerful test that can provide a variety of accurate information.

Drawbacks: The test requires specialized personnel (a petrographer).

Freeze and thaw (ASTM C67)

How it works: A sample of brick or structural clay tile is subjected to 50 freeze-thaw cycles, and its deterioration is measured and recorded.

Typical uses: Determination of resistance to freeze-thaw damage.

Strong points: A standard laboratory test is available.

Drawbacks: Specialized laboratory personnel are required.

Moisture

How it works: For brick and structural clay tile, a unit is saturated by water and the weight gain is recorded. For concrete block, a sample unit is dried and the change in its weight and dimensions is recorded.

Typical uses: Finding the rate of absorption, drying shrinkage, and other moisture-related characteristics of masonry materials.

Strong points: A standard laboratory test is available.

Drawbacks: Specialized laboratory personnel are required

Some other, less commonly performed tests include a test for water permeability, which determines the amount of water penetration through a standard masonry sample; various tests for wall finishes dealing with chemical resistance, opacity, etc.; and a brick test for resistance to efflorescence.

Like concrete, masonry can be load-tested in accordance with the general guidance provided in local building codes (see Sec. 2.5.4) and various specialized standards. Load testing can help determine not only the allowable compressive or lateral load on the masonry wall, but also flexural bond strength, and even the allowable diagonal tension (shear). The testing procedure requires removal of a large sample from the existing wall.

2.9 Building Envelope

No field investigation is complete without examination of the building envelope—the exterior walls and the roof. The preliminary examination takes place during the initial walk-through. An excellent guide to conducting a visual evaluation is Norman Becker's *The Complete Book of Home Inspection*.[24] Despite the book's title, the inspection procedures it describes apply to any building. Some common building-envelope problems likely to be detected during field investigation are discussed here.

2.9.1 Inspecting the roof

Freund and Olsen[2] recommend inspecting the chimneys first—with binoculars if necessary—and then the roof, gutters, and downspouts. (We would note that a strictly structural inspection might dispense with these items if a competent architect is already performing this work.) Following that, they suggest checking the walls in a clockwise direction, starting at the north elevation.

Chimneys are indeed common sources of problems. A poorly maintained chimney might contain deteriorated, cracked, or loose bricks in the top few courses, with crumbling mortar and damaged flue tiles. Its step flashing might be loose or missing, causing leaks. The chimney could have pulled away from the building or be leaning, creating a potential safety hazard. The pulling away and leaning could be caused by differential settlement of the foundation or by uneven deterioration of mortar (the exterior mortar ages faster and thus compresses more). Chimneys without caps typically have a sloping cement course at the top to deflect rainwater. This cement course tends to deteriorate quickly, requiring periodic replacement. The condition of this cement course can be easily checked with binoculars; a rugged outline, indicative of crumbling, is easy to see.

Roof inspection normally begins with an overall walk-through to assess the general condition of the roof and find the areas of local damage, sagging, and patching. On flat roofs, typical problems include erosion, breaks, bulges, bubbles, and blisters (be careful not to step on those and puncture the roofing). Flat roofs with local depressions full of ponded water could simply suffer from insufficient slope or clogged drains, but this condition might also indicate that the roof structure is sagging because of member deterioration.

On pitched roofs covered with asphalt shingles, the signs of damage and aging—missing shingles, cupping, a loss of granules, and excessive wear in the slots—are obvious even to a layperson. The number of layers of shingles can be counted by examining the roof edges and penetrations, or can be estimated from the age of the building, assuming the service life of asphalt shingles to be 20 years. Slate or clay tiles can be cracked, broken, chipped, loose, or simply missing. Areas patched up with roofing cement suggest past leaks. In any roof, signs of growing moss, mildew, or even small trees are a cause for concern, because they may indicate rotting roof sheathing underneath (Fig. 2.19).

The typical troublesome areas, such as flashing and penetrations, are then inspected in a more detailed manner. Flashing that has lost its embedment or is torn or deteriorated can invite leaks and cause damage to structure.

The roof eaves, scuppers, gutters, and downspouts are inspected for signs of leaks. Aluminum gutters tend to leak at the joints (Fig. 2.20) and tend to overflow at low points created by inadequate pitch.

Figure 2.19   A tree growing on the roof may indicate interior decay.

Figure 2.20   Aluminum gutters frequently leak at the joints.

Gutters are certainly not structural items, but unchecked leaks can lead to deterioration of the supporting boards and subsequent spread of the rot into the ends of the roof rafters. This decay would probably be missed if the gutters were not examined.

2.9.2 Inspecting exterior walls

Exterior walls are inspected by standing at each corner and looking down the line of sight to check for any wall bulges, out-of-plumbness, and signs of settlement. Settlement can be suspected when the walls show characteristic stepped cracks (Fig. 2.21) and doors and windows are not square in their openings or won't operate properly (Fig. 2.22). Vines, moss, and other vegetation growing on the walls can add a certain charm to the building—and can also hide cracking or other wall damage.

Many existing buildings are clad in brick veneer. This type of wall construction is susceptible to a multitude of problems, ranging from efflorescence to corrosion of metal ties between the brick and the structural backup. During field investigation, the observed signs of distress, such as wall cracks, are typically noted and photographed,

Figure 2.21 Stepped cracks in masonry walls.

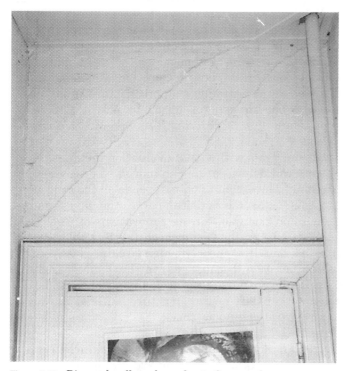

Figure 2.22 Diagonal wall cracks and out-of-square doors are typical signs of foundation settlement.

and mortar joints are inspected for signs of deterioration. Excessive efflorescence could indicate that water is seeping through the wall. Other masonry materials common in exterior walls include concrete masonry units (CMU), solid brick, and stone or stone veneer. All of these suffer from the same ailments as brick veneer—cracking, mortar deterioration, water penetration, and efflorescence. Buildings with masonry exteriors typically contain steel lintels over openings. The old lintels are frequently rusted, with associated cracking and discoloration of masonry.

Buildings clad in wood siding or shingles are ubiquitous. Wood, as already mentioned, is subject to attack by rot and insects. The most common areas of distress, which should not be missed during field investigation, are the bottom courses of siding, where water is likely to be frequently splashed and termites are likely to enter; exposed window sills; gutter boards; and sheathing around the penetrations. Again, these items are not structural, but their presence may indicate damage to the hidden structural members behind.

Many industrial facilities are covered in metal siding. Typical signs of deterioration for this wall material include rust, corroded and missing fasteners, open seams, water leakage, and physical damage (Fig. 2.23).

Some other types of exterior walls that may be encountered include glass storefronts, vinyl, asbestos and aluminum siding, natural and synthetic stucco, and proprietary systems. Each of these materials may suffer from water penetration, bulging, cracking, disintegration, and corrosion of fasteners. A systematic approach to evaluation of these systems is described in general terms in ASTM E1825-96, *Standard Guide for Evaluation of Exterior Building Wall Materials, Products, and Systems.*[29]

The foregoing material deals with visual evaluation methods. Most exterior-wall problems and deterioration can indeed be discovered by observation and probing. The extent of both depends on the condition, age, and use of the structure. In critical cases, the investigative efforts are enormous. For example, during renovations of New York City's Enid A. Haupt Conservatory—the largest U.S. public greenhouse—repairs had to be made to the 90-ft-high circular glass centerpiece. A glass mullion fell down because, as was later discovered, its fasteners had been damaged during paint-stripping operations. To make certain that no other part of the structure suffered the same fate, the project's structural engineers checked every joint and every bolt in the structure![30]

When the wall problem is obvious, but its source is unclear, testing is the only answer. For example, finding areas of water leakage

Figure 2.23 Deterioration of metal siding from aging and physical damage.

without testing may be nearly impossible, because water can penetrate through flashing locations, sealants, and at the interface
of dissimilar materials and travel within the wall. One testing
procedure includes isolating of perimeter or transition elements
and testing them by spraying water at a pressure of 30 to 35 psi.
Other wall assemblies and elements are then isolated one at a time
with a waterproof barrier seal made of polyethylene masking and

subjected to the same water spray.[31] Eventually, the area of leakage is identified.

A more detailed explanation of exterior wall problems, including water penetration, and the available repair methods is included in Chap. 13.

References

1. ASCE 11-90, *Guideline for Structural Condition Assessment of Existing Buildings,* American Society of Civil Engineers, Reston, Va, 1990.
2. Eric C. Freund and Gary L. Olsen, "Renovating Commercial Structures: A Primer," *The Construction Specifier,* July 1985, p. 36.
3. Donald Friedman, "Taking Stock," *Building Renovation,* winter 1995.
4. Carl A. Baumert, Jr. "Structural Condition Assessment of Existing Buildings," Proceedings of Structural Congress XIII, Boston, Mass., 1995, p. 4.
5. Alexander Newman, "Debating Steel-Connection Design," *Civil Engineering,* February 1994.
6. Michael R. Lindeburg, *Civil Engineering Reference Manual for PE Exam,* 6th ed., Professional Publications, Inc., Belmont, Calif., 1997.
7. Edwin H. Gaylord, Jr., et al. (eds.), *Structural Engineering Handbook,* 4th ed., McGraw-Hill, New York, 1997, p. 8-55.
8. Gary Steficek, "Creative Movement," *Modern Steel Construction,* October 1995.
9. ANSI/AWS D1.1-96, *Structural Welding Code—Steel,* American Welding Society, Miami, 1996.
10. Frank Stahl, "Bridge Rehabilitation," *Civil Engineering Practice,* Fall 1990.
11. Jerome P. O'Connor, et al., "Evaluation of Historic Concrete Structures," *Concrete International,* August 1997, p. 57.
12. ACI 364.1R-93, *Guide for Evaluation of Concrete Structures Prior to Rehabilitation,* American Concrete Institute, Detroit, 1993.
13. V. M. Malhotra and N. J. Carino, *Handbook of Nondestructive Testing of Concrete,* CRC Press, Boca Raton, Fla., 1991.
14. Mary Sansalone, "Impact-Echo: The Complete Story," *ACI Structural Journal,* November-December 1997.
15. "Methods and Equipment for Tensile Bond Testing," Repair Q&A, *Concrete Construction,* October 1998, p. 908.
16. Dipayan Jana, "Petrography: A Powerful Tool for Solving Common Concrete Problems," *Civil Engineering News,* March 1997, p. 40.
17. "In-Situ Test Systems," Catalog IST-94, Germann Instruments, Inc., Evanston, Ill., 1994.
18. Gary J. Klein, "How to Determine Floor Load Capacity," *Plant Engineering,* May 27, 1982, p. 69.
19. "Proper Load Tests Protect the Public," CRSI Engineering Data Report No. 27, Concrete Reinforcing Steel Institute, Schaumburg, Ill., 1988.
20. ACI 318-95, *Building Code Requirements for Structural Concrete,* American Concrete Institute, Detroit, 1995.
21. *BOCA National Building Code,* Building Officials and Code Administrators International, Inc., Country Club Hills, Ill. 1996.
22. *Repair Guide for Unbonded Post-Tensioned Structures,* Structural Preservation Systems Inc., Baltimore, Md., 1997.
23. Scott Greenhaus, "Post-tensioned Concrete Inspection Guide," *Concrete Repair Digest,* June/July 1997, p. 171.
24. Norman Becker, *The Complete Book of Home Inspection,* McGraw-Hill, New York, 1980.
25. NAVFAC MO-111.1, *Inspection of Wood Beams and Trusses,* Naval Facilities Engineering Command, Alexandria, Va., 1985.

26. Bruce A. Suprenant and Michael P. Schuller, *Nondestructive Evaluation and Testing of Masonry Structures,* The Aberdeen Group, Addison, Ill., 1994.

27. Norbert V. Krogstad, "Tricky Brick," *Building Renovation,* Spring 1995, p. 46.

28. *Uniform Building Code,* International Conference of Building Officials, Whittier, Calif., 1994.

29. ASTM E1825-96, *Standard Guide for Evaluation of Exterior Building Wall Materials, Products, and Systems,* American Society for Testing and Materials, West Conshohocken, Pa., 1996.

30. John Gregerson, "In the Clear," *Building Design and Construction,* May 1997, p. 39.

31. Fredrik C. Schroer, "Analyze, Observe, Test," *Restoration Direct,* STO Concrete Restoration Division, Atlanta, Ga., January 1997.

3

Renovating Steel-Framed Buildings

3.1 Steel: The Venerable Material

3.1.1 Some history

The first uses of iron date back thousands of years. Much of the iron found in prehistoric tools and weapons was probably taken from meteorites. That iron was quite strong and of excellent quality because it frequently contained hardness-imparting nickel. Native iron was so rare some 6000 years ago that it was made into jewelry by the Egyptians; it began to be used for weapons only millennia later. Tutankhamen's tomb, dated 1360 B.C., contained, among other treasures, an iron dagger.[1] There are numerous references to iron tools and nails found in the Bible.

The Iron Age is commonly assumed to have begun at about 1000 B.C., and since that time the civilizations that mastered the use of iron have consistently won their struggles for survival. For example, in the famous Battle of Marathon (490 B.C.) the Greeks demolished a vastly larger Persian army and defended their independence. They had one decisive advantage: An average Greek soldier wore 57 pounds of iron armor; an average Persian, none. The first iron production by the Greeks is traced to a large forest fire on Mount Ida in Troy (in present-day Turkey), which melted some iron ore near the surface into a primitive metal that could be fashioned into crude tools and weapons.[2]

The ancient technology relied on repeated heating and hammering of iron ore, because the furnaces then in use could not be heated to high enough temperatures to melt the iron. The semisolid iron had to be forged into shape. It is unclear when iron was first used for structural purposes, and a few isolated examples can be found throughout the world. Beckmann[3] mentions an iron pillar in India, said to have been erected in the second century, and a wrought-iron suspension bridge built in China around A.D. 600.

Widespread use of structural iron probably did not start until the industrial revolution, although invention of the blast furnace in the fifteenth century had allowed iron to be produced in liquid form and cast to any shape. Among the first documented structural uses of iron, in the sixteenth century Michelangelo encircled the great dome of St. Peter's Basilica in Rome with an iron chain to resist the outward thrust.[1]

The ancient furnaces used charcoal as fuel and could not generate enough draft to smelt iron in large enough quantities. In 1709, an Englishman named Abraham Darby developed the smelting process, which used coke as fuel and in which iron ore reacted with carbon monoxide from coke. The iron produced this way contained from 2 to 5 percent carbon, which lowered both the melting point and the viscosity of iron, making it possible to cast it into forms.[3] Some of the essentials of Darby's process are still in use today.

3.1.2   Cast iron and wrought iron

Cast iron was pioneered by the British civil engineer John Smeaton in the mid-eighteenth century. Cast iron is basically pure iron that has been extracted from ore by heating the ore layered with fuel inside a furnace. The first cast-iron Coalbrookdale bridge (England) consisted of semicircular arches and was built in 1779.[4] The first building with a cast-iron frame was reportedly the Ditherington Flax Mill, constructed in Shrewsbury, England, in 1796. Three years earlier, cast-iron columns were substituted for the usual timber in a calico mill in nearby Derby. Both these experiments were prompted by the devastating fires that frequently occurred in timber mills.[5] Cast-iron members retained high carbon content, which imparted high compressive strength—but also brittleness. Cast iron functioned well in columns and arches, but not in flexural members such as girder bridges, which proved to be unreliable and suffered many fatigue and brittle failures. Some cast-iron beams, especially lintels over masonry openings, have survived and can be occasionally found in nineteenth-century buildings.

A further improvement in iron structures took place in the mid-nineteenth century, when *malleable iron* and *wrought iron* were introduced. Both these materials contained no additives other than slag fibers and were quite pliable, but they could not be strengthened by tempering. Malleable iron was more reliable than cast iron and was widely used beginning in 1841. The first iron trusses started to appear in the mid-nineteenth century in the United States and Europe.[4] Malleable-iron structures held up better than their cast-iron brethren, but malleable iron was still not reliable enough for widespread structural use in flexural members.

The term *wrought iron* reflects its method of manufacture: repeated flattening of a piece of iron by a steam hammer and folding it over for further hammering. The objective of this procedure was purification of iron from slag, oxides, and other impurities. At the time, repeated hammering was the only known method of removing slag from metal, since the technology to melt the steel to a fluid condition, when the slag would simply float to the surface, did not exist. The hammering did not squeeze out all of the slag, and so much of it remained in the metal. These slag inclusions tended to separate the iron into layers with poor interlayer strength, almost like plywood. For this reason, welding to existing wrought iron—and to cast iron, for that matter—is problematic.[6]

The first wrought-iron beams were limited by the available steam-hammer capacities, but Henry Fielder's 1847 patent opened the way for making built-up (riveted) members from plates and angles.[3] Wrought-iron members, quite often in the shape of the rail sections, were used as purlins or light beams from about 1850 to 1885. In some industrial buildings, wrought-iron beams were supported by cast-iron columns. Both malleable iron and wrought iron were displaced by structural steel in tall buildings after about 1885 but lingered on in smaller structures for a few more decades.[1]

As shown in Table 3.1, the maximum allowable bending stresses in wrought-iron beams in the period from 1900 to 1923 were typically taken as 12,000 psi. Some engineers use an allowable bending stress of 14,500 psi in evaluation of existing structures of this period. Birkmire[7] states in his 1891 book that the elastic limit of wrought iron should be at least 25,000 psi.

Table 3.1 also contains the allowable bending-stress values for cast iron and for early rolled steel. The allowable stresses for today's cast-iron members conforming to ASTM A48 are typically taken as

TABLE 3.1   Typical Design Values for Bending Members Used in Buildings, 1900–1923

Material	Maximum allowable bending stress, psi	Average modulus of rupture, psi	Implied safety factor
Rolled steel	16,000	60,000	3.75
Wrought iron	12,000	50,000	4.17
Cast iron	3,000	40,000	13.33

SOURCE: References 1 and 11.

16,000 psi for extreme-fiber compressive stress in bending and 3000 psi for extreme-fiber tension bending stress; the allowable stresses for both shear and tension are limited to 3000 psi.[8] The allowable stresses in cast-iron columns are discussed further in Sec. 3.3.

3.1.3 Early structural steel

Structural carbon steel came into use with the invention of the Bessemer converter in 1856 and the later introduction of the Siemens-Martin open-hearth process. These two developments started the modern age of steel construction by making possible large-scale economical production of structural steel. The technology at last allowed the steel to be liquefied and slag to float to the surface, where it could be removed.

Carbon steel is a nearly ideal construction material. (By definition, it contains less than 1 percent of carbon by weight, plus small amounts of other elements.) Carbon steel is very strong, has a high modulus of elasticity (the measure of rigidity), and is isotropic (has the same properties in every direction). It can be fashioned into a variety of shapes for maximum efficiency and is noncombustible. Steel is highly ductile; it behaves elastically for more than one-half of its ultimate-strength range, and after that yields at a constant stress for a strain of 1.5 to 2 percent.[4] A steel flexural member can stretch about 20 percent before its ultimate failure—a very desirable trait that allows for timely detection and evacuation in case of an overload.

Among steel's drawbacks is its susceptibility to oxidation (rusting), so steel structures need to be protected from corrosion. Another problem of steel is reduction of its strength at elevated temperatures; it requires fire-protection measures.

Structural steel enjoyed a fast rate of acceptance. In 1884, the first steel I-beams appeared. Also in 1884, the first structural-steel frame was constructed, for Home Insurance Company Building in Chicago.[9] This building was originally designed to contain wrought-iron I-beams, but these were changed to steel during construction because of price and availability. After this project, wrought-iron beams rapidly disappeared from tall buildings.[1] In 1889, the Rand McNally Building in Chicago became the first skyscraper built with all-steel framing.[10]

As with any emerging technology, chaos reigned at first, with every steel mill producing its own assortment of steel shapes. The first attempts at introducing standardization were made in 1896 by the newly formed Association of American Steel Manufacturers, later the American Institute of Steel Construction (AISC). The first structural-steel sections were I-beams, angles, channels, and tees; wide-flange beams appeared only in the early 1900s. Continuously refining the rolling technology, the industry was capable of producing 15-in-deep members in 1884, 24-in in 1900, and 36-in by 1927.[9]

The first AISC *Specification*—less than 20 pages long—was released in June 1923. Today, the *Specification* and the accompanying *Manual of Steel Construction* encompass every aspect of steel design and contain hundreds of pages.

3.1.4   Which iron is it?

How can one tell whether the existing structural member is made of cast iron, wrought iron, or steel? Quite often, all one has to do is examine the member's configuration and the surface texture. Beckmann[3] suggests the identification tips mentioned below.

Cast-iron beams were typically cast in two-piece molds filled with tamped sand, and the sandy or gritty texture was ingrained on their sides. These beams had unequal flanges, with wider flanges placed at the bottom, in the region of flexural tension stresses. The internal corners were rounded, to reduce the buildup of shrinkage stresses, but the external corners were square. The flange width could vary along the length of the beam to correspond to the shape of the moment diagram, with the widest dimension being in the middle of the span.

Cast-iron columns were typically hollow round shapes, although some cross-shaped sections were occasionally produced. Such columns were provided with separate or integral cast-iron caps and base plates, further described in Sec. 3.3 and easily identifiable. The round columns show the vertical seams from casting at the opposing sides.

Both wrought-iron and structural steel members have a smooth surface, as both were rolled, not cast. Sometimes, wrought iron can be identified by some material delamination at the edges caused by corrosion. As already noted, wrought-iron members are typically built up by riveting together iron plates or small I-beams and flange plates. Wrought iron may also be found in tie rods for arches, roof trusses, and fasteners. Structural steel beams were usually rolled in one piece and sometimes contained a stamp of the manufacturer, listing its name and the section's weight per foot.

If the age of the structure is known, another identification clue can be found by using the dates of typical uses mentioned in this chapter. And, of course, a definitive conclusion can be reached by performing a metallurgical analysis of metal coupons, preferably removed by sawing or drilling. (The best place to take a coupon in a beam is at the tip of the flange near the beam's end, where stresses are lowest.) The laboratory analysis will not only identify the type of metal, but also provide data about its weldability and strength.

3.2 Past Design Methods and Allowable Stresses for Iron and Steel Beams

This section contains information on the typical properties of old steels and on some old design methods. For critical applications, it is important to verify the actual material properties of the metal by laboratory analysis of coupons taken from the existing members. Such analysis is described in Chap. 2. Additional information about old steels and obsolete steel sections can be found in *Iron and Steel Beams: 1873 to 1952,* a classic publication by the American Institute of Steel Construction,[11] and in FEMA-274, *NEHRP Commentary on the Guidelines for the Seismic Rehabilitation of Buildings.*[12]

3.2.1 Beams made during 1900 to 1923

The first building codes were enacted at the end of nineteenth century, because many catastrophic fires occurred in major cities during that period. The codes were based on very limited research and therefore prescribed conservative values for live loads and allowable stresses. No live-load reduction was permitted. The typical design values for iron and steel bending members used in buildings from 1900 to 1923 are given in Table 3.1.

Remarkably, as early as the 1880s, the basic concepts of beam behavior were well understood, even if the details were sometimes curious by today's standards. For example, in his 1891 book, Birkmire[7] shows tie rods framing into the webs of I-beams for lateral bracing, with so-called government anchors added for stability at supports. As now, beam-to-beam connections were made with bolted or riveted clip angles (they were called knees). The maximum size of I-beams mentioned by Birkmire is 24 in, the minimum, 4 in. Tee sections, channels, and angles with equal and unequal legs were already available at the time.

The typical steel used for buildings of that period conformed to ASTM A9; steel conforming to ASTM A7 was used for bridges. From 1900 to 1908, the medium steel category was common, with the tensile strength varying between 60,000 and 70,000 psi and the minimum yield point equal to one-half of the tensile strength. From 1908 through 1931, the structural steel category became more common, with the tensile strength varying between 55,000 and 65,000 psi and the minimum yield point equal to one-half of the tensile strength.

The reader may also encounter a variety of specifications for structural steel produced during that period. Not only were such standards promulgated by local building codes, but steel producers and construction companies published their own standards. In some instances, building designers developed their own customized design standards for major structures. Still, the vast majority of those documents incorporated the same allowable working stress for structural steel, 16,000 psi. The early structural steel beams were overwhelmingly designed as simple-span members.

Remarkably, even the early steel produced during the 1890s was not that different from the A36 steel used in the 1950s and 1960s, except that its yield point was somewhat lower.[12]

3.2.2   Beams made from 1923 to 1936

In 1923, the first AISC *Specification* increased the basic allowable working stress to 18,000 psi; it was kept at that level until 1936. From 1923 to 1931, steel conforming to ASTM A9, structural steel category, continued to be common; it had tensile strength varying between 55,000 and 65,000 psi and a minimum yield point equal to one-half of the tensile strength, but not less than 30,000 psi, a new provision. The other design requirements were essentially unchanged, except that the first cautious provisions dealing with live-load reduction for columns started to appear. The method of moment distribution was commonly used for statically indeterminate structures.

3.2.3   Beams made from 1936 to the present

In 1936, as the improved technology afforded a more uniform product quality, the AISC increased the basic allowable working stress to 20,000 psi. The tensile strength of structural steel was similarly increased to between 60,000 and 72,000 psi, with the minimum yield point specified as being equal to one-half of the tensile strength, but not less than 33,000 psi. During this time, building codes became comfortable with the concept of live-load reduction and became more prescriptive in other areas, such as wind loads.

In 1942, to conserve badly needed steel, the War Production Board approved a temporary increase in the allowable working stress from 20,000 psi to 24,000 psi. A similar increase in the allowable shear stress was mandated, but the allowable column stresses stayed unchanged.

In the 1960s, the prevalent steel used in buildings and bridges was ASTM A7, with a typical yield point of 33 ksi. Another steel used in applications requiring good weldability was ASTM A373, with a yield point of 32 ksi. Beginning with the 1963 edition of the AISC *Specification,* the minimum yield point of steel was specified as 36,000 psi, and the basic allowable working stress was increased to 24,000 psi.

Since the 1970s, the use of high-strength steels, especially ASTM A572 Gr. 50, which has a yield strength of 50 ksi, became more and more common. By the late 1990s, the premium for using that steel rather than the previously standard ASTM A36 material has essentially disappeared, and steel conforming to ASTM A572 Gr. 50 has become prevalent. This steel can be identified in the field by green and yellow markings on the unpainted pieces. The latest structural steel to come into widespread use is ASTM A992, also known as ASTM A572 Gr. 50 with special requirements per AISC Technical Bulletin 3 dated March 1997. This steel, which essentially superseded both A36 and A572 steels, has a minimum yield point of 50,000 psi and a tensile strength of at least 65,000 psi.

Table 3.2 illustrates the evolution of structural steel design standards in the twentieth century.

3.3 Early Iron and Steel Columns

As was already noted, for flexural members, cast, malleable, and wrought iron were largely displaced by structural steel after about 1885. Columns made with those materials lingered longer. Wrought-iron columns were more expensive than cast-iron and were widely used for only two decades, from 1870 to 1890.[1] Wrought-iron columns typically consisted of several I-beams, channels, angles, zees, or plates riveted together to achieve a box, a cross, or an I-beam section. The round "Phoenix" columns were made of four to eight segments riveted together. The rivet spacing was limited to 32 times the thickness of the material.[7]

Cast-iron columns outlived other applications of this material and became quite widespread. Why? Despite being weak and unpredictable in tension and bending, cast iron is very strong in compression—indeed, stronger than steel itself. One author mentions that in 1905 the ultimate strength of cast iron in compression was commonly taken as 90,000 psi, while that of structural steel was only 60,000 psi and that of wrought iron 50,000 psi. Also, fabrication of cast-iron columns required only a minimum investment, and the raw material—pig iron—was widely available.[1] Many American cities had foundries producing an assortment of cast-iron columns. Cast iron represented an economical alternative to steel, especially when used in combination with timber framing.

Another advantage of cast-iron columns was their good corrosion and fire resistance, better than those of wrought iron or steel. Wormicl[19] relates that after a huge 1956 fire that raged for 25 hours in New York City's Wanamaker department store and reduced it to a pile of rubble, the cast-iron columns remained standing—intact. That structure was built in 1864.

Not to be forgotten was the adaptability of cast iron to ornamentation and decoration. As Birkmire[7] observed in 1891, "In almost all modern buildings cast-iron columns are used, not only for their

TABLE 3.2 Typical Design Values for Structural Steel Used in Buildings

Period	ASTM specification	F_u (tensile strength), ksi	F_y (minimum yield point), ksi	F_b (maximum bending stress), ksi
1900–1908	A9 (medium)	60–70	50% of F_u	16
1909–1923	A9 (structural)	55–65	50% of F_u	16
1923–1931	A9 (structural)	55–65	50% of F_u but at least 30	18
1932	A140-32T	60–72	50% of F_u but at least 33	18
1933	A9 (structural)	55–65	50% of F_u but at least 30	18
1934–1936	A9 (structural)	60–72	50% of F_u but at least 33	18
1936–1960	A7 (structural)	60–72	50% of F_u but at least 33	20
	A9 (structural)			
1960–1963	A7 (or A36)	60–72 (58)	50% of F_u but at least 33 (36)	22
1963–present	A36	58	36	24
1970s–present	A572, Gr. 50	65	50	33

SOURCE: References 11, 17, and 26.

strength...but as a means of decoration, being ornamented to suit the style of the building and the taste of the architect." Many exterior columns were capped with ornate capitals and bases that were cast separately, even though the regular column bases were usually cast integrally.

A typical interior cast-iron column had a round shape (although rectangular and even H shapes were produced on occasion) and had simple cast brackets for girder support (Fig. 3.1). The holes for attaching the girders were kept to a minimum, because casting them was impractical and drilling them in the field was difficult. The splices were typically bolted together or could be bolted to girders.[1]

The first cast-iron columns were solid; hollow sections became available later. Sometimes, hollow cast-iron columns were used as roof drain conduits, as shown in Fig. 3.2. Such columns can be found interspersed among—and visually almost indistinguishable from—round wood columns in old mill buildings.

The technology for producing hollow cast-iron columns was at first imperfect. The biggest problem was the fact that the interior molds often shifted during column casting in a horizontal position, producing columns with uneven wall thicknesses. The spectacular collapse of Pemberton Mill in Lawrence, Massachusetts, in 1860 was blamed in

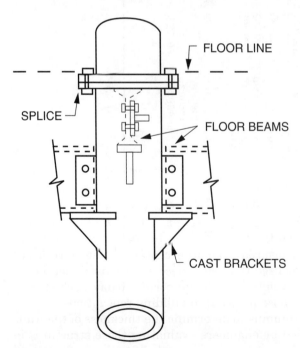

Figure 3.1 A typical cast-iron column.

Figure 3.2 This hollow cast-iron column doubles as a drain conduit.

part on such "outrageously defective," in the words of one observer, cast-iron columns. To safeguard against this defect, the Boston building law of 1885 required that the columns be drilled and the thicknesses of their walls measured prior to acceptance. New York City's building law of 1871 required that each piece of structural metal bear a stamp indicating the weight it could support.[13] (Today, such stamps can be quite helpful to those involved in building renovations.)

Drilling of cast-iron columns to determine the thickness of the walls is still in use—it is used by engineers evaluating these structures in the course of building renovations. Unless the expensive ultrasound equipment described in Chap. 2 is available, drilling small holes at the opposite faces of the column to insert a narrow steel tape is the most direct method of measuring the wall thickness. The best place to make such holes is at the bottom of the column at the opposite sides. If the measurements differ, an average can be used for analysis.[14] Once the outside diameter and the average wall thickness are known, the load-bearing capacity of the column can be analytically determined by multiplying the area by the allowable stress.

The allowable stresses can be those permitted at the time of construction (if this is allowed by the governing building code) or those permitted by today's codes. The original allowable stresses could have been a function of the manufacturing method. For example, Birkmire recommended that the maximum allowable column stress be limited to one-fifth of the ultimate strength, "provided that thickness of metal in column is uniform, with turned ends, secured top and bottom and bolted through flanges." For a more typical case of columns "secured by an uncertain method," he recommended using one-sixth of the ultimate strength. If the column ends were chipped off with a chisel (a common procedure, he noted), one-eighth of the ultimate strength should be taken.[7]

Most modern building codes allow the maximum compression stress in *new* cast-iron columns to be taken as $(9000 - 40\ l/r)$, where l/r is the slenderness ratio, not to exceed 70. In general, the maximum stresses in shear, tension, and extreme-fiber tension stress in bending in cast-iron members are limited to 3,000 psi; the extreme-fiber compression stress in bending is limited to 16,000 psi. Some engineers use these formulas for existing cast-iron columns as well, provided that the material is in good condition. The expected yield strength (or assumed proportionality limit, for purists) of cast iron produced in the late nineteenth century ranged from 15 to 21 ksi, with 18 ksi being the most common.[15]

Throughout the years, a wide variety of formulas for determination of allowable loads on steel and iron columns have been prescribed by different sources. These have been compiled by Friedman[14] for columns built from as early as 1848 to our day and are not reproduced here. A good source of information on capacity determination for old cast-iron columns can be found in Paulson et al.[16] An alternative is, of course, to use the present-day code equations and to dig deeper only if their results show an insufficient member capacity.

When steel columns eventually appeared, they were offered in a dazzling multitude of shapes and forms. Most were made of interconnected angles, plates, channels, and similar built-up sections. Latticed columns were quite popular; those were made of four angles at the corners interconnected by flat bars in a zigzag pattern. During the 1930s, these columns were largely replaced by the less labor-intensive solid sections. Since latticed columns are quite likely to be encountered during renovations and since modern textbooks rarely cover their design, we illustrate the process in Example 3.1.

Example 3.1 Check the latticed column section shown in Fig. 3.3 for an axial load of 590 kips. The column has pinned ends and is 24 ft long. The load is applied concentrically. The building was constructed in the late 1930s.

solution Determine the moment of inertia of the total section

For L 6 × 6 × 3/4,

$$A = 8.44 \text{ in}^2 \quad I_x = I_y = 28.2 \text{ in}^4 \quad r_x = r_y = 1.83 \text{ in} \quad r_z = 1.17 \text{ in}$$

The distance between the centers of gravity of the opposite L6 × 6 × 3/4 is

$$20 - (1.78)(2) = 16.44 \text{ in}$$

$$I_{ox} = I_{oy} = 4 [28.2 + 8.44 (8.22)^2] = 2400 \text{ in}^4$$

The area of the total section is

$$A_o = 4 (8.44) = 33.76 \text{ in}^2$$

The radius of gyration of the total section is

$$r_y = (2400/33.76)^{1/2} = 8.45 \text{ in}$$

The slenderness ratio is

$$KL/r_y = 1.0 (24)(12)/8.45 = 34.1$$

Determine F_a. Since the building was constructed in the late 1930s, assume that the steel has a minimum F_y of 33 ksi (see Sec. 3.2.3). Use the allowable-stress formula from the latest AISC *Specification* (see Ref. 17). The term C_c (from AISC Table 4) is 131.7 for F_y of 33 ksi. Since KL/r_y is less than C_c, use AISC *Specification* formula E2-1:

$$F_a = \frac{[1 - (KL/r)^2/2(C_c)^2] \, F_y}{5/3 + 3(KL/r)/(8C_c) - (KL/r)^3/(8C_c^3)}$$

$$= (0.966)(33)/1.761 = 18.1 \text{ ksi}$$

(Note that since F_a in the above formula is directly proportional to F_y, it can also be determined from Table 3-36 in the manual for $F_y = 36$ ksi and multiplied by the ratio 33/36. The interpolated Table 3-36 value for KL/r_y of 34.1 is 19.64 ksi; multiplying it by 33/36, we get the same 18 ksi.)

The allowable load on the column is then

$$P_{all} = A_o F_a = 33.76 \times 18.1 = 610 \text{ kips} > 590 \text{ kips} \qquad \text{OK}$$

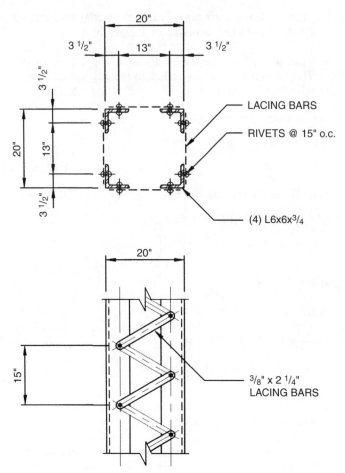

Figure 3.3 A latticed column for Example 3.1.

Check the lacing design against the provisions of the AISC *Specification* at the time.

The single line of lacing is most appropriate when the distance between the rivets in the flanges does not exceed 15 in,   OK

The inclination of the lacing bars to the axis of the column is not less than 60°.   OK

L/r_z of main angles between the lacing is $15/1.17 = 13 < KL/r = 34.1$ of the whole section.   OK

The horizontal component of the force in two lacing bars is 2 percent of the total force:

$$0.02\,(590) = 11.8 \text{ kips} \qquad \text{or} \qquad 11.8/2 = 5.9 \text{ kips on each lacing bar}$$

Axial force in lacing bar = 5.9/ (cos 30°) = 6.81 kips

For $^3/_8$-in lacing bar,

$$r_y = 0.288 \times 0.375 = 0.108 \text{ in} \quad \text{and} \quad L/r_y = 15/0.108 = 139$$

F_a=8.54 ksi (multiplying the value from AISC manual Table 3-36, secondary members, by 33/36).

$$P_{\text{lace}} = 2.25 \times 0.375 \times 8.54 = 7.2 \text{ kips} > 6.81 \text{ kips} \quad \text{OK}$$

Conclusion: The column is acceptable.

3.4 Properties of Early Fasteners

3.4.1 Rivets

From the time the first steel structures appeared up to the middle of the twentieth century, structural steel elements were connected almost exclusively by rivets. Riveting involves joining two or more pieces with a soft-steel cylinder, one rounded head of which is made during the rivet's fabrication and the other formed during the joining process. To form the second head, the rivet can be heated to a red-hot condition or can be deformed cold. Obviously, the rivet material must be soft and pliable enough to be able to deform without breaking.

A typical "hot" method used for field work involved heating a rivet to about 1800°F, inserting it in the matching holes in the joined members, and forming the second head by repeated hammering with a portable rivet gun. Shop riveting typically utilized bull riveters, much more powerful devices that were able to apply huge pressures and could form a rivet head in a single stroke. In addition, the bull riveters forced the rivets to completely fill the holes. Cold-driven rivets, as the name implies, needed no heat and relied instead on tremendous pressures to form their heads.[2]

After hot-driven rivets cooled, they shrank and squeezed the connected parts together, producing somewhat of a friction-type connection. However, the amount of the clamping force depended greatly on several variables, including the method of riveting—cold-driven rivets did not shrink at all—and friction was conservatively neglected by the specifications.

A rigorous analysis of riveted connections is far from simple. As Gasparini and Simmons[18] note, the design of riveted (or bolted, for that matter) joints is, to this day, largely empirical. The actual distribution of load among the rivets in the joint is affected by the clamping force of each rivet, by the friction and the deformation of the connected parts, and by the details of the hole edges. The number, pattern, spacing, and edge distances used in the early riveted connections were based largely on trial and error, augmented by some rough-cut computations. Not being able to calculate the stresses in rivets, the practitioners had to determine the actual strength of the connection by tests and use some percentage of it for design.

Moreover, the pioneers of riveted-connection design differed even on the basic philosophy. In 1850, British engineers Fairbairn and Hodgkinson published the results of their experiments on riveted connections. Their report essentially assumed that the rivets acted purely in shear. In contrast, Clark and Stephenson, who also published the results of their research that year, treated rivets as friction-type fasteners. The bitter rivalry between the two approaches has split engineers into two schools of thought for a long time.[19]

The consensus design specifications that finally emerged treated rivets as shear-type fasteners. To determine the strength of a riveted joint, one needed to know only the size, number, and arrangement of the rivets and the thicknesses of the connected parts.

Rivet sizes encountered in existing buildings can range from $1/_2$ in to $1^1/_2$ in, with $3/_4$- and $7/_8$-in diameters being the most common. The rivet diameter is usually shown on the original design drawings, if these are available, or can be determined by removing and measuring a rivet. (Keep in mind, however, that the design values for rivets were based on their diameters before driving.) Another way to find the rivet size is simply to measure the hole diameter and subtract $1/_{16}$ in, since the standard hole is typically $1/_{16}$ in larger than the diameter of the fastener it contains.

The most common shape for a rivet head is semicircular (button head). Occasionally, rivets with flattened or countersunk heads may be encountered; the allowable load on those has been historically taken as 50 percent of the load on rivets with button heads.[2]

The strength of the rivet material can be determined by checking the existing drawings or by testing. Typical values for old rivet steel are listed in Table 3.3. The 1949 edition of the AISC *Specification*[20] specified the allowable shear stress on rivets as 15,000 psi and the allowable tension stress as 20,000 psi.

In general, the allowable shear and tension per rivet can be established following the provisions of the current AISC *Specification.* Then, the total load on the group of rivets can be determined by the methods included in the AISC manual[17] for high-strength bolts.

3.4.2 Bolts

Structural steel members were bolted together even before riveting and welding became commonplace. Unfortunately, the first joints made with unfinished bolts were considered weak and loose.[14] It was

TABLE 3.3   Typical Design Values for Rivet Steel Used in Buildings
in 1900–1950

Period	ASTM Specification	Tensile strength, ksi	Minimum yield point
1900–1908	A9	50–60	50% of tensile strength
1909–1913	A9	48–58	50% of tensile strength
1914–1923	A9	46–56	50% of tensile strength
1924–1931	A9	46–56	50% of tensile strength, but at least 25 ksi
1932–1949	A141	52–62	50% of tensile strength, but at least 28 ksi

SOURCE: Reference 11.

only in the 1930s that bolting was given another look. By then it was realized that riveting required heating and hammering for installation, and therefore introduced a fire hazard, a great expense, and a lot of noise. These disadvantages of rivets pointed to bolts as the probable substitutes.

Rough, unfinished bolts made of low-carbon steel, also known as unturned, common, rough, or ordinary bolts, were the first to be used. When installed in the punched holes, these fasteners were considered suitable substitutes for hand-driven rivets and were assigned similar design values. Unturned bolts can be recognized by their square nuts. The tensile strength of these bolts, specified by ASTM A307-61T, was 55,000 psi. This tensile strength was to be computed on the so-called stress area, determined as

$$\text{Stress area} = 0.785 \, (D - 0.9743/n)^2$$

where D is the nominal size of the bolt and n is the number of threads per inch (quoted from Ref. 21).

The so-called turned bolts were made of stronger steel and, when installed in drilled holes, were considered equal to power-driven rivets.[21] Indeed, the 1949 edition of the AISC *Specification*[22] already specified the same allowable tension on the nominal area of any bolts and rivets (20,000 psi) and the same allowable shear on turned bolts in finished holes and on rivets (15,000 psi). In contrast, unfinished bolts were assigned an allowable shear value of only 10,000 psi.

Since the 1950s, a new breed of "high-strength" bolts conforming to ASTM A325 has been gaining popularity. In 1951 the AISC endorsed the first specification for high-strength bolts.[10] These fasteners were able to withstand high levels of tension and shear and could be installed inexpensively. According to some studies, high-strength friction bolts were more than 20 percent less expensive than rivets. Not surprisingly, high-strength bolts quickly displaced rivets in the field.

By 1963, rivets were rarely used in field connections,[14] but they lingered longer in fabricator's shops because the pneumatic machinery for rivet installation was already available there.

Table 3.4 lists the allowable shear values on various types of fasteners with $3/_4$-in and $7/_8$-in diameters that were common in the 1950s and 1960s.[23]

The Research Council on Riveted and Bolted Structural Joints originally recommended a one-for-one substitution of high-strength bolts for rivets of the same diameter.[22] Later on, these bolts were assigned much larger design values in shear and tension. ASTM A325-61T has specified their minimum tensile strength on the stress area (see the preceding definition) as 120,000 psi.

High-strength bolts could be used in so-called friction connections, which transmitted loading by means of friction forces generated by pretensioning the bolts. To develop dependable friction values, the contact surfaces had to be free from paint, oil, galvanizing, and similar coatings. In essence, high-strength bolts worked in the way some thought rivets did, except that the clamping action of the bolts could be controlled and therefore counted on in the design. The minimum stress of pretension, known as proof load, was included in the material specifications. Friction connections could be used in the design of lateral-load-resisting systems, among other applications.

High-strength bolts conforming to ASTM A325 remain a staple of structural engineers to this day. There is plenty of available literature on their design, and further discussion here is not warranted. These fasteners are identified by the mark "A325" on their heads and three radial lines. The nut is marked by three circumferential marks spaced at 120° or by the number "2."

TABLE 3.4   Some Typical Shear Values for Fasteners Used in Buildings in the 1950s and 1960s

Fastener	Allowable shear, ksi	$3/_4$-in diameter	$7/_8$-in diameter
Power-driven rivets and turned bolts in reamed holes	15	6.63	9.02
Unfinished bolts	10	4.42	6.01
High-strength A325 bolts; friction and bearing type, threads included in shear plane	15	6.63	9.02
bearing type, threads excluded from shear plane	22	9.72	13.23

NOTES:
1. Bearing value of connected members may control—check by current AISC *Specification*.
2. Values indicated are for single shear; values for double shear are twice those shown.
SOURCE: Reference 23.

3.5   Open-Web Joists

3.5.1   Introduction

Open-web steel joists, also known as bar joists, are essentially small parallel-chord trusses. Their chords are typically made of small double angles and the diagonals of round bars (hence the name *bar joists*). These structures are designed by their fabricators and are therefore proprietary in nature. The design engineer of record typically selects the required joist designation from standard tables published by the Steel Joist Institute. The designation guarantees that a joist will support a certain level of uniformly distributed loading, based on its standard designation and the design span. The designation is silent, however, on the sizes of the joist components, leaving them up to the fabricator to determine.

The first bar joists, which appeared in 1923, had the configuration of a Warren truss, with both chords and the diagonals made of round bars. The diagonals consisted of a single bent bar. The economy of this framing was readily recognizable, but its progress was slowed by the multitude of proprietary designs available, which was confusing to the specifiers. In 1928, the Steel Joist Institute was formed to bring some standardization to the industry. The institute adopted its first standard specification in 1928 and the first load table for what was later known as the SJ series in 1929.[24]

Since then, the institute has been developing a succession of open-web steel joist series, all of which can still be found in existing buildings. Figure 3.4 pictures the so-called Kalmantruss joists that were produced in the late 1920s and early 1930s. Figure 3.5 shows a few other types.

Present-day bar joists almost exclusively support metal deck floors and roofs, and it is easy to forget that the old joists were intended to

Figure 3.4   Kalmantruss joists.

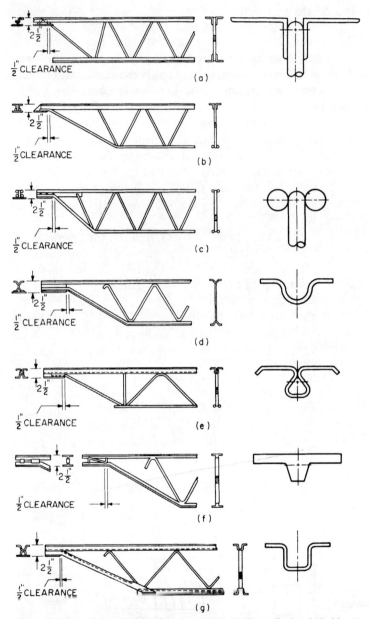

Figure 3.5 Some types of old open-web steel joists. [*From Frederick S. Merritt, (ed.),* Building Construction Handbook, *2d ed., McGraw-Hill, New York, 1965, p. 7-3. Reproduced with permission of The McGraw-Hill Companies.*]

support timber or plywood floors. Some joists had their top chords adapted to receive nails from flooring, and some had similarly adapted bottom chords as well. In most cases, special wood nailers could be attached to the regular joists, as shown in Fig. 3.6*a* and *b*, but some manufacturers made special chord sections to receive the nailers (Fig. 3.6*c*). Some joists were even advertised as capable of receiving nails directly (Fig. 3.6*d*).

3.5.2 When the original drawings are not available

Open-web steel joists are among the most economical structures around, and they have been extensively used in low- and medium-rise buildings throughout the country. The cost-effectiveness of the joists can undoubtedly be traced to their proprietary nature, as each fabricator is able to select the most economical steel sections for local conditions. The flip side of the proprietary coin, however, is great difficulty in identifying bar joist sizes in the field. Indeed, a structural engineer can face few more disappointing sights than the presence of open-web steel joists in a building of uncertain age that needs an evaluation and for which no original design drawings are available. What to do in this situation?

The first step should be an attempt to determine the age of the building; even a very approximate time frame is better than none at all. The search for information can include all the avenues described in Chap. 2, including checking the building cornerstone, where the date of construction might be carved. Once the approximate age of the building is known, the investigators proceed to measure the joists. The measurements should include the overall depth, the configurations of the diagonals, the depth of bearing ends, the type of bridging, and the sizes of chord and diagonal members. In some cases, the sizes

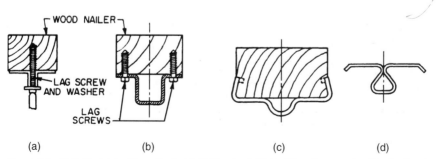

Figure 3.6 Nailable joists. [*From Frederick S. Merritt, (ed.),* Building Construction Handbook, *2d ed., McGraw-Hill, New York, 1965, p. 7-3. Reproduced with permission of The McGraw-Hill Companies.*]

of the welds connecting the diagonals to the chords also need to be measured. For critical applications, taking a steel coupon to verify the strength may be required. One often overlooked source of information is a fabrication tag, which might still be wired to a diagonal near one of the ends of the joist.

Of course, while these measurements are being made, the general condition of the framing should not be neglected. Any joists that show excessive deflection, separation of welds, advanced corrosion, or any signs of past modifications should be documented. After all, if the structure is clearly damaged, what is the point of investigating it in detail?

Armed with information about the age of the building and the sizes of the joists, the designer may be able to determine the original joist designation from the old load tables published by Steel Joist Institute. The publication *Steel Joist Institute 60-Year Manual, 1928–1988* [25] is an indispensable tool in the arsenal of any engineer involved with renovation of existing buildings. This publication includes an extensive chronology of the various bar joist series offered in the United States since 1928, as well as load tables and standard specifications for all the series.

Perhaps the most important clue to identifying the type of joist in question is the depth of its end bearing. Table 3.5 shows depths of bearing for various joist types.

Other clues that can help identify the joist series include the cross section of the web members: SJ, S, J, H, and K series typically had round bars for diagonals, while the rest of the joist series listed in Table 3.5 normally had angles. Also, the depth of the joist can help identify it. "Regular" joists—those of the SJ, S, J, H, and K series—can be from 8 to 30 in deep, and joists built between 1930 and 1952 fall into an even narrower range, from 8 to 16 in. In contrast, long-span joists of the L, LA, LJ, and LH series *start* at 18 in and can go as deep as 48 in.[25]

Sometimes, despite all the detective work, the existing bar joist cannot be precisely identified. When the depth and the series of the joist, but not the chord size, have been determined, the old load tables can be checked using the joist with the lightest possible chord size. If that joist yields an

TABLE 3.5 Depth of End Bearing for Various Joist Series

Joist series	Depth of end bearing, in
SJ, S, J, H, and K	2.5
L, LA, LJ, and LH	5
DLJ and DLH	5 for chord sizes 10–17
	7.5 for chord sizes 18–20
Joist girders	6

SOURCE: Reference 25.

insufficient loadbearing capacity, and the building cannot be "grandfa-thered," more expensive methods of analysis have to be used. One such method is to analyze the joists as trusses (this is when the sizes of the welds connecting diagonals to chords will have to be measured). To aid those involved in this thankless exercise, Table 3.6 shows typical allow-able tensile stresses that were used for various types of joists at different times. A more dependable method of capacity determination is load test-ing. One such load test is described in Chap. 12, Case 1.

3.6 Strengthening Floors

What can be done if the existing floor is inadequate for the proposed loads? This section examines a few methods of adding load-carrying capacity to existing buildings framed with structural steel.

3.6.1 Finding reserve load-carrying capacity analytically

It is much easier to upgrade the load-carrying capacity of a building by analysis than by investing in bricks and mortar. There are several methods of unlocking the reserve capacity of the framing. The first one involves harnessing the larger allowable stresses permitted by today's codes than were used originally. As was already discussed, from the 1930s until 1963, the allowable bending stress in steel beams was lim-ited by the AISC specifications to $0.6F_y$ (that is, to 60 percent of the yield stress of steel). The present codes specify a limit of $0.66F_y$, a 10 percent increase. Therefore, the total design loading on a beam can be increased by 10 percent if today's codes are used for analysis of steel members.

TABLE 3.6   Maximum Stresses in Open-Web Joists

Year	Joist series	Maximum tensile stress, ksi
1929–1958	SJ	18
1953–1961	"L," L, "S," S	20
1962–1965	J	22
1965–1978	J	22 (sometimes 30)
1961–1966	LA	20 or 22
1961–1986	H	30 (sometimes 22)
1962–1988	LH	$0.6F_y$ (22–30)
1967–1978	LJ	$0.6F_y$ (22–30)
1970–1972	DLJ	$0.6F_y$ (22–30)
1970–1988	DLH	$0.6F_y$ (22–30)
1986–present	K	30 or 22
1978–present	Joist girders	$0.6F_y$ (22–30)

SOURCE: References 24 and 25.

If the *total* design loading on a beam can be increased by 10 percent, the *live* load may be increased much more. For example, if the dead load does not change and is equal to the live load, the allowable live load may be increased by 20 percent!

A further increase in the design live load is possible if the dead load is decreased, as by removing heavy plaster ceilings and using lightweight acoustical ceilings instead, but that requires some actual construction—and disruption—as opposed to the analytical work.

The second method of increasing the theoretical live-load capacity of steel framing is to use advanced methods of analysis. The third is to check the live-load reduction provisions of the original code: None might have been permitted at the time, while today's codes allow it.

3.6.2   Adding new steel members

Here is a typical problem: The existing roof framing was designed for a uniform snow load, but the current building code specifies a much larger snow drift loading on parts of the roof. In this case, when the existing framing is grossly undersized for the proposed loading, or when the floor deck appears overstressed, the simplest course of action is to add new steel members between the existing beams and joists.

This work often requires removal of roofing and decking, so that new framing can be lowered into position by a crane, and entails some interruption of the building operations. To reduce the downtime, some preparatory work can be done in advance: The bearing seats for the proposed members can be installed, some existing attachments weakened or removed, and mechanical and electrical services disconnected. One often overlooked issue involves differences in stiffness between new and existing members. With flexible decking, this could cause uneven framing deflections under load. The best cure for this problem is to use additional beams or joists of a size and stiffness analogous to those of the existing ones.

The ultimate way of reducing the downtime for relatively small areas is a complete removal of the existing framing—beams, deck, roofing, and all—in one piece. The whole assembly can be lifted off by a crane and another preassembled section lowered into its place and attached to the walls.

Taken to the extreme, this approach may evolve into a replacement of the whole structure with new beams, decking, and lateral bracing. This radical surgery involves total removal of the interior of the building, keeping only the façade; it is sometimes performed on existing structures with exteriors of some historic or sentimental significance.

At the other extreme, the existing members and decking cannot be disturbed and must remain in place. Inserting new beams through windows and doors by hand or with the aid of some basic rigging is a challenging but not impossible task. The beams can be supplied in pieces and assembled with full-penetration welds. Complete bearing of the decking on the new beams can be ensured by means of steel shims or nonshrink grout.

But what can be done if the existing beams are incapable of carrying the proposed loading even after the new beams are inserted? This situation occasionally arises when the proposed loads are very large or the existing framing is deteriorating. In that case, removing the existing dead load in the old members can help. This can be accomplished by deflecting the old members upward with hydraulic jacks or by similar means and tying the existing and new members together by mechanical connections or by concrete encasement. Once the jacks are removed, the existing members will share their tributary dead load with their new neighbors.

3.7   Reinforcing Steel Members by Welding

3.7.1   Is the steel weldable?

Existing steel members can often be reinforced by attaching additional components to act compositely with the original sections, most commonly by field welding steel angles, channels, or bars. This work is labor-intensive and involves potential fire hazards, but it at least avoids the disruption caused by hauling the new beams into place.

The first question that comes to mind when contemplating such reinforcement is: Can we weld to the existing structure?

To answer this question, one must determine whether the existing steel is weldable. Some early steels had large concentrations of carbon, sulfur, and phosphorus, and welding them may be problematic. Too much carbon and phosphorus result in a steel that is overly brittle; high sulfur content often produces porous welds. As a rough guide, steel made prior to 1923 should be tested for weldability. Weldability of steel produced between 1923 and 1936 is generally good, and steel made after 1936 is normally weldable.[26] The reason for the qualifiers in this statement is that the ASTM A7 allowed for a wider range of chemical composition than the later ASTM A36 steel.

Unless the material specifications or mill certifications for the steel in question are available, the best method of arriving at an intelligent answer is to conduct chemical testing of the steel by removing coupons, as described in Chap. 2. This relatively inexpensive service, which can be performed by many testing laboratories, can determine the mechanical and chemical properties of steel. The critical chemical property for determination of weldability is carbon equivalent, a number that combines the percentage of pure carbon and certain fractions of several other elements present in steel. There are a number of formulas for computing the carbon equivalent. When the carbon equivalent is less than a certain number, which depends on the formula, the steel is weldable.

Of course, prior to embarking on any such testing, one should first examine the existing structure for the presence of any welded connections: If the steel was successfully welded before, the answer is obvious. If still unconvinced, the engineer may order an in-place test for weldability, which will provide information about weld ductility and hardening of the base metal. This "fillet bend test" is specified in Section 5 of AWS D1.1.[27] Essentially, it involves welding a small steel plate tab to the existing member and hitting it with a hammer. If the existing steel is weldable, the fillet weld will deform without breaking; if it is not, the weld will separate from the base material. This method is rather crude; for critical applications, there is no substitute for laboratory testing.

3.7.2 Increasing chances for successful welding

In his classic article "Field Welding to Existing Steel Structures,"[6] which should be read by everyone involved in specifying this type of work, David T. Ricker lists some of the precautions that can be taken to increase the chances of successful welding to the existing framing. Some of these steps that the design engineer can take include

- Specifying proper welding electrodes, typically E60XX or E70XX series for ASTM A7, A9, A36, and A500. Low-hydrogen welding electrodes used for manual welding can help.

- Using preheating, especially for thicker members. Both the AISC manual and AWS D1.1 contain tables for minimum preheat and interpass temperatures for various material thicknesses joined by regular or low-hydrogen electrodes. Thicker members require higher preheat and interpass temperatures because they absorb more heat from the welds and make them cool too fast, potentially resulting in increased internal stresses and embrittlement. Maximum preheat temperatures should not exceed 450°F. The interpass temperatures apply to welds made with several passes, generally those over $5/16$ in. The welds should be allowed to cool down between the passes.

- Positioning the welds properly, especially avoiding welding perpendicular to the existing stress lines, which usually run along the longitudinal axis of the member. If transverse welds are absolutely required, they should be made in stages to minimize the heating input.

- For static loads, whenever possible, specifying intermittent fillet welds instead of the more expensive and less dependable continuous groove welds. In general, do not require any more welding than is necessary, in order to reduce weld shrinkage and distortion. Similarly, partial-penetration welds should be chosen in lieu of full-penetration welds for static compression loads.

- Requiring smooth transitions between butt-welded tension members, such as tapering the width and thickness of the existing flanges welded to a narrower and thinner new plate.

- Specifying that all undercuts, notches, and gauges be filled and ground out, because most weld fractures start at such points of stress concentration.

- Specifying proper surface preparation. Existing steel members are usually covered with rust, fireproofing, or many layers of paint (or just dirt), and all these contaminants should be removed to bare metal before starting the work. Similarly, any existing galvanizing should be locally ground off. (Section 11 of AWS D1.1.[27] recommends that the contaminants be removed for a distance of 2 in from each side of the welds.)

Ricker provides many other useful bits of information. For example, welding on top of the existing welds is acceptable; E60XX and E70XX electrodes can be combined at the same weld if required; welding should be avoided at temperatures below 0°F; a minimum preheat of 70°F is needed when the temperature is between 0 and 32°F; there is no need for the cover plate material to match the material of the existing beam.

The preheat requirements mentioned here need some amplification. There are several reasons to preheat existing steel. First, preheating slows down the rate of cooling of the weld and of the base material, helping to produce a more ductile metallurgical structure and reduce the possibility of weld cracking. Second, a longer cooling time allows hydrogen to escape from the heat-affected zone (HAZ), also reducing potential cracking. Third, preheating helps reduce shrinkage stresses in the HAZ, an especially important consideration for highly restrained joints. Fourth, preheating may raise the temperature of the steel above the temperature that makes it vulnerable to brittle fracture. And finally, preheat helps preserve some mechanical properties, such as weld metal notch toughness.[28]

It is usually more economical to make longer welds of smaller size than to make shorter but thicker welds, because the volume of weld metal is proportional to the square of the weld size times the weld length. Weld cost is a linear function of weld volume. Therefore, increasing weld size increases the cost much faster than increasing length.

3.7.3 Reinforcing beams by welding

The need to reinforce existing steel beams by welding additional steel angles, channels, or bars to act compositely with the original sections is quite common. Often, this may be the only way to increase the load-carrying capacity of the framing. How is this done?

A typical reinforcing detail involves welding a cover plate to the bottom flange of the beam, as shown in Fig. 3.7a. Note that the cover plate is wider than the beam to permit horizontal fillet welding, and that intermittent welding is used to reduce shrinkage stresses, as previously discussed. Unless the existing beam acts compositely with concrete, such reinforcement is not as effective as it might seem: Increasing the effective area of the bottom flange shifts the neutral axis downward, the distance from the neutral axis to the top flange increases, and the resulting section modulus of the combined section changes relatively little. Still, a modest increase is often all that is needed. Besides, the alternative is to weld members to the top *and* bottom of the beam (Fig. 3.7b), a procedure that reduces the use of reinforcing metal but doubles the amount of the welding required.

When reinforcing existing beams in this fashion, remember to examine the existing connections and to make certain that they can carry the increased design load on the reinforced beam. The issue of reinforcing existing connections is addressed later in this chapter.

Simply welding a cover plate to the beam does not change the existing stress distribution until the additional load is applied. Indeed, a beam that is overstressed by the existing loading cannot be helped by welding additional pieces to it, because the existing stresses are already in place and are not affected by reinforcement. It is only the *future* live load that will be resisted by the combined section.

Perhaps the biggest problem of welding to existing steel is that welding generates high temperatures and temporarily reduces the load-carrying capacities of existing members. Indeed, if welding to a loaded beam elevates its temperature above the melting point, the beam may fail. For this reason, temporary shoring should be specified for members welded under load.

Shoring may be erected for another purpose as well: It is desirable to relieve the existing stresses in the beam being reinforced to the largest degree possible. Ideally, the beam should be jacked up in the middle by the amount of the existing dead-load deflection (when all live load is removed); that way, it carries essentially no stress when the welding is performed. Of course, any jacking up should be done slowly and carefully in order not to damage floor finishes and existing partitions above.

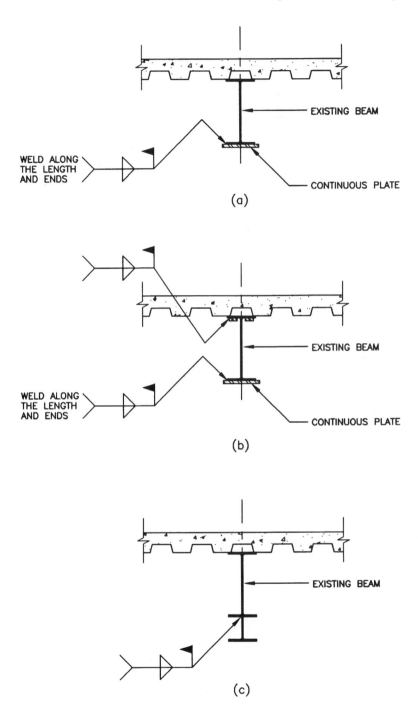

Figure 3.7 Reinforcing existing beams by welding: (*a*) Cover plate welded to the bottom flange; (*b*) cover plates welded to both flanges; (*c*) WT section welded to the bottom flange.

Example 3.2 illustrates the procedure of reinforcing an existing steel beam with a cover plate using the load and resistance factor design (LRFD) method of design.

Example 3.2   Reinforce an existing simply supported W16 × 36 beam made of A36 steel to carry a service dead load of 0.5 kip·ft and a service live load of 1.1 kips·ft. The beam spans 25 ft and is fully braced by the floor.

solution   Use the LRFD method, which assumes that the total area of the steel section is under plastic (yield) stress. The tension and compression area are separated by the plastic neutral axis (PNA). Compute:

$$w_u = 1.2 \times 0.5 + 1.6 \times 1.1 = 2.36 \text{ kips} \cdot \text{ft}$$

$$M_u = 2.36 \times 25^2/8 = 184.4 \text{ kips} \cdot \text{ft}$$

$$\text{Required } Z_x = 184.4 \times 12/\,(0.9 \times 36) = 68.30 \text{ (in}^3) > Z_x = 64.0 \text{ provided}$$

Required plastic moment to carry the load $= F_y\,(Z_{x,\text{req}})$
$$= 36 \times 68.3/12 = 204.9 \text{ kips} \cdot \text{ft}$$

Maximum plastic moment the beam can resist $= 0.9 \times 36 \times 64.0/12 =$
$$172.8 \text{ kips} \cdot \text{ft}$$

Because the beam is overstressed by only about 7 percent, try adding a bottom plate of minimal size. For ease of welding, select a plate that is wider than the beam. Try a $^3/_8$-in by 9-in plate.

Properties of W16 × 36 from AISC manual:

$$b_f = 6.985 \text{ in} \qquad t_f = 0.43 \text{ in} \qquad k = 1.125 \text{ in} \qquad d = 15.86 \text{ in}$$
$$t_w = 0.295 \text{ in} \qquad A_s = 10.6 \text{ in}^3 \qquad Z_x = 64.0$$

Properties of $^3/_8 \times 9$ plate:

$$A_{\text{pl}} = 3.375 \text{ in}^2$$

Construct the model of the steel beam following the approach of the AISC LRFD Manual, p. 4-4. Compute

$$A_f \text{ (flange area)} = b_f \times t_f = 6.985 \times 0.43 = 3.00 \text{ in}^2$$

$$A_w \text{ (web area)} = (d - 2k)\,t_w = (15.86 - 2 \times 1.125)\,(0.295) = 4.01 \text{ in}^2$$

$$K_{\text{dep}} - h - t_f - 1.125 \qquad 0.43 - 0.005 \text{ in}$$

$$A_k = (A_s - 2A_f - A_w)/2 = (10.6 - 2 \times 3.0 - 4.01)/2 = 0.295 \text{ in}^2$$

$$A_k/K_{\text{dep}} = 0.424 \text{ in}$$

$$K_{\text{dep}} + t_f = k = 1.125 \text{ in}$$

Since the section is under a uniform yield stress, the area in tension must equal the area in compression. Determine the location of the PNA; assume it falls within the web for the first try.

Area in compression = area in tension = $(A_s + A_{pl})/2 = (10.6 + 3.375)/2 =$
$$6.99 \text{ in}^2$$

$$A_f + A_k + t_w (y - k) = 6.99 \text{ in}^2$$

$$3.00 + 0.295 + 0.295 (y - 1.125) = 6.99$$

$$y = 13.65 \text{ in from top}$$

Check:

$$d - y = 15.86 - 13.65 = 2.21 \text{ in} > k$$

Therefore, the PNA is indeed located within the web.

Find plastic forces in various components of the section (see Fig. 3.8). Force = $(F_y)(\text{area})$.

$$C_1 = 36 \times 3.0 = 108 \text{ (kips)}$$

$$C_2 = 36 \times 0.295 = 10.62$$

$$C_3 = 36 (13.65 - 1.125) (0.295) = 133.02$$

$$T_1 = 36 (15.86 - 13.65 - 1.125) (0.295) = 11.52$$

$$T_2 = 36 \times 0.295 = 10.62$$

$$T_3 = 36 \times 3.0 = 108$$

$$T_4 = 36 \times 9.0 = 324$$

Sum the moments from these forces about the PNA:

Force (kips) × distance from PNA (in) =		plastic moment M_p (in · kips)
$C_1 = 108$	$(13.65 - 0.43/2) = 13.43$	1450.98
$C_2 = 10.62$	$(13.65 - 0.43 - 0.695/2) = 12.87$	137.70
$C_3 = 133.02$	$(13.65 - 1.125)/2 = 6.26$	833.04
$T_1 = 11.52$	$(2.21 - 1.125)/2 = 0.54$	6.25
$T_2 = 10.62$	$(2.21 - 0.43 - 0.695/2) = 1.43$	15.21
$T_3 = 108$	$(2.21 - 0.43/2) = 1.995$	215.46
$T_4 = 324$	$(2.21 + 0.375/2) = 2.40$	776.79
		3435.43 in · kip

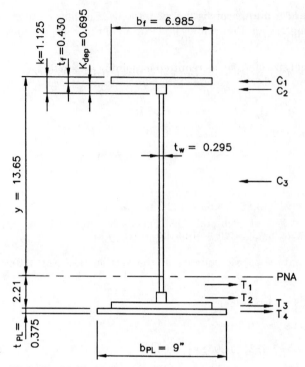

Figure 3.8 Model of the beam reinforced with welded cover plate for Example 3.2.

$$M_p = 3435.43/12 = 286.28 \text{ kip} \cdot \text{ft} > 204.9 \text{ kip} \cdot \text{ft} \qquad \text{OK}$$

Conclusion: Use a $^3/_8$- by 9-in plate.

Find the required length of the plate. First, determine the theoretical cutoff points x where the factored moment does not exceed that resisted by the beam section alone:

$$w_u = 2.36 \text{ kips} \cdot \text{ft}$$

$$R_u = 2.36 \times 25/2 = 29.5 \text{ kips}$$

$$29.5x - 2.36x^2/2 = 172.8$$

or

$$1.18x^2 - 29.5x + 172.8 = 0$$

$$x = 9.37 \text{ ft and } 15.63 \text{ ft}$$

The actual cutoff points are moved closer to the supports by the amount of the required welding.

Force in the weld = plastic capacity of plate = $A_{pl}(F_y)$

$$F \text{ (weld)} = 3.375 \times 60 = 202.5 \text{ kips}$$

Use a minimum fillet weld size of $\frac{3}{16}$ in, as required for joining a $\frac{1}{2}$-in flange and a $\frac{3}{8}$-in plate. Use two welds at each end of the plate. The plastic capacity of a $\frac{3}{16}$-in weld is $3 \times 1.39 = 4.17$ kips · in.

$$\text{Required length of } \frac{3}{16}\text{-in weld} = 202.5/\,(4.17 \times 2) = 24.28 \text{ in}$$

This is a little excessive, so try $\frac{1}{4}$-in welds. The plastic capacity of a $\frac{1}{4}$-in weld is $4 \times 1.39 = 5.56$ kips · in.

$$\text{Required length of } \frac{1}{4}\text{-in weld} = 202.5/\,(5.56 \times 2) = 18.2 \text{ in}$$

Use two $\frac{1}{4}$-in welds 19 in long at each end. Also, specify nominal welding along the length of the plate, such as $\frac{1}{4}$-in fillet welds 2@12 in.

The actual cutoff points are
$$9.37 - 19/12 = 7.79 \qquad \text{say } 7.75 \text{ ft}$$

and

$$15.63 + 19/12 = 17.21 \qquad \text{say } 17.25 \text{ ft}$$

The total plate length is $17.25 - 7.75 = 9.5$ ft. The plate should be centered between supports.

The reinforced beam is shown in Fig. 3.9.

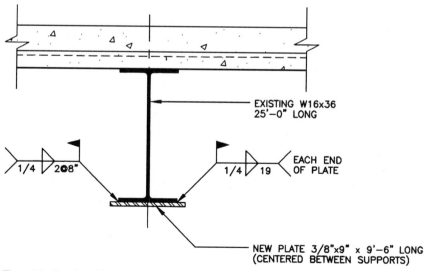

Figure 3.9 Reinforced beam for Example 3.2.

Sometimes it is impractical to relieve the existing stresses in the beams or, more commonly, the cost of doing so is much higher than whatever savings it can bring. In that case, the member is reinforced under load. As Ricker[6] observes, "The procedure is safe and feasible in most cases." Indeed, welding members under load is commonplace: In addition to buildings being so repaired, oil and gas companies routinely repair pipes carrying their products under pressure by welding, and even the eyebars in suspension bridges have been repaired by welding while carrying traffic. However, one should not needlessly create problems. For instance, if the cover plate must be spliced, it is best to splice it with full-penetration welding prior to welding it to the beam, so that shrinkage stresses created when the splice weld cools are not transmitted to the beam.

Some engineers following working-stress design methods use a conservative approach for reinforcing loaded members by limiting the stress in the added steel to the difference between the allowable and the actual stresses in the existing beams. For example, if the allowable stress in the beam is 18 ksi and it is presently stressed under a dead load to 8 ksi, the maximum stress in the cover plate should not exceed 10 ksi. This approach is recommended in Blodgett[29] and in AWS D1.1[27] for members carrying dead-load stresses in excess of 3000 psi, unless the dead-load stresses in these members are relieved or the added materials prestressed.

Others use a more common method of designing the cover plate so that the maximum stress in it does not exceed that specified by today's codes and the stress in the existing beam does not exceed that allowed by today's code for the actual grade of steel. For example, when reinforcing an existing steel beam produced in 1949 with a steel plate made of ASTM A36 material, the section should be so proportioned that the maximum stress in the plate does not exceed 24 ksi and the stress in the beam, 22 ksi. (The latter can be determined from Table 3.2, which lists the minimum yield point of the steel produced in 1949 as 33 ksi, and the 1989 AISC *Specification,* which allows the maximum bending stress for compact beams to be $0.66F_y$. Alternatively, one can use the $0.6F_y$ permitted in 1949, which results in a stress of 20 ksi in the existing beam.)

Whenever a more substantial increase in capacity than that afforded by a cover plate is desired, the beam may be reinforced by welding a WT section to the bottom flange. The attachment can be made with symmetrical fillet welds on each side of the web or by full-penetration welds. As mentioned previously, the most economical connection is made with intermittent fillet welds; these are usually supplemented by sizable end welds. A typical beam reinforced with a WT section is shown in Fig. 3.7c.

Existing composite steel beams can be strengthened using the same methodology. Further information on this subject and additional design examples can be found in Refs. 30 and 31. These references can also be used for strengthening noncomposite beams, if the thickness of concrete is taken as zero.

What about the deflection criteria for reinforced beams? The common practice is to limit the deflections to the values anticipated in the original design, so that the reinforced beams are compatible with the rest of the floor, or to follow the standard code requirements, such as limiting live-load deflections of floor beams supporting plastered ceilings to $L/360$. The situation is less clear if the target beam is visibly distorted. The choice of straightening it out before reinforcing or leaving it as is belongs to the design engineer.

3.7.4   Reinforcing columns by welding

Like beams, existing steel columns can be reinforced by welding cover plates or other sections. The basic design procedure parallels that for beams, except that columns normally have cover plates on each face. Symmetrically positioned reinforcing members may reduce the overall slenderness ratio (L/r) of the combined section and make possible higher allowable stress in compression than existed originally. Again, it is desirable to remove as much load from columns as possible before welding. For multistory structures, the effort of shoring several floors may not be cost-effective, and reinforcing them under stress may be inevitable.

If a column is reinforced to improve its bending strength in combination with compression, the procedure outlined for beams may be followed. In this instance, complete bearing of the reinforcing pieces on the existing base plate may not be needed. Complete bearing is a necessity, of course, if the reinforcement is needed to increase the column area in compression. A common technique for that is to leave a gap of about $1/_2$ in between the added material and the base plate and to fill it with weld.[6]

Another peculiarity of reinforcing steel columns is a frequent need to fix deteriorated column bases, such as that shown in Fig. 3.10. The basic approach for this kind of repair is to shore the column, remove all the deteriorated metal to sound material, and weld or bolt the reinforcing to the column, designing the connection for the full load minus the load to be carried by the column, if any can be justified.

Ricker recommends that cover plates for both beams and columns be welded starting at one end and working toward the other or starting in the middle and working toward the ends. The worst thing to do is to start at both ends and work toward the middle—this may result in the

Figure 3.10 A completely deteriorated column base.

plate's looking like a worm. For symmetrically positioned cover plates in columns, distortion caused by welding can be reduced if the welds at the opposite faces of the column are made at the same time.

Some engineers worry that welding cover plates to columns may reduce the material properties of the original sections. A program of testing of cover-plated columns has reached the conclusion that the ultimate stress of the reinforced columns is not affected, and that the influence of welding is confined to a relatively small area in the vicinity of the welds.[32] The same program noted that the heat produced by welding of cover plates resulted in temporary lengthening of the columns, followed by subsequent shortening. If the columns were prevented from such movement by other members, there was a buildup of additional stresses in the columns. Since columns are rarely truly fixed at both ends, the realistic magnitude of such additional stresses is difficult to quantify, but designers should at least be aware of this effect.

3.7.5 Reinforcing trusses by welding

Existing steel trusses can be reinforced by welding additional members, using the procedures outlined for beams and columns. If reinforcement is done to repair deterioration, new steel plates or angles may be welded to existing truss elements. For local member reinforcing, the new welded-on piece can simply overlap the damaged section by a certain distance, such as 9 in. Badly corroded members must be ground to sound metal capable of receiving the welds; if the final thickness is insufficient, the reinforcement may be designed to locally span the corroded area and rely only on the end welds.

If reinforcement is done to increase member capacity or if deteriorated members cannot support the welds, new sections may be designed to span between the truss panel points without reliance on the existing construction. The design and details would be similar to those used in the repair of light-gauge trusses, described in a case study in Chap. 10.

The repair of truss chords should be done symmetrically even if only one side of the chord is damaged. Reinforcing only one of the double angles in a chord can lead to an unbalanced chord section that transfers loading to the diagonals with an unintended eccentricity. If such one-sided reinforcement is unavoidable, the diagonal and vertical truss members should also be upgraded to resist the eccentricity. Hill[33] has investigated a collapse of steel trusses where the top- and bottom-chord angles had previously been reinforced by welding additional continuous angles in an asymmetrical fashion. He concluded that during a heavy snow accumulation, the single-angle vertical struts of the trusses became vastly overstressed from combined compression and bending caused by the eccentricity in the connection—and literally folded over. Unfortunately, many engineers might overlook this potential problem.

The foregoing discussion has dealt with reinforcing truss members, assuming that proper connections could be made to existing gusset plates. But what if the gusset plates are corroded or damaged? This situation is rather rare in buildings; it is much more common in bridges. Still, prolonged unrepaired roof leaks might damage roof trusses to a degree that could render gusset plates unsafe. The most reliable method of repair in this situation is to replace the trusses. It is possible to replace the gusset plates (and probably other truss members as well), when conditions require, by constructing extensive shoring systems. An extraordinary example of replacing gusset plates in a bridge under load is contained in Ref. 34.

3.8 Reinforcing Beams by Composite Action with Concrete

Another method of improving the load capacity of an existing steel beam is to make it act compositely with the concrete floor it carries. In composite construction, the slab becomes a part of the beam. When used in new buildings, composite action brings the well-known benefits of better economy, increased member stiffness, and decreased framing sizes. In renovation, composite action can help substantially strengthen the existing beams and increase their stiffness.

Composite action of this type is not a new phenomenon. As Griffis[35] observes, the Methodist Building in Pittsburgh, constructed in 1894, was the first in the United States to utilize concrete-encased steel beams. From that time on, composite beams have received increasingly close attention from engineers, researchers, and, eventually, building code writers. The first building code acknowledging the advantages of composite construction was the 1930 New York City Building Code, which allowed the maximum bending stress of composite steel beams to be taken as 20 ksi when 18 ksi was the norm for noncomposite beams (see Table 3.2).

There are two basic ways to achieve composite action. The first is by encasement of a steel beam in concrete, where the slab is assumed to be connected to the beam by natural bond. Proper encasement occurs, according to the AISC *ASD Specification*,[36] when concrete cover over the beam sides and soffit is at least 2 in, the top of the beam is at least 1.5 in below the top and 2 in above the bottom of the slab, and the concrete contains wire mesh or other reinforcing steel in its sides and bottom to prevent spalling. Existing beams are rarely strengthened by encasement, because of the costs involved and because the top surfaces of beams are typically located directly under the slabs rather than 2 in above.

Composite beams can be constructed with or without temporary shoring being kept in place until the concrete hardens. Shored composite-encased sections are assumed to carry all dead and live loads. For unshored beams, the AISC *ASD Specification* requires the encased steel section to carry all dead and superimposed loads applied prior to concrete hardening without exceeding a maximum bending stress of $0.66F_y$. The loading applied after the concrete has hardened is assumed to be carried by the composite section. Alternatively, the steel beam may be designed to resist all dead and live loads without exceeding the maximum bending stress of $0.76F_y$.

An advantage of this approach for strengthening existing framing is that existing beams may already be surrounded by concrete encasement for fireproofing reasons and may be considered composite even if they were not originally designed as such. Therefore, the load-carrying capacity of an encased existing beam can be analytically upgraded without investing a penny in the actual construction and even without computing the composite section properties, simply by taking the maximum bending stress in the adequately encased beams as $0.76F_y$.

The second method of achieving composite action is by providing shear connectors between steel and concrete, as specified in Sec. I4 of the AISC *ASD Specification* or Sec. I3 of the AISC *LRFD Specification*. Any reinforcing bars present in the slab that run parallel to the beam may be included in the composite section. Shear connectors of all kinds have been tried, but today almost all composite beams are built with welded-on headed studs, typically $^3/_4$ or $^7/_8$ in in diameter. Whenever concrete is placed on a formed steel deck, the maximum diameter of the shear studs should not exceed $^3/_4$ in (an AWS D1.1 requirement).

According to Griffis,[35] the first building utilizing headed studs for composite action was IBM's Education Building in Poughkepsie, New York, constructed in 1956. Since that time, composite construction has become increasingly popular, although it was not mentioned in AISC specifications until 1978.

When shear connectors are used, the AISC *ASD Specification* allows the composite section to be designed to carry all dead *and* live loads, even when steel is unshored, without exceeding the maximum allowable stresses specified in its Sec. F1.1. This liberal provision is justified by the ultimate-strength behavior of the composite section, even if the other relevant requirements of the specification are based on the allowable-stress concept. However, there is also a requirement that limits the maximum stress in a steel beam constructed without temporary shoring to $0.9F_y$, assuming that the steel beam alone carries all the loads while the concrete is still "wet" and that the composite section carries all loads applied after the concrete has reached 75 percent of its specified strength. Therefore, the sum of the wet-concrete stresses and those computed on the basis of the composite section properties from the superimposed loads should not exceed $0.9F_y$. The stresses in concrete should not exceed $0.45f'_c$.

In many cases, providing full composite action between steel and concrete is unnecessary, and a partially composite section will suffice. The flexural strength of a partially composite beam is governed by the shear strength of the shear studs: The larger the number of studs, the more the beam behaves like a fully composite member.

A practical method of strengthening existing steel beams involves introducing composite action by installing steel-stud shear connectors in the holes drilled in the slab, as shown in Fig. 3.11. After the studs are installed, the space around them is filled with high-strength grout. The length and spacing of studs must conform to the requirements of AISC specifications.

To act as a unit, the composite section must be able to internally transfer a shear force that is the smallest of the following:

$0.85f'_cA_c$ The strength of the concrete slab

A_sF_y The strength of the steel beam

nQ_n The number of shear studs between the point of maximum moment and a point of zero moment on each side times the ultimate shear capacity of a stud

The AISC *LRFD Manual*[37] lists the values of Q_n for $3/_4$-in headed studs as a function of concrete type and strength. These values are as follows:

For normal-weight concrete ($w = 145$ lb/ft^3):

$$f'_c = 3 \text{ ksi} \qquad Q_n = 21 \text{ kips}$$
$$f'_c = 3.5 \text{ ksi} \qquad Q_n = 23.6 \text{ kips}$$
$$f'_c = 4 \text{ ksi} \qquad Q_n = 26.1 \text{ kips}$$

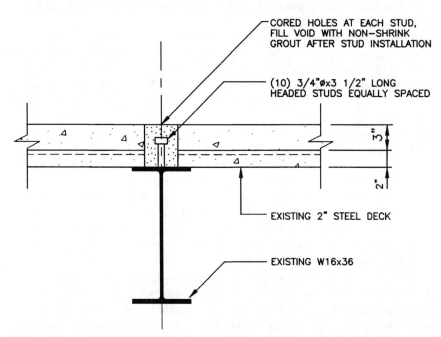

Figure 3.11   Beam reinforced by welding shear connectors for Example 3.3

For lightweight concrete ($w = 115$ lb/ft^3):

$$f'_c = 3 \text{ ksi} \qquad Q_n = 17.7 \text{ kips}$$
$$f'_c = 3.5 \text{ ksi} \qquad Q_n = 19.8 \text{ kips}$$
$$f'_c = 4 \text{ ksi} \qquad Q_n = 21.9 \text{ kips}$$

Example 3.3 further explains the procedure for strengthening existing steel beams by introducing composite action with concrete by installing shear connectors in drilled-in holes.

Example 3.3 Assume that the W16 × 36 beam in Example 3.2 cannot be accessed from the bottom. Propose a solution that does not require welding to the bottom flange. Use the following additional information:

1. The concrete is of normal weight with $f'_c = 3$ ksi.
2. The overall thickness of the concrete floor is 5 in; 2 in of that is taken by a steel floor deck running perpendicular to the beam.
3. The 25-ft-long beams are spaced 8.33 ft on centers.

solution Weld new shear studs in cored holes. The studs, if properly designed and installed, will make the slab act compositely with the W16 × 36 steel beam. Use the design procedures and load tables of the AISC *LRFD Manual*. In this example, since the required added capacity is modest, it is very probable that a partially composite section will suffice and that the expression nQ_n will control the design.

Since the beam is existing and its size is already known, proceed directly to the "Composite Beam Selection Table" in the manual (p. 4-23). A quick review of the table demonstrates that for *any* combination of Y2 (a distance from the centroid of the concrete to the beam's top flange), expression nQ_n, and location of the PNA, the resisting moment exceeds the required moment. The minimum table value of ϕM_n is 215 kip · ft, whereas the ultimate applied moment M_u from Example 3.2 is only 184.4 kip · ft.

Therefore, we can select the smallest value of $nQ_n = 95.4$ kips. From this, we can find a total number of studs (remembering that nQ_n refers to the number of shear studs between the point of maximum moment and a point of zero moment on *each* side times the ultimate shear capacity of a stud) using a stud reduction factor of 1.0 and a stud shear capacity of 21 kips:

$$n = 95.4 \times 2/21 = 9.1 \qquad \text{say 10 studs}$$

Check stud spacing:

$$25 \times 12/ (10 + 1) = 27 \text{ in} < 32 \text{ in minimum} \qquad \text{OK}$$

The minimum stud length should be such that after installation, the stud extends not less than 1.5 in above the top of the steel deck:

$$L = 2 + 1.5 = 3.5 \text{ in}$$

It is not necessary to use longer studs because their tops would then be too close to the top of the slab.

Check live-load deflection of the reinforced beam. The modulus of inertia of the composite section can be quickly determined from tables entitled "Lower Bound Elastic Moment of Inertia for Plastic Composite Sections," contained in the AISC *LRFD Manual* on p. 4-49. Again, even taking the smallest table value of 555 in^4 we can see that live-load deflections are acceptable:

$$\Delta_{ll} = \frac{5 \times 1.1 \times 25^4 \times 1728}{384 \times 29,000 \times 555} = 0.505 \text{ in} < 25 \times 12/360 = 0.83 \text{ in} \qquad \text{OK}$$

Check shear:

$$V_u = 2.36 \times 25/2 = 29.5 \text{ kips}$$

$$\phi V_n = \phi 0.6\, F_{yw}\, A_w = 0.9 \times 0.6 \times 36 \times 15.86 \times 0.295$$
$$= 90.95 \text{ kips} > V_u \qquad \text{OK}$$

Conclusion: Use 10 equally spaced headed shear studs with $^3/_4$ in diameter, 3.5 in long, in cored holes, as shown in Fig. 3.11.

(Note: In this example, we were able to avoid computing the exact values of the effective concrete flange thickness a, and the exact location of the PNA. This example is characteristic of upgrading existing beams for moderately increased loads. The precise trial-and-error procedure contained in the AISC *LRFD Manual* should be used for cases where a major upgrade is required.)

As the example demonstrates, introducing composite action by adding shear connectors works well if the bottom flange of the beam cannot be accessed (the operations in the space below cannot be interrupted, for example, or there are pipes hung from the bottom flange). A variation of this method may be used in the common case where the existing floor includes a thick nonstructural topping above a structural slab, such as several inches of cinder fill between the wearing finish.

In this situation, instead of reinforcing the bottom of the steel beam with welded plates or shapes, the *top* flange can be reinforced by removing the nonstructural topping and placing reinforced concrete in its place. The new concrete can be connected to the steel by welded-on shear studs for composite action. If the existing structural slab still remains under the new slab, and composite action between these slabs is desired, the two slabs can be tied together with shear pins. One such design utilizes short steel dowels in drilled-in adhesive anchors. In one project that has used this approach, as documented by Torello and Epstein,[38] 550 number 8 reinforcing bars were embedded into proprietary adhesive anchors to tie the slabs together.

To reduce the incidence of longitudinal cracking, which is quite common in composite-slab construction, some reinforcing steel can be placed at the top of the slab perpendicular to the steel beam. This negative reinforcement can also be provided by a double layer of welded wire fabric.

3.9 Strengthening Beam Connections

Whenever existing steel beams are reinforced, one should not forget about strengthening—or at least checking—their connections. Checking connection strength may or may not be a straightforward effort. The first task is to determine what was the original design load on the connection. In the best-case scenario, that load was larger than the proposed design reaction. Then, the problem is largely solved: All that is needed is to check the actual field condition of the connection and perhaps to verify some items analytically. The original load on the connection may be shown on the existing structural drawings, if these are available, as a number in kips near each end of the beam.

In the second-best scenario, the existing reactions are unknown, but the existing drawings include the connection details. Upon field verification, the connection capacity can be established using the original textbooks or analyzed using modern methods. For example, one of the most common beam-to-column connections found in old buildings is a riveted stiffened beam seat; a common connection between two beams is a riveted double angle. The loadbearing capacities of both of these can be checked using the appropriate tables in the current edition of the AISC manual and adjusting the numbers downward by the ratios of the shear capacities of the old rivets or bolts to those of the high-strength bolts on which the tables are based.

In the worst-case scenario, no information about the existing connections is available; this means that the connections have to be field measured and sketched out. After that, the analysis can proceed as discussed previously.

Incidentally, the uncertainty about design loading on existing connections can be traced in part to the steel industry's belief that extra connection capacity—anything that exceeds the design reaction—is a waste of money and should be discouraged. But, as the author has questioned in one of his articles,[39] why should the connections be the weakest parts of the beam? (That article won an award but did not change the industry's attitude on this issue.) Today, many engineers require steel connections for noncomposite beams to be designed for one-half of the uniform-load capacity listed in the AISC manual for the beam size and the design span indicated. The engineers involved in renovations of buildings designed under this approach would appreciate its certainty and uniformity.

What can be done if the existing connection is not strong enough to support the proposed loading? Depending on the structure, several steps can be taken. If the existing rivets or bolts are a problem, whether because of their insufficient load-carrying capacities or because of deterioration, they can be replaced with modern high-strength bolts of the same or larger diameter. (In the latter case, the existing fastener holes should be reamed to a proper size.) The drawings can specify under what conditions the rivets should be replaced. For example, existing $7/_8$-in rivets could be replaced if the protrusions of their heads are reduced by corrosion to less than 0.5 in or the maximum visible diameter of the heads to less than 1.31 in.

What are the actual methods of removing old rivets? Unlike bolts, which can be taken apart, rivet heads must be removed either by mechanical means or by burning them off with a torch. A typical mechanical method involves using a pneumatic hammer with a chisel tip to remove the head; after this, the rivet's shank is pushed out with a punch and hammer. Pushing out the shank may not be easy because, as already noted, the hot-installed rivet shank often deforms to fill the irregular shape of misaligned holes. (The falling pieces represent a safety hazard, and appropriate precautions must be taken.) The mechanical methods inflict the least amount of damage on the existing steel and are preferred for fatigue-critical applications (mostly bridges) and in combustible buildings. An alternative is burning off the rivet heads. It is less expensive and, if done properly, provides acceptable results.[40]

A common method of strengthening single shear-plate bolted connections involves welding another plate or angle at the opposite side of the beam's web and replacing old bolts that acted in single shear with new bolts that would act in double shear and provide a corresponding increase in capacity.

Welded connections can be similarly upgraded. If existing welds are overstressed by the proposed loading, the welds can be extended or removed by grinding and replaced with larger-capacity welds. A typical strengthening detail involves welding additional clip angles to extend the existing connection downward, if the space permits. Existing clip angles that are too thin or too short can be augmented by new material.

An existing seated-beam connection can be strengthened by welding additional web angles. Conversely, existing web-angle connections can be reinforced by welding new seated angles. Unstiffened seated-angle

connections can be improved by welding stiffeners. The possible reinforcing schemes are limited only by the designer's imagination.

Normally, existing connections utilizing bearing-type bolts should
not be reinforced by welding unless the welds are designed to carry the
full load on the connection. On the other hand, existing friction-type
high-strength bolts and rivets can work in combination with new
welds. AWS D1.1 Sec. 11.5.3 states that if existing fasteners in the latter case are (theoretically) overstressed by the total load, only a dead
load will be assigned to them, if they can safely carry it, and all other
loading will be carried by new welds. If, however, existing fasteners
are overstressed even by the dead load, the new welds will carry the
total load. As AWS D1.1 Sec. 11.5 cautions, the sequence of welding
should be arranged to proceed in a symmetrical fashion, especially if
the structure is reinforced under load.

The foregoing discussion has dealt with shear connections—those
transmitting only vertical loads—with or without added tension or
compression. However, many old beam-to-column connections were of
rigid or semirigid varieties—those designed for full or partial fixity.
Some examples of the old "wind" moment connections are shown in
Fig. 3.12. Verification or reinforcement of such connections should proceed in concert with analysis of the building for lateral loads.

3.10 Composite Steel-Concrete Columns

Existing steel columns can be reinforced by encasing them in concrete
rather than by welding additional members, as was discussed in Sec.
3.7.4. In some cases, encasement has a definite advantage, especially
when the column must be fireproofed or when a square or rectangular
column section is desired for architectural reasons. In addition, encasing a wide-flange column in concrete helps prevent it from buckling
under load and increases its rigidity.

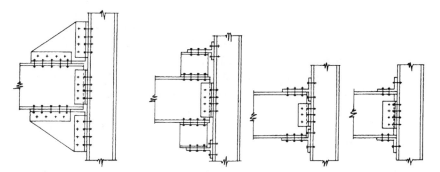

Figure 3.12 Typical "wind" connections. [*From John Hancock Callender (ed.),* Time-Saver
Standards for Architectural Design Data, *6th ed., McGraw-Hill, New York, 1982, p. 2-94.
Reproduced with permission of The McGraw-Hill Companies.*]

According to the AISC *LRFD Specification,* included in *Manual of Steel Construction, Load and Resistance Factor Design,*[37] structural steel components of composite columns may consist of rolled or built-up steel shapes (including laced columns). Concrete-filled steel pipe or tubing also belongs in the composite-column category, but filling the inside of an existing tubular column with concrete is a method rarely used for strengthening. The AISC *ASD Specification* does not cover composite columns, and the *LRFD Specification* has to be followed. It requires that certain design limitations be met in order for the encased column to qualify as a composite section:

- The area of the steel column must be at least 4 percent of the total. (If it is not, the column should be designed as a reinforced-concrete member.)

- Concrete encasement must be reinforced with longitudinal rebars and lateral ties to prevent spalling under load and during fire.

- The minimum clear cover to ties or longitudinal bars must be 1.5 in.

- The 28-day compressive strength of the encasement concrete must be between 3 and 8 ksi for normal-weight concrete and at least 4 ksi for lightweight concrete.

- The specification states: "The specified minimum yield stress of structural steel and reinforcing bars used in calculating the strength of a composite column shall not exceed 55 ksi." As explained in the *Commentary on the AISC LRFD Specification* (a part of the *LRFD Manual*), the stress of 55 ksi corresponds to the maximum strain of 0.0018, below which concrete is considered to remain stable without spalling.

Section I2.1.b of the AISC *LRFD Specification* differentiates between load-carrying longitudinal bars, which must be continuous at framed levels, and restraining longitudinal bars, which may be added to satisfy the spacing requirements for longitudinal bars—at least 0.007 in^2 of steel area per inch of bar spacing. The restraining longitudinal bars may be interrupted between the floors (otherwise, they would have to run through the intersecting steel beams). To allow the load carrying longitudinal bars to be continuous at framed levels, the bars are typically placed in column corners. Indeed, the most common arrangement of the longitudinal bars is four, one in each corner. If more bars are used, the designer must be careful to provide lateral support for the additional bars, so that no bar is farther than 6 in from a laterally supported longitudinal bar. ACI 318[41] considers a bar to be laterally supported if it is located in a corner of a tie with an inclusive angle of not more than 135°.

The ties must be spaced no more than two-thirds of the smallest dimension of the composite section and must provide at least 0.007 in² of steel area per inch of bar spacing. Additionally, ACI 318 requires that spacing of ties in reinforced-concrete columns not exceed 16 longitudinal bar diameters or 48 tie-bar diameters.

The lateral ties in composite sections obviously cannot be of the regular closed shape typically used in reinforced-concrete columns. To fit around a steel column, two-piece U-shaped ties are used, overlapped to provide the required splice length. The minimum size of the ties depends partly on the size of the longitudinal bars. ACI 318 allows #3 tie bars for #10 and smaller longitudinal bars and #4 ties for larger bars. Figure 3.13 illustrates a typical composite column.

Splices in longitudinal bars should be located carefully. For single-story encased columns, it is better to avoid any splices. For multistory columns used in composite frames, splices are best made at the middle of their clear heights, where the inflection points for flexure are typically located. Making splices in this location could permit the use of less expensive compression lap splices or compression butt splices in lieu of the tension lap splices that would be required if the splices were made at the floor line.[35]

The specification does not require shear connectors between steel and concrete encasement, just as it does not require them for fully encased beams. In axially loaded composite columns, the load transfer to the concrete section is developed by direct bearing of steel beams at connections. *Specification* Sec. I2.4 prescribes that the maximum ultimate design strength of concrete in bearing be taken as $1.02 f'_c A_b$, where f'_c is the specified compressive strength of concrete and A_b is the loaded area. This value applies to cases in which the supporting concrete area is

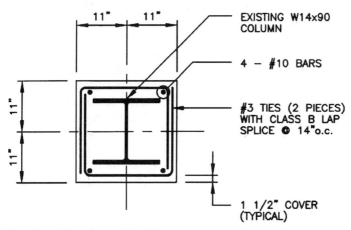

Figure 3.13 Typical composite column.

wider than the loaded area at least on one side and the remaining sides are restrained against lateral expansion.

In addition to direct bearing of beams, at least some load transfer between the steel column and the concrete takes place by bond. Some engineers question whether load transfer by bond is reliable and specify shear studs welded to the steel column as an alternative means of sharing the load. Others do not. Similar questions remain in the minds of some about the transfer of a moment and shear at the joints between steel beams without concrete encasement and composite columns. In the absence of definitive research, Ref. 42 recommends that shear connectors with the maximum spacing of 36 in be provided. It proposes that the shear connectors be designed for the larger of

1. The sum of all beam reactions at the floor level.

2. Whenever the expression $P_u/\phi_c\, P_n$ is less than 0.3, the force computed by multiplying the tension steel area (including rebars) by F_y. This force is intended to resist a moment equal to the nominal flexural strength of the composite section. (The resistance factor for composite columns is 0.85.)

If specified, shear connectors are typically welded to the column flanges—on the outside if possible, on the inside if space does not permit it. Shear connectors welded on the inside are staggered about the web.

The existing column base plate typically covers most, but not all, of the composite column's footprint. The base plate might interfere with the drilled-in dowels that are typically required to transfer a compression load into the foundation. If this problem exists, consideration should be given to reducing the base plate size to allow the dowels to be placed.

Regardless of the design details, one must remember that the existing dead load carried by the steel column is not automatically transferred to the concrete encasement. Therefore, if the steel section is already overstressed by the dead load, the excess load must be removed by shoring and jacking the beams. If the steel column alone can safely support the dead load but not the total load, the composite section may be assumed to carry the total loading. Recall that a similar assumption was made in the case of steel beams made composite with concrete by shear connectors (Sec. 3.8).

LRFD Specification Sec. I2.2 establishes the ultimate design strength of axially loaded composite columns by modifying the basic column equations E2-1 through E2-4. The designer can simply follow the modified equations. Alternatively, the *LRFD Manual* contains tables for axial design strengths of composite columns for various combinations of steel column and concrete encasement sizes and concrete strengths. The tables are provided for steels with F_y of 36 and 50 ksi. For existing steel columns of other yield strengths, the longhand solution is needed.

The foregoing discussion has dealt with the design procedures for columns subjected only to axial loads. Whenever combined compression and bending is present—in columns with eccentricities, for example—a more complicated analysis is needed. The AISC *LRFD Specification* Sec. I4 refers designers to its basic interaction formulas H1-1 through H1-6, with certain modifications. To simplify these efforts, the AISC has published its Design Guide No. 6, *Load and Resistance Factor Design of W-shapes Encased in Concrete,* by the already quoted Lawrence G. Griffis.[43] The guide contains numerous design tables and is a "must-have" for anybody involved in this challenging task.

At this time, research sponsored by the structural steel industry is underway to produce an economical composite column section that does not require complete envelopment of steel by concrete. But for existing construction, given the complexities of the composite design, in some cases it is simpler to neglect the composite action and rely on the concrete encasement as the sole load-carrying element. This approach may be especially handy when cast-iron (or even wood) columns are encased in concrete to strengthen them.

3.11 Openings in Existing Steel Beams

During the course of renovations, it is often necessary to make openings in existing steel beams for the passage of piping, small ducts, and conduits. When the existing headroom is low, there may not be another choice, even though it is always preferable not to cut into steel beams. How can one determine what size openings in existing framing can be safely made? At what point does a beam require reinforcing? How should the required reinforcing be designed? All these questions must be resolved before any holes are made in existing framing.

Assuming that the structural engineer is faced with a fait accompli about the presence of new openings, he or she may still be able to influence the opening locations. Mercifully, most openings are made through the beam's web; a proposal to run a pipe horizontally through the beam flange might be greeted with a threat of bodily harm. Making holes in the web is unlikely to seriously affect the moment capacity of the wide-flange beam but can significantly reduce its shear capacity.

Many engineers follow the rule of thumb that holds that relatively small openings—those not exceeding one-third of the beam depth, for example—can be safely made near the neutral axis of the beam if they are located in the middle third of the beam, where shear stresses are usually minimal. In the noncomposite steel wide-flange beams typically found in old buildings, the neutral axis is located at middepth; in composite beams or in those with cover plates, the neutral axis can be located by a straightforward analysis. The rule also holds that the adjacent holes should not be spaced closer than three hole diameters. If located within the end thirds of the beam span, any penetration requires reinforcement by a plate. The common approach for dealing with small openings in the web is illustrated in Fig. 3.14.

When the parameters of web openings exceed those indicated in Fig. 3.14, special analysis is needed to verify that the shear capacity of the web with the opening is adequate. Quite often, the analysis concludes that the web must be reinforced. Reference 44, a publication by United States Steel, contains design tables and outlines a rather lengthy method of analysis intended to determine if web reinforcement is needed. It does not include the design of the reinforcement.

When reinforcement is required, the procedure for analyzing and strengthening the beam with a proposed large web opening can become quite involved. One good source of information is AISC Steel Design Guide No. 2,[45] which contains a wealth of theory, reference tables, and design examples. Those finding themselves dealing with this issue on a regular basis might consider investing in a specialized computer program for the design of web openings, such as WEBOPEN, offered by the AISC.

In general, the most cost-efficient reinforcement of web penetrations consists of welded horizontal bars above and below the openings. The bars should extend beyond the hole edges to lessen the stress concentrations and to provide spaces for welding.

3.12   Steel Corrosion: Evaluation and Protection

3.12.1   General issues

Here is a common situation: An existing steel member in a structure scheduled for renovation is covered with heavy rust. What should be done about it?

Unfortunately, steel does rust, and this question is as old as steel construction itself. The first order of business in this situation is to determine the actual loss of section. It is widely known that the products of rusting expand to occupy several times the volume of the original material. Thus, appearances can be deceiving, and the rust may have actually destroyed much less metal than it seems. Unless a steel member has been in truly unfortunate circumstances—exposed to constant wetting and drying, for example—the damage could be moderate. The actual dimensional loss can be determined by scraping away the rust and

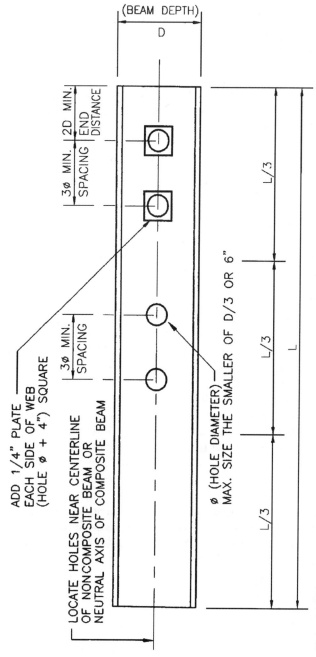

Figure 3.14 A common approach to making small (less than one-third of the beam's depth) openings in the web of a wide-flange beam.

measuring the remaining thickness of the flanges and the web (a small hole has to be drilled to measure the web). If the member designation is known from the existing drawings, the properties of the remaining section can be compared with those of the original section. If the original drawings are not available, the actual properties of the remaining section can be analyzed for the proposed loading. Corroded members that are found to be deficient for the proposed loads must be reinforced.

If the extent of rusting cannot be readily assessed, a preliminary assumption could be made that the loss of section equals $1/_8$ in. In fact, some building codes allow using steel members $1/_8$ in thicker than those required by analysis in lieu of other means of corrosion protection. This assumption is based on the fact that expanded rust tends to act as a coating of sorts that insulates the remaining section from the supply of oxygen necessary for continuing corrosion.

Should existing steel be protected from further corrosion? The answer depends on the anticipated severity of the environment surrounding the steel members. The available protective measures range from applying no coatings at all to taking the members down and sending them to be galvanized. The latter is rarely done, of course; if corrosion protection is desired, it is usually achieved by application of various coatings.

A typical coating system for structural steel consists of primer, intermediate coat, and topcoat. The three components should be compatible, even though the primer is typically specified by structural engineers and the other two components by architects. The primer is the first line of defense against corrosion. Its protection has to last until the other two coats are applied. Another function of the primer is to provide adhesion of the coating system to the steel. The intermediate coat, which gives added protection, must be fully compatible with the primer. The topcoat provides overall durability and the final level of protection.[46]

Even the best primer will do no good if the surface is poorly prepared. Indeed, as painters well know, the key to paint durability is careful surface preparation. The various degrees of surface preparation are described in Steel Structures Painting Council (SSPC) standards. The minimal degree of surface preparation, SSPC-SP2, "Hand Tool Cleaning," requires only removal of loose mill scale, paint, and rust; the highest degree, SSPC-SP5, "White Metal Blast Cleaning," requires a full abrasive blast cleaning.

3.12.2 Coating types

Some designers tend to specify "only the best" and do not care what the particular circumstances of the job are. True, the best finishes are usually the most cost-effective in challenging environments: Even if only one repainting job can be avoided, the product will have paid for itself handsomely. One such finish combines inorganic zinc-rich primer, epoxy or urethane intermediate coat, and urethane topcoat. Such three-coat systems, called zinc-epoxy-urethane and zinc-urethane-urethane, are frequently used for repainting bridges. But what if the steel is located inside an occupied building without any significant sources of corrosion? Requiring top-of-the-line coatings and SSPC-SP5 surface preparation will not only waste resources but also interrupt the building operations for a longer period. Instead, this steel could quite often be left unpainted.

Others attempt to go in the opposite direction and fix their sights on "surface-tolerant" coatings, which have become popular because they do not require extensive surface preparation and can "tolerate" some surface irregularities. When properly applied, these coatings can generate substantial savings. Some view surface-tolerant coatings with suspicion because of a few widely publicized failures, which have later been attributed to material misuse and not to product defects.

Many designers are intrigued by so-called rust reformers that are supposed to turn old rusty steel into steel covered with a black impenetrable coating that can be painted (recall the TV ads of some time ago targeted at home users?). Most rust converters contain tannic acid, which indeed reacts with rust and turns it into iron oxides. The problem is, the acid rarely penetrates rust deeply enough for a complete chemical reaction. The SSPC recommends that all loose rust be removed before applying rust reformers.[46]

Our recommendation is to go for the coating system that is the most cost-effective for the proposed use of the existing structure, no more and no less. Table 3.7 is a simplified summary of recommended coating types as a function of environmental conditions, adapted from Ref. 46. Where a severe chemical exposure is anticipated, a special investi-

TABLE 3.7   Suggested Painting Systems for Various Environmental Conditions

Environmental conditions	Painting system
Dry interiors where steel is protected	None
Interiors, normally dry	None or latex
Exteriors, normally dry	Oil base or latex
Frequent fresh-water wetting, spray, immersion	Vinyl, coal-tar epoxy, epoxy
Frequent salt-water wetting, spray, immersion	Zinc-rich, vinyl, coal-tar epoxy, epoxy
Fresh-water or salt-water immersion	Vinyl, coal-tar epoxy
Chemical exposure, acid (pH 2 to 5)	Vinyl, coal-tar epoxy, chlorinated rubber
Chemical exposure, neutral (pH 5 to 10)	Vinyl, chlorinated rubber, zinc-rich
Chemical exposure, alkali (pH 10 to 12)	Coal-tar epoxy, chlorinated rubber
Chemical exposure to mild solvents	Epoxy

SOURCE: Reference 46.

gation should be made. Some basic material choices for recoating existing steel, in addition to the zinc-rich coatings discussed in the next section, include (from Refs. 46 and 47):

- *Oil-based and alkyd coatings.* These organic materials are among the oldest coatings in existence. Oil-based coatings contain natural fish and plant oils; alkyds are based on synthetic resins. Both are familiar to most homeowners, who appreciate their moderate cost, ease of application, and liberal surface-preparation requirements. Still, as with most coatings, better surface preparation leads to longer service life. Oil-based coatings are typically used in renovations as primers; they cure by air oxidation. While they behave reasonably well in mild atmospheric environments, their resistance to moisture and chemicals is not very good. For this reason, they are most effective in combination with zinc-rich primers. Their heat resistance is about 250°F. Oil-based coatings now include acrylics and vinyls. When combined with zinc-rich primers in three-coat systems, they are called zinc-acrylic-acrylic and zinc-vinyl-vinyl.

- *Latex.* These coatings are also familiar to most people. In latex coatings, the resin is suspended among water particles. After application, the water evaporates and the resin eventually fuses together. In service, latex coatings behave similarly to alkyds. Acrylic latex coatings can be applied over existing oil-based paints because their low shrinkage during curing assures that little stress is placed on the existing base.

- *Epoxy.* Unlike oil and latex coatings, epoxies cure by a chemical reaction that requires mixing of two components, the resin and the hardener. In general, epoxy coatings have good resistance to water, solvents, and chemicals but tend to disintegrate by chalking when exposed to sunlight. For this reason, they are normally used for interior applications and when covered with other materials. Epoxies usually require extensive surface preparation to achieve their full potential. Their heat resistance is also about 250°F, but modified epoxies can resist 400 to 450°F. Recently, some thin-film epoxy coatings have been specifically marketed for recoating existing oil-based paints.

3.12.3   Repairs of galvanized coating

Galvanizing works on the principle of a sacrificial mechanism. As is widely known, when two dissimilar metals in close contact are exposed to an electrolyte, a less noble metal will become an anode and will corrode first, protecting the other metal—the cathode. If zinc-coated steel is placed in a corrosive environment, the zinc will be sacrificed first and the steel will be protected until the zinc is dissolved. At that point, the zinc has to be reapplied, or the steel will rust. Galvanized coatings can also be destroyed during the reinforcing of existing framing by welding or simply by abrasion.

Essentially, the challenge involved in repairing existing galvanized steel is how to deposit an additional layer of zinc on its surface in small areas. Fortunately, zinc is known to actually migrate to the adjoining uncoated areas and, to some degree, protect the steel there. For example, the edges of a hole drilled in a galvanized steel beam will be protected by zinc on the beam's surface. Unfortunately, zinc is consumed much faster when it attempts such rescue operations.

The acceptable methods of repairing galvanized coatings are listed in ASTM A780, *Repair of Damaged and Uncoated Areas of Hot-Dip Galvanized Coatings*. These methods include zinc-rich paints, zinc-based solders, and zinc metallizing. For any of these to be effective, the repaired surfaces must be clean, dry, and free from oil, grease, and similar contaminants that might interfere with the adherence of repair materials.

Applying zinc-rich paints is the most common method of galvanizing repair. To be effective (and to conform to ASTM A780), zinc-rich paints should contain no less than 65 percent zinc. Despite its ease of application, this repair method is visually different from hot-dip galvanizing and is not as resistant to scratches and abrasion.

There are two basic types of zinc-rich coatings: organic and inorganic. Organic resins are based on carbon particles in a polymer base, which may contain such binders as epoxy polyamide, urethane, and vinyl. Inorganic coatings are mostly based on an ethyl silicate resin, although water-based zinc-rich primers are now available. Both these types have advantages and disadvantages. Organic zinc-rich paints have better compatibility with topcoats and can better tolerate surface irregularities, but are less tough and abrasion-resistant than the inorganic variety. Inorganic zinc-rich paints are often used as primers but can also be specified for single-coat paints.[47] Zinc-rich primers are often used with epoxy intermediate coats and either epoxy or urethane topcoats. The SSPC recommends a minimum surface preparation level of SSPC-SP6, and preferably SSPC-SP10.[46]

If aesthetics are important, zinc solders can be used. This repair method usually assures a good visual match and excellent adherence to the substrate, but is more difficult to apply than zinc-rich paints. For example, the steel has to be preheated prior to the solder application. This may be difficult to do in existing buildings.

Zinc metallizing involves spraying droplets of molten zinc onto the steel, a process that is even more operator-dependent than zinc soldering. However, zinc metallizing can provide the best level of corrosion protection of the original galvanized coating.[48] This repair method is uncommon in renovations of existing structures.

In general, zinc-rich coatings are most useful for exterior applications, such as for recoating the undersides of existing steel lintels.

3.13 Thermal Prestressing of Steel Structures*

3.13.1 When to use thermal prestressing

Thermal prestressing is applying thermally preelongated or preshortened steel reinforcement to existing members that are loaded in tension or compression. In many situations, especially where no other strengthening method can be economically used, thermal prestressing is a valuable and practical way of safely achieving the desired results.

Thermal prestressing is especially valuable for strengthening structures that are already overstressed by the existing loads or are close to being so; in such cases, unrealistically large reinforcement areas would otherwise be required.

For example, imagine an axially loaded bar that is already loaded to 90 percent of its capacity and that must be reinforced to resist a load that is increased by another 25 percent. The available 10 percent of total allowable stress cannot be exceeded on the combined system. To keep the stresses in the original member from exceeding the allowable limits, this would require an area of reinforcement of 125 percent of the original area! (To illustrate this with numbers, assume that a 1-in^2 steel bar with an allowable stress in tension of 22 ksi is already loaded by a 19.8-kip tension load. To increase the load by 25 percent, or 4.95 kips, the allowable stress in the whole system must be limited to 2.2 ksi. The total area, including reinforcement is then 4.95/2.2 = 2.25 in^2.)

There are also cases in which the existing levels of stress already exceed those allowable. Here, no amount of reinforcement will help.

In both of these cases, there are two options for upgrade. The most common option is to temporarily relieve the existing stresses by jacking up the overstressed member. The second, less common option is to use external prestressing reinforcement that reduces the existing stress level while assuming a share of the existing loads and new loads. The second option, prestressing, can be achieved through mechanical means (the most common way) or by application of temperature, the subject of this section.

*The author gratefully acknowledges the contribution of John M. Wathne, P.E., of Structures North, Salem, Mass., for providing material for this section.

3.13.2 Thermal prestressing: The basics

Preelongation or preshortening of the reinforcement can be achieved by varying the temperatures of the new and the existing metal to introduce differential thermal strains. Such strains are analogous to elastic strains that result from changes in stress levels. When a new heated reinforcing member is rigidly attached to the existing member, both members will have the same stress when the new member cools down.

For example, if an existing tension member is stressed to 15 ksi, it would have an elastic strain of 0.052 percent. Since the coefficient of thermal expansion of steel is 0.00065 per 100°F, a change in temperature of 80°F will also result in a strain of 0.052 percent. (If the existing member was at a temperature of 70°F, then the new member would need to be heated to 150°F.) As the new member cools, its thermal strain becomes the equivalent of elastic strain. If the new member were attached to unyielding supports, it would develop an elastic strain of 0.052 percent, and 15 ksi of stress. If the new member is attached to the existing member, the new member will shorten as it cools and cause the existing member to shorten, too. This would reduce the stress level in each member by the same amount. The larger the ratio of reinforcement area to base steel area, the larger the amount of such axial shortening.

Consider two cases.

Case 1. A 1-in² bar is loaded to 20 kips in tension. The resulting tension stress is 20 ksi. Imagine that the bar is unloaded by external means and that a second 1-in² bar is added to it. The 20-kip load is then put back on the combined section, resulting in a 10-ksi stress in both members.

Case 2. A 1-in² bar is stressed in tension by a 20-kip load at room temperature. It now has a stress of 20 ksi. Another 1-in² bar is thermally elongated to 20 ksi equivalent strain. These two bars are then attached at their ends, and the heated bar is allowed to cool to room temperature. As the added bar reaches room temperature, the 20-ksi equivalent strain becomes 20 ksi of stress. The 20-ksi strain in the new member results in the equivalent of a 20-kip load that is applied to the existing bar; this restrains the shortening of the added bar. As the existing bar shortens, the added bar shortens by the same amount, reducing the applied level of stress in the added bar by the same amount as the level of stress in the existing bar is increased, since the two bars have the same areas. By basic mechanics, they equalize to a 10-ksi compression stress in the existing bar and a reduction of 10-ksi stress in the added bar. As a result, the tension stress in the original bar is reduced from 20 to 10 ksi and the stress in the added bar is reduced from 20-ksi equivalent stress to 10-ksi real stress.

Both cases have the same initial conditions and the same end results, although these results are achieved by different methods. The first case required that temporary external support be applied to the system in order to relieve the system of its original load. In the second case, no external support was required, as all work was internal to the system.

3.13.3 Thermal prestressing:
The procedure

Thermal prestressing of axially loaded members can be done by the following procedure:

Step 1: Determine the stress level in the existing member by dividing the axial load by the cross-sectional area.

Step 2: Select the desired stress level and determine the required total cross-sectional area by dividing the applied load by the desired stress. Subtract the area of the existing steel from the total required area to get the required area of steel reinforcement.

Step 3: Determine the required differential temperature by dividing the existing stress level by the modulus of elasticity and the coefficient of thermal expansion. Add this to the ambient temperature to get the thermal prestressing temperature.

Step 4: Heat or cool the reinforcement to the thermal prestressing temperature and instantaneously rigidly attach it to the loaded member. Let the temperatures equalize; the stresses will equalize, too.

Remember, it is critical to preheat reinforcement for tension and precool for compression. The converse application of these techniques could be disastrous, as it would increase the existing stresses rather than reduce them. Alternatively, the temperatures in the existing members can be varied in opposite directions to achieve the same results: Precool existing tension members and preheat existing compression members. The results will be the same as long as the differential temperatures are the same.

Extreme care must be taken to evaluate the effects of the heating and cooling on the existing members. The load on preheated compression members may actually increase as these members elongate with respect to the surrounding structure and attract a larger percentage of the overall load.

Attention must also be given to maintaining the dimensional and geometric properties of axially loaded members so as not to introduce bending stresses or buckling. Forces between the reinforcement and the original members must be transferred along a path that avoids eccentricities and follows the original load path as closely as possible. Otherwise, additional bending stresses may result.

3.13.4 Some practical tips

Some practical adjustments to the theoretical procedure follow.

Temperature loss during attachment. During the time that it takes to actually make the rigid attachments between the reinforcing and the base members, the temperature differential diminishes. A reinforcing member preheated to an equivalent stress of 15 ksi might have a reduced equivalent stress of only 10 ksi by the time the attachments are actually made. To compensate for this, one should

- Minimize the amount of time it takes to make attachments by having both ends attached simultaneously. Also, preheat or precool the reinforcing member within a few feet of the installation to minimize transit and adjustment times. However, do not let the precooling or preheating operations affect the temperature of the base member.

- Make the primary attachment first. Attach the ends and then make the stitch welds along the length of the reinforcement. Stress can then begin to equalize between the members while the final attachments are underway.

- Overheat or overcool for the application. This makes up for inaccuracies in the pyrometer readings and compensates for losses during handling and attachment. On one project, the engineer specified preheat and precool temperatures 20°F above those theoretically required; the temperature at the time of attachment was within 5° of that required. In that project, the engineer specified overheating or overcooling and continuously checked temperatures with a pyrometer. The welding of the end connections was completed before the design temperature differentials were reached.

- Use reinforcing steel with a yield point at least 30 percent higher than that of the existing steel. While overcompensation reduces the equalized stress in the main members, it increases it in the reinforcement.

- Leave some reserve strength in the overall design to account for inaccuracies in thermal prestressing. In this type of work, even as much as 10 to 20 percent additional capacity leads to only a small increase in material cost; it is labor costs that account for most of the total.

Design of end connections. Overdesign the end connections of the reinforcement by a margin of at least 30 percent greater than the anticipated transfer loads. Consider the effects of all eccentricities on both the main members and the reinforcement. Locate end connections where the eccentric stresses, when combined with the reinforced stress in the main member, are within acceptable limits.

Design welds to reduce the amount of warping due to weld cooling and out-of-plane bending and to provide as much rigidity as possible.

Consider the temporary section losses of all sequential weld lines and weld parallel to stress paths. Because of the magnitude of the forces being transferred, improper end connection design could cause sudden failure of both the connection and the main member.

Reinforcing trusses. Reinforcement of truss systems by thermal prestressing uses this technique in its most basic form. The bottom-chord members can be reinforced by preheating cover plates or angles and anchoring them at their ends. The ends should be located within panels, where stresses from the anchorage will not damage the chords' performance. Staggered stitch welds between end attachments are sufficient to hold the cover plates in place. Heavier welding should be added at panel points to ensure proper load transfer. Figure 3.15 illustrates termination details for thermal prestressing.

The top-chord members can be reinforced in a similar fashion, except that reinforced members must be precompressed (precooled). One way of doing this is to use a tub of dry ice.

Tension diagonals can be reinforced by thermal pretensioned plates or angles with properly adapted anchorages. Compression diagonals can be reinforced by heating the base metal and adding cold reinforcement. To avoid buckling of the diagonals, consider loosening diagonal bridging, which would tend to restrain elongation of the diagonals.

Nonuniform stress levels along the length of the member. Nonuniformity of stresses along the length of a member, such as a truss chord, can actually be used to facilitate the attachment of reinforcement: It can be located at points of low or opposite stresses. The reinforcement would be designed for the loads in the critical panel and run through the lighter loaded panels until termination. Such overreinforcing should not cause problems as long as the basic distribution of stresses is not changed by the reinforcement. The shear connections between reinforcing and main truss panel points should also be designed to preserve the existing structural scheme.

Reinforcing flexural members. Some special considerations apply for flexural members. For reinforcement without eccentricity, the total

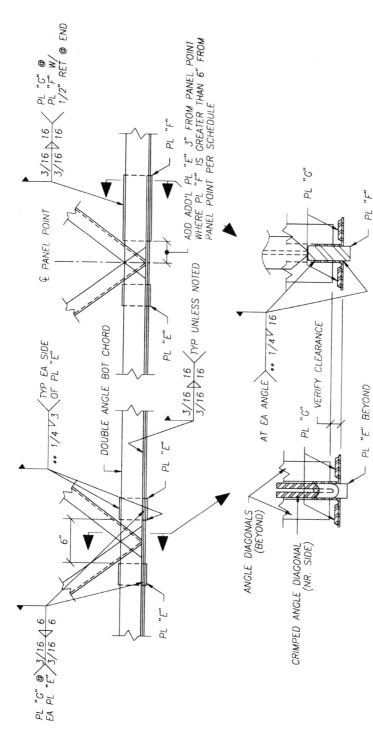

Figure 3.15 Termination details for thermal prestressing. (*Courtesy of John M. Wathne, Structures North, Salem, Mass.*) NOTES: 1. Preheat PL "G" as shown on joist heating schedule. Immediately weld ends and then weld length after PL cools. Weld all other work before "G." 2. Preheat PLs "E" and "F" as required for proper welding, complete and cool before adding PL "G."

axial and bending loads can be applied to the composite section and the stresses calculated directly. For reinforcement with eccentricity, the moments from eccentric prestressing force must be added to the bending stresses. Do not simply assume that eccentricities do not exist, neglecting them can result in failure.

3.13.5 Understand the limitations of thermal prestressing

The overall cost of using thermal prestressing must be checked against that of other methods, such as shoring (especially in relatively light unobstructed structures) or adding reinforcing (when possible and practical). Thermal prestressing demands relatively high labor cost and supervision levels, and should not be specified indiscriminately. In some cases, however, there is no other choice, as was mentioned in Sec. 3.13.1.

It is very important, given the inherent risks involved in working with existing structures under load, that the designer understand all aspects of the work before proceeding. An experienced contractor is also critical to the success of this system.

References

1. Neal FitzSimmons, "Iron and Steel Structures," BSCES/ASCE Structural Group Lecture Series at MIT, Cambridge, Mass., 1979.
2. Jack C. McCormac, *Structural Steel Design,* 3d ed., Harper & Row, New York, 1981.
3. Poul Beckmann, *Structural Aspects of Building Conservation,* McGraw-Hill International (UK), London, 1995.
4. Charles Head Norris et al., *Elementary Structural Analysis,* 3d ed., McGraw-Hill, New York, 1976.
5. "MBMA: 35 Years of Leading the Industry," a collection of articles, *Metal Construction News,* July 1991.
6. David T. Ricker, "Field Welding to Existing Steel Structures," *AISC Engineering Journal,* 1st quarter 1988.
7. William H. Birkmire, *Architectural Iron and Steel, and Its Application in the Construction of Buildings,* Wiley, New York, 1891.
8. *BOCA National Building Code,* Building Officials and Code Administrators International, Country Club Hills, Ill., 1996.
9. William McGuire, *Steel Structures,* Prentice-Hall, Englewood Cliffs, N.J., 1968.
10. Leslie H. Gillette, *The First 60 Years: The American Institute of Steel Construction, Inc., 1921–1980,* American Institute of Steel Construction, Chicago, 1980.
11. *Iron and Steel Beams: 1873 to 1952,* American Institute of Steel Construction, Chicago, 1953.
12. FEMA-274, *NEHRP Commentary on the Guidelines for the Seismic Rehabilitation of Buildings,* FEMA, Washington, D.C., 1997.
13. Sara Wermiel, "Rethinking Cast Iron Columns," *Building Renovation,* Winter 1995.
14. Donald Friedman, "Historical Building Construction," Norton, New York, 1995.
15. FEMA-273, *NEHRP Guidelines for the Seismic Rehabilitation of Buildings,* FEMA, Washington, D.C., 1997.
16. Paulson et al., "Modern Techniques for Determining the Capacity of Cast Iron Columns," in *Standards for Preservation and Rehabilitation,* S. J. Kelly (ed.), ASTM Special Technical Publication 1258, ASTM, Philadelphia, 1994.
17. *Manual of Steel Construction, Allowable Stress Design,* American Institute of Steel Construction, Chicago, 1989.

18. Dario Gasparini and David Simmons, "American Truss Bridge Connections in the 19th Century," *Journal of Performance of Constructed Facilities,* August 1997, p. 137.
19. Richard DeJonge, *Riveted Joints,* American Society of Mechanical Engineers, New York, 1945, quoted in Ref. 18.
20. *Specification for the Design, Fabrication and Erection of Structural Steel for Buildings,* American Institute of Steel Construction, Chicago, 1949.
21. Lambert Tall (ed.), *Structural Steel Design,* Ronald Press, New York, 1964.
22. Linton E. Grinter, *Design of Modern Steel Structures,* 2d ed., Macmillan, New York, 1960.
23. *Manual of Steel Construction,* 5th ed., American Institute of Steel Construction, Chicago, 1961.
24. *Standard Specifications, Load Tables and Weight Tables for Steel Joists and Joist Girders,* Steel Joist Institute, Myrtle Beach, S.C., 1992.
25. *Steel Joist Institute 60-Year Manual, 1928–1988,* Steel Joist Institute, Myrtle Beach, S.C., 1988.
26. Charles H. Thornton et al., "Vertical Expansion of Vintage Buildings," *Modern Steel Construction,* June 1991, p. 35.
27. ANSI/AWS D1.1-96, *Structural Welding Code: Steel,* American Welding Society, Miami, 1990.
28. R. Scott Funderburk, "Fundamentals of Preheat," *Welding Innovation,* vol. 14, no. 2, 1997.
29. Omer Blodgett, *Design of Welded Structures,* James F. Lincoln Arc Welding Foundation, Cleveland, 1982.
30. John P. Miller, "Strengthening of Existing Composite Beams Using LRFD Procedures," *AISC Engineering Journal,* 2nd quarter 1996, pp. 65–72.
31. Peter Kocsis, discussion on "Strengthening of Existing Composite Beams Using LRFD Procedures," by John P. Miller, *AISC Engineering Journal,* 3d quarter 1997, pp. 110–111.
32. N. R. Nagaraja Ras and Lambert Tall, "Columns Reinforced Under Load," *AWS Welding Journal,* vol. 42, April 1963. Quoted in Ref. 6.
33. Howard J. Hill, "Building Failures," BSCES/ASCE Lecture Series at MIT, Cambridge, Mass., October 1997.
34. Arthur A. Huckelbridge, Jr., et al., "Grand Gusset Failure," *Civil Engineering,* September 1997.
35. Lawrence G. Griffis, "Motivation for and Fundamentals of Composite Building Frame Construction," BSCES/ASCE Structural Group Lecture Series at MIT, Cambridge, Mass., Fall 1987.
36. *Specification for Structural Steel Buildings, Allowable Stress Design and Plastic Design,* American Institute of Steel Construction, Chicago, 1989.
37. *Manual of Steel Construction, Load and Resistance Factor Design,* American Institute of Steel Construction, Chicago, 1986.
38. George Torello, Jr., and Howard I. Epstein, "Composite Action Crucial for Renovation," *Modern Steel Construction,* May-June 1990, p. 48.
39. Alexander Newman, "Debating Steel-Connection Design," *Civil Engineering,* February 1994.
40. David T. Ricker et al., "Discussion in the Steel Interchange," *Modern Steel Construction,* September 1998, p. 9.
41. ACI 318-95, *Building Code Requirements for Reinforced Concrete,* American Concrete Institute, Detroit, 1995.
42. Committee on Design of Steel Building Structures of the Committee on Metals, Structural Division, ASCE, "Compendium of Design Office Problems," *Journal of Structural Engineering,* vol. 118, no. 12, December 1992.
43. Lawrence G. Griffis, *Load and Resistance Factor Design of W-shapes Encased in Concrete,* AISC Design Guide No. 6, American Institute of Steel Construction, Chicago, 1992.
44. "Rectangular Concentric and Eccentric Unreinforced Web Penetrations in Steel Beams: A Design Aid," United States Steel, Pittsburgh, Pa., 1981.

45. David Darwin, *Steel and Composite Beams with Web Openings,* Steel Design Guide No. 2, American Institute of Steel Construction, Chicago, 1990.
46. Simon Boocock, "High Performance Coatings for Steel," *Modern Steel Construction,* September-October 1990, p. 57.
47. "Introduction to Generic Coating Types," *Journal of Protective Coatings & Linings,* July 1995.
48. *Galvanizing Insights,* American Galvanizers Association, Aurora, Colo., vol. 2, no. 4, Fall 1997.

Strengthening Concrete Buildings

4.1 Historical Perspective

4.1.1 Introduction

At its simplest, concrete is an artificial stone made of three ingredients: aggregate, cement, and water. The aggregate components, coarse and fine, are the meat and bones of concrete, giving it strength. Cement is the glue that solidifies the mixture, making concrete more than a pile of rubble. Water is required to hydrate ("activate") the cement particles and bind together all of the components. But, if the principle is so simple, why are there thousands of publications devoted to various aspects of concrete design and construction, and why have many dissertations been written about the minutest details of concrete's behavior?

The answer: Concrete is not a fully predictable material. Specifying certain mixture proportions and a level of workmanship does not guarantee the exact outcome. Concrete will invariably be a little stronger or weaker than anticipated, it may crack where it shouldn't, and, worst of all, it may not last as long as intended. When an existing concrete building is being renovated, its load-carrying capacity is uncertain, but can be determined by the investigative efforts described in Chap. 2. If the structure is found not to be strong enough for the proposed loading, it can be strengthened.

In this chapter we explore the history of concrete and design practices of the past, to give the reader an idea of what to expect in an old concrete structure. We then discuss some typical challenges involved in strengthening concrete buildings and describe the methods of upgrading various concrete structural elements. The specific challenges of repairing deteriorated concrete, rehabilitation of slabs on grade, and renovation of posttensioned structures are addressed in the three subsequent chapters.

4.1.2 Some history of concrete

Of the three concrete components, stone and water were always familiar to humans; it is cement that makes modern concrete different from its predecessors. The ancient Assyrians and Babylonians were probably the first people to mix lime, clay, and water for wall making.[1] The Phoenicians mixed ground bricks with lime before 700 B.C., and the Egyptians used gypsum as a binder to produce the early versions of concrete and masonry.

But it was the Romans who built numerous structures of true concrete by mixing stones, lime, and "pozzuoli" or "pozzolana" binders—fine brownish-red volcanic deposits. When mixed with hydrated lime, these deposits formed a cementlike material that was capable of binding sand with various kinds of coarse aggregate, such as travertine, tuff, brick, and even marble. The Romans used this version of concrete in most of their public buildings, including the Pantheon and the Colosseum. The Pantheon, built in the second century A.D., has a 142-ft-diameter dome supported by concrete ribs and arches bearing on 20-ft-thick concrete walls. The Pantheon's dome was the largest until the twentieth century. This remarkable result can be attributed to the excellent quality of the Roman concrete and to painstaking selection and grading of the aggregate. Incredibly, the ancient builders varied the density of the aggregate in the dome—the material was made progressively lighter with increasing height.[1]

Harries[2] observes that Roman methods of concrete construction substantially differ from modern methods only in the manner of placing aggregate and cement: We mix them together, while the Romans placed them in layered courses, like a cake. Instead of erecting formwork for concrete walls, the Romans built stone facing walls and placed concrete between them. Even the basics of seismic resistance were not altogether foreign to Roman builders. Rome is located in a region of severe seismicity, but both the Pantheon and the Colosseum have survived numerous earthquakes, a fact that Harries credits to their immensely massive foundations.

After the fall of the Roman Empire, the art of concrete making was lost. New experiments with concrete started during the Middle Ages, around A.D. 1200. The first hydraulic cement—cement that hardens under water—was used around 1756 by John Smeaton, an English engineer who was experimenting with various binding materials for the third Eddystone Lighthouse. Smeaton's discovery is now called hydraulic lime. (Incidentally, Smeaton was the first to use the term *civil engineer*, to distinguish himself from the military engineers of the time.[3]) Smeaton's compatriot James Parker made the first natural cement in 1796.[4]

A much better product, called portland cement, was patented by an English mason, Joseph Aspdin, in 1824. The name was derived from the resemblance of the hardened cement to a natural limestone found on the Isle of Portland off the English coast. To make his cement, Aspdin combined crushed limestone powder from local road pavement with fine clay and burned the mixture in his kitchen. To keep others from duplicating his extremely successful invention, he also sprinkled some bogus "secret" salts in the mix.[3]

In a few decades, the use of portland cement spread throughout Europe and eventually to America, where the first building made with this material is said to be Milton House in Milton, Wisconsin (1844), which survived for more than 100 years.[5] The early plain concrete approached stone in strength, but it had a very serious limitation: poor tensile strength. Plain concrete could resist only about one-tenth as much load in tension as in compression. Worse, the failures were brittle and occurred without warning, meaning that this material could not be used in flexural members. The use of concrete could have been confined to foundations and short walls, had it not been for the introduction of steel reinforcement.

The era of reinforced concrete was announced in 1850, when French builder Jean-Louis Lambot made a reinforced-concrete boat. In 1853, his compatriot François Coignet reportedly built a 20-ft reinforced-concrete roof. However, the inventor of reinforced concrete is considered to be still another Frenchman, F. Joseph Monier, who in 1861 started making flower tubs of concrete reinforced in two directions with grids of primary and distribution bars. In 1865 he patented the system. Being a gardener, not an engineer, Monier did not understand the mechanics of what he was doing and did not place reinforcement where it was most needed—near the tension surfaces. Concrete with rationally positioned steel reinforcement was pioneered in Germany by Gustav Adolph Wayss and J. Bauschinger.[6]

The structures built during all this experimentation typically used no coarse aggregate and therefore were really made of reinforced mortar rather than what we would call reinforced concrete.

In 1854, the first true reinforced-concrete structure was patented in Newcastle upon Tyne (England) by William Boutland Wilkinson, a plasterer and concrete manufacturer.[4,7] It is believed that Wilkinson was in contact with Aspdin, the inventor of portland cement, who had established a cement factory in a nearby town. Wilkinson's innovation remained relatively unknown until François Hennebique developed reinforced concrete into a structural system some 20 years later and patented the system in 1892. The so-called Hennebique-Mouchel system became dominant in France, while other systems were established elsewhere. For example, in 1907 the Trussed Concrete Steel Company was established in England to produce its proprietary system of reinforcement.[7]

The first American reinforced-concrete building was a house in Port Chester, New York, built in 1875 by W. E. Ward.[6] The American concrete systems were different from European framing.

4.1.3 Design practices at the turn of the twentieth century

The last quarter of the nineteenth century and the first quarter of the twentieth were times of extensive development of reinforced-concrete structures. A virtual explosion of proprietary systems and patented reinforcing bars made reinforced concrete an exciting, but also rather confusing, area of construction. One observer counted as many as 144 proprietary systems in 1909 Europe.[8] A comprehensive list of these systems has been compiled by Reid.[9]

The main reason for this abundance of proprietary systems was uncertainty about the most effective way to transfer stresses between steel reinforcement and concrete. Early experiments had shown that plain metal bars, especially if contaminated with oil and grease, tended to slip and destroy the composite action on which the system was based. Each of the proprietary reinforcing bars attempted to solve this problem in its own way. Some, such as the Hyatt bar, contained short cross bars inserted into holes in rectangular main bars. The Thacher system used rivets instead, while others relied on cross angles riveted to main bars.[6]

A different approach was to indent the rebar section to improve the bond. Attempts to do so included making circular indentations in flat bars (the Staff system), crimping the bars (the De Man system, 1898), or twisting the bars (the Ransom bar, 1884). The last became quite popular in this country for a time. Another rebar system had shear reinforcing attached to main bars. Still other bars, both square and round in section, utilized deformations of all sorts. There were also reinforcing products based on triangular mesh or expanded metal. All these systems are discussed by Loov.[6]

Those involved in renovations of old concrete structures may be surprised to find several kinds of reinforcing bars used on the same project. Figure 4.1 shows various square, round, and twisted types that were found in a Boston subway station constructed around 1917. Since the theory was lagging the innovation, new structural systems were often validated by full-scale load tests.[10]

Figure 4.1   These various kinds of reinforcing bars were all used on the same project.

4.2   Design Methods of the Past

The information in this section is based largely on the definitive article by George Winter.[11] Other sources are acknowledged separately.

4.2.1   The first U.S. concrete specifications

Given the vast number of proprietary reinforcing-bar systems that existed at the turn of the twentieth century, the process of comparing them was difficult. The theory of reinforced concrete design was still in its infancy, and the first design specifications started to appear in Europe between 1904 and 1907. Those recommendations were based on the elastic theory, which assumes that stresses vary across the section in a straight-line fashion, proportional to the distance from the neutral axis.

As Winter observes, during the first two decades of the twentieth century the use of structural concrete was considered experimental and was severely restricted. Essentially, the only concrete structures that were deemed acceptable for framed floors were shallow arches placed on sheet metal forms between the bottom flanges of steel beams. Even there, concrete faced strong competition from hollow terra cotta tiles (described in the beginning of Chap. 9). The concrete industry realized that lack of acceptance by building codes was hampering the development of the material's theory and practice.

To help reach a common understanding on reinforced-concrete design in the United States, the first design specification, entitled "Requirements for Reinforced Concrete or Concrete-Steel Constructed Buildings," was published in 1908 by the National Association of Cement Users, a predecessor of the American Concrete Institute (ACI). The six-page specification was not formally adopted and was considered an advisory document.

Two years later, however, a new and vastly different edition of the specification was approved by the NACU. The 1910 edition was called "Standard Building Regulations for the Use of Reinforced Concrete." Many stipulations first made by this document were kept in the editions that followed. The 12-page 1910 specification expressed allowable stresses in concrete in terms of f'_c—its cylinder strength—as shown below.

Compressive stress:

   $0.225 f'_c$ for columns "without hoops"

   $0.270 f'_c$ for columns "with hoops"

   $0.325 f'_c$ for flexural compression (i.e., 650 psi for 2,000-psi concrete)

Shear: $0.02 f'_c$

The regulations recognized only one strength of concrete, 2000 psi, and placed a severe limit on the type of structures in which concrete could be used. The 1910 specification already recognized two-way and continuous construction and provided some rudimentary guidance for the design of such members.

The efforts of NACU, which in 1913 became the ACI, were paralleled by another, broader, group called the Joint Committee on Standard Specifications for Concrete and Reinforced Concrete. The joint committee was composed of representatives from six major engineering groups: American Society of Civil Engineers (ASCE), American Society for Testing and Materials (ASTM), Portland Cement Association (PCA), American Railway Engineering Association (AREA), American Institute of Architects (AIA), and ACI. The joint committee was developing its own recommendations and in 1916 published its first report. The report recommended allowable stresses that differed little from those in the 1910 NACU regulations, but it contained many additional provisions for the design of flat slabs and columns.

Also in 1910, the first material specifications for reinforcing steel were developed by the Association of American Steel Manufacturers. In 1911, the American Society for Testing and Materials (ASTM) adopted its first standard specification for billet steel reinforcing bars. The rebars could be plain or deformed and were available in structural, intermediate, and hard grades. A structural grade was typically used unless other grades were specified. Cold-twisted steel was also produced. Later, the ASTM developed standards for rerolled rail steel (ASTM A16) and axle steel (ASTM A17), which had the same yield strength as billet steel.[12]

The yield strength of various kinds and grades of reinforcing steel available in 1935 is shown in Table 4.1. (The listed values are representative for bars produced from 1911 through the mid-1960s, as discussed in Sec. 4.3.)

The 1920 edition of ACI "Standard Building Regulations for the Use of Reinforced Concrete" essentially kept the same allowable stresses in concrete as the 1910 NACU specification. Like all previous specifications, it did not address the issues of column slenderness and columns subjected to bending.

In 1924, the second report of the joint committee was published. To convey a more official stance, it consistently used the term *shall* instead of the previous *may*. The report raised the allowable working stresses above those recommended previously. For example, flexural compression was limited to $0.4f'_c$ within the span of bending members

TABLE 4.1   Grades of Reinforcing Steel Available in 1935

ASTM designation	Material	Yield point, ksi	Tensile strength, ksi
A15-35	Plain and deformed bars, structural grade	33	55–70
(billet steel reinforcing bars)	Plain and deformed bars, intermediate grade	40	70–90
	Plain and deformed bars, hard grade	50	80
	Cold-twisted bars	55	—
A16-35	Plain bars	50	80
(Rail-steel reinforcing bars)	Deformed and hot-twisted bars	50	80

SOURCE: *ASTM Standards*, 1936 issue.

and to $0.45f'_c$ at continuous supports. The 1924 joint committee report also included many other new design provisions. Among them were design formulas for maximum bending moments in beams with various support conditions, for flat slabs, and for columns with or without end moments. Significantly, the report introduced a column-slenderness formula that remained in the later documents up to and including the 1956 edition of the ACI code:

$$P' = P\,(1.33 - h/120R)$$

where R was the least radius of gyration of the column section, P was the allowable load on a short column, and P' was the allowable load on a slender column with height h.

Of interest to those analyzing old rectangular column footings, the report prescribed finding the maximum moments in them by dividing a footing into four trapezoidal parts, with the critical section being at the column (or pier) face and the effective width resisting bending being the dimension of the column or pier. This approach was later changed.

The 1924 joint committee report also recommended the following allowable working stresses in steel reinforcement: 16 ksi for structural-grade and rail-steel rebars; 18 ksi for intermediate- and hard-grade bars, twisted bars, and cold-drawn wire. The report listed round bars with $^3/_8$-, $^1/_2$-, $^5/_8$-, $^3/_4$-, $^7/_8$-, and 1-in diameters; square bars were available in $^1/_2$-, 1-, $1^1/_8$-, and $1^1/_4$-in sizes. In addition to these, some suppliers also stocked $^1/_4$-in round and square bars. This list excluded some other sizes that had been produced before. According to CRSI, square bars were normally deformed or, if plain in structural grade, twisted.[12]

In 1928, the ACI issued a new edition of its standard, entitled "Reinforced Concrete Building Regulations and Specifications," that for the first time was specifically intended to be a part of the general building code. It included many of the 1924 joint committee's recommendations and also contained some new provisions, such as a method of distributing the load in two directions for two-way slabs.

4.2.2 Concrete specifications from the 1030o through the 1950s

The next (1936) edition of the ACI regulations kept many of the prescriptive requirements of the 1928 document but also added new provisions. Those included a recognition of the design principles for statically indeterminate continuous structures and a new complex and lengthy section on two-way slab design. The text also contained a column formula in which the allowable compressive loads on concrete and steel reinforcement were simply added together, an approach normally found in ultimate-strength design:

$$P = 0.22 f'_c A_c + f_s A_s$$

where f_s was taken as 20,000 psi.

The third, and last, report of the joint committee was published in 1940. For reasons that are not entirely clear, the report consisted of two parts, "Recommendations for the Use of Concrete and Reinforced Concrete" and "Standard Specifications for Concrete and Reinforced Concrete." Some topics were covered in both sections.

In many areas, the report simply rearranged the information contained in the previous edition. It significantly simplified the formulas for two-way slab design by providing a table of moment coefficients for slabs with various span ratios and clearly differentiated between column and middle strips. The design of flat slabs was also streamlined. The report used essentially the same approach to column design, with some simplifications, as the 1936 ACI regulations. The quaint design provisions of the 1924 joint committee report for two-way footings— those that divided the footings into four trapezoidal parts—were changed to the approach used today.

The 1940 joint committee report recommended larger allowable stresses in steel rebars than the earlier editions. For bars in tension, the allowable stresses were 18 ksi for structural-grade and rail-steel rebars, 20 ksi for intermediate and hard grades and rail-steel bars, and 16 ksi for web reinforcement. For bars in compression, the allowable stresses were 16 ksi for intermediate-grade bars and 20 ksi for hard-grade and rail-steel bars.[12]

The title of the 1941 edition of the ACI code was shortened to "Building Regulations for Reinforced Concrete." It was prepared by ACI Committee 318, and from that edition on the ACI code was known as ACI 318. This document kept much of the 1936 ACI design provisions and contained a few slight increases in the allowable stresses, made possible by the advances in reinforced-concrete construction practice. Unlike previous specifications, this edition allowed the use of two methods for designing flat slabs: one that resembled today's equivalent frame method, and another based on the tables of moment coefficients for specified dimensional and support conditions. As Winter observes, the 1941 ACI regulations may be considered the first comprehensive modern concrete code.

The ACI code underwent relatively minor revisions in 1947 and in 1951. (The 1951 revision attempted an ill-fated liberalization of the allowable shear stresses carried by concrete, as discussed in Sec. 4.2.4.) The term *regulations* disappeared from the title, the code was now called "Building Code Requirements for Reinforced Concrete." Perhaps the most important change in the next edition, that of 1956, was an explicit recognition of the ultimate-strength concept as a valid method of design and analysis. However, the ultimate-strength design provisions were relegated to an appendix—and largely ignored. No serviceability provisions were developed at the time.

4.2.3   The 1963 and later editions of ACI 318

The 1963 edition of ACI 318 was a major step forward. Not only did the code contain twice as many chapters and pages as the previous edition, but it also included a commentary that allowed the users to understand the rationale behind code provisions. For the first time, the code clearly treated both working-stress design and ultimate-strength design as equally acceptable.

The ultimate-strength provisions of ACI 318-63 were nothing short of revolutionary. Winter, who was in charge of drafting those provisions, recalls that at that time, no Western country allowed ultimate-strength methods of design, although the U.S.S.R., as he later learned, had a somewhat similar method on the books. The basic tenets of ultimate-strength design (USD) were retained in the later editions of ACI 318.

As generations of engineering students have since learned, the major advantage of USD lies in its ability to deal with the issues of structural safety in an explicit and rational manner. This is done by limiting the maximum effects of various loads to the minimum probable strength of the member. The ultimate-strength design method makes possible a quantitative evaluation of the member's reliability, and therefore the probability of its failure. In contrast, working-stress design (WSD), also known as allowable-stress design (ASD), strives to keep the maximum stresses in concrete and steel reinforcement below arbitrarily selected allowable values. The USD approach provides a better correlation with the actual behavior of reinforced-concrete structures under load. In the late 1980s and 1990s, design specifications for other materials started to emphasize USD methods, but the advantages of USD over WSD first became apparent in concrete.

ACI 318-63 contained many other innovations. One of the most important was its treatment of columns, which now were explicitly required to be designed for axial loads and bending moments. For cases where no calculated end moments were present, the code required designing for minimum eccentricities to account for possible construction misalignment. For the first time, the strength of slender columns was to be reduced by a slenderness factor.

Some other innovations included sections on allowable deflection limits and on methods of deflection computations—the serviceability provisions that had been lacking in the previous editions. For the first time, the code extended its umbrella over prestressed concrete construction, a development that was prompted by the publication in 1960 of the first prestressed concrete code by the Prestressed Concrete Institute (PCI). (The PCI was founded in 1954.) Importantly, the 1963 code significantly reduced the excessively liberal allowable shear values specified in the 1950s, which were blamed for several beam failures in shear. The code also improved design provisions for two-way slabs and footings.

The 1971 edition of ACI 318 took the final step toward embracing USD. The 1963 code included USD provisions in an appendix, with WSD still ruling the body of the code, but the situation was reversed in 1971. USD became the preferred approach, and WSD was now called the "alternate design method." Still, simple inertia has kept many designers from changing to USD. In any case, WSD had to survive, if only because it still had to be used for making deflection calculations and for satisfying the crack-control provisions of the code. (Indeed, one of the main arguments against USD is that it requires making structural calculations twice—once for strength with factored loads in USD format, and a second time for serviceability, with actual steel stresses computed under service loads.)

In addition to announcing the triumph of USD, the 1971 code contained many other important changes. A major modification was an introduction of the moment magnification method, the newest way to account for column slenderness effects. Another new provision allowed the designers to redistribute up to 20 percent of the negative moments in bending members into the positive bending regions. This was an acknowledgment of sorts that, since USD was now king, the elastic indeterminate methods of analysis did not really apply at failure, although there was nothing else to replace these methods.

Another novelty introduced in ACI 318-71 was a provision relating to controlling the width of tension cracks in concrete. Why was that necessary? Older low-strength concrete was reinforced with 33- or 40-ksi steel and properly designed in accordance with conservative WSD methods rarely produced wide cracks. In contrast, the newly available higher-strength concrete reinforced with high-grade bars and designed in accordance with USD tended to undergo substantial strain under load, and therefore some means of restraining the size of cracks was required.

Still another change that has altered the way all concrete members are designed was a switch from checking bond stresses to computing the development lengths of steel reinforcement. A few new chapters appeared in the code, such as those devoted to the design of shells and folded plates, plus an appendix on seismic design. Of special interest to the readers of this book, the 1971 code also included a chapter on strength evaluation of existing structures.

Later editions of the code appeared in 1977, 1983, 1989, 1995, and 1999, and these, along with interim revisions, continued to improve the theory and practice of reinforced-concrete construction. As with most other trade building codes, however, the general tendency seems to be toward more complex provisions that are thought to better represent the latest research efforts—at the expense of simplicity.

Some practitioners already complain that the codes are becoming incomprehensible, and that a point is near where designing the simplest of structures will require computer software. The fact that the cumbersome concrete code is counterproductive has been recognized even by one of its principal writers, who has spent over 40 years on ACI committee 318. In a published interview,[13] Chester P. Siess, ACI's past president, acknowledges that "because the Code is so long, so complicated, and written in such bad language, at times, it doesn't always tell you when you're getting into trouble."

4.2.4 The question of shear in beams designed during the 1950s

The 1947 and earlier editions of the ACI 318 code—beginning with the 1910 edition of the NACU regulations—specified an allowable value of $0.02 f'_c$ for shear stresses carried by concrete. As long as this stress level was not exceeded, web reinforcement was not required. Where the stress limit was exceeded, the maximum spacing of stirrups was limited to d, the distance from the extreme surface in compression to the tension reinforcement. Remarkably, this limit was still used in the last (1963) WSD specifications. The only exception occurred in the 1950s and deserves a brief mention.

For some reason, the 1951 edition of ACI 318 allowed much larger shear stresses carried by concrete, $0.03 f'_c$—a 50 percent increase. No maximum absolute value for shear stresses was given. So, if beams made of 3000-psi concrete were allowed a maximum shear stress of 60 psi in 1947, four years later the allowed shear stress was 90 psi! This liberalization was blamed for several highly publicized beam failures that occurred shortly before the new 1956 edition of the code was to be published. All the failures were in long, slender beams without web reinforcement.[14]

Based on the results of forensic investigations, the shear requirements were quickly tightened, even though the issue was not yet fully understood. The 1956 code introduced an explicit maximum value of 90 psi for shear stress carried by concrete, and reduced the maximum stirrup spacing to $\frac{1}{2}d$. The code also specified a minimum percentage of web reinforcement of 0.15 percent. Unfortunately, these changes did not address the core issue of the excessive allowable shear values.

The conclusion that a shear stress of 0.03 f'_c was too liberal was reached by ACI Committee 326, Shear and Diagonal Tension. In 1962, the committee produced several papers that formed a basis for the revisions contained in the 1963 and later editions of the ACI code. As already mentioned, ACI 318-63 restored the maximum allowable shear stress in concrete of 0.02 f'_c. Since the 1963 edition, most concrete beams have incorporated shear reinforcement.

Engineers involved in renovations of concrete buildings constructed in the 1950s are wise to pay particular attention to checking shear stresses. However, the mere fact that the maximum stirrup spacing in the existing framing is twice that allowed today (d vs. $\frac{1}{2}d$) does not necessarily mean that the building is unsafe. Elstner[14] mentions that a report of ACI Committee 326 considered beams with stirrups spaced more than $\frac{1}{2}d$ to be quite safe. He also observes that during his 40 years of investigating shear problems ranging from diagonal cracks to catastrophic failures, he has never seen a shear failure in a beam with stirrups outside the laboratory.

4.3   Properties of Old Concrete and Reinforcing Steel

Practitioners involved in the renovation of existing concrete structures need to make some assumptions about the design values for old concrete and reinforcing steel. The most reliable way of determining these properties is by destructive or nondestructive field testing, discussed in Chap. 2, but for preliminary investigations the information included here should help.

As was mentioned in the previous section, the strength of concrete structures was limited by the 1910 design specifications to 2000 psi. The 2000-psi concrete was typically made with a 1:2:4 mixture, i.e., 1 part cement to 2 parts sand and 4 parts of coarse aggregate, all measured by volume (by shovelfuls, perhaps). This mixture consistently produced concrete strengths in excess of 2000 psi, as was extensively proved by testing at the turn of the century. However, the in-place strength could be compromised by adding too much water—a bit of knowledge that was not widely available at the time. The reinforcing of this period ranged in yield from 30 to 35 ksi for soft steel to 50 to 60 ksi for hard steel. The typical working stress was taken as 12 to 14 ksi for the widely available steel reinforcement with a yield point between 35 and 40 ksi.[6]

In contrast, cast-in-place concrete produced at the end of the twentieth century could easily have a strength of 5000 psi or more, and reinforcing steel can have a yield strength of 60 or even 75 ksi. Concrete produced sometime between the 1910s and the 1990s is likely to fall in the middle.

Table 4.1 lists various grades of reinforcing steel available around 1935; these values are characteristic for many buildings constructed throughout the first sixty-odd years of the twentieth century. Unless a detailed investigation is performed, as illustrated in Example 4.1, reinforcing bars placed prior to about 1968 may be conservatively presumed to have a minimum yield point of 33 ksi; those from 1968 to the late 1970s, 40 ksi; and bars produced later, 60 ksi. (Steel with the yield point of 40 ksi lingered longer in some sanitary structures and when used as web reinforcement.)

Example 4.1: Determining Strength of Twisted Reinforcing Bars During renovation of an old underground subway station in Boston (c. 1917), several types of reinforcing bars were encountered. To assess the fitness of the existing concrete for new loading, it was necessary to check the yield and tensile strength of the $^7/_8$-in-diameter twisted bars shown in Fig. 4.1.

Without the benefit of being able to look in Table 4.1, the designers consulted some technical literature of the period. According to the 1911 edition of Sweet's Catalogue,[15] Inland Steel produced hot-twisted reinforcing bars with a yield strength of 50 ksi. According to ASTM A16-1935, hot-twisted rail-steel reinforcing bars had a design yield strength of 50 ksi. The tensile strength from both sources was 80 ksi. The existing bars could also be cold-twisted, with a yield strength of 55 ksi. (In cold-twisted bars, the yield point is raised by cold-drawing.) Conservatively, hot-twisted bars were assumed.

Next, bar specimens taken from the less-stressed areas of the structure were tested in a laboratory. The actual yield strength was determined to be 63 ksi, and the actual tensile strength 83 ksi.

This information allowed the design engineers to check the existing structure using a yield strength of 50 ksi, instead of the 33 or 40 ksi that would probably have been assumed by someone without access to this reference information or unable to afford the services of a testing laboratory. The higher allowable stress values permitted the existing structure to remain in operation, rather than being demolished and rebuilt at great cost.

If finding the yield strength of existing bars is relatively straightforward, determination of their allowable bond values or development lengths may be difficult. Modern code provisions apply only to "high bond" deformed bars, which have been produced by all mills only since 1947, although many suppliers offered them earlier. Concrete reinforced with plain bars behaves differently from that containing modern deformed bars. Without reliable means of transferring stresses between steel and concrete, the whole elaborate theory of reinforced-concrete behavior is invalid. What can be done?

The Concrete Reinforcing Steel Institute (CRSI)[12] recommends the following procedure for analyzing old reinforcing bars. For *all* varieties of pre-1947 bars, the bars can be considered 50 percent effective. Also, the development length specified in today's codes can be doubled (and multiplied by the ratio of existing to tabulated yield stresses). For example, for Grade 33 bars, 10 percent can be added to the values in development length tables computed by the modern code methods for 60-ksi steel. The advantage of this very conservative method is that it allows evaluation of old bars without regard to whether they are plain, deformed, or twisted. The big disadvantage is that much of the existing construction may be deemed unsafe if its bars are assumed to be only 50 percent effective and the existing development and splice lengths too short. This method often declares deficiencies in lap splices for bars larger than #6.

4.4   Some Early Proprietary Systems

As mentioned in Sec. 4.1.3, at the turn of the twentieth century, more than a hundred proprietary bar and framing systems had already been developed. Proprietary two-way, three-way, and even four-way concrete flat slabs were common; these construction systems are especially difficult to analyze. Some pre-World War II systems were known by their trade names, and sometimes the system's name is the only structural information indicated on the existing drawings. A few systems that were especially popular are shown here, reproduced with permission from the CRSI's booklet *Evaluation of Reinforcing Steel Systems in Old Reinforced Concrete Structures.*[12]

The so-called standard system illustrated in Fig. 4.2 was referenced in the joint committee reports. This four-way system was well engineered; it was ultimately abandoned, probably because of high cost.

Figure 4.3 illustrates the proprietary Turner system, also based on four-way action. The system was invented and first used by C. A. P. Turner of Minneapolis in 1906 and is considered the earliest system that was not modeled after familiar post-and-beam construction.[4] Turner's "mushroom floor" was one of the most economical concrete floor systems at the time.

Figure 4.4 shows the S-M-I flat slab system, which was very popular in the eastern states. The system includes a combination of radial and circumferential bars, an intriguing manifestation of the engineering thought about stresses around the columns. Figure 4.5 describes some details of the system.*

---

*Slabs designed by this method could undergo excessively large deflections. Some consider this design conceptually flawed because there are locations where no bars cross a potential failure plane.

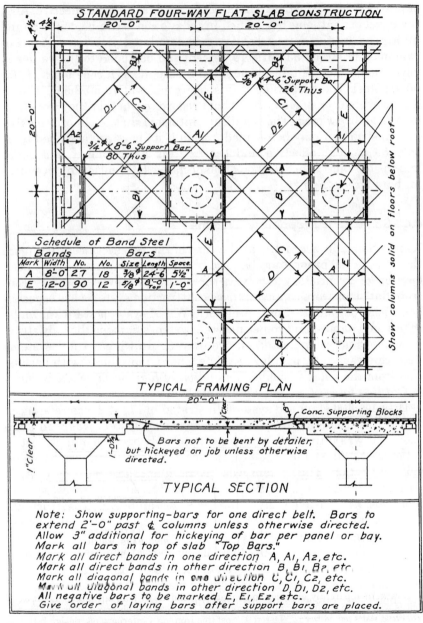

Schedule of Band Steel

Bands			Bars			
Mark	Width	No.	No.	Size	Length	Space.
A	8'-0"	27	18	3/8"	24-6	5½"
E	12-0	90	12	5/8"	8'-0" Top	1'-0"

TYPICAL FRAMING PLAN

TYPICAL SECTION

Bars not to be bent by detailer, but hickeyed on job unless otherwise directed.

Conc. Supporting Blocks

Note: Show supporting-bars for one direct belt. Bars to extend 2'-0" past ₵ columns unless otherwise directed. Allow 3" additional for hickeying of bar per panel or bay. Mark all bars in top of slab "Top Bars." Mark all direct bands in one direction A, A₁, A₂, etc. Mark all direct bands in other direction B, B₁, B₂, etc. Mark all diagonal bands in one direction C, C₁, C₂, etc. Mark all diagonal bands in other direction D, D₁, D₂, etc. All negative bars to be marked E, E₁, E₂, etc. Give order of laying bars after support bars are placed.

Figure 4.2 "Standard" four-way flat slab construction. (*CRSI*.)

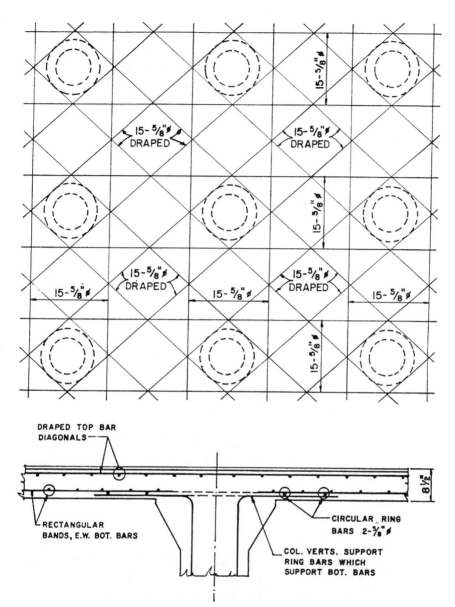

Most bars were apparently intended to be made continuous by long lap splices. Location of splices was left to detailer or placer apparently (from results observed after breaking out concrete slabs tested here). *Tested 913 psf (pig iron) on 4 interior bays; design L.L. = 250 psf; f'_c = 3000 psi; f_y = 60,000 psi. (*University of Illinois, Bulletin No. 106, May, 1918.*)

Figure 4.3 The patented Turner "mushroom" flat slab system utilized four-way action. (*CRSI*).

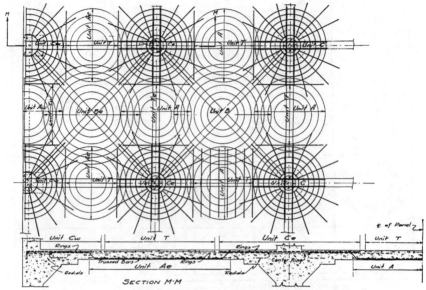

Typical S-M-I Flat Slab Reinforcement of Interior and Exterior panel. (Patented.)

(a)

Figure 4.4 S-M-I flat slab system. (*CRSI.*)

In addition to the three examples illustrated here, there were many other types of patented two-way, three-way, and four-way systems. Some of them utilized sound engineering knowledge; some were based on questionable assumptions. Sutherland and Reese mentioned in their 1943 book that even at that time, flat slabs were still largely designed by rule-of-thumb methods.[4]

4.5   Strengthening Concrete Beams

4.5.1   General methods of strengthening

The remaining sections in this chapter focus on methods of strengthening various concrete members. We will assume that structural eval uation of the existing framing has already been performed and the decision to upgrade reached prior to the start of any strengthening operations.

Our first topic is beams, the elements of concrete buildings that are most frequently upgraded, whether because of their deterioration or because of additional loading. Many techniques for upgrading beams are applicable to other structural members as well. Unfortunately, there is precious little available information on which to base our recommendations. In the words of Jay Thomas,[16] "There are no direct design

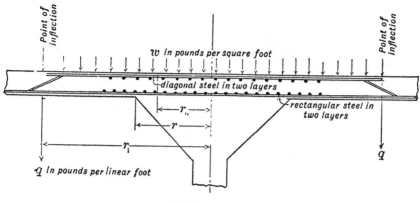

SECTION OF FLAT SLAB.

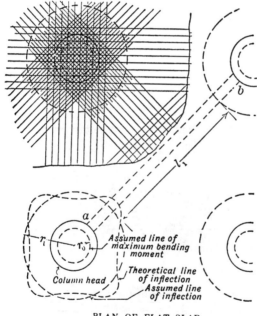

PLAN OF FLAT SLAB.

(b)

Figure 4.4   (*Continued*)

guidelines, no codes, no standards, and no practices for strengthening technology in the United States." The solutions we suggest are based on successful prior practice.

The easiest way to upgrade the load-carrying capacity of any existing structure is by analysis. As argued in Chaps. 1 and 3, it is much easier

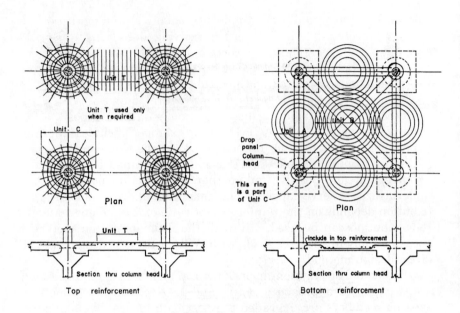

Top reinforcement   Bottom reinforcement

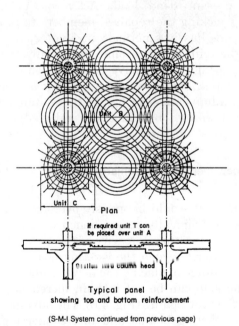

Typical panel
showing top and bottom reinforcement

(S-M-I System continued from previous page)

Figure 4.5 Details of S-M-I flat slab system. (*CRSI.*)

to work with paper and pencil than with real bricks and mortar (or concrete and steel in this case). Prior to specifying alterations to the structure, the engineer should be certain that the structure is indeed deficient. To help make this determination, a program of material testing may be undertaken to find the actual properties of the concrete and the steel reinforcement, to be used in structural analysis.

Methods of redistributing bending moments among positive and negative spans may help. ACI 318,[17] for example, includes a formula for moment redistribution in Section 8.4. As the Commentary points out, since negative moments are usually determined for one loading arrangement and positive for another, each section possesses some underutilized reserve capacity. The amount of allowed moment redistribution depends on the reinforcement ratio of the member. Moment redistribution can be used only with the ultimate-design method; it does not apply to analyses made with WSD or to slabs designed by the direct-design method.

When actual strengthening of structural members is required, there are two general approaches: active and passive.[18] In active methods, existing members are upgraded to resist both future (live and superimposed) and present (dead) loads. Active methods typically involve prestressing or jacking the repaired members to remove the existing stresses in them. Passive methods involve repairs that resist only future loads; the reinforcing becomes active only after some deformations of the existing section have occurred. This issue was already broached in Chap. 3, where we observed that a currently overloaded steel beam or column cannot be helped by welding additional sections (passive strengthening), but must be relieved of some load (the approach of active strengthening).

4.5.2 Adding new members and enlarging sections

Concrete beams that would be grossly overstressed by the proposed loading can be economically strengthened by adding new members. Depending on the structural capacity of the slab, additional members can be placed either between the existing beams or alongside them. The obvious advantage of placing the new members at midspan is that the span of the slab can be cut in half, increasing its load-carrying capacity as well as the beam's (Fig. 4.6). The advantage of placing new beams next to existing beams and interconnecting them is that a future load on the existing beams will be shared by the new neighbors.

With a passive approach, new beams resist only superimposed loads, and the existing slab and beams remain stressed by their dead loading. With an active approach, the slab and beams are jacked up by the

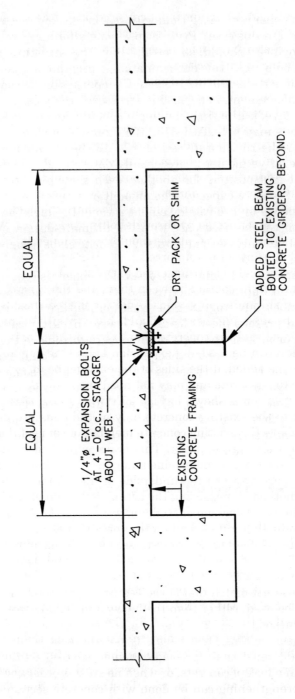

Figure 4.6 Adding steel beams at midspan of the slab.

EQUAL

EQUAL

DRY PACK OR SHIM

ADDED STEEL BEAM BOLTED TO EXISTING CONCRETE GIRDERS BEYOND.

1/4"⌀ EXPANSION BOLTS AT 4'-0"o.c., STAGGER ABOUT WEB.

EXISTING CONCRETE FRAMING

amount of dead-load deflection, a delicate and often unnecessary operation. Or, new members can be introduced that apply upward load on the existing structure; the load is generated by jacking or shimming, as explained later in this section.

It is often easier and quicker to add structural steel rather than concrete members, because new concrete beams would require formwork and shoring and are difficult to build with the slab in place. To be effective, steel beams have to maintain deformational compatibility with concrete beams they are intended to help. The load will be distributed among the new and existing beams in accordance with their relative rigidities (EI). Adding a steel channel on each side of an existing concrete beam is a common solution that allows the channels to be attached to the existing concrete columns. To share the load, the three beams can be interconnected by through-bolting (Fig. 4.7a). While the work is being done, the area above should be cordoned off to remove live load on the beam being reinforced.

In order for the load to be transferred to the steel beams, it is important to maintain tight contact between them and the concrete slab by drypacking or shimming any voids. Additional steel beams located away from the existing beams should have positive connections between the steel and the concrete slab to brace the beams laterally. This connection can be made by expansion bolts staggered about the web. Steel beams added at the sides of the existing beam are laterally restrained by the concrete and may not need such bolting.

A different solution is shown in Fig. 4.7b, where flexible steel channels fastened to the existing concrete only at the ends are used. The intent is to relieve the existing concrete member of some of the load by introducing upward forces into it. This is accomplished by deflecting the beams downward a predetermined amount, by jacking them, or by wedging the space between the underside of the slab and the beams. (One simple method of measuring the actual deflection of the channels is to use piano wire.) Jacking or wedging applies load on the channels and correspondingly reduces load on the concrete beam, allowing it to avoid either present or future overstress. One big advantage of this strengthening method over that of Fig. 4.7a is that no strain compatibility needs to be achieved in the combined system. The rigidity of the channels is need not match that of the beams, so that light and flexible channels can be used, rather than the heavy and rigid beams required by the other method.

In some cases, adding steel beams is not the best solution. When the final section must be fireproofed, when aesthetics dictate that the framing retain a look of concrete, or when no steel is otherwise present on the job, strengthening can be done with concrete section enlargement. This procedure involves unloading the existing beam as much as

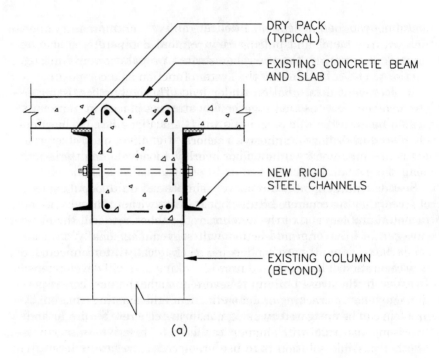

DRY PACK
(TYPICAL)

EXISTING CONCRETE BEAM
AND SLAB

NEW RIGID
STEEL CHANNELS

EXISTING COLUMN
(BEYOND)

(a)

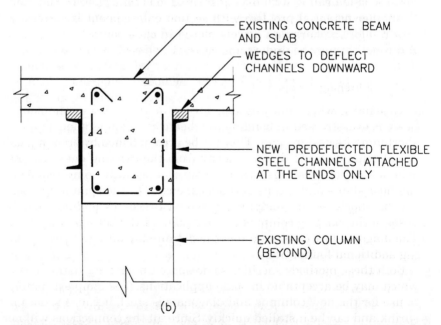

EXISTING CONCRETE BEAM
AND SLAB

WEDGES TO DEFLECT
CHANNELS DOWNWARD

NEW PREDEFLECTED FLEXIBLE
STEEL CHANNELS ATTACHED
AT THE ENDS ONLY

EXISTING COLUMN
(BEYOND)

(b)

Figure 4.7   Adding steel beams on each side of an existing concrete beam. (*a*) Rigid channels designed for strain compatibility; (*b*) flexible channels deflected downward to remove some load from the existing beam.

possible, roughening its surface to remove contaminants and to improve the bond, and placing new reinforced concrete or shotcrete around the existing beam. Proper surface preparation and interconnection is critical to making the system function as a composite whole and to prevent delamination under load. The new and existing concrete sections can be tied together by stirrups placed in horizontally drilled holes in the web of the existing beam (Fig. 4.8a), by short dowels placed in drilled-in adhesive anchors (Fig. 4.8b), or, if strengthening is accompanied by a new floor overlay, by enveloping the existing beam (Fig. 4.8c).

Section enlargement is a relatively simple and quite popular method of strengthening concrete beams, but it is not without drawbacks. One potential problem is that the new concrete or shotcrete will shrink as it cures, while the original section will stay dimensionally the same. Since the two sections are bonded and mechanically interconnected, the new concrete will be prevented from shrinking and will develop tension stresses. If the stress buildup is severe enough, the new concrete section may end up cracking or debonding from the existing concrete. The problem can be mitigated by using nonshrink concrete for repair and by specifying concrete with higher capacity to resist tension strains. Another possible solution is to use preplaced-aggregate concrete. The discussion of dealing with drying shrinkage is continued in Sec. 4.7.1.

Another potential problem with section enlargement is corrosion of new reinforcing bars and dowels placed in close contact with existing concrete that may be undergoing corrosive processes.[19]

4.5.3 Shortening the span

Occasionally, when analysis indicates that the existing simple-span beam is overstressed in bending, the beam can be upgraded simply by shortening its design span. This can be accomplished by erecting additional columns some distance away from the existing ones. The new columns require footings, which in turn necessitate some removal of the floor slab—and a considerable expense. Or, the span of the beam can be shortened by installing diagonal braces extending from the bases of the existing columns to some point at the bottom of the beam. The diagonal-brace solution has the definite advantage of not requiring additional foundations.

Both these methods sacrifice some space under the existing beams, which may be acceptable in some applications. The simplest material to use for the new columns and diagonals is steel, because it does not shrink and can be installed quickly. Since all the connections will normally be in compression, a pair of steel rods installed in drilled-in adhesive anchors might suffice.

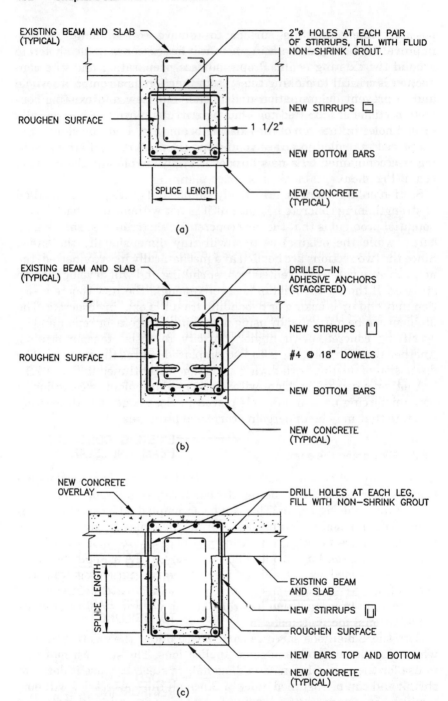

EXISTING BEAM AND SLAB
(TYPICAL)

2"∅ HOLES AT EACH PAIR
OF STIRRUPS, FILL WITH FLUID
NON-SHRINK GROUT.

NEW STIRRUPS ⊐

ROUGHEN SURFACE

1 1/2"

NEW BOTTOM BARS

SPLICE LENGTH

NEW CONCRETE
(TYPICAL)

(a)

EXISTING BEAM AND SLAB
(TYPICAL)

DRILLED-IN
ADHESIVE ANCHORS
(STAGGERED)

NEW STIRRUPS ⊔

ROUGHEN SURFACE

#4 @ 18" DOWELS

NEW BOTTOM BARS

NEW CONCRETE
(TYPICAL)

(b)

NEW CONCRETE
OVERLAY

DRILL HOLES AT EACH LEG,
FILL WITH NON-SHRINK GROUT

SPLICE LENGTH

EXISTING BEAM
AND SLAB

NEW STIRRUPS ▯

ROUGHEN SURFACE

NEW BARS TOP AND BOTTOM

NEW CONCRETE
(TYPICAL)

(c)

Figure 4.8 Enlarging a section of an existing concrete beam. The new and existing concrete can be tied together by (a) stirrups placed in horizontally drilled holes, (b) short dowels placed in drilled-in adhesive anchors, or (c) enveloping the existing beam.

When some owners see the proposed sketches showing diagonal bracing, they tend to ask why knee braces are not used instead, a solution that would result only in minimal headroom reduction. Of course, the problem with knee braces is that they exert horizontal reactions on the columns and could overstress them in bending or shear. In contrast, a diagonal brace transfers its load to the column base near the floor line, where the horizontal load can be resisted by the slab.

The most radical method of shortening the span of a failing beam is to construct a loadbearing wall under it, supported by its own foundation, if this is acceptable from a space-planning standpoint.

4.5.4 Adding bolted steel tension reinforcement

If the existing beam lacks positive moment capacity, it can be reinforced in place by adding structural-steel tension plates or built-up members bolted to the beam (Fig. 4.9). The welded U-bracket shown can be used if substantial additional steel area is needed. This is a passive design; the new steel does not become effective until the concrete deforms under some additional load. However, in the ultimate-design philosophy, both the existing bars and the added steel reinforcement are assumed to be yielding at the maximum factored load, and the combined section may be assumed to resist the total load (see Example 4.2).

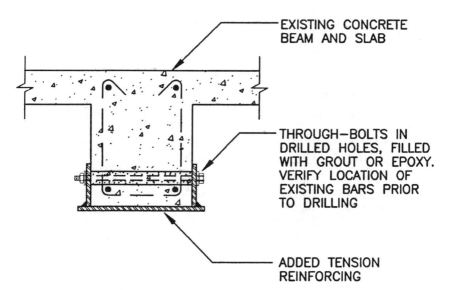

EXISTING CONCRETE
BEAM AND SLAB

THROUGH-BOLTS IN
DRILLED HOLES, FILLED
WITH GROUT OR EPOXY.
VERIFY LOCATION OF
EXISTING BARS PRIOR
TO DRILLING

ADDED TENSION
REINFORCING

Figure 4.9 Bolting a built-up steel member to improve the positive moment capacity of an existing concrete beam.

The size and spacing of connecting bolts is determined by the forces needed to transfer the factored loads from the steel members into the concrete by bolt bearing or shear. The procedure is reminiscent of reinforcing existing steel beams by welding cover plates, as described in Chap. 3. Through-bolting next to the existing bottom reinforcing bars should be approached with caution; it is best to locate the exiting beam bars in advance.

A variation of this scheme involves bolting steel plates to the underside of the beam with vertical adhesive anchors. This solution demands even more attention to locating and avoiding existing bottom bars.

Yet another version of this design involves adding *two* plates—at the top and at the bottom of the beam. The plates can be interconnected by bolting vertically through the whole depth of the beam. This reinforcing detail can be used when substantial loading deficiency has to be overcome. Unfortunately, it requires not only a lot of difficult drilling through the beam, but also covering the floor above with an overlay to conceal the top plate and its anchors.

All these kinds of beam strengthening, and some of those discussed later, can also be effective in improving the beam's stiffness. When large added members are connected by bolting, the number and diameter of the bolts may become excessive, and other upgrade options could then be explored.

Example 4.2: Concrete Girder Externally Reinforced with Structural-Steel Channel* The existing cast-in-place concrete floor framing system shown in Fig. 4.10 was designed for a superimposed live loading of 50 psf. It is desired to increase the flexural capacity and stiffness of the interior girder shown so that it will support an additional 50 psf. (Strengthening of the other beams is not addressed in this example.)

The following assumptions are made:

1. The existing member is capable of supporting the original design loads.
2. The girder can be represented by a simple-support analytical model—a conservative assumption for positive moment calculations, made for simplicity.
3. The existing shear capacity of the girder is sufficient to support the added loading.

The example uses a strength design approach, which assumes that the existing internal reinforcing steel can reach its yield-stress level at the ultimate loading and will continue to yield under the new superimposed live loading. Buckhouse[20] has found this assumption to be justified for the design of exter-

*The author is grateful to Dr. Christopher M. Foley, Ph.D, Assistant Professor, Department of Civil and Environmental Engineering, Marquette University and to Mr. Evan R. Buckhouse for contributing this example.

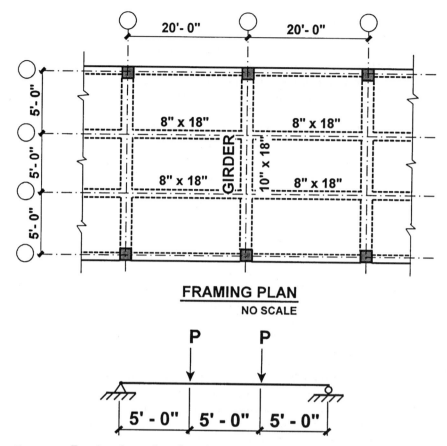

Figure 4.10 Framing plan and analytical model for Example 4.2.

nally reinforced members, but it should be carefully addressed on a case-by-case basis. Many older concrete structures are reinforced with plain bars and rely on bond stresses to transfer forces between the reinforcement and the concrete. If the bond fails before the bars yield, this assumption becomes invalid. (Fortunately, real-life beam failures related to loss of bond are rare.)

The existing beam has the following characteristics:

$$b = 10 \text{ in} \qquad h = d_2 = 18 \text{ in} \qquad d_1 = 15.31 \text{ in}$$

$$f'_c = 4000 \text{ psi} \qquad f_y = 60,000 \text{ psi} \qquad A_s = 1.32 \text{ in}^2 \text{ (3 #6 bars)}$$

The proposed channel reinforcement is shown in Fig. 4.11a. The design methodology is based on the procedures outlined in Refs. 20 to 22. The procedure is equally valid for allowable-stress design. Although the design procedure in this example assumes wedge-type expansion anchors,

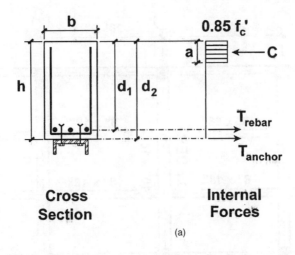

Cross
Section

Internal
Forces

(a)

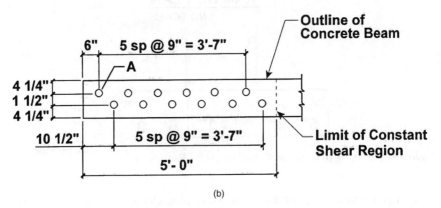

(b)

Figure 4.11 Design details for Example 4.2. (*a*) Cross section and internal forces in the reinforced member; (*b*) layout of expansion bolts.

a similar procedure may be used with epoxy-adhesive anchors. (As noted elsewhere in this book, epoxy's effectiveness is reduced at elevated temperatures; it starts to soften at 130 to 150°F. Some engineers feel that even spray-on fireproofing does not provide enough protection to ensure that the load capacities of epoxy adhesive anchors are not reduced during fire.)

Solution The first step is to compute the factored bending moment resulting from the original design loading on the beam. The existing service loads acting on the beam to be reinforced are computed as follows:

Dead load

4-in existing concrete slab: (50 psf)(5 ft)(20 ft) $\qquad$ = 5000 lb

8-in by 18-in beam: (8 in)(18 in − 4 in)(1/144)(20 ft)(150 pcf) = 2333 lb

10-in by 18-in beam: (10 in)(18 in − 4 in)(1/144)(150 pcf) = 46 lb · ft

Original live load: (50 psf)(5 ft)(20 ft) $\qquad$ = 5000 lb

The *existing* factored bending moment acting on the unreinforced section is therefore

P_u,exist = 1.4 (5.0 + 2.33) + 1.7(5.00) = 18.76 kips

$$M_{u,\text{exist}} = \frac{1}{8}(1.4)(0.146 \text{ kip} \cdot \text{ft})(15 \text{ ft})^2 + \frac{1}{3}(18.76 \text{ kip})(15 \text{ ft}) = 99.5 \text{ kip} \cdot \text{ft}$$

The next step is to calculate the *additional* factored superimposed live load moment from the beams in the framing system.

$$P_u = 1.7 \left[\frac{50 \text{ psf } (5 \text{ ft})(20 \text{ ft})}{1000 \text{ lb/kip}} \right] = 8.50 \text{ kips}$$

$$M_{u,\text{LLa}} = \frac{8.50 \text{ kips } (15 \text{ ft})}{3} = 42.5 \text{ kip} \cdot \text{ft}$$

The *total* factored bending moment acting on the externally reinforced concrete member is

$$M_u = M_{u,\text{exist}} + M_{u,\text{LLa}} = 99.5 \text{ kip} \cdot \text{ft} + 42.5 \text{ kip} \cdot \text{ft}$$

$$M_u = 142 \text{ kip} \cdot \text{ft}$$

Next, the internal tension force to be developed in the anchors is determined. It has been shown that the controlling mode of failure, on which the reinforced member strength is based, is shear failure of the anchors. The expression for the nominal capacity of the externally reinforced cross section is given by Buckhouse[20]:

$$M_n = A_s f_y \left(d_1 - \frac{a}{2} \right) + T_{\text{anchor}} \left(d_2 - \frac{a}{2} \right) \tag{1}$$

where T_{anchor} is the tension force carried in shear by the bolted anchors, A_s is the area of the existing internal flexural reinforcement, f_y is the yield stress of the internal flexural reinforcement; d_1 is the depth from the extreme compression fiber to the centroid of the internal flexural reinforcement, and d_2 is the depth from the extreme compression fiber to the shear plane of the anchors. The depth of the compressive stress block a in the externally reinforced member is a little more difficult to determine, as it depends on the anchor force. The internal equilibrium shown in Fig. 14.11a gives

$$a = \frac{A_s f_y + T_{anchor}}{0.85 \, f'_c b} \tag{2}$$

It should be noted that the force in the anchors T_{anchor} and the depth of the compression block a are both unknown as posed in Eqs. (1) and (2). Therefore, an iterative approach is needed to compute them for the required nominal moment capacity. Equation (1) can be rearranged as follows to facilitate the iterative calculations:

$$T_{anchor} = \frac{M_n - A_s f_y \left(d_1 - \dfrac{a}{2}\right)}{\left(d_2 - \dfrac{a}{2}\right)} \tag{3}$$

An initial estimate of the depth of the rectangular stress block a in the reinforced member needs to be made. This estimate should be larger than the depth of the compressive stress for the existing unreinforced section; a value of $2a_{original}$ is a good first estimate. Few iterations are now required in order to converge on the correct internal equilibrium configuration.

Equation (3) requires that the nominal section capacity be determined or known. This may seem a simple matter until we recall that the controlling mode of failure is assumed to be shear of the anchors, while the externally reinforced member's behavior is essentially flexural. The conflicting reduction factors ϕ for these two failure modes ($\phi = 0.90$ for flexure and $\phi = 0.85$ for shear) require some judgment. In this example, $\phi = 0.85$ was used for the calculations because of the relatively indeterminate material properties of the existing concrete member and the relatively difficult field installation procedures.

(A note of caution: It can be argued that a much lower value of ϕ should be used for designs involving expansion anchors. These fasteners are typically designed with a capacity reduction factor of 4, owing to their lack of reliability. There is a difference between the ultimate load values given in the manufacturers' catalogs and the ultimate load capacities of these anchors that can be used in design. A smaller degree of reduction can be justified if the anchor capacity is governed by anchor shear—a reliable property. A larger degree of reduction—lower ϕ—is prudent when the mode of failure involves concrete crushing. For example, a ϕ of 0.5 could be used in these circumstances.)

The required nominal moment capacity is then

$$(M_n)_{req} = \frac{M_u}{\phi} = \frac{142}{0.85} = 167.1 \text{ kip} \cdot \text{ft}$$

With the required nominal moment capacity determined, Eqs. (3) and (2) can be used for the iterative solution. Substituting the known quantities into Eq. (3) gives

$$T_{anchor} = \frac{12(167.1) - 1.32(60.0)\left(15.31 - \dfrac{a}{2}\right)}{\left(18 - \dfrac{a}{2}\right)} \tag{4}$$

The depth of the compressive stress block in the existing member is $a = 1.60$ in. Therefore, the initial estimate for a is $2(1.6) = 3.20$ in. Solving for the anchor force using Eq. (4) gives

$$T_{anchor} = \frac{12(167.1) - 1.32(60.0)\left(15.31 - \dfrac{3.20}{2}\right)}{\left(18 - \dfrac{3.20}{2}\right)} = 56.1 \text{ kips}$$

The revised compressive stress block depth is

$$a_{revised} = \frac{1.32(60.0) + 56.1}{0.85(4.0)(10)} = 4.00 \text{ in}$$

Substituting $a_{revised}$ into Eq. (4) to compute a new T_{anchor} and continuing the iterations (four in total) results in

$$T_{anchor} = 56.0 \text{ kips} \qquad a = 3.18 \text{ in}$$

The anchors should be designed for 56 kips. Typical design values for the ultimate shear capacities, embedment depths, and guide torque values of expansion anchors can be found in the manufacturers' literature. In this example, the values of Ref. 23 were used. For $1/2$-in-diameter carbon-steel wedge-type expansion anchors in 4000-psi concrete, the ultimate shear strength is 7.36 kips. (This value assumes that the shear plane passes through the anchor body; if the shear plane passes through the anchor threads, the nominal strength should be reduced 20 percent.[23]) The minimum required bolt embedment is 2.25 in.*

Assume that $1/2$-in-diameter anchors with $2\ 1/2$-in embedment in 4000 psi concrete are used and that the shear plane passes through the anchor body. The nominal capacity for a single anchor is then

$$V_n = 7360 \text{ lb}$$

The number of anchors required can be initially determined assuming that all anchor edge distances and anchor spacing requirements are met. Therefore, using the nominal strength previously determined, the number of anchors required is computed as

$$N = \frac{T_{anchor}}{V_n} = \frac{56 \text{ kips}}{7.36 \text{ kips} \cdot \text{anchor}} = 7.60$$

*See the note of caution involving capacity reduction factors for expansion anchors.

These anchors must be placed in a zone of constant shear. The anchors should be located to develop the tension force in the channel at the point of maximum bending, which in this example is at the middle third of the span.

The edge distance and spacing of anchors are two critical issues that pertain to both wedge-type expansion anchors and epoxy-adhesive anchors. Therefore, the initial estimate of 7.6 anchors is just a starting point. A preliminary anchor layout based on the existing dimensions for the beam soffit and also the length of the constant shear zone is shown in Fig. 4.11b. To enhance anchor spacing, given the limited width of the beam soffit, the anchors are staggered. This initial layout will be used to determine revised anchor strengths based on reduction factors that take into account limited edge distance and anchor spacing.

The reduction factors for small edge distances and spacing are contained in the anchor manufacturers' design manuals.[23,24] From Ref. 23, for $1/2$-in-diameter bolts placed as shown, the anchor spacings (AS), edge distances (ED), and corresponding reduction factors (R) are

Anchor A:
$$AS = 4.75 \text{ in} \qquad R_{AS} = 1.00$$
$$ED = 4.25 \text{ in} \qquad R_{ED} = 0.63$$

Remaining anchors:
$$AS = 4.75 \text{ in} \qquad R_{AS} = 1.00$$
$$ED = 4.25 \text{ in} \qquad R_{ED} = 0.63$$

The nominal strength of the anchors can then be computed as

$$T_n = (1.0)(0.63)(7.4) + 11(1.0)(0.63)(7.4) = 55.9 \text{ kips}$$
$$\approx 56 \text{ kips}$$

Therefore, use twelve $1/2$-in-diameter wedge-type expansion bolts with $2^{1}/_{2}$-in embedment and the layout shown in Fig. 4.11b.

The usual checks pertaining to the structural steel channel member must be made prior to acceptance of this design. These checks (such as block shear, tension on net section, etc.) are not made in this example for the sake of brevity. Also, the total area of steel should be checked to ensure that it does not exceed the maximum (75 percent balanced) required to avoid brittle failure. The designer should also ensure that the installation procedures contained in the anchor manufacturers' design handbooks are followed during construction. Careful specification writing should be performed to ensure proper in-field service of the retrofit scheme.

One additional point: Since the holes drilled in steel are typically slightly larger than the outside bolt diameters, some plate slippage could occur under load. To avoid slippage, the small gaps between the edges of the holes and the bolts could be filled with epoxy.

In lieu of the trial-and-error method of Example 4.2, a simpler approach can be used to find the required size of an added member (a plate is used for illustration purposes). The first step in this procedure is to calculate the forces in the existing reinforcing bars, the depth of a compression stress block, and the moment capacity of the original section in a conventional way, using a capacity reduction factor ϕ of 0.90. In the next step, the required additional moment capacity is found by subtracting the existing moment capacity from the applied maximum bending moment. The added capacity is produced by a couple consisting of the tension force in the plate and the resultant of the compression block located below the compression block of the original section. The required factored force in the plate can be found using a conservative capacity reduction factor ϕ of, say, 0.5, to account for the relative lack of reliability provided by drilled-in anchors (Fig. 4.12a).

One general difficulty with adding bolted steel tension reinforcement is that the exact position of the existing reinforcing bars at the bottom of the beam must be known. Care should be taken not to damage the bars by drilling during anchor installation. (The bar locations can be found by the nondestructive investigation techniques described in Chap. 2 or by chipping the concrete cover and exposing the bars. The chipped areas should be patched.)

When the existing bars are few and far between, it may be possible to dispense with added steel plates. Instead, new reinforcing bars can be placed in longitudinal slots cut at the bottom of the beam (Fig. 4.12b). The slots are later filled with a suitable repair material, such as epoxy mortar. If the bottom legs of the existing beam ties are cut during this operation, their shear capacity should be investigated.

Sometimes the bars are spaced so closely or so erratically that drilling from the bottom is risky. In this case, a solution similar to that shown in Fig. 4.9, where the external reinforcement is placed at the sides of the existing member near the bottom, can be used. One potential difficulty with using side-attached members: They could interfere with pipes and conduits running alongside the beam.

4.5.5   Adding bonded steel plates

Instead of being bolted, steel reinforcing plates or other shapes can be attached to concrete by two-part epoxy adhesives. "Gluing" reinforcing plates is certainly easier than drilling and bolting them. Also, repairs can be done quickly, with minimum disturbance of the existing construction or the occupants. The obvious disadvantage of this system is its complete reliance on the quality of workmanship. It is easy to visualize everything that could go wrong with epoxy application: The surface is not properly prepared, the epoxy material is deficient or simply does not deform at the same rate as the beam, the system is applied in

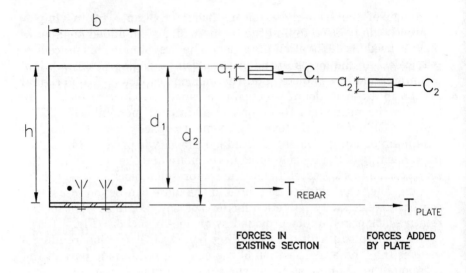

FORCES IN
EXISTING SECTION

FORCES ADDED
BY PLATE

(a)

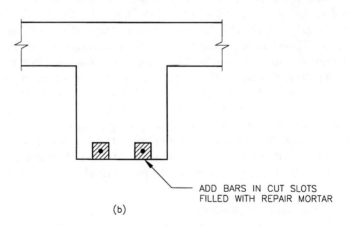

ADD BARS IN CUT SLOTS
FILLED WITH REPAIR MORTAR

(b)

Figure 4.12 A method of determining forces in the reinforcing member.

a careless manner, and so on. The final result would be the same: a loose cover plate that does no good at all.

Still, if specified and applied properly, the bonded-plate system offers an excellent alternative to mechanical methods of attachment. Three factors are critical to success of the system:[19]

1. The surface must be thoroughly prepared, with mill scale and contaminants removed and the surface roughened; abrasive blasting of both concrete and steel works well.

2. The epoxy should have a bond strength equal to or exceeding that of the concrete and be suitable for the environmental conditions.

3. The reinforcing plate should be long and thin enough to avoid brittle plate separation from the concrete.

The epoxy can be placed by pressure injecting flowable resin or by applying gel to the mating surfaces. As with any glue, connecting the members with a two-part epoxy requires pressing them together temporarily, so erection anchors and temporary falsework are needed. The support during installation—and some additional shear capacity—can be provided by grouted-in threaded rods (Fig. 4.13).[18] It is a good idea to provide some supplemental anchors at the ends of the plate. This can prevent debonding caused by the high local bond stresses that develop as the plate takes up load. Another possible approach to avoiding abrupt transition at the ends of the plate is to gradually taper (feather) them.

The Achilles' heel of the system in exterior applications is corrosion of the steel plate at the interface with the concrete. This can lead to a loss of bond. Protecting added steel by coating it with paint is not the

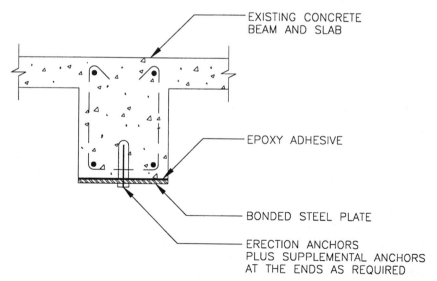

EXISTING CONCRETE
BEAM AND SLAB

EPOXY ADHESIVE

BONDED STEEL PLATE

ERECTION ANCHORS
PLUS SUPPLEMENTAL ANCHORS
AT THE ENDS AS REQUIRED

Figure 4.13 Adding a bonded steel plate to improve the positive moment capacity of an existing member.

answer, because coatings might interfere with adhesion. Another issue to keep in mind is that added exterior reinforcement should not be used as a long-term solution for structures undergoing active corrosion. The new steel member can not only conceal the signs of corrosion until it is too late, but also contribute to galvanic action and make the situation worse, as described in Chap. 5. And, as already stated, there is a problem with using epoxies at elevated temperatures.

4.5.6 Adding bonded FRP plates and wraps

In Fig. 4.13, plates of fiber-reinforced plastic (FRP) can be used instead of steel. FRP is made of small but very strong fibers bonded with resin, and its strength sometimes surpasses that of steel. The fiber material commonly used in the past was glass, and the familiar acronym GFRP stands for glass fiber–reinforced plastic. Because some FRPs are resistant to corrosion caused by acids, alkalies, and salts, reinforcement with FRP plates is most popular for rehabilitation of bridges. For enclosed buildings, corrosion resistance is less important, but even here FRP plates are finding acceptance. The main disadvantage of FRP materials in general and GFRP in particular is that their modulus of elasticity—the measure of the material stiffness—is very low. The stiffness of GFRP is about one-fifth that of steel, which means that a beam made from GFRP will deflect five times as much as a steel beam of the same size.

The epoxy materials needed for bonding steel are different from those required to attach FRP, and proper selection of epoxy material is critical to the success of the project. The epoxy should be sufficiently stiff and strong to transfer the forces between the plate and the concrete. Rubber-toughened epoxies seem to be good candidates for this application.[25]

FRP reinforcement can fail in two ways: flexural and local. The flexural mode of failure includes crushing of concrete or rupture of the plate. The local mode of failure can involve debonding of the FRP plate or a shear failure of the concrete layer between the plate and the longitudinal reinforcement. A flexural mode of failure suggests ductile behavior of the repair and is desirable; the local failure must be avoided. Indeed, Ross et al[26] conclude that the bond strength between the plate and the concrete is the most important factor affecting the response of beams reinforced with FRP plates. (They also conclude that the most significant improvement of flexural strength is attained in lightly reinforced concrete beams and with very thin FRP plates.)

Malek et al.[27] state that shear and normal stress concentrations near the ends of FRP plates, as well as flexural cracks in the beams, must be considered in the design of FRP reinforcement. They provide general design methods for computing shear and normal stresses at various points of the interface. The location of cutoff points of the plates is critical in ensuring a flexural, rather than local, mode of failure. To avoid local stress concentrations and to ensure a flexural mode of failure, the longest possible bonded lengths of FRP plates should be specified.[28] Emmons et al.[29] mention a rule of thumb that limits the width-to-thickness ratio of FRP plates to 50 to ensure ductility at failure.

Fiber composites have other limitations as well. Popular aramid fibers, despite their excellent impact resistance (one aramid, Kevlar, is used in bulletproof vests), have poor compressive strength, which limits their application to tension members. In addition, aramid costs almost as much as carbon, a much stronger material. Glass fibers have their own shortcomings; perhaps the most important is their poor resistance to alkalis, so that GFRP fibers must rely on the protection afforded by the resin. The resin thickness, however, is quite small—a few millimeters at best—and its long-term protective ability is suspect.[19] Since alkalis are always present in concrete, concerns about the durability of glass fibers should be addressed before specifying GFRP reinforcement. Another possible problem with glass fibers is creep rupture, or failure under sustained load at levels lower than those the material can support instantaneously. Therefore, GFRP is best used for strengthening members to resist quickly applied loads—seismic, for one—rather then long-term loading. Also, there are limits on how much added strength FRP reinforcement can provide. Some authorities suggest 50 percent as a reasonable upper limit.[30]

The most attractive material used for FRP strengthening is carbon-fiber-reinforced plastic (CFRP). CFRP possesses many desirable traits: It is strong, resistant to corrosion caused by both acids and alkalis, lightweight, and nonmagnetic. Some carbon materials used for structural strengthening have strengths over 500 ksi—8 times more than Grade 60 reinforcing bars—and stiffnesses comparable to that of steel.[31] The disadvantages? They are relatively minor: Carbon, like steel, conducts electricity, so it could become a part of an electromagnetic chain of corrosion or even promote corrosion of steel.[30] In most cases, carbon is the FRP material of choice.

A wider acceptance of CFRP and other fiber composites is hampered by the scarcity of long-term experience with these products, coupled with a lack of design standards and specifications for their use. When these standards are developed, FRP materials are certain to gain wide acceptance in rehabilitation of concrete building structures.

A state-of-the-art report on the theory and design of both steel and FRP bonded plates is ACI Special Publication No. 165, *Repair and Strengthening of Concrete Members with Adhesive Bonded Plates.*[32] The publication includes a number of design papers dealing with all aspects of bonded-plate design and construction, such as details of anchorage, performance of various adhesives, and temperature limitations. It includes papers devoted to construction-related issues, such as surface preparation and methods of epoxy application. Another relevant recent publication is *State-of-the-Art Report on FRP for Concrete Structures* by ACI Committee 440.

Polymer resins soften and lose their strength at elevated temperatures. Therefore, the use of epoxy adhesives for bonding steel and FRP reinforcing plates should be avoided in applications where high service temperatures might be present. As Gaul[33] points out, when the temperature rises above a certain point—the so-called glass transition temperature—many of the structural properties of epoxies, among them strength, stiffness, and creep resistance, rapidly decrease. For epoxy adhesives used in construction, this point falls between 70 and 170°F. When the temperatures eventually decrease, the epoxies regain their structural properties, but in the interim some damage could have occurred.

To reduce this degradation of properties during fire, proper fireproofing has to be provided for critical repaired elements. Fireproofing can include intumescent coatings and drywall enclosures, both of which are rather expensive. The cost of this fireproofing, on top of the already high cost of the FRP materials, can make the system less economical than other strengthening options. Also, as already noted, it is questionable whether fireproofing—especially by a thin coating—will prevent degradation of epoxy properties.

One rationalization offered for not only allowing the use of FRP materials at elevated temperatures, but even avoiding fireproofing altogether, is that FRP should be used only as supplemental reinforcement. That is, even if the FRP is totally lost, the existing building structure will not collapse.[30] In this approach, the role of FRP is secondary—to improve the margin of safety or to meet some detailed code provision, such as lateral confinement for columns and minimum stirrups for beams—rather than critical for the immediate safety of the structure. This argument might make intuitive sense for some, although building codes do not differentiate between "important" and "less important" provisions. Also, this raises some interesting philosophical questions about the probabilities of experiencing the maximum level of loading during fire.

How can one verify that fiber reinforcement was properly bonded? The bond can be checked by sounding, which can reveal delaminations, and by pull-off testing. In a well-bonded application, the failure occurs in the concrete and is of a flexural, rather than local, nature, as discussed previously. A more extensive verification can involve load testing, described in Chap. 2.

A variation on the theme of adding FRP bonded plates is wrapping the member with bonded CFRP sheets. This design, while relatively new, is sponsored by some of the most prestigious names in the industry and is meeting with phenomenal success. One of the world's largest projects involving carbon-fiber reinforcement is strengthening of a 2100-space parking garage at the Pittsburgh International Airport in 1998. The 1992 garage was framed with precast double-tee beams, which started to crack and spall. More than 1250 precast double-tees were strengthened for both flexure and shear by applying CFRP sheets to the first 8 ft of the precast tee stems. Prior to strengthening, epoxy was injected into more than 44,000 linear ft of surface cracks.[34] The process of using carbon FRP reinforcement to increase the shear capacity of beams is further described in Sec. 4.5.8 and in Example 4.3.

4.5.7 Reinforcing by post-tensioning (external prestressing)

A popular method of strengthening concrete beams is by post-tensioning, also called external prestressing. In this method, the members are reinforced by applying external forces to counteract the effects of the design loads. The forces are delivered by means of prestressing tendons located outside the reinforced section (external prestressing) or inside it (internal prestressing). External prestressing is more common, but internal tendons can also be used in rehabilitation of existing post-tensioned structures, as described in Chap. 7.

The tendons are connected to the structure at the anchor points, which are typically located at the member ends. The desired jacking force is provided by deviation blocks, or deflectors, fastened at the high or low points of the structure. A common approach to strengthening concrete beams by external prestressing is shown in Fig. 4.14. The deflectors help position the cables in a manner that produces net forces in the desired direction, as shown in Fig. 4.15.

External prestressing has been extensively used in bridge rehabilitation, but it is also a very cost-effective method of strengthening building components. The advantages of this reinforcing method are its simplicity, ease of inspection, and the ability to replace tendons if needed. The disadvantages are a need for corrosion protection in exterior and aggressive environments, a need to protect the system from fire, and vulnerability to acts of vandalism. All three potential drawbacks can be mitigated by encasing the post-tensioning system in concrete or shotcrete.[19] Obviously, this substantially increases the sizes of the reinforced members, a situation that may or may not be acceptable. Additionally, the use of this system requires unfettered access to the sides (and perhaps to the ends) of the members; this may be difficult in existing buildings.

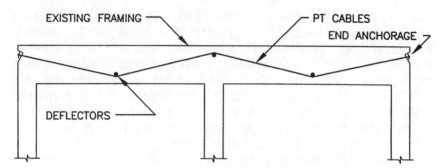

Figure 4.14 Strengthening concrete beams by external prestressing.

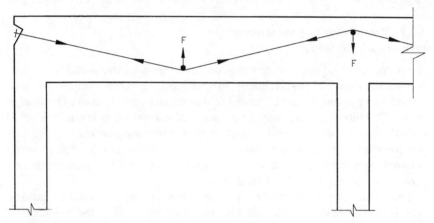

Figure 4.15 The resultant of the forces applied by external prestressing counteracts the applied loads.

Which wires should be specified for post-tensioning? For interior applications, high-strength steel wire strands of $\frac{1}{2}$-in diameter are typically specified. For exterior uses, some occasionally specify FRP wires instead, because of their corrosion resistance. However, as just mentioned, the standards for FRP design are far from established, and there are questions about the effects of the alkalis in concrete on FRP. Once the standards for testing and installation of FRP fibers arrive, post-tensioning with this material is likely to increase in popularity.

The deviators typically consist of structural steel brackets or saddles bolted to the sides or bottom of the existing beams. An alternative simple deviator detail, mentioned by Vejvoda,[35] is to drill a hole in the beam's web and insert a grouted pipe protruding on both sides. It is a good idea to ease the metal edges to avoid damaging the tendons. The tensioning is usually applied by hydraulic jacks, but in cases of tight clearances, turnbuckles or nuts and threaded rods have been used.[36]

Whenever external prestressing is proposed, consideration should be given to the effect of the newly introduced forces on the structure. The geometric ends of the member may not be accessible, and the load is transferred via the eccentric forces from the side- or bottom-mounted brackets. The member must be checked for these forces, and the strength of the segments between the brackets and the ends should be checked as well. Prior to external prestressing, the member should be repaired, with all cracks filled and spalls patched, so that the new forces are distributed uniformly across its section.[36]

Like FRP reinforcing plates attached with resin, external prestressing tendons must be protected from fire, unless they are used only as supplemental reinforcement and their failure would not jeopardize safety.

4.5.8   Reinforcing beams for shear

The previous discussion concerned the need to upgrade the flexural capacities of existing concrete beams, but similar methods can be used to improve the shear capacities as well. For example, two steel reinforcing plates can be used to improve the shear resistance of an existing concrete beam. The new plates cover the sides of the existing beam and are through-bolted to it in at least two places (Fig. 4.16). Depending on the circumstances, both shear and tension plates could be used on the same beam in various regions of the beam's length.

Another method of upgrading the shear capacity of an existing concrete beam is to add new stirrups. One technique is to place U-shaped bars in adhesive anchors drilled vertically into the bottom of the beam. There is usually not enough concrete clear cover to place the new stirrups outside the existing tension bars, and U-shaped bars are drilled in between the bottom bars. Naturally, the bottom bars must be located first, and great care must be taken to avoid damaging them during drilling.

Concrete beams with insufficient shear capacities may exhibit diagonal cracks emanating from the interior edges of the supports toward the middle. If the cracks are too wide, the aggregate interlock mechanism may no longer be able to hold the separated pieces in place (Ref. 37 places the limit on crack width at 0.02 in). Added side plates or grouted stirrups may not do much to repair and strengthen beams with such severe shear damage. Instead, some sort of clamping action can help. In the design illustrated in Fig. 4.17, several pairs of bolts are placed on each side of the cracked beam, as close to it as possible, and tensioned to somewhat tighten the crack. These clamps may physically prevent the concrete from separating. Another approach is to reestablish the aggregate interlock mechanism by injecting epoxy into the cracks. Alternatively, the span of the beam with visible shear

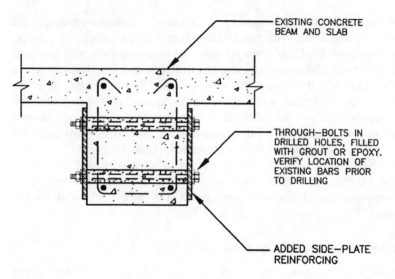

Figure 4.16 Adding steel plates to improve the shear resistance of an existing concrete beam.

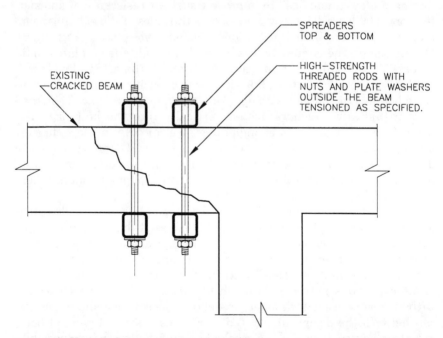

Figure 4.17 Upgrading the shear capacity of existing concrete beams by through-bolting. (*After Ref. 37.*)

damage can be shortened by adding columns or diagonal braces, or perhaps by steel brackets bolted to the columns.

A new method of shear strengthening uses epoxy-bonded FRP composites to wrap the ends of the beam in thin laminates or FRP fabric. Describing the history of this method, Triantafillou[38] acknowledges the scarcity of research on using FRP for shear strengthening and the contradictory nature of the available recommendations. He proposes a comprehensive design procedure for reinforcing concrete members using ultimate-design methods and a concept of effective FRP strains, in an approach analogous to conventional design of beams with internal stirrups. According to his research, strains in FRP are not constant but decrease with an increase in the axial rigidity of the FRP reinforcement. This promising strengthening method awaits additional research and testing.

Example 4.3 illustrates the procedure of increasing shear capacity of a beam using carbon FRP reinforcement.

Example 4.3: Accommodating a New Load Pattern by Increasing Shear Capacity of a Beam with FRP Reinforcement* The beam shown in Fig. 4.18a was originally designed to carry two point loads from mechanical equipment spaced 6 ft apart. New equipment was installed that resulted in the same load magnitude, but with a smaller footprint—the point loads from the new equipment were spaced 3 ft apart (Fig. 4.18b). In the original construction, stirrups were left out of the 6-ft region at midspan because of the low shear demand. The new load pattern may, therefore, result in a shear deficiency in this region. In order to accommodate this new load pattern, MBrace FRP shear reinforcement may be designed to correct the deficiency. The beam cross section is shown in Fig. 4.18c. The following material properties have been determined: $f'_c = 4000$ psi, $f_y = 60$ ksi.

- *Assess the current condition.*

$$V_c = 2\sqrt{f'_c}\, b_w d = 2\sqrt{4000 \text{ psi}}\ (12 \text{ in})(18 \text{ in}) = 27{,}322 \text{ lb} = 27.3 \text{ kips}$$

There are no stirrups in the portions of the beam that require strengthening, because V_u was less than $\frac{1}{2}(\phi V_c)$:

$$(\phi V_c)/2 = (0.85 \times 27.3)/2 = 11.6 \text{ kips} > V_u = 9.81 \text{ kips}$$

But in the new condition, V_u becomes 36.8 kips $> \frac{1}{2}(\phi V_c)$, so additional shear reinforcement must be provided (Fig. 4.18d).

---

*This example is reprinted with permission from *MBrace Composite Strengthening System: Engineering Design Guidelines,* by Master Builders, Inc., Cleveland, Ohio, 1999. MBrace is a registered trademark of MBT Holding A.G.

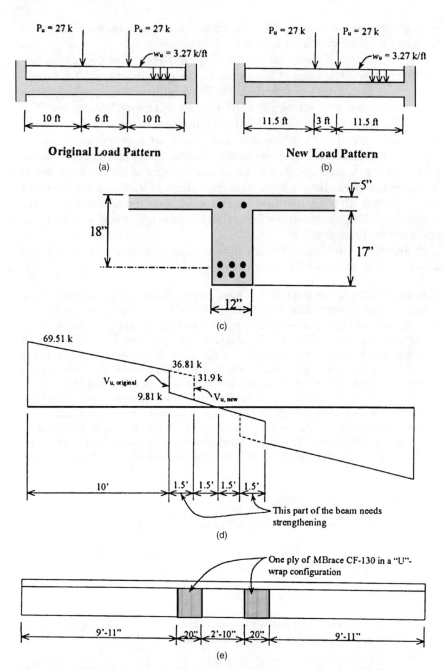

Figure 4.18 Increasing the shear capacity of a beam with FRP reinforcement for Example 4.3. (*a*) Original load pattern; (*b*) new load pattern; (*c*) cross section of beam at midspan; (*d*) shear requirements; (*e*) beam elevation showing the location and configuration of the designed FRP shear reinforcement. (*Reprinted with permission from* MBrace Composite Strengthening System: Engineering Design Guidelines, *Master Builders, Inc., Cleveland, Ohio, 1999.*)

- *Determine the shear contribution that must be provided by the FRP.*

$$V_u = \phi(V_c + 0.85V_f)$$

where V_f is the shear contribution that must be provided by the FRP.

$$36.81 \text{ kips} = 0.85(27.3 \text{ kips} + 0.85V_f)$$

$$V_{f,\text{ req'd}} = 18.8 \text{ kips}$$

- *Select materials and geometry.* MBrace CF 130 reinforcement is chosen for the shear retrofit. Its ϵ_{fu}, ultimate strain, is 0.017 in/in, and f_{fu}, design strength, is 550 ksi. Because of geometric considerations, it is desired to use a 20-in-wide U-wrap to cover each of the two 1.5-ft lengths of the beam that are deficient in shear. Assuming one ply, the shear contribution may be computed.

- *Determine the effective bond length.*

L_0 (effective bond length of one ply of FRP) = 2 in   for MBrace CF 130

$$L_e \text{ (effective bond length of the FRP strip)} = \frac{1}{\sqrt{n}} L_0 = 2 \text{ in}$$

$$\text{for one ply } (n = 1)$$

- *Determine the reduction factor on the ultimate strength of the sheet.*

k_1 (multiplier on the effective bond length to account for

$$\text{the concrete strength)} = \left(\frac{f'_c}{4000}\right)^{2/3} = 1$$

The depth of the FRP shear reinforcement d_f is equal to the depth to the tension steel centroid minus the slab thickness:

$$d_f = d - h_s = 18 - 5 = 13 \text{ in}$$

The effective depth of the FRP shear reinforcement is d_f less the effective bond length:

$$d_{fe} = d_f - L_e = 13 - 2 = 11 \text{ in}$$

$$\text{Multiplier } k_2 = d_{fe}/d_f = 11/13 = 0.846$$

$$R \text{ (reduction factor)} = (k_1 k_2 L_e)/(468 \epsilon_{fu}) = [1(0.846)(2)]/[468(0.017)]$$

$$= 0.213$$

- *Determine the stress level in the fiber at ultimate load* (f_{fe}).

$$f_{fe} = Rf_{fu} = 0.213(550 \text{ ksi}) = 116.9 \text{ ksi}$$

- *Find the shear contribution of the FRP and compare it to the required value.*

$$V_f = A_{fv} f_{fe}(\sin \beta + \cos \beta) d_f/(s_f) \leq 4 \sqrt{f'_c}\ b_w d$$

where A_{fv} is the total area of one strip, β is the orientation angle with respect to the longitudinal axis of the beam, and s_f is the strip spacing (taken here as 12 in).

$$V_f = 2(1)(0.0065\ \text{in})(20\ \text{in})(116.9\ \text{ksi})\ (1 + 0)13\ \text{in}/(12\ \text{in})$$

$$\leq 4\sqrt{4000\ \text{psi}}\ (12\ \text{in})(18\ \text{in})$$

$$= 32.9\ \text{kips} < 54.6\ \text{kips}$$

$$V_f = 32.9\ \text{kips} > V_{f,\text{req'd}} = 18.8\ \text{kips} \qquad \text{so one ply is sufficient}$$

The final design is summarized in Fig. 4.18e.

4.6   Strengthening Structural Slabs

In some cases, existing structural slabs can be strengthened using the methods for upgrading beams. In this section, we examine some additional reinforcing methods that are applicable specifically to structural slabs. Rehabilitation of slabs on grade is covered in Chap. 6.

4.6.1   Strengthening one-way slabs

One-way slabs function similarly to shallow concrete beams, but some methods of beam reinforcement cannot be practically applied to them. Solutions involving addition of external or internal plates, whether bolted or bonded, and those depending on grouted shear reinforcement are obviously inappropriate; enlarging section is also rarely practical.

Perhaps the easiest method of strengthening one-way slabs is by shortening their span. The most practical approach is to add steel beams at midspan, as illustrated in Fig. 4.6. If for some reason this upgrade technique cannot be used, steel plates can be added to improve the flexural resistance of the existing slab, using the methods outlined for concrete beams. Since most concrete slabs in need of upgrade are rather thin, the most reliable method of adding plate reinforcement is by through-bolting (Fig. 4.19a).

How often should the steel plates be spaced? Obviously, the larger the spacing, the less the field labor and the more economical the upgrade. But plates spaced too far apart end up working as discrete beams rather than as components of a uniform slab. ACI 318 does not allow primary flexural slab reinforcement to be spaced farther apart than 18 in or three times the slab thickness. But this sensible provision does not seem to be applicable to postinstalled external reinforcement. In the absence of definitive code provisions, some engineers use a plate

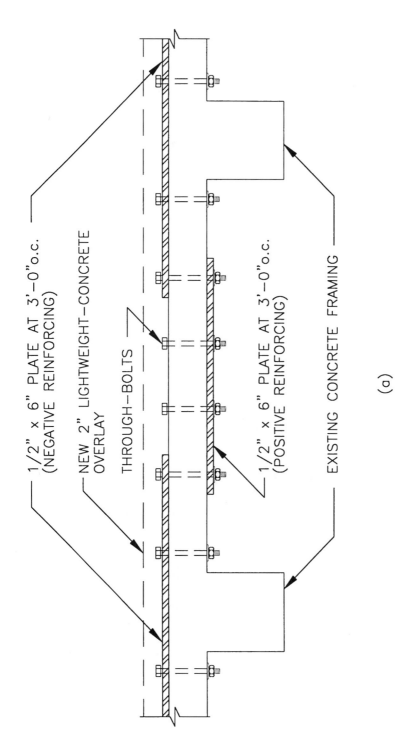

Figure 4.19 Methods of improving the flexural resistance of concrete slabs. (*a*) Adding bolted steel plates.

1/2" x 6" PLATE AT 3'−0"o.c. (NEGATIVE REINFORCING)

NEW 2" LIGHTWEIGHT−CONCRETE OVERLAY

THROUGH−BOLTS

1/2" x 6" PLATE AT 3'−0"o.c. (POSITIVE REINFORCING)

EXISTING CONCRETE FRAMING

(*a*)

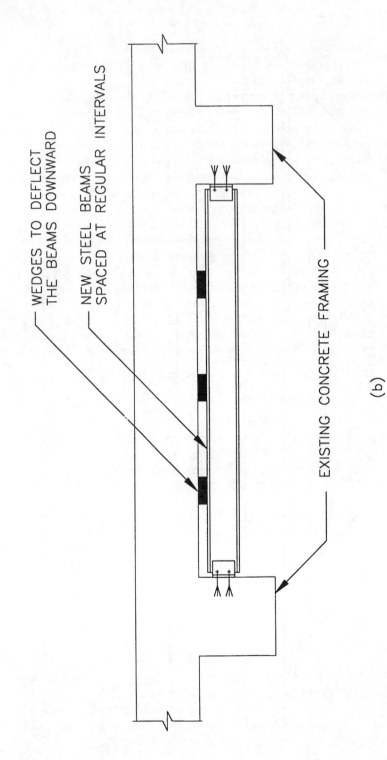

WEDGES TO DEFLECT
THE BEAMS DOWNWARD

NEW STEEL BEAMS
SPACED AT REGULAR INTERVALS

EXISTING CONCRETE FRAMING

(b)

Figure 4.19 (*Continued*) Methods of improving the flexural resistance of concrete slabs. (*b*) adding steel beams deflected downward by wedges.

spacing of 3 to 4 ft if secondary ("temperature") reinforcing bars are present in the slab. These bars help distribute loading to the steel plates in the direction normal to the span.

An alternative approach, shown in Fig. 4.19b, uses the already discussed idea of predeflecting flexible beams located underneath concrete. After steel beams, placed at close intervals, are attached at the ends, wedges are driven between the beams and the slab. The downward deflection can be computed to produce the desired upward force level. As explained in Sec. 4.5.2, this design does not require strain compatibility between the steel and the concrete sections, but merely introduces a counteracting load into the slab.

Another method of strengthening slabs, discussed in Sec. 4.5, involves the application of various fiber composites, particularly carbon-fiber-reinforced concrete. In recent years, some new products have been introduced. In one system, a low-viscosity, high-solids epoxy primer is applied first, and the surface is leveled with a special epoxy putty. A coat of resin saturant is applied next, followed by a sheet of carbon fiber—the heart of the system. A second coat of resin saturant is then brushed on, and an additional (or more than one if necessary) sheet of carbon fiber is applied. The system can be covered with an optional corrosion-resistant topcoat.[31]

4.6.2   Strengthening two-way slabs

Two-way slabs can be strengthened by some of the same general methods used for reinforcing concrete beams and one-way slabs. However, using bolted or bonded steel reinforcing strips becomes problematic, because of intersections. It is possible to weld the strips at each intersection, but the procedure becomes so complicated that it is probably easier to cover the whole top and bottom surfaces of the slab with large steel plates or with bonded overlays containing reinforcing bars.

Two-way slabs rarely fail in bending, because they tend to develop a membrane action that substantially increases their capacity to carry loads. For example, Rangan and Hall[39] reported the results of three large-scale tests of flat-plate floors. (Flat-plate slabs are those without beams or column capitals.) The floors were designed for a service load of 113 psf and a design ultimate load of 188 psf. The maximum load actually carried by the floors was between 471 and 498 psf, and even that was limited by a punching-shear failure at the edge columns, not by flexural capacity. Indeed, the main weakness of this system is insufficient shear or moment capacity of the slab at column locations, especially at edge columns. It should be noted that membrane action becomes effective only after substantial deflections have taken place. Therefore, the ultimate capacity of the slab may be adequate, but serviceability failures—damage to finishes, excessive sag, etc.—can still occur.

When the shear capacity of a flat-plate floor needs to be upgraded, the usual strengthening method involves using shear collars—fabricated steel brackets placed under the slab around the columns. The brackets, which are essentially postinstalled column capitals, are bolted to the columns, leaving some space between their tops and the bottom of the slab. The gap is then filled with nonshrink grout to assure uniform bearing. By providing a wider area for slab bearing, shear collars can reduce punching-shear stresses to acceptable limits.

Fabricated steel brackets can also be used to increase the moment-resisting capacities of existing flat-plate slabs. In this case, the brackets are bolted both to the columns and to the slabs. The intent is to provide an additional means of transferring bending moments between the two in the negative bending regions (at the top of the slab near the columns).

When steel brackets are undesirable from an aesthetic or maintenance standpoint, cast-in-place concrete shear collars can be used. Instead of being bolted, these are connected to the columns by drilled-in dowels.

Another, less common method of increasing the moment-resisting capacities of two-way slabs is by external post-tensioning. The procedure follows the principles mentioned in Sec. 4.5, except for the location of tendon deviators (deflectors): In slabs, these are bolted to the underside at predetermined locations. The deviators consist of special steel saddles made of steel tubular members, which may be laterally braced for stability by diagonal outriggers bolted to the underside of the slab (Fig. 4.20). When the tendons need to pass through a column, a drilled hole, usually reinforced by grouted-in steel members, is provided. The hole location should be adjusted to avoid damaging vertical bars in the column.

Barchas[40] describes a project in which a specialized engineering/construction firm used an extremely sophisticated method of external post-tensioning to reinforce all the floor slabs of a 15-story condominium building in Los Angeles. In that unique project, a system of post-tensioned tendons could be placed only in one location, under the second floor, because all other floors had insufficient headroom. Instead of reinforcing each floor separately, the designers had to rely on one system of post-tensioned tendons running in two directions under the second floor.

The forces generated in the tendons at that one location had to be sufficient to reinforce all the floors above. To transfer the upward forces to the higher floors, steel tubular columns were erected above the deviator locations at each story, from the second floor to the roof. The upward force of 150 kips had to be produced at each building bay to transmit a 10-kip force to each of the 15 levels. The tendons were placed in bundles wrapped in metal lath and covered with 1.5-in-thick vermiculite plaster for fire protection.

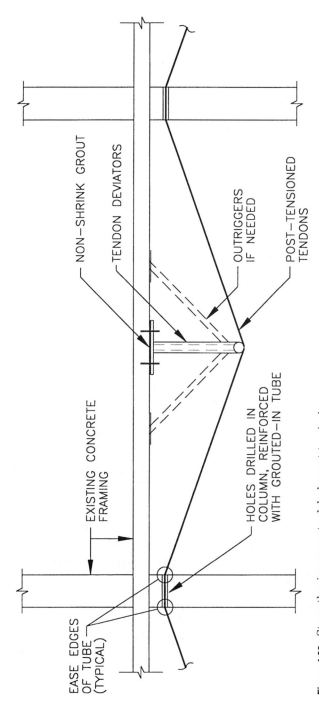

Figure 4.20 Strengthening concrete slabs by post-tensioning.

NON-SHRINK GROUT

TENDON DEVIATORS

OUTRIGGERS
IF NEEDED

POST-TENSIONED
TENDONS

EXISTING CONCRETE
FRAMING

HOLES DRILLED IN
COLUMN, REINFORCED
WITH GROUTED-IN TUBE

EASE EDGES
OF TUBE
(TYPICAL)

4.6.3   Strengthening sagging slab cantilevers

Cantilevered slabs of all kinds are architects' perennial favorites. But concrete cantilevers require a lot of design attention if they are to function properly over the long term, because they are not as forgiving of construction irregularities as regular slabs. In cantilevered slabs, if the top slab reinforcing bars are impaired in any way, moment redistribution is not possible, and the load cannot be shifted to other parts of the structure.

What can happen to the top bars? They can rust, being close to the surface in a location where free drainage is prevented. They can be damaged during construction or subsequent alterations. Or, most frequently, they can be simply placed too low, thus reducing the available moment arm resisting the load. Any one of these occurrences can seriously degrade the ability of a cantilevered slab to carry loads.

A typical manifestation of the problem is a sagging and cracked slab. Sagging can be caused not only by rebar weaknesses, but also by creep or by excessive long-term deflections. Transverse cracks may form at the top surface near the support, an easily noticeable problem.

There are several methods of dealing with drooping slab cantilevers, ranging from basic and inexpensive to elegant and costly. The strengthening approach can be passive, with the slab reinforced only for future loads and the sag left in place, or active, with the slab jacked up and then strengthened for both present and future loads.

The basic methods include reducing the span of the cantilever by adding steel brackets or by diagonal and knee braces. Another simple but not always acceptable solution is to install new columns and supporting beams under the exterior edge of the slab. Cash-strapped owners might elect the cheapest solution—to remove the cantilevers, or at least to shorten their span. All these methods are self-explanatory and do not require further discourse.

The elegant and costly techniques, on the other hand, require a certain expertise on the part of both designers and contractors. These methods involve either drilling in new reinforcing bars to supplement the existing bars or drilling in post-tensioned tendons. Examples 4.4 and 4.5 illustrate the two approaches.

Example 4.4: Strengthening Cantilevered Balconies*   A relatively new 22-story residential high-rise in south Florida featured cantilevered wraparound balconies that were sagging and cracking. The balconies were framed by cantilevering 8.5-in flat slabs about 8 ft. A field investigation determined

---

*This example is a summary of a published case study by Anderson and Holland[41] of Eugene Holland and Associates, Ltd., Schaumburg, Ill.

that the top reinforcing bars were placed too low—up to 3.5 in deep instead of the 1 in required. The moment resistance of the slab was therefore greatly reduced, and strengthening was required.

Some possible rehabilitation options involved installation of additional columns or brackets; others involved shortening the span of the cantilevers. All these were ruled out because they would drastically change the look of the building and present logistical problems. Instead, the design team decided to reinforce the cantilevers in place by grouting in #10 reinforcing bars.

But how to get the rebars into the slab? The reinforcement could be inserted by cutting slots at the top of the slab and grouting the bars in place. The bars would have to be 16 to 19 ft long, extending inside the building from 8 to 11 ft. This normally economical option was rejected because it would involve significant disruption of the building's use and would require replacing the expensive perimeter floor finishes.

Another method would rely on coring horizontal holes in the slab and placing rebars in an adhesive resin, similar to the detail shown in Fig. 4.21. This method was eventually chosen because it was the least disruptive, even though technically challenging.

The most obvious challenge was that of coring the slabs for 16 to 19 ft with sufficient precision. The designers specified holes of $1^5/_8$-in diameter that had to be cored with a maximum vertical tolerance of 0.24 in at the middle and 0.5 in at the end of the hole. Another challenge was not damaging the existing slab reinforcement. Prior to coring, the contractor was required to locate the existing bars and to determine the exact configuration of each balcony slab; this major effort required the services of a professional engineer. If the core still encountered a top bar parallel to it, the coring was stopped and moved to an adjacent location.

The next challenge was how to insert the bars and bond them into the holes. The bars could be bonded either with epoxy or with polyester-resin cartridges. Using epoxy would have allowed placing the bars in the holes and pressure injecting the space around them. However, this method would require constant cleanups and would be expensive. Instead, polyester-resin cartridges were selected. In a typical procedure, about one-half of the hole was filled with the cartridges before the bar was inserted. The bar could be spun into the hole with drilling equipment, the bar's end being beveled to simplify the insertion. Once the bar was inserted halfway, it was connected to the power drill to complete the placement.

The strengthening procedure was so new that a lot of field experimentation and fine-tuning had to take place before the work started. At the bidding stage, one of the bidders performed an on-site trial repair for all the others to see. This trial run revealed a complication: The core cut into an embedded electrical conduit. During construction, the conduits had to be monitored and promptly filled to prevent the water used in the drilling operations from

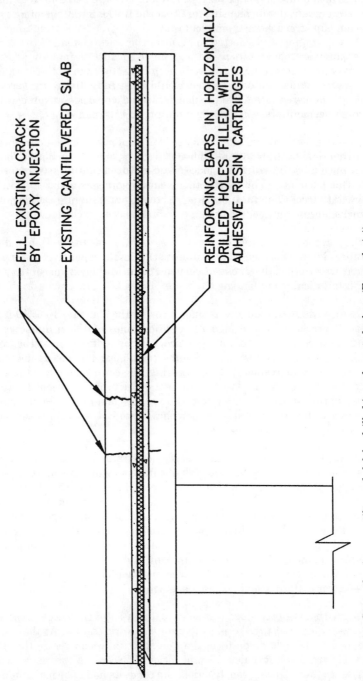

FILL EXISTING CRACK BY EPOXY INJECTION

EXISTING CANTILEVERED SLAB

REINFORCING BARS IN HORIZONTALLY DRILLED HOLES FILLED WITH ADHESIVE RESIN CARTRIDGES

Figure 4.21 Strengthening a cantilevered slab by drilled-in and grouted rebars (Example 4.4).

entering the building. Some other innovations: The selected contractor made a sample concrete slab in the shop to train the workers; the resin manufacturer came up with a revised product that had an extended setting time; and a 10-ft-deep staging was selected to perform the work instead of the typical 30-in swing stage.

The workers installed a total of about 2100 bars. After the work was completed, all the cored holes were patched with concrete and covered with a thin coat of cement plaster, and the building was repainted.

Example 4.5: Strengthening Cantilevered Roof Slabs* A high-visibility cross-shaped office building in Rhode Island featured cantilevered roof slabs around the perimeter. The length of the cantilevers was about 9.33 ft; the slab was 8 in thick. The cantilevers were cracked and slightly sagging, with the cracks largely confined to areas extending 20 ft on each side of the eight exterior corners.

A field investigation revealed that the cracks had a repetitive pattern. At least one crack typically extended perpendicular to the perimeter from the corner column to the outer edge of the slab (Fig. 4.22). The cracks originated at the previously filled-in rectangular roof openings. They often ran through the thickness of the slab and were from 1 to 6 mm wide. There was evidence of previous repairs, but no rust marks (these would suggest corrosion of the existing reinforcement). The top and bottom surfaces of the cantilevers were covered with a cementitious coating, which had cracked and spalled in some places. The cantilevered areas were poorly drained, and some had standing water.

As in the previous example, the strengthening options involving changes to the building's appearance were rejected, and it was decided to reinforce the cantilevers by grouting in bonded post-tensioned bars. However, in this case it was possible to avoid full-length horizontal drilling. The reinforcement could be placed by cutting slots at the top of the slab and grouting the bonded tendons in place.

The tendons consisted of 1.25-in-diameter bars conforming to ASTM A722, Type II, with a yield strength of 150 ksi. Six threaded bars were installed at each corner, as shown in Fig. 4.23. The bars were placed in bonded ducts that were filled with grout after post-tensioning.

Prior to cutting, the existing concrete was shored and the designated cracks in the roof overhang were filled with epoxy by pressure injection. (Since sagging was relatively minor, no attempt to jack the slab upward was made.) Then, slots were cut in the existing slab, and the concrete beams were cored

---

*This example is based on a project designed by Maguire Group Inc. of Foxborough, Mass., which supplied the accompanying illustrations.

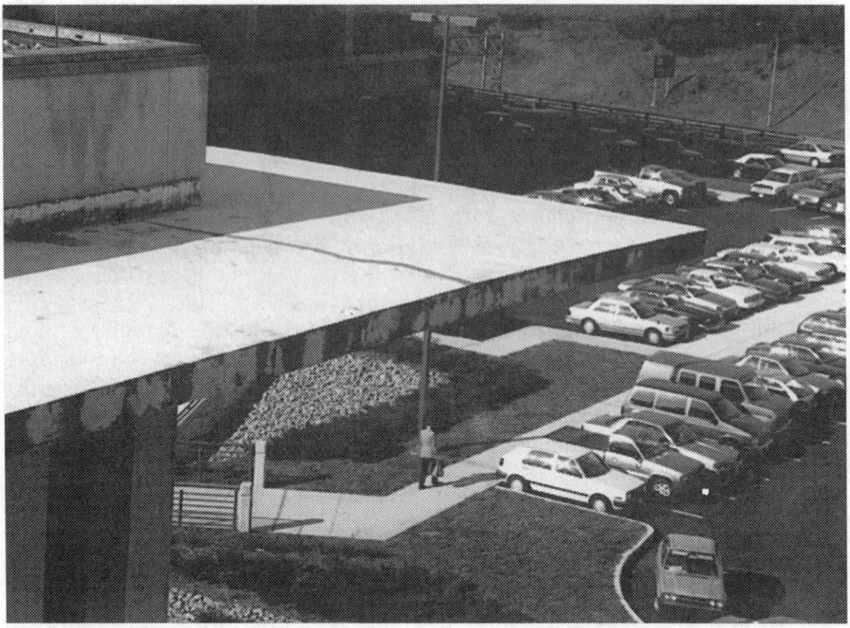

Figure 4.22 Cracked and sagging cantilevered roof slab for Example 4.5.

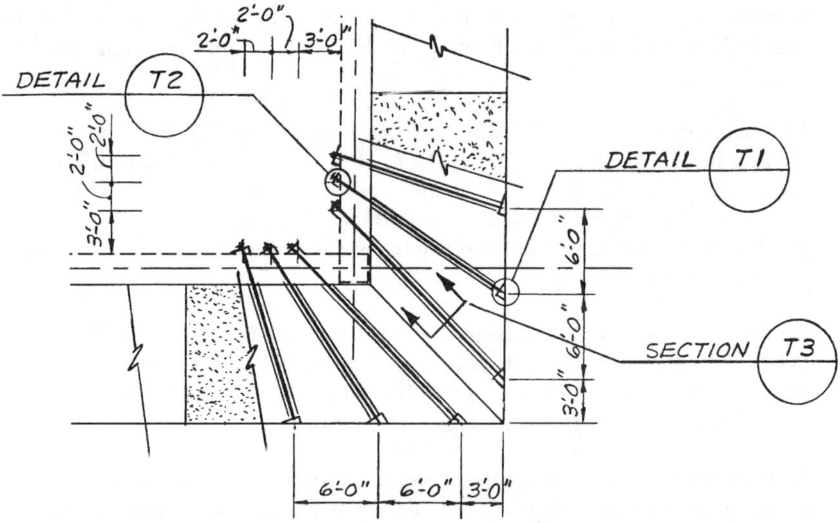

Figure 4.23 Typical layout of post-tensioned bars at a corner for Example 4.5

for installation of the post-tensioned system. The post-tensioned bars were placed at a slight slope—6 in from the bottom of the slab at the exterior surfaces and 8 in at the interior.

The top surface of the slabs received a new concrete overlay of varying thickness to provide slope for drainage and to protect the repairs. The overlay was tied to the existing concrete by $^3/_4$-in stainless steel anchor bolts located in a 2-ft by 2-ft grid. Prior to tensioning of the bars, the top of the existing slab was sandblasted to improve the bond, the anchors were installed, an epoxy bonding agent was applied, and the new overlay was placed and cured. To reduce shrinkage cracks in the overlay, welded wire fabric was placed there. Figure 4.24 shows the cross section at a typical post-tensioned bar and overlay.

As in the previous example, the workers had to deal with some nonstructural interference during installation of tendons. In this case, it was the mechanical ducts that had to be removed and later reinstalled. To make the work less visible from the outside and to provide corrosion protection, the end anchorages of the tendons were recessed into concrete and the space around them filled with nonshrink grout (Fig. 4.25). At the interior, such measures were deemed unnecessary and even dangerous because there the tendons were anchored to the sides of existing concrete beams. Instead, tapered bearing plates were provided (Fig. 4.26). Once the tendons were tensioned, installation of the new roofing, insulation, and flashing completed the job.

4.7 Strengthening Concrete Columns

A need to reinforce existing concrete columns often arises when the proposed loads exceed the available capacities. Column strengthening is typically needed in projects involving vertical additions, where the existing columns have to support one or more added floors. Sometimes columns need strengthening when it is discovered that their concrete is much weaker than anticipated.

4.7.1 Section enlargement

Existing concrete columns can be strengthened in a variety of ways. Perhaps the most common is section enlargement, a solution that allows the upgraded framing to remain all-concrete, provides built-in fireproofing, and preserves a rectangular column configuration. Another benefit of encasing existing columns in concrete is a reduction in slenderness and increase in rigidity. As already discussed, the added section will resist only future superimposed loads; it provides no help when the column is overloaded by the present loads, unless these loads are removed first.

Fig. 4.27*a* shows a method of encasing an interior isolated column in additional concrete, while Fig. 4.27*b* provides a detail for an exterior

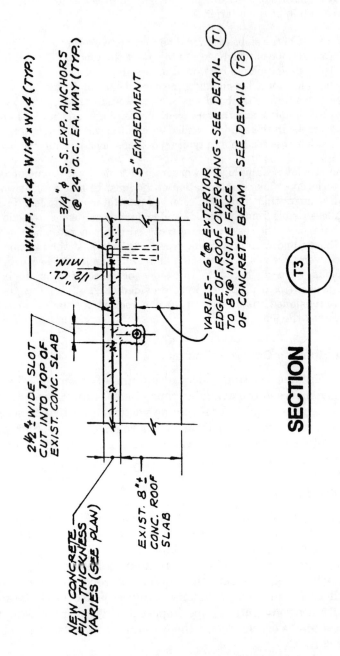

NEW CONCRETE FILL - THICKNESS VARIES (SEE PLAN)

2½" ± WIDE SLOT CUT INTO TOP OF EXIST. CONC. SLAB

W.W.F. 4×4 - W1.4 × W1.4 (TYP.)

3/4" ⌀ S.S. EXP. ANCHORS @ 24" O.C. EA. WAY (TYP.)

1/2" CL. MIN.

EXIST. 8" ± CONC. ROOF SLAB

5" EMBEDMENT

VARIES - 6" @ EXTERIOR EDGE OF ROOF OVERHANG - SEE DETAIL (T1) TO 8" @ INSIDE FACE OF CONCRETE BEAM - SEE DETAIL (T2)

SECTION

(T3)

Figure 4.24 Cross section at a typical post-tensioned bar for Example 4.5.

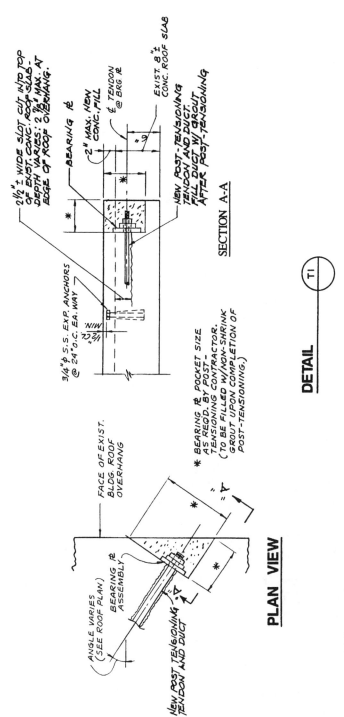

SECTION A-A

DETAIL (T1)

PLAN VIEW

Figure 4.25 Anchorage of a post-tensioned bar at the exterior end for Example 4.5.

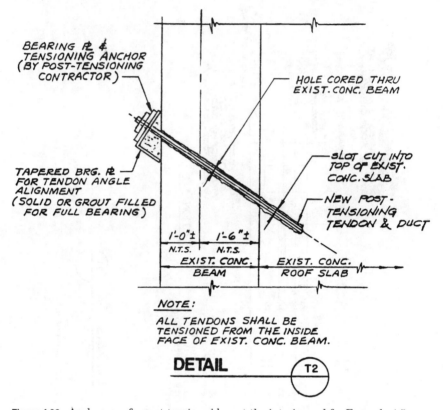

BEARING ℝ &
TENSIONING ANCHOR
(BY POST-TENSIONING
CONTRACTOR)

HOLE CORED THRU
EXIST. CONC. BEAM

TAPERED BRG. ℝ
FOR TENDON ANGLE
ALIGNMENT
(SOLID OR GROUT FILLED
FOR FULL BEARING)

SLOT CUT INTO
TOP OF EXIST.
CONC. SLAB

NEW POST-
TENSIONING
TENDON & DUCT

1'-0"±   1'-6"±
N.T.S.   N.T.S.
EXIST. CONC.   EXIST. CONC.
BEAM   ROOF SLAB

NOTE:

ALL TENDONS SHALL BE
TENSIONED FROM THE INSIDE
FACE OF EXIST. CONC. BEAM.

DETAIL   T2

Figure 4.26 Anchorage of a post-tensioned bar at the interior end for Example 4.5.

column that forms a part of the wall. In both cases, composite action between the new and the existing concrete is achieved by a combination of bond and anchoring dowels. Good bonding is promoted by roughening and cleaning the existing surfaces and by applying a bonding agent. The usual rules regarding tie spacing and longitudinal bar support in concrete columns are applicable for the encased ones as well.

Unless added longitudinal bars are located only in column corners, some of them will be interrupted by intersecting beams and will have to stop at the floors. If more than four longitudinal bars are used, the bars located farther than 6 in away from the corners must be laterally supported. The support may be provided by drilled-in dowels, as shown in Fig. 4.27.

Some concerns with section enlargement parallel those for encased structural steel columns, discussed in Chap. 3, Sec. 3.10. ACI 318 requires that spacing of ties in reinforced-concrete columns not exceed 16 longitudinal bar diameters or 48 tie-bar diameters. The ties in composite sections are made of two pieces, hooked at the ends or overlapped

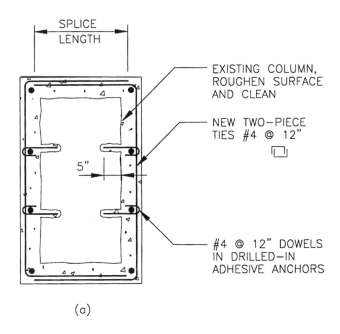

(a)

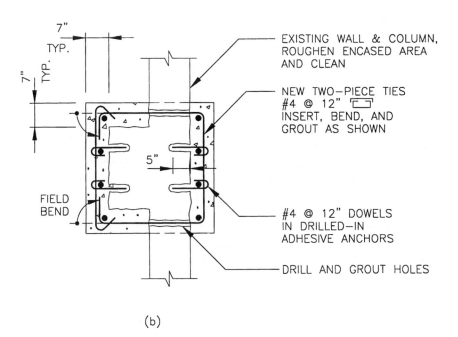

(b)

Figure 4.27 Strengthening concrete columns by section enlargement. (a) Interior isolated column; (b) exterior column forming part of a wall.

to provide the required splice length. As with regular concrete columns, the minimum size of the ties is #3 for #10 and smaller longitudinal bars and #4 for larger ones.

When an exterior column cast integrally with the wall needs to be reinforced, the new encasement can be divided into two halves, one on each side of the wall. It is then necessary to drill through the wall to extend the ties between the two halves. To keep the size of the drilled holes reasonable, field bending the ends of the tie bars as shown in Fig. 4.27b can be specified.

The biggest problem with column enlargement is drying shrinkage. It takes months, and perhaps even years, for concrete to hydrate and complete the shrinkage process, as discussed in Chap. 6. The few days or even weeks allowed by construction schedules are certainly not enough. When a green concrete enclosure is interconnected with the existing fully hydrated concrete, the shrinkage is restrained, and tensile stresses develop in the new material. If these stresses are large enough, the new section will crack perpendicular to the member's length.

So, in order for the section-enlargement methods of Figs. 4.8 and 4.27 to be successful, the new concrete must have an extremely low rate of drying shrinkage or be made with shrinkage-compensating cement.

Preplaced-aggregate concrete usually offers the lowest rates of drying shrinkage and is therefore an excellent candidate for this application.[18] The construction procedure for this type of concrete involves placing coarse aggregate into forms and then pumping fluid grout into the voids. Since the pieces of the coarse aggregate are already in contact with one another, drying shrinkage is kept to a minimum. Specific information about proportioning and gradation of the aggregate can be found in ACI 304.1R.[42]

Although simple in theory, the actual process of building column enclosures with preplaced-aggregate concrete requires experience. Placing grout into a narrow area within the existing building may be rather difficult. Typically, the grout is conveyed from the top through pipes extending the full height of the column; the pipes are gradually withdrawn as the bottom part is filled. The obvious problem here is the placement and withdrawal of the pipes. This often requires drilling holes in the concrete beams above. A simpler method is to pump the grout from the bottom through holes in the formwork.[43]

4.7.2 Adding columns

A concrete column can also be reinforced by constructing a new column next to it. The two columns can be totally separate or can be interconnected by drilled-through and grouted rods or similar means. Adding a column is especially advantageous when the existing column receives a large part of its load from a single beam. As in the section-enlargement method, the new concrete will resist only future superimposed loads. It cannot relieve the existing column of the loads it already carries, unless the loads are removed.

The new column can be made of structural steel or concrete. Concrete might be preferred because of fire-rating considerations, but from a space-saving standpoint a structural steel column is better. A good compromise may be achieved by encasing a steel column in concrete. If steel is used, many engineers prefer visually pleasing tubular shapes. If concrete is used, a rectangular section is typically selected, but it is also possible to build a C-shaped concrete column enveloping the existing on three sides and made composite with it by drilled-in dowels.

The obvious problem with adding adjacent columns is what to do with their foundations. The existing column foundations are typically not wide enough, and there is probably insufficient space for new foundations next to the existing ones. Therefore, the method of adding columns works best when the existing foundations or supports are wide enough to accommodate both columns.

Assuming that an adequate foundation exists, how should the new column be attached to it? Positive attachment is needed for lateral bracing of the column ends; it is also needed to transfer compressive forces, if concrete bearing alone is inadequate for that purpose, and to prevent shifting during an earthquake.

In brand-new construction, the column vertical bars are spliced with the rebar dowels cast into the supports. In our case, no such dowels are available, but it is possible to postinstall foundation dowels at each longitudinal column bar by drilling and grouting them into existing concrete. Obviously, this solution works only if the longitudinal bars are few (say, four, one in each corner). If there are many column bars, it might be more practical to place a steel bearing plate under the new column, to which the dowels would be welded. The plate itself can be connected to the support by drilled-in anchors (Fig. 4.28). The anchors should extend deep into the support (ideally, into the bottom of the foundation) to prevent brittle failure under load and to make column erection safer.

Whether made of structural steel or concrete, the new column has to bear tightly not only on the bottom supporting surface, but also at the top, against the underside of the beam. For this purpose, a gap is left above the column, to be filled with nonshrink grout after installation. In the case of a concrete column, the grout should be applied as late as possible, to allow much of the drying shrinkage to take place. This could take many months, as explained in Chap. 6. To minimize the effects of drying shrinkage, the two adjacent concrete columns can be intentionally separated by a bond breaker, with no interconnecting dowels used.

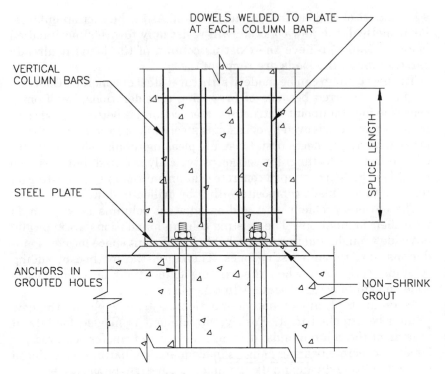

Figure 4.28 Attachment of an added concrete column to the foundation.

4.7.3 Reinforcing with structural steel

A third method of strengthening existing concrete columns is by reinforcing them with steel shapes. This can be done by placing steel angles in the corners and attaching them to the existing concrete with adhesives or drilled-in anchors (Fig. 4.29a), or by totally enveloping the column with steel plates (Fig. 4.29b). Designers attempting the latter course of action are wise to review the paper by Rangan,[44] which contains a proposed calculation method and other useful information on the subject.

Reinforcing with four angles is less expensive, although this method has not been widely used. One example of reinforcing columns using this technique is described by Oey and Aldrete,[45] who state that they approximated Rangan's methodology for the design of concrete-filled tubular steel columns by lacing the angles together using 2.5-in-wide plates of the same thickness as the angles. The spacing of the lacing plates was such that the slenderness of each angle was less than the overall slenderness of the composite column. Each lacing element was designed for a shearing force acting perpendicular to the member and equal to 2 percent of the compressive force in the column.

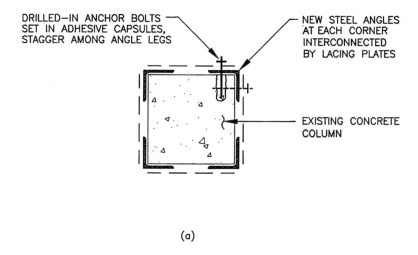

DRILLED–IN ANCHOR BOLTS
SET IN ADHESIVE CAPSULES,
STAGGER AMONG ANGLE LEGS

NEW STEEL ANGLES
AT EACH CORNER
INTERCONNECTED
BY LACING PLATES

EXISTING CONCRETE
COLUMN

(a)

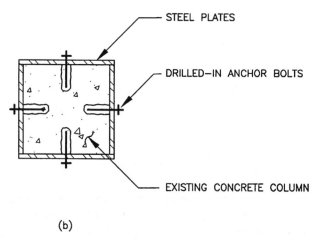

STEEL PLATES

DRILLED–IN ANCHOR BOLTS

EXISTING CONCRETE COLUMN

(b)

Figure 4.29 Strengthening concrete columns with steel shapes. (a) Steel angles placed in the corners; (b) enveloping the column with steel plates.

The angles were attached to existing concrete with $\frac{1}{4}$-in-diameter, 2.5-in-long anchor bolts set in epoxy. (Incidentally, an example of all-steel latticed column design can be found in Chap. 3, Example 3.1.)

Another possible approach is to encase the existing concrete column in a loosely fitted welded steel "jacket" and to fill the resulting voids with cement grout. The jacket produces a confining effect on concrete—as ties and spirals do—that not only increases column capacity in compression and bending but also imparts better ductility. Shams and

Saadeghvaziri[46] provide an overview of the latest developments in concrete-filled tubular columns, the behavior of which is similar to that of concrete columns reinforced by plate jacketing. They point out that at present there are three different design codes dealing with concrete-filled tubular columns: ACI 318, the AISC *LRFD Manual,* and a procedure offered by the Structural Specification Liaison Committee. All three, according to the researchers, seem to underestimate both the flexural stiffness and the confining effect of the steel tube on concrete.

Shams and Saadeghvaziri conclude that a new design method is needed to properly design concrete columns encased by steel. This method should be able not only to predict the ultimate strength and stiffness of such columns, but also to account for their slenderness effects, various aspect ratios, the contribution of shear connectors to bond values, and other important issues. In the meantime, those using steel plates or angles to reinforce existing columns are well advised to stick to conservative design approaches.

Steel jackets for reinforcing concrete columns against shear and flexure have been extensively employed in bridge construction, especially in California, where Caltrans until recently listed column jacketing as the only approved method of column strengthening. However, as Cercone and Korff[47] observe, installation of steel jackets is labor-intensive and time-consuming, and heavy equipment is required in order to lift the tons of steel involved. Moreover, the thickness of plates is often determined by practicality of installation—the need to keep the plates from buckling during lifting and grouting, for example— rather than by structural requirements.

4.7.4 Reinforcing with fiber wraps and fiberglass jackets

One emerging and promising technology that could replace steel jacketing for flexural and shear reinforcing is encasing critical regions with continuous epoxy fiber wraps. Wrapping columns with fiber composites is widely used in seismic strengthening of bridges, but it also has its place in building renovation. This relatively inexpensive technique not only prevents the repaired columns from spalling and protects them from further damage from the elements, but also increases their shear and flexural strength as a result of the confinement it provides.

One unusual example of epoxy fiber wrapping is described by Dial.[48] During an adaptive reuse of a former cement plant, four 200-ft-tall concrete smokestacks, cracked and spalled in some areas but otherwise in sound condition, were restored. The rehabilitation process included removing loose and deteriorated concrete, applying a polymer repair mortar, sealing the cracks with epoxy, and wrapping 4.5-ft-wide fiberglass strips saturated in epoxy around the stacks' perimeter. Badly deteriorated areas were wrapped twice. Once the epoxy in the wrap had hardened, a layer of fabric with a concretelike texture was placed, also attached with epoxy. The wrapped areas received a final coat of colored epoxy 24 h later.

In another variation of the wrapping method, carbon-fiber composite wraps are used. As already mentioned, carbon-fiber-reinforced plastic possesses many desirable traits: It is strong, resistant to corrosion, lightweight, and nonmagnetic. When the standards for its design are developed, validated by experience, and widely accepted, carbon wrapping will probably become the prevalent method of column strengthening. Example 4.6 illustrates the procedure of increasing the ultimate load capacity of a column.

Example 4.6: Increasing the Ultimate Load Capacity of a Column with FRP Wrap*

A circular column, 16 in in diameter, with 10 #7 bars was originally designed to carry a factored axial load of 570 kips and a factored moment of 135 kip · ft. The column has 1.5 in of clear cover and #3 spiral transverse reinforcement. Design the number of plies of MBrace CF 130 carbon fiber needed to support a 20 percent increase in factored loads. The concrete and steel reinforcement properties are $f'_c = 5000$ psi and $f_y = 60,000$ psi.

- Compute the factored axial force and bending moment for the 20 percent live load increase.

$$P_u = 1.2(570 \text{ kips}) = 684 \text{ kips}$$

$$M_u = 1.2(135 \text{ kip} \cdot \text{ft}) = 162 \text{ kip} \cdot \text{ft}$$

To use the nondimensional interaction diagrams given in Fig. 4.30 (from Appendix A of Ref. 31), the following values must be calculated:

- Compute the existing steel reinforcement ratio.

$$\rho_g = \frac{A_s}{A_g} = \frac{6.0}{201} = 0.03$$

- Compute the diameter of the circle defining the reinforcement centroid.

$$\gamma h = 16 \text{ in} - 2(1.5 \text{ in}) - 2(3/8 \text{ in}) - (7/8 \text{ in}) = 11.375 \text{ in}$$

$$\gamma = (\gamma h)/h = 11.375/16 = 0.71$$

---

*This example is reprinted with permission from *MBrace Composite Strengthening System: Engineering Design Guidelines,* by Master Builders, Inc., Cleveland, Ohio, 1999. MBrace is a registered trademark of MBT Holding A.G.

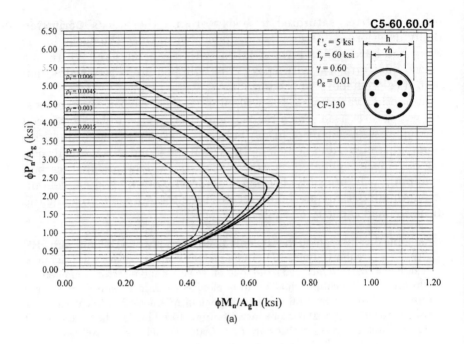

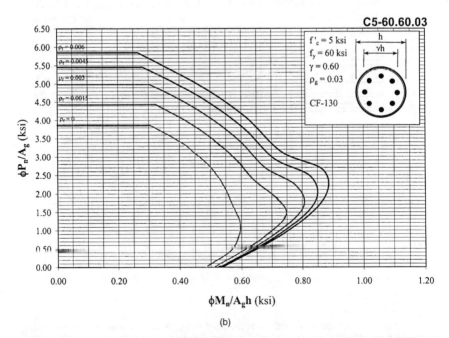

Figure 4.30 Column interaction diagrams for round columns wrapped with FRP hoop reinforcement. (*Reprinted with permission from* MBrace Composite Strengthening System: Engineering Design Guidelines, *Master Builders, Inc., Cleveland, Ohio, 1999.*)

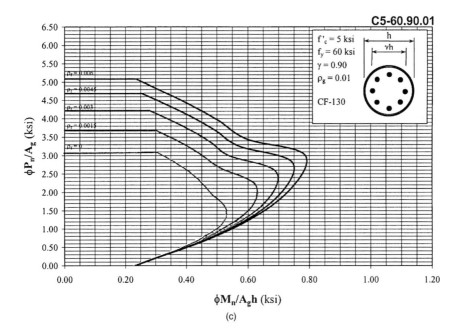

(c)

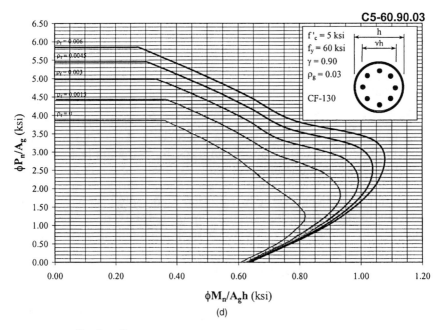

(d)

Figure 4.30 (*Continued*)

▪ *Find the factored unit axial force and bending moment.*

$$P_u/A_g = (684 \text{ kips})/(201 \text{ in}^2) = 3.40 \text{ ksi}$$

$$\frac{M_u}{A_g h} = \frac{162 \text{ kip} \cdot \text{ft} (12 \text{ in/ft})}{201 \text{ in}^2 (16 \text{ in})} = 0.60 \text{ ksi}$$

With these values, ρ_f, the required FRP reinforcement ratio (the ratio of the volume of fibers to the volume of encased concrete), may be determined from the nondimensional interaction diagrams given in Fig. 4.30 (Appendix A of Ref. 31).

For $\gamma = 0.6$, $\rho_g = 0.03$:

$$\rho_f = 0.003$$

For $\gamma = 0.9$, $\rho_g = 0.03$:

$$\rho_f = 0.0015$$

From linear interpolation, if $\gamma = 0.71$, then $\rho_f = 0.0024$.

▪ *Compute the required jacket thickness.*

$$nt_f = \rho_f h/4 = 0.0024(16 \text{ in})/4 = 0.010 \text{ in}$$

▪ *Compute the required number of plies.*

$$n = (0.010 \text{ in})/(0.0065 \text{ in/ply}) = 1.6 \text{ plies} \qquad \therefore \text{Use 2 plies}$$

Thus, 2 plies of CF 130 will be adequate to allow a 20 percent increase in factored loads.

Yet another alternative to heavyweight steel jackets and FRP wrapping is encasing existing columns in fiberglass jackets. The jackets are custom made to fit the column profiles and are applied in layers "snapped" onto columns and secured with layers of high-strength urethane adhesive. In one project, where concrete bridge columns were encased in these jackets, the installation reportedly took less than 2 h per column. Like the other strengthening methods, fiberglass jackets work by increasing column resistance to spalling during earthquakes.[49]

4.7.5 Shortening the length of the columns

Occasionally, the existing columns are too slender for the imposed loading. This situation arose in a building in Florida with a two-way flat-plate system that had to be strengthened. The engineers elected to shorten the unbraced length of the columns by adding 4-ft square drop panels, 1 or 1.5 ft deep. To facilitate concrete placement into the drop panels, 3-in openings were made in the existing slabs. The new and existing concrete were interconnected with six #4 dowels drilled into the underside of the existing slab.[50] Another method of shortening the unbraced length of the columns is four sturdy steel knee braces bolted to the column and to the slab.

These methods present some conceptual problems in columns with "pinned" ends, which would have to be converted into columns with fixed ends at the top. A careful analysis of the relative rigidities of the floor and the column needs to be made, to ensure that fixity can be achieved.

4.8   Openings in Existing Slabs

4.8.1   General issues

It is difficult to imagine a renovation project that does not involve making openings in slabs for plumbing and fire-protection pipes, HVAC ductwork, or stairs and elevators. The sizes of the required openings can vary from small cores needed for piping to removal of whole framing bays when the architect desires to "open up" the building.

Like any modification of existing framing, cutting openings in slabs should be approached with caution and avoided if possible, perhaps in favor of making wall penetrations. Unlike steel beams, which can be visually inspected and reinforced if needed, concrete framing presents only a flat surface, impenetrable to view and difficult to reinforce. Making a wrong cut can jeopardize the integrity of the framing and even lead to collapse. For example, some old concrete floors are supported by flat tile arches spanning between the webs of steel beams, with or without tie rods. Carelessly cutting through the tile without investigation destroys the integrity of the system. (As stated in Chap. 2, this can happen even *during* the investigation when probes are made in framing of unknown composition.)

Some of the challenges involved in making slab openings in various structural systems are examined here. Quite often, the structural engineer has some say in the selection of the opening locations. Knowing where to make the openings and how to reinforce their edges can spell the difference between a successful renovation project and a troublesome one.

What is the best way to make an opening? Small holes are usually cored to the desired diameter; large openings are cut with circular saws. The problem with a circular saw is that it cuts a longer slot at the top than at the bottom of the slab. One way to avoid this overcutting is to make circular cores at the corners and connect them by saw cuts (Fig. 4.31). Lately, concrete chain saws with plunge cutting capabilities have become available. These saws do not overcut and do not require core holes. Some engineers are concerned that making square-cornered openings produces stress concentrations, resulting in cracking at reentrant corners. This is probably true, but the slab reinforcement is usually able to keep any cracks from opening up and becoming visible.

4.8.2 Making openings in one-way slabs

One-way concrete slabs typically span between concrete or steel beams, less commonly between loadbearing walls. Except in very small buildings, the slabs have multiple spans and are structurally continuous. Slab continuity is a double-edged sword when openings have to be made. On the one hand, it is beneficial, because it allows for some moment redistribution. For example, weakening the slab at midspan by cutting rebars reduces its positive moment-resisting capacity, but that could be acceptable if there is excess capacity in the negative regions (at the supports). On the other hand, because of continuity, a large opening can upset the structural balance in the slab, as explained here.

To find the least critical area in which to make an opening, the engineer should review the original construction documents to see whether positive or negative flexural reinforcement controls the slab design. If the amount of positive reinforcement is barely acceptable or worse, the holes can be made near the supports (such as opening 1 in Fig. 4.32).

Figure 4.31 Avoiding overcutting by coring holes at the corners and sawcutting.

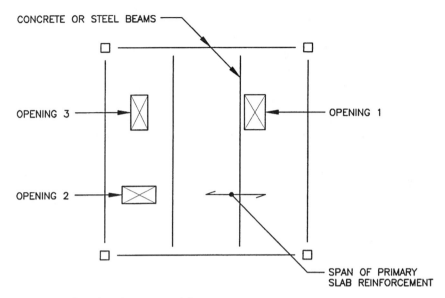

CONCRETE OR STEEL BEAMS

OPENING 3

OPENING 1

OPENING 2

SPAN OF PRIMARY
SLAB REINFORCEMENT

Figure 4.32 Openings in one-way slabs.

Conversely, top bars with insufficient area may indicate that openings are best located near the midspan. A proper orientation of rectangular openings is also important: A hole with the long side parallel to the span of the slab, such as opening 2 in Fig. 4.32, will interrupt fewer bars than an opening with the long side in the perpendicular direction, such as opening 3.

Large openings in continuous slabs may be reinforced on all sides with steel beams, two of them spanning between the adjacent floor beams. When the exact opening location can be established, the need for steel beams may be obviated by careful analysis. In this case, the opening has to be made in such a manner that the slab overhang is long enough to provide the proper development length for the top bars in the next panel, but not so long that the slab cantilever is overstressed.

Naturally, examining the flexural capacity of the slab is only the first step, and other factors may prove more important. For instance, the slab's shear capacity might be critical. Also, the supporting members could have been designed as concrete T-beams or composite steel beams. In both of those cases, making openings near the beams might be acceptable from the standpoint of slab well-being but could weaken the beams.

Sometimes, an opening of the largest practical size—short of cutting the beams—is desired, meaning that the slab has to be totally removed between the edges of the supporting members. As was just said, making openings like that in continuous slabs can upset their structural balance. Why? The interior panels in continuous slabs are designed assuming large negative moment capacities at slab supports, and the top rebars in the interior spans are designed for larger bending moments than the bottom bars. If the top bars are cut, their contribution is eliminated. The positive (bottom) reinforcement in the slab areas adjacent to the cut-out panel will have to pick up the slack, but unless the bottom bars were originally grossly overdesigned, they do not have the required capacity. As a result, these slab panels may deflect excessively, and their cut edges, unless reinforced, may curl up.

Openings in cast-in-place concrete joists should be located to fit between the joist stems. Again, the joists should be checked for a T-beam action. If a joist has to be cut, headers at the edges of the opening spanning between the adjacent joists are required. Naturally, the supporting joists must be able to support the loads or be reinforced by one of the methods used for beams (Sec. 4.5).

4.8.3 Making openings in precast floor slabs

Hollow-core precast floor planks function as one-way slabs, and the process of making openings in them should be guided by the same principles, with some variations. When small vertical holes are made in prestressed floor panels, care should be taken not to cut the strands located between the voids in hollow-core slabs. Small holes in hollow-core planks are made through the open cells, where there are only thin concrete sections to go through. The holes can be made by core drilling or by using a hammer and chisel.[51]

When large openings need to be made, severing some strands may be unavoidable. This is not necessarily disastrous, because the strands are typically bonded along their full length, and a cut does not totally negate their structural contribution. (In contrast, cutting a post-tensioned unbonded tendon results in a total loss of its capacity. Making openings in post-tensioned floors is a big challenge, and this difficulty is a major drawback of the system.) When a strand has to be cut, the surrounding concrete is carefully removed first, and then the strand is burned off with a torch or cut off with a concrete saw.

Hollow-core planks are normally constructed with grouted keys that are able to transfer superimposed loads from one plank to another. Therefore, the planks can be designed assuming some kind of support at all four edges. Making a sizable opening near one edge may not only weaken the actual cut plank, but also affect the flexural or shear capacities of its neighbors. According to the *PCI Design Handbook*,[52] the following guidelines are generally applicable in this situation:

1. Slab openings located near the panel ends may be neglected in flexural design if they extend not more than the smaller of one-eighth of the span length or 4 ft toward the midspan.

2. The remaining development length of the cut prestressing strands, measured from the edge of the opening to the point of maximum moment, should be checked.

3. For flexural design, panels adjacent to those with openings located near the midspan or larger than one-quarter of the span should be analyzed as having a free (unsupported) edge at that side.

4. For shear design, panels adjacent to openings located closer than three-eighths of the span to supports can be analyzed as having a free edge.

So where is the least disruptive place to make a small penetration in a precast plank? The answer depends on the available plank capacity. If the panel has a large shear capacity relative to the loading (as when the span is short and the plank is rather deep), the opening is best made near the support on the centerline of the plank. If the panel's shear capacity is borderline but the unused flexural capacity is substantial, the opening is best made near the midspan (not closer than three-eighths of the span to the supports) on the centerline of the plank, not closer than one-fourth of the span toward the edges.

Where can one get information about the flexural and shear capacities of the existing plank? If the plank designation is known from the existing drawings, the capacities can be determined from the *PCI Design Handbook* or from the manufacturer's literature. If the exact type of the plank is unknown and the planned opening is of substantial size, it might be prudent to reinforce the cut edges with steel frames rather than to rely on the general rules of thumb.

4.8.4 Making openings in two-way slab systems

In a two-way slab, the primary bars extend in both directions, perpendicular to each other. The slab is divided into column and middle strips running both ways, with the column strips having a width of one-quarter of the span length on each side of the column centerline, and the middle strips taking up the areas between the column strips. The column strips include any beams located between the columns. This type of framing presents some complications from the standpoint of cutting holes: It is not easy to find the least critical area for an opening or to reinforce the slab, should that be needed.

For new two-way slab construction, a few rules regarding openings are contained in ACI 318.[17] It allows making openings of any size in the slab area at the intersection of middle strips, provided that the total required cross section of reinforcing bars required for strips in both directions is maintained. As far as the area of intersecting columns strips is concerned, the code allows a maximum of one-eighth of the column strip in either span to be interrupted by an opening. In the area common to one middle strip and one column strip, not more than one-quarter of the bars in either strip can be interrupted. In all these cases, the solution is relatively simple: The bars that would be interrupted by an opening are simply moved beyond the cut area into the remainder of the slab.

This rule helps little when making openings in existing floors. What can be done then? An obvious solution is to analyze the existing slab system in the hope that it contains some excess capacity that can be sacrificed, or that perhaps some redistribution of moments into the uncut sections can take place. For preliminary decision making, some rules of thumb can be followed. These rules are different for various slab types.

In two-way slabs with beams at the column centerline (shown in Fig. 4.33a), the beams act as column strips, but the slab alone acts as the middle strip. The corners act primarily in shear, which is rarely a problem. Here, holes with sizes of up to one-quarter of the span can be safely cut in the corners (area 1), but not through the beams, of course. As Rice[53] points out, four square openings in the corners might actually lighten the load on the panel. Area 2, common to one middle strip and one column strip, is less desirable for making penetrations. (Check whether the beams were designed as composite or T-beams before allowing any holes there.) Area 3, at the intersection of middle strips, is the least preferred, but even here, as Rice suggests, square openings not exceeding one-eighth of the span can usually be safely placed without any strengthening. Larger holes require analysis.

Here is another point for two-way slabs with beams: Do not remove the entire slab between the beams; some slab overhang should be left to allow development of top rebars from the adjoining panels, as discussed previously. Also, if the entire slab in a bay is removed, the existing beams around it will receive loading only from one side and could fail in torsional overstress. If removal of the entire panel cannot be avoided, the beams and the adjacent slabs may have to be strengthened.

In flat-plate slabs (Fig. 4.33b), the situation is reversed. As stated already, the design of flat-plate slabs is typically governed by the punching-shear capacity of the slabs around the columns. Therefore, no openings larger than, say, 12 in should be cut around the columns (area 3). It is important to grasp that any holes cut in this area reduce the critical section available to resist punching shear, as explained in ACI 318 Sec. 11.12.5. (If column capitals are present, making small

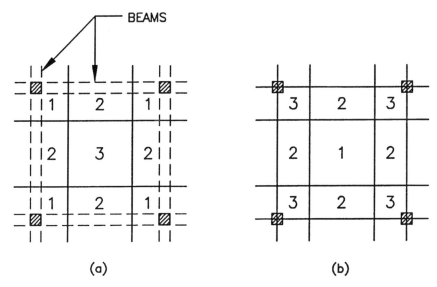

Figure 4.33 Suggested areas for making openings in two-way slab systems, numbered in order of preference. (*a*) Two-way slabs with beams; (*b*) flat slabs. (*After Ref. 54.*)

holes through them for placement of drainage pipes is acceptable, because shear stresses within the capitals are low.)

Area 2, at the intersection of middle and column strips, is less critical. Small openings—perhaps less than 15 percent of the span length—may usually be made there without incident. Still, the best place to make an opening in a flat-plate slab is in the center of the bay (area 1), where middle strips intersect. Some suggest that this whole area may even be removed without ill effects,[53] because this would simply reduce the load on the two-way column strip system.

What about the cases where the openings are inconveniently large or in the wrong place? Then, the cut edges must be reinforced, usually with steel channels or wide-flange sections. In two-way slabs with beams, the added members can be anchored directly into the beams. In flat-plate slabs, the only point of attachment is columns. The presence of column capitals or dropped slabs (such as those shown in Fig. 4.2) makes for some interesting detailing. The easiest solution is to place the beams below the obstructions and to provide another level of beams around the opening, to be supported by the lower members.

To help develop stresses in the cut reinforcing bars, Rice recommends that the cut ends of the bars be welded to side plates attached to the steel beams at the edges (see Fig. 4.34). While this welding is not always required, it is desirable in some cases, especially when top slab bars are interrupted. Some space should be allowed between top of the beam and the underside of the slab for drypack grout or steel shims.

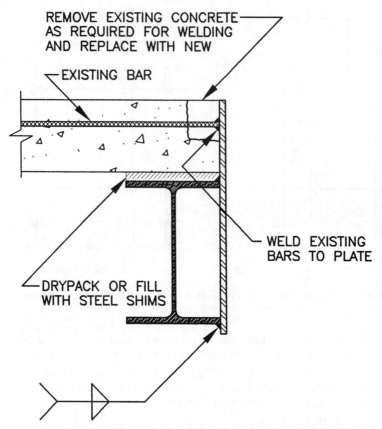

REMOVE EXISTING CONCRETE
AS REQUIRED FOR WELDING
AND REPLACE WITH NEW

EXISTING BAR

WELD EXISTING
BARS TO PLATE

DRYPACK OR FILL
WITH STEEL SHIMS

Figure 4.34 Detail at a slab opening reinforced by a steel beam. (*After Ref. 53.*)

References

1. Stella L. Marusin, "Ancient Concrete Structures," Concrete International, January 1996, p. 56.
2. Kent A. Harries, "Concrete Construction in Early Rome," Concrete International, January 1995, p. 58.
3. Engineers of the Millennium," CE News, December 1999.
4. Hale Sutherland and Raymond C. Reese, Introduction to Reinforced Concrete Design, New York, Wiley, 1943.
5. W. Gene Corley, "Protecting the Public from Fools and Rascals: Building Codes for the Millennium," Concrete International, September 1992, p. 57.
6. Robert E. Loov, "Reinforced Concrete at the Turn of the Century," Concrete International, December 1991, p. 67.
7. Rutter Carroll and J. G. Cabrera, "Reinforced Concrete in Northeast England," Concrete International, June 1998, p. 59.
8. Albert L. Colby, Reinforced Concrete in Europe, Chemical Publishing Company, Easton, Pa., 1909. (Quoted in Ref. 11.)
9. H. A. Reid, Concrete and Reinforced Concrete Construction, 2d ed., Myron C. Clarke Publishing Co., New York, 1908. (Quoted in Ref. 6.)
10. PCI Committee on Industry Handbook, "Background and Discussion of the PCI Design Handbook, Fourth Edition," PCI Journal, November–December 1996, p. 30.

11. George Winter, "Development of a National Building Code for Reinforced Concrete, 1908–1977," Concrete International, December 1982, p. 27.
12. Evaluation of Reinforcing Steel Systems in Old Reinforced Concrete Structures, Concrete Reinforcing Steel Institute, Chicago, 1981.
13. Nancy L. Gavlin, "Writing the Code—More than 40 Years on Committee 318," An interview with Chester P. Siess, Concrete International, November 1998, pp. 37–42.
14. Richard C. Elstner, "Concrete Beam Shear Design," Concrete International, September 1992, p. 70 (letter).
15. Sweet's Catalogue of Building Construction, The Architectural Record Co., New York, 1911.
16. Jay Thomas, "FRP Strengthening—Experimental or Mainstream Technology?" Concrete International, June 1998, p. 57.
17. Building Code Requirements for Structural Concrete (ACI 318-95) and Commentary (ACI 318R-95), American Concrete Institute, Farmington Hills, Mich., 1995.
18. Peter H. Emmons, Concrete Repair and Maintenance Illustrated, R.S. Means Co., Inc., Kingston, Mass., 1993.
19. Peter H. Emmons et al., "Strengthening Concrete Structures, Part I," Concrete International, March 1998, p. 53.
20. Evan R. Buckhouse, "External Flexural Reinforcement of Existing Reinforced Concrete Beams Using Bolted Steel Channels," MS thesis, Marquette University, Milwaukee, Wis., 1997.
21. Christopher M. Foley and Evan R. Buckhouse, "Strengthening Existing Reinforced Concrete Beams for Flexure Using Bolted External Structural Steel Channels," Structural Engineering Report No. MUST-98-1, Marquette University, Milwaukee, Wis., 1998.
22. Christopher M. Foley and Evan R. Buckhouse, "Method to Increase Capacity and Stiffness of Reinforced Concrete Beams," ASCE Practice Periodical on Structural Design and Construction, vol. 4, no. 1, February 1999, pp. 36–42.
23. Powers Rawl, Fastening Systems Design Manual, 2d ed., Powers Fastening, Inc., New Rochelle, N.Y., 1997.
24. Hilti Product Technical Guide, Hilti Corp., Tulsa, Okla., 1997.
25. Hamid Saadatmanesh and Mohammad R. Ehsani, "Fiber Composite Plates Can Strengthen Beams," Concrete International, March 1990, p. 65.
26. C. Allen Ross et al., "Strengthening of Reinforced Concrete Beams with Externally Bonded Composite Laminates," ACI Structural Journal, March–April 1999, pp. 212–220.
27. Amir M. Malek et al, "Prediction of Failure Load of R/C Beams Strengthened with FRP Plate Due to Stress Concentration at the Plate End," ACI Structural Journal, March–April 1998, p. 142.
28. Marco Arduini and Antonio Nanni, "Parametric Study of Beams with Externally Bonded FRP Reinforcement," ACI Structural Journal, September–October 1997, p. 493.
29. Peter H. Emmons et al., "Strengthening Concrete Structures, Part II," Concrete International, April 1998, p. 56.
30. "Composites: Coming on Strong," Concrete Construction, January 1999, pp. 120–124.
31. MBrace Composite Strengthening System: Engineering Design Guidelines, Master Builders, Inc., Cleveland, Ohio, 1999.
32. Repair and Strengthening of Concrete Members with Adhesive Bonded Plates, ACI Special Publication No. 165, American Concrete Institute, Farmington Hills, Mich., 1996.
33. Robert W. Gaul, "A New Generation of Epoxy Adhesives," Structural Engineering Forum, December/January 1995, pp. 32–37.
34. "Parking Garage Repair Completed with Carbon Fiber," Civil Engineering, March 1999, p. 24.
35. Miroslav F. Vejvoda, "Strengthening of Existing Structures with Post-Tensioning," Concrete International, September 1992, p. 38.
36. "Repair Q&A: Avoiding Replacement of Prestressed Beams," Concrete Construction, August 1998, p. 716.

37. Theodor Krauthammer et al., "Effects of Short Duration Dynamic Loads on RC Structures," Concrete International, October 1994, p. 57.
38. Thanasis C. Triantafillou, "Shear Strengthening of Reinforced Concrete Beams Using Epoxy-Bonded FRP Composites," ACI Structural Journal, March–April 1998, p. 107.
39. Vijaya B. Rangan and A. S. Hall, "Moment and Shear Transfer between Slab and Edge Column,"ACI Structural Journal, May–June 1983, pp. 183–191.
40. Karen J. Barchas, "Repair and Retrofit Using External Post-tensioning," Concrete Repair Digest, February 1991, p. 4.
41. Lindsay M. Anderson and Eugene P. Holland, "A Unique Way to Repair Flat Plates," Concrete Construction, March 1990, p. 298.
42. ACI 304R-92, Guide for the Use of Preplaced Aggregate Concrete for Structural and Mass Concrete Applications, American Concrete Institute, Farmington Hills, Mich., 1992.
43. "Repair Q&A: Preplaced-Aggregate Concrete for Column Enlargement," Concrete Construction, December 1998, p. 1109.
44. Vijaya B. Rangan, "Design of Slender Hollow Steel Columns Filled with Concrete," Proceedings, International Conference on Steel and Aluminum Structures, 1991, pp. 104–112.
45. Hong Sioe Oey and Carlos J. Aldrete, "Simple Method for Upgrading an Existing Reinforced-Concrete Structure," ASCE Practice Periodical on Structural Design and Construction, Vol. 1, No. 1, February, 1996, pp. 47–50.
46. Mohammad Shams and M. Ala Saadeghvaziri, "State of the Art of Concrete-Filled Steel Tubular Columns," ACI Structural Journal, September–October 1997, p. 558.
47. Larry Cercone and James Korff, "Putting the Wraps on Quakes," Civil Engineering, July 1997, p. 60.
48. Stephen Dial, "Smokestack Rescue," Civil Engineering, May 1998, p. 62.
49. "Fiberglass Jackets Used in California Column Retrofit," CE News, June 1998, p. 26.
50. "Engineer's Troubles Spur Reviews, and Repairs at Condo," ENR, June 8, 1998, p. 15.
51. "How to Make Openings in Flexicore Decks," Durastone Flexicore Corp., Lincoln, R.I.
52. PCI Design Handbook, Precast and Prestressed Concrete, 4th ed., Precast/Prestressed Concrete Institute, Chicago, 1992.
53. Paul F. Rice, "Determining Safe Locations for Holes in Concrete Floors," Plant Engineering, April 1977.
54. Edwin H. Gaylord, Jr., Charles N. Gaylord, and James E. Stallmeyer, eds., Structural Engineering Handbook, 4th ed., McGraw-Hill, New York, 1997.

Repairing Deteriorated Concrete

5.1 Overview

5.1.1 Introduction

It was once said that concrete is an ancient but not an eternal material. Concrete strength tends to increase with age, but only up to a point. Eventually, concrete deteriorates and requires maintenance and repair, just like any other construction material. Modern science has made enormous progress in understanding the reasons for the deterioration of concrete, and today's concrete structures—if properly designed and built—should prove quite durable. Unfortunately, most of the building stock in this country and worldwide is of an earlier vintage, much of it constructed at the turn of the twentieth century, when concrete theory was still in its infancy. Now, many decades later, these aging buildings need rehabilitation.

In this chapter, we discuss strategies for repairing concrete, focusing on restoration of its structural capacity, appearance, and durability. The topic of strengthening concrete buildings is addressed in Chap. 4, upgrading slabs on grade in Chap. 6, and renovation of post-tensioned structures in Chap. 7.

The general approach to the evaluation of distressed concrete buildings and assessment of their structural capacities is described in Chap. 2. This necessary process includes information gathering, document review, site investigation, and office analysis. For the purpose of our discussion here, we assume that all these steps have already been taken and that the causes of the problems are known, so that we can concentrate on the solutions.

5.1.2 Options other than repair

The discussion in this chapter assumes that the repair option has already been chosen, except in some case studies where the relative merits of all the available options are contemplated. Yet, it is important to keep a broad perspective and to know when to repair concrete and when not to. In many cases, options other than repair will better meet the owner's needs.

A case in point: Some owners neglect proper maintenance for years and decide to repair only when the signs of problems are unmistakable. Repairing concrete in such circumstances is expensive and disruptive to the occupants, and might even be counterproductive. It is also not without risk: As mentioned in Chap. 1, botched repair projects are among the common reasons for undertaking another round of renovation efforts.

In lieu of repair, other possible courses of action can be taken. A "do-nothing" approach, despite all its negative connotations, is sometimes appropriate. For example, concrete with reinforcing bars that are in the early stages of corrosion (with only some rust stains) may have many years of useful life left before repairs must be done. A building in which the structure is not readily accessible—it is covered with conduits, pipes, and finishes, for example, all of which would require costly rerouting—is another prime candidate for delaying repairs until absolutely necessary. A structure for which the disruption caused by renovations would be economically severe is in the same category.

Regardless of its technical and economic advantages, the do-nothing approach may be hard to defend if political or bureaucratic considerations enter into the picture, as they usually do. A need to include the repairs in the next year's budget—the use-it-or-lose-it mentality—is a typical reason to accelerate the repair schedule before the optimum time (who knows when the funds for the repairs will be available again?). Similarly, a desire to "do something" may be high on the agenda of those controlling the purse strings.

Sometimes it is better to replace, rather than repair, a dilapidated concrete structure. The relative merits of renovation vs. tearing down and replacement are debated in Chap. 1. Heavily deteriorated concrete members that are easy to replace, such as the pipe support pedestal in Fig. 5.1a, should indeed be replaced, not repaired.

Another possibility for dealing with deteriorated concrete buildings involves reducing the level of loading they carry, in lieu of renovations. This could include closing off the areas with the worst deterioration, making them inaccessible to the occupants; relocating the operations that produce the heaviest live loads; and removing some of the existing dead loads, e.g., heavy ceilings and floor finishes.

(a)

(b)

Figure 5.1 (a) This deteriorated concrete pedestal should be replaced rather than repaired; (b) this exposed concrete framing could be fixed.

5.1.3 Causes of concrete deterioration

Concrete deterioration can take many forms and be caused by several factors. Some of the most common ones are listed here.

1. *Destructive influence of the environment.* Environmental damage includes chemical attack by acids or alkalis, oxidation, carbonation, fire, and freeze-thaw cycles. A typical result of these is shown in Fig. 5.1*b*.

2. *Structural failure.* The causes of structural failure can include excessive loading, impact, or fatigue; all of them result in cracking, crushing, abrasion, and dimensional deformation of concrete. Structural failure can also occur as a result of errors in design and construction.

3. *Restricted movement.* When concrete members are unable to expand and contract freely, cracking, crushing, or buckling typically follows. (A separate case of structural distress is concrete creep, an increase of strain under a sustained stress.)

All of these factors are aggravated by poor workmanship. Similar signs of concrete distress can have different causes, and it is important to isolate the proper cause before attempting the repair. Repairing a misdiagnosed problem is likely to have the same rate of success as pulling a tooth to heal a broken bone.

The main types of concrete damage and methods of repair are discussed in this chapter, with the exception of some surface defects in slabs on grade, which are covered in Chap. 6. Various kinds of concrete defects are also described and illustrated in the *Guide for Making a Condition Survey of Concrete in Service,* (ACI 201.1R-92), by the American Concrete Institute.[1]

5.2 Repairing Cracks

5.2.1 Why concrete cracks

Cracks are the most frequent and easily recognizable sign of concrete problems. Because the tensile strength of concrete is only about one-tenth of its strength in compression, concrete tends to crack and break when subjected to large tensile loads. In reinforced-concrete structures, the role of resisting tensile stresses belongs to steel reinforcement.

Cracks can occur for a variety of reasons. Minor flexural cracks are a part of normal concrete behavior. Heavy cracks occur when the reinforcement does not do its job, either because the bars are placed incorrectly or because the member sustains damage from one of the causes listed previously. Also, many cracks occur while concrete is still wet and has not yet formed a monolithic whole with the steel reinforcement. The various kinds of cracks are described in the next sections.

5.2.2 Shrinkage cracks

Concrete starts to shrink within the first hour after placement, as the water is absorbed by the cement particles and by the subgrade, and some water evaporates from wet concrete. The shrinkage process can continue for months. The mechanics of concrete shrinkage are not fully clear, but it is generally accepted that the component that shrinks is the cement paste. According to one explanation, shrinkage is caused by the surface tension of water within the capillary pores formed in the paste during evaporation. As a capillary pore starts to dry out, the remaining water forms a meniscus between the adjacent cement particles, and the forces of surface tension pull the particles together. As a result, the cement paste, and ultimately the concrete, shrinks. In addition to the shrinkage properties of the cement paste, which are obviously important, the total amount of shrinkage also depends on the paste-to-aggregate ratio, the stiffness of the aggregate, and the bond strength between the paste and the aggregate.[2]

This understanding of the shrinkage mechanism has resulted in the introduction of shrinkage-reducing admixtures that can be added to the concrete mix water. These materials work by lessening the forces of surface tension in the pore water. One of the most popular admixtures is derived from propylene glycol. Shrinkage-reducing admixtures have been shown to significantly decrease the rate of concrete shrinkage.[2,3]

In theory, unrestrained concrete should not crack as it shrinks. In practice, some form of external restraint is always present, whether steel reinforcement, formwork, anchorage into the adjoining construction, or friction with the subgrade. This restraint prevents free shrinkage and subjects concrete to tensile stresses that it is ill equipped to resist. Eventually, these stresses overcome the modest tensile resistance of concrete, and it cracks.

What factors aggravate concrete shrinkage? The most important are mix design and curing. Concrete that is not properly cured or that contains too much water will almost certainly crack. Proper curing includes continuous wetting or submersion in water for a period of several days (curing for seven days is often specified). It is often stated that shrinkage is more pronounced in concrete mixes that are cement-rich, because they tend to contain more water than usual to maintain the same water-cement ratio and produce more internal heat during hydration. In reality, it is the actual water content, regardless of the amount of cement, that is the main factor affecting shrinkage. Some types of cement cause more volume changes than others. According to Ropke,[4] calcium chloride admixtures in 2 percent concentration may significantly increase concrete shrinkage, but entrained air in the usual doses does not seem to affect it. Superplasticizers have also been found to increase the rate of shrinkage.

There are two major types of shrinkage cracks: drying shrinkage cracks and plastic shrinkage cracks. *Drying shrinkage cracks* start to form when concrete loses its water sheen and can take several months to fully develop. Most often, these are hairline, or barely perceptible, cracks. [A hairline crack is typically defined as being less than 0.003 in (0.08 mm) wide.] Drying shrinkage cracks are often straight or ragged, forming at random locations (Fig. 5.2), but they can also appear as crazing on the surfaces of walls and slabs. These cracks are usually shallow and cosmetic in nature. Drying shrinkage cracks often occur in slabs on grade with inadequate joints.

Plastic shrinkage cracks, on the other hand, are often (but not necessarily) wide and deep, sometimes even extending through the whole thickness of the slab. These cracks typically run parallel to one another. They may form when wind or heat causes concrete to lose water rapidly, usually 30 min to 6 h after the concrete placement. Because of their size, plastic shrinkage cracks can be structurally significant and are often repaired. The repairs, if applied early enough during construction (at the time the cracks start to form), often need be nothing more elaborate than a fine spray from a garden hose.[4]

Concrete tends to crack when its shrinkage is restrained. Shrinkage cracks are especially common at reentrant corners of slabs on grade, but they can also occur over reinforcing bars and embedded conduits placed too close to the surface, around anchor bolts in concrete footings, and on tops of foundation walls. Cracks in this category can appear as early as within a day or as late as several months from the date of placement. Shrinkage cracks may or may not be of structural importance.

Cast-in-place beams and elevated slabs may also crack if they are restrained against shrinkage. These "restraint" cracks are typically perpendicular to the length of the member and extend through its thickness; they are evenly spaced, are of uniform width, and can have a torn appearance. Restraint cracks usually form within 3 or 4 months after placement and remain dormant after that.[5]

The stresses from slab shrinkage can cause cracking in the supporting members. For example, shrinkage of a floor slab placed on top of a freshly constructed concrete wall might result in diagonal cracks near the wall ends, as the top of the wall is pulled toward its center.[6] Being

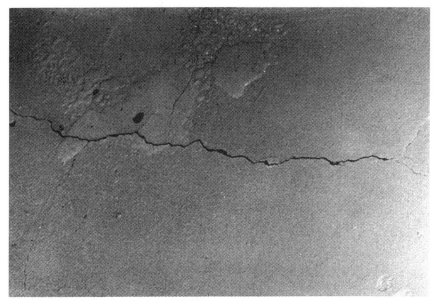

Figure 5.2   A drying shrinkage crack.

pulled perpendicular to its plane by the shrinking slab, the wall might also develop horizontal cracks at midheight.

5.2.3   Cracks caused by structural distress

Structural cracks tend to occur at the points of maximum stress in the member. The characteristic diagonal cracks in concrete walls are usually caused by settlement of foundations (Fig. 5.3a). Closely spaced cracks running perpendicular to the main reinforcement and located near the midspan at the bottom surfaces of beams and elevated slabs are typically caused by flexural stresses (Fig. 5.3b). In two-way slabs, the top surface above the column (Fig. 5.3c) can show either common flexural cracks (those radiating from columns) or rare and dangerous cracks indicating punching-shear failure (those that are circumferential). In basement-floor slabs, large cracks near the middle of the floor could be caused by hydrostatic uplift (Fig. 5.3d), especially if water stains are also present there.

Cracks at the side surfaces of beams near their supports can be caused by a shear failure. In long precast girders welded at both supports, this type of cracking reflects restrained temperature movements.

A torsional overstress is typically manifested by a series of diagonal cracks parallel to one another. Both torsional and shear cracks are inclined, but shear cracks are inclined in the same direction on both

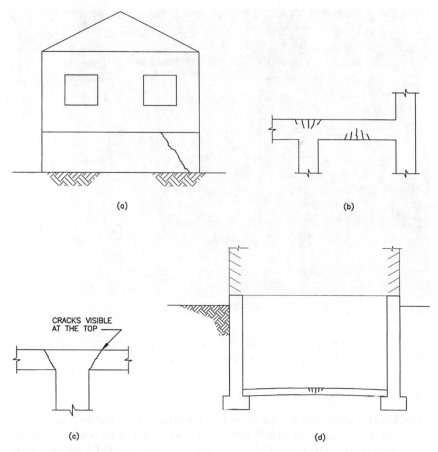

Figure 5.3 Examples of cracks caused by structural distress: (a) diagonal cracks in foundation walls caused by settlement; (b) flexural cracks running perpendicular to the main reinforcement and located at the points of maximum stress; (c) flexural or punching shear cracks at the top of two-way slabs above the columns; (d) cracks in basement-floor slabs caused by hydrostatic uplift.

sides of the beam, while torsional cracks run in opposite directions and form a spiral pattern.[5]

Cracks that seem structural in nature but are located beyond the points of maximum stress might stem from impact, blast, or some other unusual loading. Structural cracks often do not extend through the thickness of the member and are widest at the exterior surfaces. Obviously, these cracks can appear at any time after concrete is cured.

5.2.4 Other types of cracks

Some cracks run directly above the reinforcing bars. Most probably, these are *plastic settlement cracks,* which appear during construction when concrete consolidation is locally restrained by the reinforcement. Similar cracks include those caused by lateral shifting of the formwork during placement or by premature stripping of the forms, before the concrete has gained adequate strength. Formwork-related cracks are easy to spot, as they run near and parallel to the forms and are formed soon after placement. A similar type of cracking can be caused by movement of supports, including excessive settlement of subgrade and vertical deflection of the forms.

Another, more ominous cause of cracks running directly above the bars is corrosion of reinforcement, a major problem discussed in Sec. 5.3. Corrosion cracks appear years after construction. And finally, cracking can be the result of chemical deterioration, such as alkali-silica reaction.

Concrete can crack if its thermal expansion and contraction is not accommodated by properly designed and spaced expansion and contraction joints. Typical examples include cracking at reentrant corners in doors and windows and at dapped ends of beams.[7] Diagonal cracks in concrete columns can be caused by excessive movements of the floor as a result of large thermal changes or by lateral loads.

And, cracks can be caused by mechanical damage. In one case, the author had to decide what to do about the cracks that formed around anchor bolts in recently cast circular piers (Fig. 5.4). The cracks could have been caused either by abuse during handling or by shrinkage, probably the former. In that case, the cracks were repaired, because they compromised the holding capacity of the anchor bolts. As recent research[8] has demonstrated, the design capacity of headed anchor bolts (and other fasteners, for that matter) is significantly reduced by cracks in surrounding concrete.

5.2.5 Dormant vs. active cracks

Cracks in concrete can be divided into two major classes: dormant and active. Once formed, dormant cracks do not increase in size and length. Active cracks, true to their name, change their width or length under load. Dormant cracks typically result from shrinkage, initial movement of supports, or previous structural overload. Depending on their size and location, they may or may not need repair.

Active cracks, on the other hand, are usually formed in response to a continuing movement or to present overload; they open and close as a result of external forces. Repairing active cracks is difficult. While it might seem a good idea to repair them to prevent their uncontrolled propagation, this is easier said than done: If the underlying cause is not fixed, a new crack will form next to the "repaired" one.

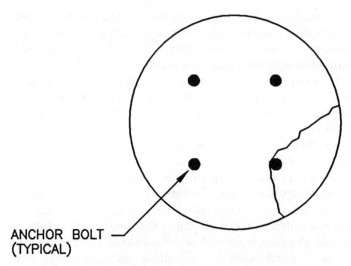

ANCHOR BOLT
(TYPICAL)

Figure 5.4   Cracks around anchor bolts in a circular pier.

How can one determine whether a crack is active or dormant? One simple way is to bridge the surface of the crack with a rigid but non-shrink material, such as plaster or even paint, and inspect the patch periodically. Depending on the circumstances, the measurements can be taken daily, weekly, or monthly. The longer the waiting period, the more chances there are to detect an infrequent movement. An active crack will, of course, reflect through the coating and become visible, a dormant one will not. The actual magnitude of the movement can be determined, if that is needed, by periodically measuring the crack width in a few places. To measure the crack width, a crackscope (illustrated in Chap. 2) or a crack comparator is used. Another helpful tool is a crack monitor, a simple two-piece device glued to both sides of the crack.

The origins of some cracks are difficult to determine. In any case, cracking caused by one factor will contribute to further deterioration and instigate other kinds of cracks. For example, shrinkage cracks facilitate corrosion of the reinforcing steel, which eventually rusts and causes further cracking and spalling of concrete. Similarly, cracking induced by corrosion may lead to weakening of the member and might be followed by structural cracking.

Additional information about various types of cracks and methods of repairing them is contained in ACI 224.1R.[7]

5.2.6   Which cracks to repair?

Which cracks should be repaired? Certainly cracks that significantly reduce the structural capacity of the member or of anchor bolts embedded in it should be. Accordingly, the first step in assessing the need to repair is to diagnose the cause and to identify cracks that are overly wide or are structural in nature. Any cracks in flexural members should be investigated especially carefully. Transverse cracks in beams and frame columns reduce their stiffness and durability, even if the strength is not affected, and should in most cases be fixed. The fact that these cracks originated from shrinkage or some other nonrecurring cause is irrelevant.

Any crack that is wide enough to lead one to question whether the aggregate interlock still exists—that is, whether the stresses can be transferred across the crack—might be considered a candidate for repair. In practice, this usually means cracks wider than 0.035 or 0.04 in. To visualize these numbers, consider the following rough equivalents given by Fricks:[9] A major credit card has a thickness of 0.030 in, while a dime is 0.053 in thick. (Also, a new dollar bill is 0.0043 in thick.)

A crack that leaks water usually gets filled. As discussed in Sec. 5.3, it is fruitless to repair cracks caused by rusting reinforcement without fixing the underlying problem first.

Beyond these clear situations, the choice is up to the owner. Cosmetic cracks are often left unrepaired, not only because of the expense involved, but also because the repair might look worse than the original damage. Some of these cracks, especially those not exceeding a width of 0.012 in, may heal autogenously (by themselves) as unhydrated cement in their edges is exposed to moisture. However, too much moisture is harmful: A continuous flow of water through a crack washes away the unhydrated cement and should be prevented by sealing the crack.[10]

Not all cracks in structural concrete are harmful. Some cracking in reinforced-concrete flexural members (beams and slabs) is to be expected. Indeed, concrete adjacent to primary reinforcing bars is *presumed* to be cracked in both working-stress and load-factor design methods. The question is not whether concrete should be allowed to crack at those locations, but whether the crack width is acceptably small to protect the steel reinforcement from corrosion. Also, the cracks should not be so wide as to frighten the occupants into believing that collapse of the building is imminent. One method of quantifying these admittedly "soft" criteria is to follow the crack-width limits contained in the previous edition of ACI 318, *Building Code Requirements for Structural Concrete,*[11] which are 0.013 in for exterior and 0.016 in for interior applications.

Another source on allowable crack widths is ACI 224R,[12] Table 4.1, "Tolerable Crack Widths." For exposure conditions of dry air or protective membrane, the table's crack limit is 0.016 in (0.41 mm). For conditions of humidity, moist air, or soil, the suggested limit is 0.012 in

(0.3 mm). For concrete exposed to deicing chemicals, the crack width should not exceed 0.007 in (0.18 mm). Even tighter limits are given for concrete exposed to seawater and for water-retaining structures. ACI 224R cautions that the values in the table are intended only as a general guide and may not apply to all conditions.

Crack width well in excess of the limits contained in either ACI 318 or ACI 224R might indicate structural distress, although wide cracks can also reflect the effects of long-term creep. In any case, cracks of this magnitude are the prime candidates for repair.

5.2.7 Methods of crack repair

As stated already, the cause of distress must be eliminated prior to repairing the cracks. Thus, filling a constantly widening crack caused by settlement of supports without first fixing the foundations is futile. Similarly, filling flexural cracks in a beam that is not strong enough or is overly flexible under load will not be productive until the beam is strengthened. Cracks that leach efflorescence to the underside of slabs (Fig. 5.5) may indicate ongoing corrosion of reinforcement and should not be simply sealed. When floor cracks are traced to thermal expansion and contraction, it may be necessary to install additional expansion joints. Cracking is usually a symptom, not a problem.

There are three basic methods of crack repair: to "glue" the cracked concrete back together by epoxy injection or grouting, to "stitch" the cracked concrete with dowels, or to enlarge the crack and "caulk" it with a flexible or semirigid sealant. Unfortunately, there is as yet no ideal multipurpose material that can not only structurally repair the crack, like an epoxy, but also stretch a significant amount, like a flexible sealant. Semirigid epoxies are popular for cases where performance compromise is tolerable; however, these can neither fully restore the section's rigidity nor stretch a lot.

The gluing option is appropriate for fixing dormant cracks that significantly reduce the strength or stiffness of the member. Stitching may work for arresting crack growth and for strengthening the member in shear. When this technique is used for shear reinforcement, mild steel dowels are diagonally drilled and grouted with epoxy into the cracked girder perpendicular to shear cracks.[13]

The caulking option may be used on moving (active) cracks to prevent water intrusion and therefore reduce the potential for corrosion. It also prevents accumulation of debris within the crack that restricts contraction and leads to further crack widening. In this method, the crack is enlarged to an opening of at least 0.25 in at the mouth by routing, grooving, or chipping; cleaned by water jet, compressed air, or

Figure 5.5   Random cracks with efflorescence at the underside of an elevated slab.

sandblasting; and filled with a sealant. If the sealant is placed in a rectangular shape, the bottom surfaces should be covered with a bond breaker to prevent adhesion and allow proper deformation of the sealant.

Repairs to leaking nonstructural cracks typically consist of sealing the cracks by injecting a two-part polyurethane grout. This low-viscosity material foams in contact with moisture and is well suited for stopping the leaks. To repair numerous leaking floor cracks, it is often advantageous to cover the whole floor surface with an elastomeric membrane, coating, or floor topping.

Some fine points of the polyurethane grouting procedure are given by Baker,[14] who describes a project involving filling leaking cracks in underground concrete vaults. In this project, injection was done via $1/2$-in drilled holes spaced about 1 ft apart. For wide cracks, the holes were drilled directly into them. For narrow cracks, injection holes were drilled at an angle to intersect the crack near the middle of the wall; the holes were staggered about the crack. The injection ports ("packers") were then placed into the holes and the cracks flushed with water using high-pressure hand pumps. After the cleaning, the actual injection with a hydrophobic polyurethane grout mixed with an accelerator was performed. The injection typically started at the lowest point of the vertical crack, where resistance to grouting was the largest, and proceeded to the adjacent packers. After injecting two or three packers, workers returned to the first one, which could frequently accept more grout.

5.2.8 Epoxy injection of cracks

Epoxy injection is one of the most common methods of crack repair. It is also one of the most misunderstood, and for this reason deserves further discussion. During this procedure, epoxy resin is injected into cracks and microfissures in concrete and into voids between concrete and reinforcing steel. The bond between the epoxy and the concrete is stronger than the concrete itself, and so, if no further distress occurs, the repaired material should be as good as new.

This repair option is typically used for fixing dormant cracks that significantly reduce the strength or stiffness of the member. Epoxy injection is commonly prescribed for repair of cracks caused by earthquakes, during which the violent bursts of seismic energy are dissipated by large structural movements—and cracking. Epoxy injection is not recommended for cases in which the chloride content of the concrete is significant: It would mask the symptoms of corrosion without addressing the cause.

The injection technology is well developed, although, as Gaul[15] observes, even after decades of use, epoxy injection remains partly an art rather than a precise science. One good source of how-to information on the topic is a book by Trout.[16]

The injection procedure consists of placing the ports through which the epoxy flows into the crack, capping (sealing) the crack to contain the resin, and pumping the epoxy under pressure. A crack can be sealed by covering it with hydraulic cement, thixotropic* epoxy, thermosetting wax, or a special polymer sealing tape. Experienced operators know where to place the ports for best performance, how long to inject the material, and under what pressure to inject it.

Because of their ability to flow into thin cracks, low-viscosity (light and fluid) resins are normally used for injection. Alternatively, when complete sealing is not required, a nonsag epoxy gel can be applied. The gel will not be able to penetrate as deeply as a liquid epoxy, because it resists the capillary forces that help the liquid get into the finest cracks.[15] Some contractors occasionally attempt to use ultra-low-viscosity (less than 100 cP) resins. These resins flow into cracks faster but are also absorbed into concrete more readily; therefore, they require some continuing injection after the crack has been technically filled.

---

*Thixotropic epoxies are normally viscous (heavy) but become fluid after shaking or stirring. They eventually return to their original condition.

According to Trout, low-viscosity resins are typically applied at pressures between 200 and 300 psi, although higher pressures, from 300 to 500 psi, might be required to fill fine cracks. He recommends beginning injection at the widest point of the crack, from which the epoxy can flow easily into the other parts. If the opposite were done, the epoxy would flow much more slowly and might not fill the narrow cracks completely, as it would flow instead into the wider areas that offered the path of least resistance. A crack intersection is also a good place to start the injection. (Note that this procedure is the opposite of the approach described for polyurethane grouting.)

When to stop the injection? A common rule is to consider the crack segment between two adjacent ports well filled when the epoxy injected in one port exits through the other. Trout, however, disagrees, explaining that this "port-to-port" mentality does not work for injecting typical shrinkage cracks—the most common type—which are widest at the surface and gradually decrease in thickness. In this case, the epoxy could simply flow near the surface and appear at the next port well before penetrating deeper. The situation is even worse in overhead applications, since the epoxy naturally gravitates downward and flows to the neighboring ports. Stopping the injection at that point would result in an only partly filled crack.

The port-to-port rule actually works fairly well with the gel injection, but for more common repairs utilizing low-viscosity resins, Trout recommends the following approach. For cracks over 0.007 in wide, the injection should continue until the refusal—a point beyond which no more resin can be accepted by the port—and then be carried on for another minute or so. The fact that resin flows from the adjacent ports should be taken simply as evidence that the work is proceeding properly. For cracks narrower than 0.007 in, the situation is more complicated. Since the flow volume is so small, filling these cracks takes a lot of time, and it is difficult to determine the point of refusal. To be successful, such work requires experienced personnel.

After the crack is filled, the ports are removed and, if desired, the surface is ground smooth. If sealing tape was used, it is peeled off, and the job is complete.

5.3 Corrosion of Reinforcement and Its Effects on Concrete

Corrosion of steel reinforcement is the most common cause of concrete failures. The cost of repair and replacement of concrete structures in North America that have deteriorated as a result of rebar corrosion is estimated to be in the tens of billions. The situation has worsened over the last few decades, and the frequency of corrosion-related failures has skyrocketed.[17] Let us briefly examine why corrosion happens, how it damages concrete, and what can be done to stop it.

5.3.1 Why steel reinforcement corrodes

Steel rusts—this fact is familiar to everyone. But why doesn't the steel reinforcement in concrete normally rust? The answer is that concrete possesses natural corrosion-inhibiting properties. In chemical terms, the cement in hydrated (cured) concrete is basically a saturated solution of calcium hydroxide [$Ca(OH)_2$]. Being very alkaline, it neutralizes any acids that might attack the steel reinforcement. An initial spurt of corrosion does indeed happen when the steel is exposed to the moisture in concrete, but this quickly results in the steel reinforcement's being coated by a thin *passivating layer* of iron oxide (Fe_2O_3), which acts as a barrier to moisture and oxygen and prevents any further corrosion.

This passivating layer remains stable in an alkaline environment, and as long as the pore water in concrete retains its high alkalinity, the steel does not rust. This protection is lost, however, if the degree of alkalinity is lowered by some external factor, chiefly an attack by corrosive chemicals. The most harmful are chlorides, specifically calcium chloride ($CaCl_2$), which is used as a deicing agent or a concrete setting accelerator. At first, the loss of passivity occurs in a few isolated spots, meaning that the severely corroded rebar areas are separated from one another by areas of steel that are unaffected by corrosion.[17]

Another mechanism by which the alkalinity of concrete is reduced is *carbonation,* a chemical reaction between carbon dioxide (CO_2) contained in the air and calcium hydroxide [$Ca(OH)_2$] in cement. Carbonation results in the formation of calcium carbonate ($CaCO_3$) and water. (A similar reaction can occur between sulfur dioxide and cement.) Carbonation destroys the passivating layer by reducing the alkalinity of the pore water from a pH of between 12.6 and 13.5 to one between 8 and 9.[18]

Carbonation is a relatively slow process that starts at the surface of concrete, and it may take many years for it to reach the steel reinforcement. The penetration rate of the carbonation "front" is nearly proportional to the square root of the exposure time.[19,20] So, if after 10 years the depth of carbonation is 0.5 in, a depth of 1 in may be expected to be reached only after 40 years. These numbers are for illustration purposes, as the rate of penetration rarely exceeds $\frac{1}{8}$ to $\frac{1}{4}$ in in 30 years. Still, the process can be surprisingly fast in some industrial buildings where the air contains high concentrations of carbon dioxide and where the concrete is of poor quality.

Carbonation seems to proceed faster in some types of concrete. Chandler[21] states that sulfate-resisting cement increases the depth of penetration by up to 50 percent and portland blast-furnace cement by up to 200 percent over the regular portland cement. Also, concrete with some lightweight aggregates seems to be particularly vulnerable to rapid onset of carbonation.

Carbonation is most likely to occur at a level of relative humidity between 50 and 70 percent. It diminishes in very dry environments, because the reaction requires moisture in order to proceed, and in extremely moist conditions, because concrete pores filled with water act as barriers to carbon dioxide. Owing to variable concrete moisture, carbonation levels can be quite different even within the same building.[18,20]

Carbonation does not advance in a straight line: Its path is distorted by coarse aggregate, which is not affected by carbonation, and by cracks, which offer shortcuts. When carbonation eventually reaches the reinforcing steel in the presence of both oxygen and moisture, rusting is likely to start. The exposed building corners, open to attack from two fronts, are most vulnerable to both chlorides and carbonation.

5.3.2 How corrosion begins

Corrosion of reinforcement embedded in concrete is an electrochemical reaction, involving both chemical processes and the flow of electricity between various areas of steel and concrete. Because of the variability of protective film and mill scale thickness, concrete imperfections, cracking, and uneven depth of carbonation, the same piece of reinforcing steel can have areas with different electric potentials. These areas become electric (galvanic) cells that act as anodes and cathodes, as in an electric battery. The role of electrolyte—the liquid that conducts electricity—is taken by the water contained in concrete pores and the chemicals dissolved in the water. (The theory of corrosion is quite complicated and is grossly simplified here for the sake of brevity.)

Corrosion starts when an electrochemical circuit is established that allows passage of hydroxyl ions $[(OH)^-]$ and oxygen through the electrolyte. The electrons flow from the anode (in this case, a negatively charged cell) via the steel reinforcement to the cathode (in this case, a positively charged cell) and back through the electrolyte (Fig. 5.6). The terms *anode* and *cathode* and the convention concerning the way the electricity flows are somewhat confusing, because on the way back—through the electrolyte—the direction of current is from cathode to anode.

A simpler way to make sense of it is to remember the expressions "sacrificial anode" and "protected cathode," because in a galvanic cell the anode is consumed first, breaking up into positive ferrous ions and electrons. The following reaction takes place at the anode:

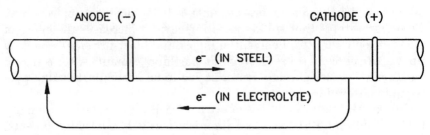

Figure 5.6   The flow of electricity in a galvanic cell.

$$Fe \rightarrow Fe^{++} + 2e^-$$

The positive ions are dissolved in the electrolyte, while the electrons move through the steel to the cathode. At the cathode, they combine with oxygen and water to create hydroxyl ions:

$$2H_2O + O_2 + 4e^- \rightarrow 4OH^-$$

Eventually, the positive ferrous ions coming off the anode and the hydroxyl ions from the cathode meet and react to form ferrous hydroxide:

$$Fe^{++} + 2OH^- \rightarrow Fe(OH)_2$$

If the anode and the cathode are isolated by an impervious barrier, the circuit cannot be completed and corrosion cannot occur. Concrete cover over the reinforcement can act as a barrier of sorts by delaying the reaction: It prevents the aggressive chemicals from dissolving in the pore water and forming an electrolyte. This barrier is not perfect, however, and its effectiveness depends on the thickness and permeability of the concrete. The thinner and more permeable the cover and the higher the temperature and the level of humidity, the earlier the start of corrosion. The relationship of the cover thickness to the speed of corrosion is nonlinear. According to some estimates, reducing the cover by one-half may shorten the time to corrosion by some 86 percent![19]

The porous mill scale does not function as a protective barrier. On the contrary, it presents an additional complication because it has a lower electric potential than concrete and acts as a cathode to steel. This means that if most of the rebar surface is covered with mill scale, but this scale is cracked or abraded in a single spot, corrosion of that spot will begin immediately, manifesting itself as very dangerous deep pitting.[22] Pitting corrosion can also occur in concrete that is deprived of oxygen but contains plenty of chlorides, such as saturated concrete.

The danger of this type of corrosion is that it is nearly impossible to detect visually, even when the steel section is severely degraded.[20]

The rate of corrosion depends on the speed of the ions traveling back from the cathode to the anode, which is a function of the electric potential of the reinforcing bars and the electrical resistance of concrete. From the basic theory of electricity, the larger the potential and the lower the resistance, the higher the current. Low resistance levels, which encourage corrosion, occur when concrete pores contain a lot of electrolyte, i.e., moisture and chlorides.

How can chlorides get into concrete structures? Deicing salts containing chlorides are responsible for the majority of chloride contamination. Corrosive chemicals can also get into concrete as admixtures, mainly in the form of calcium chloride used as a cold-weather accelerator. Other potential sources of chloride contamination include unwashed (and therefore containing sea salt, which is essentially sodium chloride) beach sand, harmful fumes from chemical processes rising to the underside of the structure, and spilled corrosive liquids used in operations or maintenance. Sometimes, corrosion can even be precipitated by acid etching used, ironically, in renovations to remove efflorescence from the surface of concrete. Muriatic acid, which is commonly used for this purpose, is nothing but diluted hydrochloric acid (HCl). There are many other avenues of possible chloride entry, including chloride-containing aggregates (a problem common in the Middle East) and salt-water spray on waterfront buildings. Buildings located on soils with high saline-water levels, as in some desert regions, can get chlorides by capillary action through footings and columns.[20]

Corrosion can be precipitated not only by harmful chemicals, but also by electric currents passing through concrete. As Pfeifer[23] points out, corrosion can be started by a leakage of direct electric current from an external source. Indeed, corrosion frequently originates at embedded nonstructural metal conduits made of aluminum or galvanized steel. Furthermore, if the surface of a conduit is cracked, it can act as both galvanic cell and water carrier. When exposed to moisture, nonstructural metal items partly embedded in concrete can also be a problem, with their exposed parts acting as anodes and their cast-in parts as cathodes. The larger the size of the embedded part relative to the exposed part, the more rapidly the latter will corrode.[19] To safeguard against this problem, metallic embedded items could be coated with epoxy or made of stainless steel.

The fact that both water and oxygen are required in order to initiate corrosion is illustrated by the phenomenon of concrete piles placed in salt water. Despite the fact that plenty of aggressive chemicals (chlorides) and, obviously, water are present, the immersed portions of the piles rarely corrode, because there is little oxygen to sustain the reaction. (One exception is pitting corrosion, discussed previously.) In the splash zones, however, all the ingredients for trouble are present, and these areas indeed often deteriorate.

5.3.3 The process of rusting and spalling

Once corrosion is under way, it begins to produce porous iron oxides, which accumulate in the anodic areas. These rust products occupy up to six times the volume of the original steel, pushing the concrete apart from within. The surface cracks at the bar locations become noticeable. As more and more rust accumulates, the cracks continue to grow. The crack width increases especially fast if water can get into the cracks and freeze there, acting as a wedge. Figure 5.7*a* illustrates this process for the underside of a concrete beam.

Eventually, the cracked concrete cover spalls off, exposing the rusted bars and leaving them totally unprotected from further deterioration (Fig. 5.7*b*). Once the bars lose their bond with the concrete because of spalling, the section no longer functions as a reinforced-concrete member, but behaves more like a damaged plain-concrete beam. At this point, failure is not far off. Cantilevered beams and slabs with corroded top reinforcement are especially vulnerable and should be treated with extreme caution. (This is not to say that any loss of bond because of rusting automatically renders the structure unsafe. There are some reports that considerable loss of bond may result in only a small reduction of section capacity.)

In slabs, the process of spalling proceeds in a similar fashion, albeit with a slightly different shape of cracking. As in beams, the deterioration starts with cracks along the rebar locations. As rust products accumulate, two additional cracks are formed near each bar. In exposed exterior slabs, water freezing in the cracks aggra-

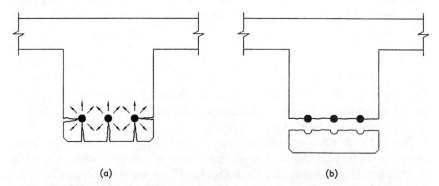

(a)   (b)

Figure 5.7 Spalling resulting from rusted reinforcement: (*a*) Rust products exert splitting forces on concrete from within, leading to cracking; (*b*) eventually, the cracks widen and the cracked section spalls off.

vates the situation. At some point, concrete spalls in two wedge-like pieces[5] as shown in Fig. 5.8 for relatively widely spaced top bars. Obviously, the spalls act as water collectors and exacerbate corrosion in the adjacent areas.

If the bars are spaced close together, the concrete may spall as one whole sheet (Fig. 5.9). Now that the reinforcement is exposed, it will continue to corrode until nothing but rust remains (Fig. 5.10).

The effects of corrosion are least noticeable in lightly reinforced slabs, especially those containing only welded wire fabric and those covered with floor finishes or overlays. These slabs offer very few clues to the rusting underneath until reinforcement in the nearby beams

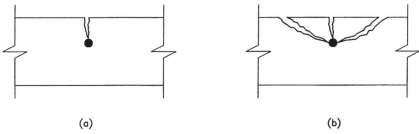

(a)   (b)

Figure 5.8   The process of spalling in top slab rebars: (a) Cracks form above the bar locations; (b) two additional cracks are formed near each bar, and eventually, the concrete spalls in two wedgelike pieces. (*After Ref. 5.*)

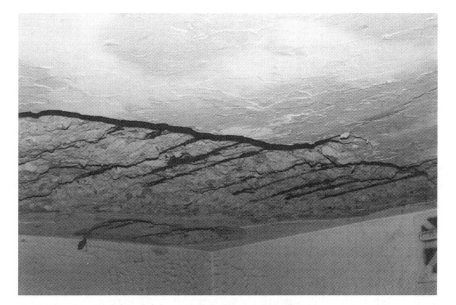

Figure 5.9   A part of this slab has spalled off as a single sheet.

Figure 5.10 Totally corroded steel reinforcement. The remains of these bars could be removed by hand.

starts to be seriously affected. Since the top bars in the beams are hidden within the slab, the damage may not be apparent until the slab is literally lifted off the beams by the expanding rust products. When that happens, it is too late for repair, and the only effective remedy is removal and replacement of the whole floor area.

A similar phenomenon is illustrated in Fig. 5.11. Here, corrosion of reinforcing bars in a concrete balcony exposed to a saltwater environment has caused major cracking of the slab edge and essentially split it in half, while the slab surface shows few signs of distress.

Figure 5.11   Cracking of concrete balcony slab edge caused by corrosion.

In structures consisting of slabs, beams, girders, and columns, corrosion usually starts in places where access by water and salts is easiest and reinforcement is the most congested. The most vulnerable place is the top of the beams or slabs around columns—an intersection point for reinforcement running in three different directions. Figure 5.12 illustrates leakage-induced deterioration of concrete around a column. Characteristically, the corroded top reinforcement is hidden in the slab; corrosion becomes obvious only when it metastasizes to the bottom bars, which are exposed to view.

Some differentiate between the so-called microcell corrosion, which affects individual bars, and macrocell corrosion, which affects the whole concrete structure. The most damaging situation occurs when the electrochemical reaction takes place between the top and bottom layers of reinforcement, with the top bars acting as cathodes and rapidly corroding.[24]

5.3.4 Assessing corrosion activity in concrete

Which structures are most susceptible to corrosion? Usually those that are exposed to weather, which supplies plenty of water and oxygen, or

Figure 5.12 Deterioration of concrete around a column caused by water penetration through the depressed slab at right.

to deicing salts containing chloride ions. Sulfate ions, found in some chemicals such as battery acid, can also cause corrosion of steel. Among the most vulnerable buildings are parking garages and underground structural slabs supporting ramps and roadways. A typical scenario of corrosion initiation involves repeated applications of chloride-containing salts, which are absorbed into the exposed concrete slab, penetrate deep enough to reach the rebars, and eventually accumulate in major concentrations.

Can the signs of corrosion be detected visually before it is too late? The investigation usually starts when some cracking, spalling, or rust stains are noticed. As was just discussed, for corrosion initiated from the top surface (as by deicing salts), such visual clues may appear too late, after the process has spread to the bottom bars or a horizontal delamination occurs between the slab and the beams. Corrosion initiated from the bottom is easier to detect, because it can be betrayed by rust stains or by leaching salts that leave a white residue (efflorescence). But visual signs are not always present. In a saturated chloride-laden environment, the volume of rust may not increase all that much if the rust products are able to diffuse into concrete pores, leaving no visible signs of cracking and spalling.[19]

So how can one detect corrosion reliably? One way is to remove some concrete cores and visually check them for delaminations and rusted rebars, but nondestructive testing methods are more practical. The extent of a deteriorated or delaminated area can be determined by chain drag, pulse velocity, thermography, ground-penetrating radar, and other nondestructive testing methods described in Chap. 2. Chemical-analysis tests measure the actual concentration of chloride and sulfate ions and can help determine with reasonable certainty whether unseen corrosion is taking place.

Chloride concentration is found by taking concrete cores and chemically analyzing them or the concrete dust generated in the process. The cores can also be tested for specific gravity, concrete compressive strength, percent of voids, and absorption, all of which contribute to the picture of structural condition. The cores can also be analyzed petrographically to determine whether microcracking, aggregate-silica reaction, or other internal problems exist.

There is no magic number for the chloride concentration at which corrosion of reinforcement must start. Research conducted in Great Britain concluded that a chloride ion content of less than 0.4 percent by weight is unlikely to cause damage for some time.[25] (This translates to about 15 lb/yd^3.)

In the United States, the typically suggested chloride ion thresholds are much lower. A minimum contamination amount of 1.2 lb/yd^3 is generally considered necessary for initiation of corrosion. When chloride concentration levels approach 3.0 lb/yd^3, corrosion can be expected to rapidly accelerate and severe rusting of steel and spalling of concrete to take place. Chloride levels in excess of 7.0 lb/yd^3 are likely to cause major loss of steel section and disintegration of concrete.

Some testing laboratories report chloride concentrations in pounds per cubic yard, some in percentage by weight of concrete, and some in parts per million (ppm). One part per million is 0.0001 percent by weight, and conversion from ppm to percent by weight is easy, but conversion to pounds of chloride ion requires knowledge of the actual concrete density. If the concrete density is unknown, an approximation of 3800 lb/yd^3 can be used. Thus 700 ppm equals approximately 700 × 0.000001 × 3800 = 2.6 lb/yd^3. (It also equals 0.07 percent by weight.)

To complicate matters more, chlorides and carbonation can work together. As carbonation lowers the pH of concrete, smaller concentrations of chloride ions are needed to cause corrosion. For example, if it takes a chloride ion concentration of 7000 to 8000 ppm (approximately 26 to 30 lb/yd^3) to start corrosion of embedded steel in concrete with a normal pH level of 12 to 13, it may take only 100 ppm (approximately 0.4 lb/yd^3) in concrete with a drastically lowered pH of 10.[26]

In addition to measuring chloride concentration, one can learn about the probable source of the chemicals by constructing a profile of the chloride distribution over the depth of the core. Typically, as a result of wetting and drying effects, chlorides deposited by deicing salts tend to concentrate near the surface at first and then gradually diffuse into the concrete. Therefore, if the chloride concentration *decreases* from the surface down, it is likely that the chlorides have migrated from the outside. If the concentration is *uniform,* the chlorides could have been introduced into the concrete by admixtures, unwashed beach sand, or salt water used for mixing. Occasionally, chlorides come from multiple sources.[27]

Another method of detecting the signs of corrosion is the half-cell potential test, described in Chap. 2. This test includes electric potential mapping augmented by some coring to the surface of reinforcement for measurement calibration. If the measured electric potential is below a certain value, corrosion is unlikely to occur, and any observed signs of distress are probably caused by other factors. Many engineers consider corrosion probable if the measured potential is at least 0.30 V for parking deck structures and at least 0.20 V for exterior concrete walls and columns.[23] ASTM C-876 considers a potential in excess of 0.35 V as certain evidence of corrosion.

Example 5.1 Gallegos and Quesada[28] describe a case of a 10-yr-old shopping center in Lima in which widespread signs of corrosion suddenly appeared. The building was framed with two-way waffle slabs. It was located 3 miles away from the sea and was not subjected to any unusual chemical attack. Nevertheless, concrete cracks parallel to the bars and rust spots were observed, with new signs of corrosion appearing every day.

Field and laboratory investigation included taking concrete samples from a variety of areas, including those with no evidence of corrosion, and analyzing chloride-ion concentrations and chemical products of corrosion on rebars. Water-soluble chloride-ion concentrations varied drastically between 50 and 500 ppm. After determining that the worst-case probable concentration of chloride ions from the background sources should not have exceeded 100 ppm, and in the absence of deicing salts or other sources of contamination, the investigators concluded that chlorides must have been added to the mix during construction to speed up placement and maximize the reuse of expensive ribbed-slab formwork. This conclusion was supported by the fact that the columns had a chloride-ion content of less than 100 ppm and exhibited no signs of corrosion. The investigators also noted that "the presence of large localized amounts of chloride originating from a poor dispersion of the additive was unmistakable."

To find the areas of corrosion, the electric-potential measurements were taken and calibrated against chloride-ion content. It was found that corrosion occurred at a potential of only 0.2 V.

As for carbonation, this process can be reliably detected by petrographic examination, which can discover the appearance of carbonates, the by-products of the reaction, deposited in concrete. Carbonation can sometimes be betrayed by the presence of fine surface dust. The depth of its penetration can be determined by applying phenolphthalein to a core removed from concrete or to the edge of a fresh cut.

The state-of-the-art techniques for evaluating corrosion damage in reinforced-concrete structures are further described in *Techniques to Assess the Corrosion Activity of Steel Reinforced Concrete Structures* (STP 1276), published by ASTM.[29]

5.3.5 Repair of corrosion-related deterioration

Successful repair of deteriorated concrete caused by corroded reinforcement requires a good understanding of the reasons for the observed corrosion and of its extent. All too often, repairs are confined to the areas of visible rust, without addressing the rest of the structure. This approach not only fails to stop but might even accelerate the rate of corrosion in the areas adjacent to the repairs. A successful repair project starts with a thorough investigation of the whole structure, proceeds with a delineation of the damaged areas, and concludes with selection of an effective and cost-efficient repair method. It also includes a program of follow-up inspections to assess the effectiveness of the repair.

In general, the two most common methods of dealing with concrete deterioration caused by corrosion of steel reinforcement are removal and replacement and patching. Less common, but perhaps more promising, methods of repair include cathodic protection and the use of corrosion inhibitors. The following sections describe these methods in some detail, except for removal and replacement, which was already discussed.

It is important to grasp that no single method of repair is appropriate in every situation, and that it is impossible to prescribe a rigid procedure for even such a relatively simple method as patching. Moreover, there seems to be no perfect technology to fix damage caused by corrosion of reinforcement. The practical repair methods available today— those that cost less over the long term than replacement—can only prolong the service life of the building by a few years or, at the very best, decades. Emmons and Vaysburd,[30] who represent a premier concrete restoration contractor, regretfully observe that in their experience of over 20 years and 3500 repair projects, they never stumbled upon a panacea or were even able to formulate the right way of protecting reinforcement in repaired structures.

5.4 Patching Spalls and Deteriorated Areas

5.4.1 Introduction

Patching is typically used to restore concrete sections damaged by delaminations and spalls. This repair method calls for removing all deteriorated concrete, exposing and cleaning rusted steel reinforcement, coating the bars with a corrosion-inhibiting primer (and adding new bars where required), patching the areas of removed concrete, and applying a protective sealer, membrane, or floor topping to prevent further ingress of corrosive chemicals. The size of the patch can vary from a few inches to many feet. Patching is the most popular repair method and is familiar to most practitioners, but its ultimate effectiveness depends on the quality of design and execution, as well as on the rate of corrosion activity in the structure.

5.4.2 Methods of concrete removal

A lot has been written about which equipment works best for removing concrete, an operation that accounts for a large portion of the repair cost. Heavy jackhammers can obviously do the work faster, but at the price of removing too much and damaging ("bruising") the surrounding sound concrete. The least amount of damage occurs when light chipping hammers, at most 30 lb in weight, are used. For repair of elevated slabs, even lighter hammers (20 lb maximum) are recommended.[31] The use of heavy jackhammers can be justified for initial chipping and for removal of large quantities of concrete, with light hand-held hammers reserved for final chipping.

When large thicknesses of concrete must be taken out, the cost-effective demolition methods include using diamond saws, which cut concrete with a steel wire containing diamonds set in steel beads; stitch drilling, which involves drilling overlapping cores through concrete; scarifying; and planing. Like heavy jackhammers, some of these mechanical demolition methods may result in bruising of existing concrete and have been blamed for delamination of patches below the scarified surface. Scabblers and bush hammers produce the largest amount of bruising, and both have been implicated in a number of cases where slab overlays failed because of delamination of repairs.[32] Conversely, shotblasting and waterblasting tend not to bruise concrete. A reasonable compromise can involve removing most of the concrete with mechanical equipment and finishing the demolition by blasting.

Another method, hydrodemolition (removing concrete with high-pressure water jets), offers a dust-free operation. In addition, it produces the cleanest surfaces ready to receive new concrete. However, it brings problems of handling and disposing of large quantities of water and wet flying debris. Hydrodemolition can also damage embedded conduits and sometimes results in excess removal of concrete. If the slab is cracked, this demolition method can completely remove parts of the slab between cracks.[33] In addition, water can damage suspended ceilings, lights, and other finishes below.

5.4.3 When to use patching

The answer depends in part on the amount of concrete deterioration that has already occurred. Patching is appropriate when the concentration of aggressive chemicals is small to moderate and when the areas of visible deterioration are relatively few. For example, a concrete floor system can be patched if the average concentration of chloride ions is less than 1.2 lb/yd^3 and spalling caused by corrosion is confined to a few slab or beam soffits (Fig. 5.13). If, on the other hand, most beams and large areas of slabs are badly deteriorated, this repair method is best avoided in favor of removal and replacement. The cost of the latter might be comparable to, or only slightly in excess of, the cost of patching, but the final result would be much better.

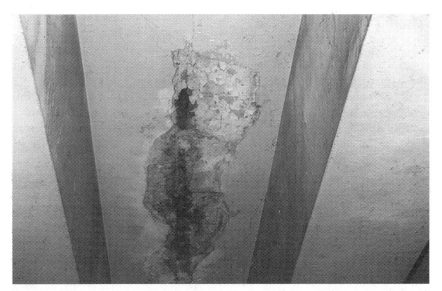

Figure 5.13 This isolated spall can be repaired by patching, provided that chloride levels in the concrete are not excessive.

There are conceptual problems with patching contaminated concrete. Patching does nothing to help the areas adjacent to the spall—usually the areas in advanced stages of corrosion. Worse, if only the part of the structure undergoing serious active corrosion is replaced, a new concrete patch may introduce one more galvanic cell into the concrete and speed up the rate of corrosion in the surrounding areas. An electrochemical imbalance is introduced when a dense patch is used to repair a porous substrate, resulting in less oxygen reaching the bars within the patch than elsewhere. The rebars within the patch begin to act as anodes and corrode. Similarly, if a chloride-free material is used to repair chloride-contaminated concrete, as is commonly done, the bars within the patch act as cathodes and cause the anodic bars in the surrounding concrete to corrode.[34]

5.4.4   The patching method

Here is a discussion of some salient—and sometimes arguable—points of patching.

How much concrete to remove?   Typical project specifications call for removal of "all unsound concrete," recognizing the fact that the actual extent of deterioration can be reliably determined only at the job site. The approximate outline of the deterioration can be established by chain drag or other concrete investigation methods listed in Chap. 2; after that, the boundary is sawcut and the concrete within it chipped out. The contractor would typically remove all concrete that is cracked and that easily crumbles or falls apart under the hammer blows. To prevent localized stresses caused by uneven shrinkage, concrete should be removed in a manner that allows placement of the repair material in a roughly uniform thickness.

All concrete surrounding heavy and moderately rusted reinforcement is also removed until the exposed bars show only light rust. Concrete is usually specified to be removed around all exposed reinforcing bars so that a distance of at least 0.75 or 1.0 in, popularly known as a finger gap, is provided around each bar. This allows the repair material to coat the entire perimeter of the bars. Removal of concrete from behind the bars is admittedly difficult and expensive, as the contractor has to be careful not to damage the existing reinforcement during the work.

Some decry the practice of removing sound concrete from behind the bars as wasteful, saying that rusting often takes place only at the exterior face of the bars, while the back face is frequently rust-free. Others retort that, on the contrary, in many cases severe corrosion occurs at the underside of the top layer of reinforcement.[35] In any event, we view full-perimeter concrete removal as clearly beneficial for rusted reinforcement, because the new concrete provides a full passivating layer around the bars, which the old concrete has lost. It also prevents a new galvanic cell from being formed between the patch and the chloride-contaminated concrete in contact with the same bar. (To be sure, when the repair is performed to fix a spall caused solely by mechanical damage, the surrounding concrete is in sound condition, and the reinforcement is not heavily rusted, completely exposing the bars in this fashion is not necessary.)

What should be done if the concrete is strong but contaminated with chlorides? Many engineers, including this author, specify removal of all concrete with significant chloride contamination. If this concrete is not removed, chlorides in the surrounding concrete will migrate toward the repaired area, where a new galvanic cell will form, inviting a new burst of corrosion. Unfortunately, removing all contaminated concrete can sometimes mean removing the whole structure, and some compromises must be made if that is not acceptable to the owner. As already stated, the onset and rate of corrosion depend on a variety of factors, and there is no established "safe" limit of chloride contamination that can be "forgiven." Engineering judgment weighs in heavily in these circumstances. In some cases when all chloride-contaminated concrete cannot be removed, using a cathodic protection system, described later in this chapter, could arrest further corrosion.

To aid in estimating the cost of the repair effort, the contract documents could include preliminary estimates of quantities of work, determined on the basis of the field investigation. Quite often, the actual quantities will be substantially different. Apart from the legitimate questionable areas, some contractors may be motivated to remove too little concrete if the contract includes a lump-sum payment, or too much if the payment is made per volume removed. Proper inspection and monitoring of the demolition work can help mitigate this problem.

How to prepare the surface? Once the unsound concrete has been removed, the existing surface may need further preparation to provide an appropriate platform for the repair. Surface preparation should include removal of all dust particles and loose material by pressure washing, high-pressure air, or light abrasive blasting.

Whenever deterioration is caused by chemical ingress, it is wise to check the pH level of the prepared concrete surface. If the pH level is below 10, some recommend carefully washing the concrete surface with a weak solution of sodium hydroxide, perhaps repeatedly, to raise the pH level above 11. Conversely, if concrete containing reactive aggregates has been attacked by alkalis, the surface can be washed with water to lower the pH level below 12.[36]

The usual methods of concrete removal—jackhammers, scarifiers, planers—often leave feathered edges. Although some patching materials are theoretically capable of being applied at feathered edges, many others are prone to spalling under concentrated loading, especially when the thin sections are poorly cured and dry out too fast. Therefore, surface preparation should include making square saw cuts at least 1 in deep at the edges of the repair. A good sawcutting practice is not to overcut the concrete at the intersections but to stop short of them, making the transition between the perpendicular cuts with a chipping hammer.

An issue about which there is some controversy is whether the surface of existing concrete should be mechanically roughened. Most engineers believe that roughening the surface improves the bond and increases resistance to sliding. Indeed, the shear-friction provisions of ACI 318[11] provide for an increased coefficient of friction for concrete surfaces that have been intentionally roughened by $1/4$ in or more. However, roughening the existing concrete surface by bush-hammering or chipping is not the same as giving fresh concrete a broom finish. As mentioned previously, these methods of concrete removal tend to bruise concrete.

Schrader[37] argues that a mechanically roughened surface is weakened by microfractures that lower its bond strength, even though these are undetectable by the naked eye. He allows that roughening and cleaning the surface is justified when concrete is "fractured, oil contaminated, severely carbonated, [or] smooth polished," but states that even in these cases the bond strength will be decreased. One method of concrete removal that does not produce microfractures is hydrodemolition, which uses water jets with pressures in excess of 20 ksi. But hydrodemolition is not without problems, as discussed previously, and is best used in large concrete-removal jobs. Also, as Sarvinis[38] points out, hydrodemolition should not be used on post-tensioned concrete, because it tends to force water into the protective sheathing and thus invites future corrosion.

As already noted, chipping followed by sandblasting is probably the most cost-effective compromise for surface preparation. Investigations[39] have confirmed that surfaces prepared that way produce very good bond strengths, ranging between 200 and 300 psi.

An important aspect of surface preparation is wetting the concrete prior to application of cementitious repair materials. Why? A dry substrate tends to absorb some water from the patch and thus weaken it. However, overwetting can also be harmful, because it can dilute the cement paste in the patch where it counts the most—at the interface— and lead to increased shrinkage of the patch material. Schrader suggests that the optimum condition exists when the surface is slightly drier than saturated, a condition that occurs when a surface that had been saturated changes color from dark to light.

A good overview of various methods of surface preparation and some specific recommendations for dealing with various surface contaminants are provided by Holl and O'Connor.[40]

To coat the reinforcing steel or not, and if yes, with what? Engineering opinions differ in this regard. Here are the options.

Epoxy coating. One school of thought favors epoxy coating, at this time the dominant method of protecting reinforcing bars in concrete structures. Under this approach, all exposed rebars are coated with an epoxy primer, forming an impenetrable barrier to any further chemical attack; any added reinforcing is factory-coated with epoxy. In addition to protecting the bars, the epoxy also helps to electrically isolate them and to remove them from the chain of corrosion.

But one of the problems with epoxy coating is that it decreases the bond between the reinforcing bars and the concrete; as a result, ACI 318[11] requires 30 percent longer splice lengths for epoxy-coated bars than for uncoated bars. Also, some experiments suggest that epoxy-coated bars may suffer from *undercutting*—progressive corrosion creeping under the coated area.[35]

If used, epoxy coating should be applied very carefully. The most critical areas to be covered are at the edges of the repair. During this work, care should be taken not to coat the surrounding concrete with epoxy, which would then act as a bond-breaker, unless the coating is specifically formulated to go over both steel and concrete. The epoxy should be allowed to dry before patching.

Zinc-rich primers. Another school advocates using zinc-rich primers. Zinc, acting as a sacrificial anode, can protect both the coated bars and, to some degree, the uncoated bars in the surrounding concrete. The zinc coating must be consumed by corrosion before the steel deteriorates. Unlike an epoxy coat, zinc remains effective even when scratched. In addition to field-applied primers, zinc coating can be introduced by specifying galvanized reinforcing bars conforming to ASTM A767. In theory, using zinc-rich primers makes a lot of sense, but they seem to be underappreciated by the engineers involved in concrete rehabilitation, perhaps because their effectiveness is not fully known.

Emmons and Vaysburd[30] cite several studies on the effectiveness of galvanizing for protection of reinforcing bars in concrete and conclude that its performance has been mixed. They attribute the conflicting results to variability of application, including different coating thicknesses (obviously, the thicker the zinc layer, the better the protection), the condition of the steel substrate, the quality of the concrete protective cover, and the degree of exposure to corrosive chemicals. We hope that future research will shed more light on the effectiveness of zinc-rich primers.

Going bare. The third school prefers leaving the bars bare, arguing that the passivating film provided by the new concrete or mortar encasement offers enough protection. The proponents of this method add that if the bars are zinc-coated or electrically isolated by epoxy, the corrosion will simply accelerate at the interface with the existing concrete. There is some truth to that statement. As already discussed, a sacrificial coating applied on only one part of the reinforcing bar, while the rest is unprotected and surrounded by chlorides, introduces another electric cell in the concrete. The result: Corrosion is facilitated at the closest unprotected material—the edges of the patch. (This adds to our previous argument that all chloride-contaminated concrete should be removed.)

Coating with cement slurry. In the past, yet another solution was popular: coating the bars with cement slurry (sometimes with polymer-modified cement slurry), on the assumption that it provides enough passivating protection. Laboratory evaluations proved, however, that while the coated steel was indeed well protected, corrosion accelerated in the surrounding concrete.[41] As just discussed, such a coating is a highly alkaline element, and introducing it in the middle of the repair zone makes for one more galvanic cell. In that cell, the coated bar acts as a cathode, and the bars in the surrounding concrete act as anodes, subjected to rapid corrosion.

What to specify? At this time, there are unanswered questions about all these approaches. Future development may bring other coating products that can protect the steel reinforcement within the patch, where it is most vulnerable, without causing deterioration elsewhere. For example, the latest variation on the zinc-rich primer theme, zinc-epoxy primer coating, holds some promise. This material consists of zinc particles and an epoxy binder that provides adhesion. There are some preliminary indications that this coating may be able to protect the steel in the patch without causing corrosion in the surrounding areas.[30]

Quality of application is critical. In general, the quality of application is at least as important as, and probably more important than, the choice of a coating. Using rust-inhibiting primers, for example, may lead to spalling if the material is carelessly sprayed on the rebars instead of being brushed on. Spraying is obviously cheaper, but it unintentionally coats not only the bars, but also the surrounding concrete, and acts as a bond breaker. According to Haydon,[42] this type of improper application was responsible for the failure of patches made during an extensive repair effort in a high-rise residential complex in New York City. The patches spalled off within a few years, and a new renovation program was needed.

Whichever coating is applied—if any material is applied at all—it is important to remove all loose and heavy rust from reinforcement. Rust removal is best accomplished by sandblasting or grit blasting rather than by the less expensive, but also less effective, wire brushing. The latter does not adequately remove rust from the back side of the bars and in general does little more than polish the bars.

When to use added reinforcing steel? If the existing reinforcement is corroded, it has lost some of its cross section almost by definition. Rust-removal procedures further diminish the area. Is the remaining section sufficient, or should supplemental steel be added? The answer depends on the circumstances. Are the corroded bars the primary reinforcement, or are some of them temperature and distribution bars, or perhaps even carrying (support) bars? At the bottom of slabs and beams, the carrying bars have the smallest concrete cover and typically are the first victims of corrosion. Fortunately, those bars are the least structurally significant and are needed only during concrete placement. Temperature and distribution bars are often acceptable even with heavily reduced cross-sectional areas.

For structural reinforcement—this includes ties and stirrups—the question is whether the remaining steel area is still adequate for the present loads. (It does not hurt to mention that the thickness measurements must be made *after* rust removal.) One rule of thumb holds that reinforcement should be added when the existing bars have lost more than 20 percent of their area. When the loss of section is under 10 percent, no added steel is needed. When between 10 and 20 percent of the bar section is lost, the answer requires some judgment and analysis.

The added reinforcement can be lap spliced to the existing bars—a solution that obviously is practical only for large-size repairs—or spliced by welding or mechanical couplers. Alternatively, the new bars can be drilled and grouted into the sides of the cut next to the corroded ones, maintaining the same concrete cover.

To use a bonding agent or not? Bonding agents, as the name implies, are intended to improve the bond between the patch and the existing concrete. The most common bonding agent is cement slurry, perhaps with some added polymers or latex, such as styrene, butadiene, and acrylic. Bonding agents of this kind allow for passage of moisture and water potentially containing dissolved chloride or sulfide ions. Accordingly, cement slurry will not protect the patch from these chemicals. Other bonding agents are based on epoxy resins; these form barriers to water.

Repairs with cementitious patching materials rarely need bonding agents, except when the intent is to isolate the patch electrically and chemically from surrounding concrete that contains high concentrations of chlorides. While intuitively appealing, such "isolationism" is fraught with problems. Removing the patch from the electrochemical chain of corrosion may sound like a good idea, but the bonding agents that can make it possible are typically epoxies. Epoxy-based bonding agents provide excellent bonding to the substrate, but enveloping the patch with a water barrier can lead to a buildup of pressure behind the patch, caused by water or ice within the surrounding concrete, and eventually cause spalling. In horizontal exterior or parking-garage applications, water barriers will simply collect water infiltrated from above. If this water freezes, the patch will spall out. Also, epoxies expand and contract with changes in temperature more than concrete, again making spalling a possibility.

Schrader[37] advises against using epoxy-based bonding agents for vertical or unformed sloped shotcreted repairs. In his words, "Fresh epoxy acts like a slippery grease, causing the new material to slough or slide. If the contractor waits until the epoxy hardens to avoid this problem, the epoxy becomes a bond breaker." The window of epoxy effectiveness— when it is fresh enough to develop a good bond but still sufficiently tacky—is impractically short for anything but small patches.

There are times when bonding agents *are* needed. Certain types of polymer patching materials, especially those applied as dry mortars, require special steel primers and bonding agents for proper adhesion, especially for vertical and overhead applications.[36] Also, bonding agents are typically required for repair mixes with very low slump.[37] When used, bonding agents should be applied in accordance with the manufacturer's instructions, some of which require a certain minimum level of pull-off strength, such as 100 psi, in the existing concrete. (A pull-off strength test is described in Chap. 2.) The failure should occur in the parent concrete, not at the interface, if the substrate is to be considered acceptable for bonding.

The main argument for using bonding agents is that this forces the contractor to address the issue of surface preparation, including roughening concrete, cleaning it to remove dust and other contaminants, and wetting it. Indeed, if the surface is not properly prepared, the new work will most likely fail. But a contractor who is reckless or ignorant enough to neglect surface preparation might also apply the cement slurry incorrectly by placing it too thick or even allowing it to dry. Then, what was intended as a bonding agent will instead become a bond breaker, making the situation worse than if it had not been applied at all. Conversely, the slurry might be too fluid, without proper cement content and with an overly high water-cement ratio.

Schrader cautions against using latex emulsions as bonding agents. He argues that, while these milky-white products feel very sticky on one's hands and seem like good bonding agents, they are not. Latex emulsions can actually reduce bond, especially if they are allowed to dry out prior to the patch placement, which could take only a few minutes in hot weather. The effectiveness of latex emulsions can be somewhat improved by mixing them with cement in a slurry. Latex *can* improve bond if it is used as an admixture to produce latex-modified concrete or mortar. Schrader states that bonding agents are not required for shotcrete work. Shotcreting usually requires only air-blowing for surface preparation. If the surface is heavily contaminated, it can be cleaned with air, water, or even sand.

Which repair material to use and how to apply it? The most common repair materials have traditionally been concrete or portland cement–based mortars. Their obvious advantages are low cost, compatibility with the substrate, proven technology, and familiarity to contractors. Depending on the size of the area to be repaired, these cementitious materials can be applied by various methods. For large areas of thin patches, shotcrete—a mixture of aggregate, cement, and water sprayed onto the repaired surface by a pneumatic gun at high velocity—is usually cost-effective. For deeper patches, including full-depth repairs, cast-in-place concrete applied by forming is the most practical. Where shrinkage must be minimized at all costs, concrete with preplaced aggregate may be used. Small and thin repairs are best performed by patching with specially formulated repair mortars. Pressure grouting involves pumping grout through the access ports into an area confined by tight formwork.

Lately, these materials have been augmented and even supplanted by polymer-based mortars, resin-based or magnesium-phosphate repair products, and various prepackaged mixes. Concrete used for repair may now contain silica fume, high-alumina cement, and corrosion inhibitors.

The more these new materials proliferate, the more important the issue of their thermal compatibility with concrete becomes. If the patch has a different coefficient of thermal expansion from the rest of the concrete structure, the two will tend to move differently with changes in temperature. [The two materials that come to mind are epoxy mortars and methylmethacrylate (MMA), both discussed in Sec. 5.8.] As a result, stresses will develop at the interface, and, if the patch is not properly anchored, cracks will follow. With the help of water getting into these cracks and freeze-thaw action, the patch could eventually spall.

Many prepackaged repair materials have much higher shrinkage rates than regular concrete. As Schrader[43] observes, even those products that are sold as low-shrink formulations may have up to twice the rate of shrinkage of conventional concrete. While the typical rate of concrete shrinkage ranges between 350 and 650 millionths of an inch per inch of length (also called microstrain), or 0.035 to 0.065 percent, some prepackaged repair materials have shrinkage rates of 1000 microstrain, or 0.1 percent. At that rate, a 10-ft patch will theoretically shrink 0.12 in, although the actual amount will be less because of the restraint offered by the surrounding concrete, and some of it will be made up by cracking.

The various repair materials and methods for their application, together with their advantages and disadvantages, are further discussed in Sec. 5.8.

Should the repair be cured? As a general rule, the answer to this question is yes, although this point is often forgotten. Cementitious materials in particular should be moist cured just like any other concrete, lest they suffer in the same manner as uncured concrete: by developing shrinkage cracking owing to rapid evaporation, resulting in a porous and weak patch surface. In the worst-case scenario, shrinkage stresses in the repaired section would be large enough to completely debond the patch from the parent concrete. All this can lead to water penetration and saturation of the repair area with corrosive chemicals. Instead of protection, the patch becomes a liability.

A few cement-based repair products are "self-curing." These do not require moist curing, except perhaps in unusually dry conditions, when some brief curing might be needed. Certain rapid-setting and polymer-modified materials also do not need moist curing.[36] As always, whenever proprietary materials are involved, the manufacturer's instructions should be strictly followed.

Is additional protection necessary? Depending on the application, the surface of the repair may need to be protected from water or corrosive chemicals by a sealer, membrane, or floor topping. A good coating applied to the patched surface will keep out moisture and, if pigmented, will mask the inevitable differences in color. Coatings are especially helpful in exterior repairs and for traffic-bearing surfaces, such as floors of parking garages. Coatings are generally unnecessary for soffits of interior slabs and beams that are not exposed to corrosive chemicals.

Concrete in a corrosive environment can benefit from the chemical barrier provided by epoxy, although epoxy is a double-edged sword, as we have seen. Epoxy coating could be particularly well suited for cases in which the repaired framing still contains chlorides. There, the impermeable barrier may help "starve" corrosion by preventing oxygen from getting to the reinforcing steel. A case in point: The repaired structure in Example 5.1 was coated with a 10-mil epoxy-based paint for this exact purpose.[28]

An uncritical use of epoxies, surface sealers, and membranes can lead to problems. If the coating stops both water *and* vapor, it will trap moisture inside the concrete, leading to spalling when the pressure builds up sufficiently. An example of delamination caused by a buildup of water and ice pressure is shown in Fig. 5.14. To avoid this problem, the best coatings—typically acrylic or acrylic-polymer based—are impenetrable to exterior moisture but allow water vapor to escape.

In any case, the protection offered by sealers and coatings is not absolute, since they cannot prevent water penetration through cracks and joints. The edges of the patch, where narrow cracks typically develop owing to shrinkage of the patch material, are especially vulnerable. Some experts recommend routing and sealing the patches with flexible sealants.[39]

5.4.5 The details

The patching method for repairing spalls of limited areas in slabs, walls, and beams is illustrated in Fig. 5.15. Note that the existing partly corroded reinforcement is specified to be augmented by new epoxy-coated rebars when the existing bars have lost more than 20 percent of their area. An additional note cautioning the contractor not to damage the existing reinforcement during concrete removal could also be included in the drawings.

To repair slabs with major concrete delaminations and serious corrosion of reinforcement, the method shown in Fig. 5.16 can be used. In this detail, both the delaminated concrete and the corroded reinforcement are removed and replaced with new formed concrete and epoxy-coated rebars. The major challenge here is to make certain

Figure 5.14 Debonding of concrete coating caused by buildup of water pressure.

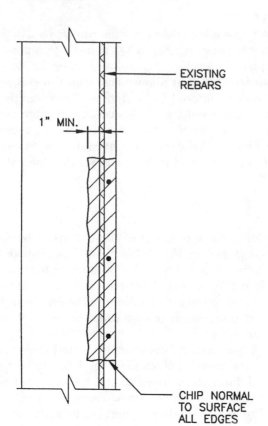

REPAIR PROCEDURE:

1. REMOVE ALL LOOSE CONCRETE TO SOUND SURFACE.

2. SANDBLAST CONCRETE AND REBARS, REMOVE DUST AND DEBRIS.

3. ADD EPOXY–COATED REBARS TO ALL EXISTING REBARS WITH 20% OR MORE LOSS OF SECTIONAL AREA.

4. COAT EXISTING REBARS WITH PROTECTIVE COATING PER SPEC.

5. APPLY CONCRETE BONDING AGENT PER SPEO.

6. IF THE PATCH THICKNESS IS LESS THAN 1 1/2", FILL PATCH WITH REPAIR MORTAR PER SPEC. IN 1/2" LAYERS, OTHERWISE FILL WITH CONCRETE IN FORMS. (WET SURFACE PRIOR TO CONCRETE APPLICATION)

7. FOR UNREINFORCED CONCRETE ADD TIE WIRE ANCHORS (AT LEAST 2 PER PATCH) AND STAINLESS STEEL TIE WIRE.

Figure 5.15 Typical method of repairing spalls of moderate size.

that the new concrete does not debond from the existing. For this reason, U-shaped support dowels are drilled and grouted into the parent concrete.

Another method of anchoring the patch by mechanical means is to use stainless-steel tie wire supported by tie-wire anchors drilled into existing concrete. Yet another design relies on a grid of thin stainless-steel rods held in place by stainless-steel eye bolts grouted into holes drilled into sound concrete.[23] In patches where existing closely spaced reinforcing bars extend into the surrounding concrete, no additional anchorage is usually needed.

5.4.6 Minor patching

Truly minor patching involving repair of surface defects, such as filling in tie holes and honeycombing, follows the same general procedures outlined in the previous section but does not need to be as elaborate. These surface defects can be repaired in a straightforward manner. First, the area of the defect is chipped to a minimum depth of 1 in, and its perimeter is given square edges to avoid feathering. The surface is wetted. If desired, a bonding agent is brushed on; it can be made of one part grout and one part cement, with just enough water for a paintlike consistency. Then a patching mortar is immediately rammed in (dry packed).[44]

The patching mortar may consist of a sand-cement mixture (typically 1 part cement to at least 3 parts sand, although some specifications call for 1 part cement to $2\frac{1}{2}$ parts sand). It can include white cement as needed for color matching; the optimum proportions of white and gray cement are determined by trial installations. To minimize shrinkage, the mortar should have only enough water to be molded into a ball. The ball should neither slump nor crumble when placed on a flat surface. The packed mortar should be left slightly protruding above the adjacent existing surface, left to harden for 1 to 2 h, and screeded flush. Finally, the repair is finished to match the existing texture and cured, as discussed previously.[45]

5.4.7 Patching columns

Our discussion of the patching method was illustrated mostly by repair of slabs and beams. What about columns? Concrete columns with modest-size spalls may be repaired by patching, provided that the concrete is not excessively contaminated by chlorides. Columns with more serious damage should be strengthened rather than repaired.

Why? The main difference between the repairability of flexural members—beams and slabs—and of columns is that beams and slabs can be easily shored or at least cordoned off to remove the live load from the

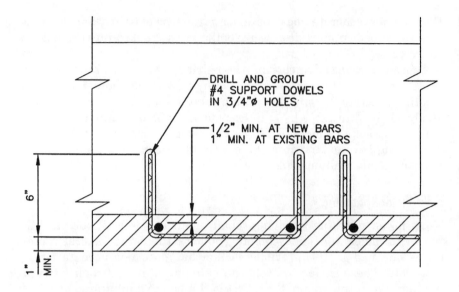

DRILL AND GROUT
#4 SUPPORT DOWELS
IN 3/4"∅ HOLES

1/2" MIN. AT NEW BARS
1" MIN. AT EXISTING BARS

6"

1" MIN.

REPAIR PROCEDURE:

1. SHORE THE WALL OR THE SLAB BEING REPAIRED FOR AT LEAST 14 DAYS AFTER REPAIR.

2. REMOVE ALL LOOSE CONCRETE AND TOTALLY RUSTED REBARS.

3. KEEP ANY REBARS WITH AT LEAST 50% OF THE AREA LEFT. REMOVE RUST AND APPLY PROTECTIVE COATING PER SPEC.

4. PROVIDE NEW EPOXY-COATED REBARS. INSTALL #4 SUPPORT BARS BY DRILLING AND GROUTING.

5. REMOVE ALL LOOSE CONCRETE AND DUST.

6. APPLY BONDING AGENT TO CONCRETE AND REBARS.

7. APPLY PORTLAND-CEMENT CONCRETE IN FORMS.

Figure 5.16   Repair of walls and slabs with major delaminations.

repaired area. Then, these members will carry only their own weight and nothing more. The situation is different for concrete columns, because it is rather difficult to unload from them the weight of all the framing they support. This means that any repair material will not carry any part of the existing dead load, as explained in Chap. 4. Worse, removal of deteriorated concrete around reinforcing will leave the exposed loaded bars unbraced and vulnerable to buckling. The more concrete is removed, the bigger the share of the dead load carried by the bars, and the more precarious their condition becomes. Even if the column ties are preserved during concrete removal, the main column bars can still buckle inward—with disastrous consequences.

For this reason, it is best not to repair concrete columns with major deterioration, but to strengthen them by adding supplemental framing as described in Chap. 4, being careful to isolate the new members from the existing chain of corrosion.

5.5 Cathodic Protection and Electrochemical Chloride Extraction

If corrosion is caused by electric current flowing through the electrolyte, it will be stopped if a current flowing in the opposite direction is introduced. The reactions at the anode and the cathode described in Sec. 5.3.2 are reversible, and a supply of electrons to the anode will bring positively charged ferrous ions back from the electrolyte and restore iron:

$$2e^- + Fe^{++} \rightarrow Fe$$

This is the principle behind the cathodic protection method.[46] Cathodic protection is the only repair method that can stop the corrosion process in chloride-contaminated slabs; however, it cannot reverse existing deterioration. Cathodic protection is best used for structures where corrosion has just begun.

There are two kinds of cathodic-protection systems: *active,* relying on externally supplied electric currents to counteract those generated by corrosion, and *passive,* using the sacrificial action of zinc or other metal to protect steel reinforcement.

Active systems, also known as impressed-current systems, are made up of electrically interconnected wire mats embedded in concrete topping and attached to a source of electricity. The electric current overcomes the corrosion-causing current and converts steel bars from sacrificial anodes into protected cathodes. Active cathodic-protection systems do work, but they suffer from some inherent problems: They are expensive to install, they require continuing maintenance, and they depend on external power supplies. Worse, applying excessive levels of electricity in prestressed structures has been linked to hydrogen embrittlement of steel. These drawbacks limit the use of active systems in buildings to rehabilitation of an occasional parking garage. Cathodic-protection systems of this type are much more common in renovation of bridge decks; the first such systems were installed in California in 1974.[47]

Passive cathodic-protection systems do not require external power supplies. One new product, marketed by a major manufacturer for stopping corrosion of deteriorated concrete balconies, consists of 10-in-wide rolls of zinc metal foil coated with a hydrogel adhesive. The foil is connected to the slab rebars by means of stranded copper wire mechanically tapped into them. The manufacturer recommends making one such connection per 150 ft^2 of the floor area, with a minimum of two connections per unit of work, such as a balcony. The product can be applied to the top of the slab and covered with carpeting, or it can be adhered to the underside of the slab and painted. The system functions akin to galvanizing, except that the anode is postinstalled and does not physically cover the steel. Products of this kind hold a lot of promise for stopping the deterioration of concrete framing without resorting to patching. However, some studies have indicated that such products are most effective in applications that involve periodic wetting of concrete, such as balconies and seawater splash zones. The reason is that dry concrete may not be able to provide a good enough electrolyte for sacrificial zinc action.

Cathodic protection can stop corrosion in chloride-contaminated concrete structures but not necessarily in carbonated concrete, perhaps because of its large electrical resistance. At this time, only conventionally reinforced (not prestressed) structures can be cathodically protected. Manufacturers of some recently developed systems claim 20-year service lives for their products with only minor routine maintenance.[48]

An alternative to cathodic protection is offered by proprietary systems for *electrochemical chloride extraction* and replenishment of alkalis that have been used in Europe, Hong Kong, and the Middle East. Like active cathodic protection, these systems involve external power sources. The purpose of the externally applied electricity is not so much to make anodes and cathodes switch places as to drive chloride ions away from reinforcement and to improve the alkalinity of concrete (i.e., raise its pH) by generating hydroxyl ions near the steel. Like cathodic-protection systems, these methods require attaching a metal mesh to the protected concrete. The mesh is covered with an overlay, typically sprayed cellulose fiber saturated with an alkaline solution, to which chloride ions migrate and are neutralized.[20]

Recently, these methods of electrochemical chloride extraction (ECE) have been successfully tried in the United States. A companion process, realkalization, is intended to stop corrosion caused by carbonation, something cathodic-protection systems cannot yet do. According to Whitmore,[49] the main difference between cathodic-protection and ECE systems is the amount of time they need to accomplish their goals: Cathodic protection is permanent, while ECE systems, with a higher current density, require only 4 to 8 weeks and are removed after that. (Realkalization takes less time—3 to 6 days.[26]) Whitmore also states that the cost of the two installations is about the same, but unlike cathodic protection, ECE requires virtually no maintenance.

5.6   Corrosion Inhibitors

The challenges of protecting concrete from corrosion are not only of interest to engineers. The R&D departments of the major construction-chemical manufacturers are busy developing the elixir of youth for concrete structures that could stop corrosion and prevent further deterioration. If such a product could be invented and proven effective, it would revolutionize the practice of concrete repair.

In recent decades, most major manufacturers have unveiled products intended to inhibit corrosion in newly constructed concrete, and some promote these products for stopping corrosion in existing concrete as well. Without getting into trade names, which are likely to have become obsolete by the time you read this book, let us briefly review the available products in generic terms.

Corrosion inhibitors are reported to form protective barriers and maintain the passivating layer of iron oxide around the steel. They can be classified by the target of their protection as anodic, cathodic, and anodic-cathodic. The most-studied corrosion inhibitors contain *calcium nitrite,* an anodic inhibitor that has proved effective for deterring corrosion in new concrete. Some calcium nitrite products also contain setting retarders, since the chemical has a side effect of acting as an accelerator. Obviously, setting retarders are not needed for existing concrete.

Other corrosion inhibitors contain amines, esters, and other chemicals capable, like calcium nitrite, of migrating through concrete and attaching themselves to reinforcing steel. The resulting coating is reported to prevent chlorides from reacting with the steel. Some of these products are intended to protect both anodes and cathodes in the electrochemical chain.

While the most common way to introduce corrosion inhibitors into new concrete is by using admixtures, for existing concrete (other than the new patch material), this option is obviously precluded. The only practical method of applying them in concrete repair is to coat the concrete surface and hope that the chemicals migrate deep enough to reach the steel, as they are supposed to do. It is also possible to remove the concrete around the bars and coat them directly, but that brings us back to patching. The best time to apply corrosion inhibitors, according to the manufacturers, is at the first signs of corrosion (and even then it could be already too late, as just discussed, because by then rusting could be widespread and serious).

To address the difficulties of bringing corrosion inhibitors to the embedded steel reinforcement, a new product called *migrating corrosion inhibitor* (MCI) has been introduced. MCI is an anodic-cathodic corrosion inhibitor that can reportedly penetrate concrete cover by a variety of methods, including diffusion in pore water, high vapor pressure (allowing MCI to protect the steel even at the surrounding air pockets), and moving through small cracks in concrete. Essentially, this concrete protector is designed to use the same entryways used by the offenders—chlorides and carbonation. MCI can be applied in a variety of ways: either mixed into the patch material, brushed on the surface, or placed into drilled holes.

When specifying corrosion inhibitors, it is important to select the proper product to fit the actual conditions and avoid electrochemical imbalance. Such an imbalance can be introduced when a dense patch is used to repair a porous substrate, or when a chloride-free material is used to repair chloride-contaminated concrete. In dense patches, anodic corrosion inhibitors should be added; the cathodic corrosion inhibitors are better suited for patching chloride-contaminated concrete. A dense patch used in concrete full of chlorides will benefit from a product that offers both anodic and cathodic protection.[34]

How well do corrosion inhibitors work in existing concrete? It is not yet clear. After all, the products are carefully named "corrosion inhibitors" and not "corrosion stoppers." The consensus of three experts interviewed by *Concrete Repair Digest* in 1997 was that too little factual information existed at that time to make a judgment. One expert stated that if sufficient amounts of chlorides and moisture could get to the bars, no corrosion inhibitor would help for long. The main unanswered question was—and still is—whether corrosion inhibitors are effective in arresting the active corrosion of rusted steel.[34]

Emmons and Vaysburd[30] point out that, according to one theory, nitrite and chloride ions engage in competing reactions with steel reinforcement, and that their relative concentrations around the steel bars determine which has the upper hand. Therefore, calcium nitrite is effective only as long as its concentration overwhelms that of chlorides; its protective power is gradually lost as it is consumed in the chemical reaction. There are other unanswered questions about corrosion inhibitors, especially with regard to adding them to repair patches for protection of steel reinforcement. Will they stay in the patch, or will they move elsewhere with pore water? And, again, what will happen to the reinforcing bar if only a part of it is protected and the rest is still embedded in chloride-containing concrete?

5.7   Other Types of Damage to Concrete

5.7.1   Attack by chemicals other than chlorides

Chlorides are said to cause the largest amount of damage to concrete, but there are other chemicals that can invite deterioration. Corrosive chemicals are most harmful in liquid form and therefore tend to attack floors and foundations first, but corrosive vapors can also cause plenty of damage to overhead framing in industrial plants. Unprotected concrete in aggressive environments can deteriorate within a few years.

The most common—and least harmful—manifestation of chemical activity in concrete is *efflorescence,* or leaching of water-soluble calcium hydroxide from concrete. In some instances, such as that shown in Fig. 5.17, heavy efflorescence in concrete framing can look rather alarming, but the problem is usually cosmetic.[50] Still, concrete exposed to continuing leaching and loss of some of its vital components becomes a cause for concern, if only because its strength is reduced and its porosity increased, facilitating further entry of water.

In new concrete, efflorescence can be minimized by reducing permeability and restricting water ingress. Permeability can be lowered by various concrete admixtures, among them water reducers, superplasticizers, and air entrainers, and by pozzolans, such as silica fume. The best results are achieved when these materials are used in combination with good proportioning, placing, and curing. In existing buildings, the only step that can usually be taken to control leaching is to eliminate, or at least reduce, the entry of water by coatings and to channel the water away.

Sulfate attack is a more serious threat to the integrity of concrete framing. Sulfates can be present in groundwater or can be introduced during

Figure 5.17   Heavy efflorescence in concrete framing.

building operations as liquids or vapors. In contrast with chlorides, which corrode steel reinforcement, sulfates damage concrete by chemical reaction with calcium hydroxide and other cement components. Like that of chlorides, however, the action of sulfates results in products that occupy much more volume than the original materials. This expansive chemical reaction can be supplemented by a physical action of sulfate-salt crystal growth, which can occur during frequent wetting and drying cycles.[50]

Either way, the resulting deterioration is debilitating. Figure 5.18 illustrates a disintegrated concrete column subjected to chemical attack, in this case by repeated leakage of battery acid that was routinely poured into a drain pipe embedded into the column. This type of damage is relatively rare. In most cases, sulfate attack occurs in the ground through the sulfates present in the groundwater. It often results in cracking and disintegration of building foundations.

Concrete with severe sulfate deterioration is usually replaced. Foundations and wall bases with only minor damage can be allowed to remain unrepaired but carefully monitored. Even then, some remedial steps can be taken. For example, groundwater can be redirected away from the affected area by pumping and drainage. Another possible remedy is ground improvement by injecting lime or other alkaline materials to lower the acidity level of the soil. Ground improvement may be useful in cases where groundwater is stagnant; it is not effective where it is flowing.

Acid attack is another danger to concrete in industrial plants. Sulfuric, hydrochloric, and nitric acids are especially harmful. Generally, strong acid concentrations and high temperatures increase the severity of the attack. The damage is most pronounced when the acid solution is moving, so that fresh acid is constantly available to attack concrete. Another unfortunate situation arises when the resulting product is soluble and cannot form a protective barrier against further deterioration.

Acids cause deterioration by dissolving cement paste (especially calcium hydroxide) or by chemical reaction with the aggregate.[50] The typical result is a concrete surface that crumbles to the touch and loses sand particles. Acid attack rarely causes structural failure, but, as with sulfate damage, the best remedy for the uncommon cases of severe deterioration is to replace the affected concrete.

The issue of increasing the chemical resistance of existing industrial floors is further discussed in Chap. 6.

5.7.2   Damage from frost and freeze-thaw cycles

Concrete placed in cold weather can suffer frost damage if it is not properly protected. Water freezing in concrete pores or in small

Figure 5.18   Disintegration of concrete subjected to sulfate attack.

pockets exerts internal pressure on concrete and can lead to its partial disintegration. The damaged surfaces appear to be covered with ice crystals that are visible with the naked eye or with the aid of a magnifying glass. Ice crystals are easiest to detect on wet concrete surfaces.

The damage occurs most frequently at the corners, where the concrete surface exposed to freezing is twice that in the flatwork.[51] Most commonly affected is concrete that is not air-entrained; this includes much existing concrete. The concept of adding uniformly distributed tiny air bubbles that can absorb some of the internal expansion caused by freezing and thus reduce the effects of frost damage is relatively recent.

Frost-damaged concrete is rarely a problem that must be fixed during building renovations, unless it is in a critical area such as a beam or elevated slab with borderline adequacy. Also, the presence of ice-crystal marks does not prove that concrete has sustained permanent damage—it only means that its surface was frozen sometime before it attained its full strength. The actual strength of a section with signs of frost damage can be verified by coring and testing.

Unlike frost damage that occurs during concrete curing, freeze-thaw damage can happen to concrete of any age. It starts when water enters small cracks in concrete and freezes there, expanding in the process and enlarging the cracks. During the next freezing cycle, the cracks are expanded still more, and the repeated cycles of freezing and thawing eventually lead to substantial deterioration of the concrete surface. Another problem involves absorptive aggregates, such as sandstone and some cherts, that can be damaged by freezing.

The freeze-thaw resistance of concrete can be improved by air-entraining admixtures and by proper proportioning and curing. Making concrete less permeable by adding pozzolans also helps. Still, if concrete is frequently saturated with water because of its exposure—in waterfront buildings, for example—damage from freeze-thaw cycles is difficult to avoid.

In renovation of existing buildings, any evidence of freeze-thaw damage should be carefully evaluated to assess its severity. More often than not, the affected concrete can be left in service, perhaps with a protective coating applied to stop further entry of water. Sections with significant damage can be restored by patching. In the worst cases, the deteriorated areas are replaced.

5.7.3 Alkali-aggregate reaction

Some minerals are known to react with alkalis in the presence of water, expanding in the process. When these minerals are used as concrete aggregates, they are vulnerable to attack from two alkali metals, sodium and potassium, that are always present in small amounts in cement. The resulting expansion, frequently accompanied by the formation of silica gel, can cause cracking, section disfigurement, and loss of strength. Most problems occur in exterior applications within several years of construction.[52] The alkali-aggregate reaction can manifest itself as alkali-silica reaction (ASR) between alkalis in cement and silica in aggregate, and as alkali-carbonate reaction involving certain carbonate aggregates, such as limestone.

Not every theoretically reactive aggregate will deteriorate: The reaction requires moist conditions, a critical concentration of alkalis, and certain aggregate particle sizes. The total alkali content of concrete is commonly expressed as equivalent sodium oxide content; it typically ranges between 0.4 and 1.6 percent. The aggregate sizes that are reportedly most susceptible to ASR are those from $\frac{1}{16}$ to $\frac{3}{16}$ in (1 to 5 mm).[20] The reaction may be hastened if concrete is attacked by deicing salts containing sodium and potassium ions.

The signs of alkali-aggregate reaction in unreinforced concrete include characteristic three-legged cracks that eventually join into a crazing-type network. Sometimes, a gel can be seen exuding from the cracks. In reinforced concrete, the cracks run parallel to the main bars, which provide some restraint against cracking in the other direction. Thus, in concrete columns, these cracks run longitudinally. Alkali-aggregate reaction inside concrete sections can be detected by petro-graphic examination of cores. The reaction rarely causes a major loss of strength in properly reinforced concrete structures, but the result-ing cracking and dimensional instability not only is unsightly, but—like any cracking—also invites water entry and facilitates corrosion of reinforcement. In severe cases, chunks of concrete can spall off and become falling hazards.[53]

The only reliable way to stop advanced alkali-aggregate reaction that has already caused severe deterioration is to replace the damaged concrete. In less critical cases, it may be possible to stop the reaction by cutting off the source of moisture—for example, by placing an impermeable overlay or a membrane. If one area is affected, it is important to determine whether other locations that do not yet exhib-it signs of alkali-aggregate reaction are in danger. If they are, a long-term testing program should be instituted to keep a close watch on the rest of the structure.

Alkali-aggregate reaction does not occur very often, and other possi-ble causes of cracking should be investigated before this diagnosis is offered.[25] Also, the reaction may eventually cease when the reactive ingredients are consumed, and the uncovered damage may have already stabilized.[20] Researchers who have studied the effects of sim-ulated alkali-silica reaction on the structural behavior of reinforced-concrete beams have found that, despite visible cracking and reduced compressive strength, the flexural capacity of the beams affected by the reaction was nearly the same as that of the control beams.[54]

5.7.4   Delayed ettringite formation

This type of damage to reinforced concrete was diagnosed and has become a cause for concern relatively recently. Ettringite is another name for calcium sulfoaluminate hydrate, a product of early hydration of concrete that results from calcium aluminate in cement reacting with sulfate. For some not entirely clear reasons, possibly including inadequate curing conditions, excessive heat, cement with too much clinker sulfate, and certain types of aggregate, ettringite production is occasionally delayed until after concrete has hardened.

When exposed to moisture, the material formed during the delayed ettringite formation may expand and crack, breaking the concrete from the inside. Damage caused by delayed ettringite formation has been reported relatively infrequently, either because the problem is rare or because it is routinely misdiagnosed as an alkali-silica reaction. The results of this chemical reaction indeed resemble those of a sulfate attack or an alkali-silica reaction—small, severe map cracking. Modern tools of petrographic analysis, including the scanning electron microscope with attachments allowing identification of atoms in crystals, now allow reliable identification of whether delayed ettringite formation is present.[55]

5.8   Materials for Concrete Repair

5.8.1   An overview

The advantages and disadvantages of some patching materials were briefly mentioned in Sec. 5.4.4. Here, we take a broader look at the options available to the engineer.

The multitude of available repair materials is staggering. Concrete used for repair may contain silica fume, high-alumina cement, magnesium phosphate, and corrosion inhibitors. Repair mortars can be polymer- or resin-based and come in a variety of prepackaged mixes. Sulfur concrete is used in some critical applications calling for extreme corrosion resistance.

Technical magazines are full of studies debating the relative advantages of various concoctions produced by the manufacturers. It seems at times that designers try to prove their sophistication by specifying the latest wonders. The majority of the new materials are indeed beneficial in some specific circumstances. But sometimes their claimed advantages are only marginally helpful; at other times, these exotic materials can do more harm than good because of their side effects.

A typical application requires the repair material to be at least as strong as the existing concrete—but not significantly stronger—and the repair to be durable under various loading and environmental conditions. The strength criterion is easily satisfied, because most formulations are designed with that in mind. The durability is more difficult to predict: It depends on many variables, such as the area and depth of the repair, the permeability of the repair material, and methods of surface preparation and curing. The degree of exposure to aggressive chemicals also matters a lot.

The characteristics that seem to be most important in determining how well a repair material performs are the amount of shrinkage after curing and the coefficient of thermal expansion. If one or both of these are excessively large, the repair may debond—fail by delamination—because of large stresses that develop at the interface with the existing concrete. Also, cracks may develop around the edges of the repair and allow water penetration. After a few freeze-thaw cycles, spalling can occur.

Repair materials fall in one of two categories: those based on portland cement and those based on synthetic resins. The former make use of the passivating properties provided by cement; the latter rely on erecting a protective barrier against oxygen and moisture.

5.8.2 Portland cement–based materials

Portland cement concrete is still the most common, best understood, least expensive, and perhaps most reliable repair material. Portland cement concrete may contain various admixtures for improved workability in various circumstances or for better durability and corrosion resistance. Among its main advantages is a theoretical thermal compatibility with existing concrete.

In practice, the shrinkage and thermal expansion characteristics of concrete vary, depending on the mix design, water-cement ratio, and aggregate gradation. The greater the cement content, the more thermal movement and shrinkage can be expected. Even if the proportions of the repair material are identical to those of the original concrete, if the aggregate was from a different source, the repair material may behave differently. For this reason, Schrader[43] recommends using, if possible, the exact same aggregate for repairs as was used in the original structure. This is rarely feasible, of course, in renovation of old buildings, which generally contain concrete of unknown origin. A more practical way to control shrinkage is to use shrinkage-compensating (Type K) cement or some of the newly developed shrinkage-reducing admixtures.

As the name implies, shrinkage-reducing admixtures (SRA) are intended to reduce drying shrinkage in concrete—and, presumably, the cracking associated with it. These admixtures work by reducing the surface tension of the water in concrete pores and thus reducing the forces pulling cement particles together (the mechanism of drying shrinkage is briefly explained in Sec. 5.2.2). A good overview of SRAs is given by Holland.[56] He notes that test results show SRAs to be effective in reducing shrinkage, but urges caution in interpreting the manufacturers' test results and observes that SRAs' effectiveness depends on concrete composition. He also brings up some areas of concern: The admixtures tend to reduce the compressive strength of concrete (by up to 15 percent, according to some data), there may be some limitations on using them in freeze-thaw environments (they shouldn't be used there, according to some suppliers), there are specific requirements for air-entrainment content (7 percent, according to some manufacturers), and they have a very high cost. Also, SRAs are not intended to combat thermal cracking, which is often misidentified as drying-shrinkage cracking and is probably as prevalent.

Plain-vanilla portland cement concrete is not appropriate in all circumstances. In particular, it may not provide a good enough barrier against aggressive chemicals. To mitigate this problem, proper curing and keeping the water-cement ratio low will help produce a dense and uncracked concrete surface by reducing the harmful effects of drying shrinkage. Lowering the permeability of the concrete may also strengthen its defenses against entry of chemicals.

ACI 318 contains recommendations for the maximum water–cementitious materials (w/c) ratio and the minimum 28-day compressive strength (f'_c) for special exposure conditions. To help achieve low permeability, a w/c ratio of 0.5 and f'_c of 4000 psi are suggested; for concrete exposed to freezing and thawing in moist conditions or to deicing chemicals, the w/c ratio is further reduced to 0.45 and f'_c increased to 4500 psi. For the most severe cases, where reinforcement must be protected from corrosion caused by chlorides in deicing salts, seawater, or brackish water, the w/c ratio is limited to 0.4 and the required concrete strength is increased to 5000 psi.

Another disadvantage of regular portland cement concrete is its relatively slow setting time. It is rarely necessary to wait the full 28 days for the theoretical strength to be reached before stripping concrete forms or being able to place the building in service. Still, several days need to elapse for concrete to reach even 50 percent of its design strength. For example, it usually takes three days for a mix with Type I portland cement and a w/c ratio of 0.4 to gain a compressive strength of 2000 psi at 70°F.[57] If the owner cannot wait that long, concrete with high-early (Type III) cement can shorten the curing time to perhaps 4 to 6 h. If even that is too long a time, a prepackaged rapid-set cementitious repair material that cures in as little as 2 h may provide the answer.

5.8.3   Placement of cementitious materials

Depending on the size of the repaired area, cementitious materials can be placed as cast-in-place concrete, shotcrete, concrete with preplaced aggregate, repair mortar, and grout applied under pressure. These methods of application are briefly described.

Shotcrete, also known as gunite and pneumatically applied concrete, involves spraying a mixture of aggregate, cement, and water onto the repaired surface, typically with high-velocity pneumatic guns. There are "dry-mix" and "wet-mix" techniques of application, both of which are capable of producing material with a compressive strength of 4000 psi.

In an older and more established dry-mix process, premixed aggregate (mostly sand) and cement are carried by compressed air to the end of the hose into the nozzle, where water is added from a separate reservoir. The mixture is then sprayed onto the repaired surface. One advantage of this method is that the water can be precisely adjusted at the nozzle and its usage minimized. As a result, dry-mix shotcreting normally produces mixtures with low water-cement ratios and low shrinkage. Applied under high pressures, these mixtures are dense and adhere well to the surface. Since the material is delivered by a hose, there is some flexibility in overcoming access restrictions.

In a relatively new wet-mix process, introduced in the 1950s, water is mixed with aggregate and cement at the outset, prior to spraying. The resulting paste is more difficult to move through a hose than a dry mix and requires a higher water content for workability. The result: In the past, wet-mix shotcrete used to shrink more than dry-mix, did not bond to the substrate as well, and tended to sag. For these reasons, the wet-mix process was rarely used for renovation work. However, wet-mix technology has been much improved and its disadvantages largely overcome. Being a less expensive alternative, wet-mix shotcrete is now dominant.

One problem with shotcrete is rebound—the material bouncing off the target area. According to ACI 506,[58] shotcrete placed on sloping and vertical walls by the dry-mix process is likely to lose 15 to 25 percent of its mass to rebound, with as much as 25 to 50 percent lost on overhead surfaces. The rebound losses are smaller in wet-mix shotcrete: 5 to 10 percent and 10 to 20 percent, respectively. Because coarse aggregate particles rebound more easily, the material that is left contains a lot of fine aggregate and cement—a richer mix than intended. When the exact composition of the repair is important, the mix can be adjusted to compensate for rebound, and the placed material can be tested.

Cast-in-place concrete is the most practical repair method for relatively deep patches, including full-depth repairs. It is less suitable for shallow repairs because of the coarse aggregate content. Cast-in-place concrete generally requires a minimum thickness of 1.5 in; repair mortars are preferred for smaller thicknesses.

Repair mortars are often used for thin patches of relatively small size. Horizontal and some vertical patches can be made with basic portland cement mortars, but these materials are at a disadvantage in overhead applications, where the weight of the mortar tends to pull it away from the substrate. The maximum single-application thickness for portland cement mortars is about 0.5 in; for repairs thicker than that, the work must stop while the mortar cures. When the work resumes, the hardened mortar has to be roughened to assure a bond with the next layer—an additional step in the process. A prepackaged rapid-setting repair material would speed up the process in such circumstances.

Concrete with preplaced aggregate is useful for reducing shrinkage and in structures with congested reinforcement. As the name implies, coarse aggregate is first placed in the forms, then grout consisting of fine sand, cement, and expansive admixtures is pumped into the voids. The main advantage of this method is negligible shrinkage, because the pieces of preplaced coarse aggregate are already touching each other and cannot move as the concrete sets. Still, grout can shrink away from the aggregate, resulting in cracking. Expansive admixtures are normally added to prevent this from happening. Specific information about proportioning and gradation of the aggregate can be found in ACI 304.1R.[59]

Pressure grouting involves pumping grout through access ports into an area confined by tight formwork, where it is fully cured prior to removal of the formwork. Pressure grouting is typically specified for vertical and overhead surfaces.[33] Regular cementitious grout is prone to shrinkage, which can cause delamination and failure of the repair, so it is best to use nonshrink grout or shrinkage-compensating (Type K) cement. There are also some fast-setting pumpable concrete repair materials on the market, such as the microconcrete RENDEROC LA by Fosroc Inc.

5.8.4 Modified portland cement materials

When reduced permeability of concrete is desired, adding microsilica, fly ash, or other *pozzolans* can help. These admixtures fill the pores in concrete and help make it stronger, denser, less permeable, and more cohesive. Pozzolans can replace some of the cement in the mix, but their dosage must be carefully controlled when large repair areas are involved. Concrete with high concentrations of these materials tends to suffer from increased plastic shrinkage cracking and requires careful curing. Also, pozzolans affect concrete workability by reducing bleed water and sometimes confuse concrete finishers, so that they start the finishing operations too early.

Polymer-modified concrete is an excellent and popular repair material. Its most familiar version is *latex-modified concrete,* frequently used for slab overlays. Other polymers include acrylic and polyvinyl acetate (PVA). Unlike pozzolans, polymers not only reduce the permeability of the repair concrete, but also increase its bond strength with the substrate. Polymer-modified concrete is typically stronger, more durable, and better at resisting freeze-thaw cycles than plain concrete.

In addition, it has a reasonably good chemical resistance to alkalis and diluted acids—better than that of plain concrete but worse than that of epoxy-based materials. And finally, it is easy to apply and is less expensive than many other repair options.[4]

Acrylic-modified concrete can be very effective in repair of vertical members. Marusin[60] describes a project involving the repair of concrete columns, spandrels, and balconies on a high-rise building in Chicago. After evaluating and testing many repair products, the design team had chosen acrylic-modified concrete for large-size repairs of columns.

Still, polymer-modified concrete is not without warts. Emmons[13] observes, quoting Plum,[61] that one reported side effect of using latex additives is a dramatic increase in flexural creep under conditions of high humidity. Another problem is that latex modifiers tend to reduce the modulus of elasticity of the repair concrete. Emmons also cautions against applying latex-modified concrete by means of dry-mix shotcreting, because this might turn the latex into a bond breaker. And finally, repairs with latex concrete require especially prompt finishing and curing.

5.8.5 Prepackaged polymer-based repair mortars

When used for patching, prepackaged polymer-based repair mortars frequently produce very good results. These materials are typically applied as trowel-on mortars, although some can be placed as flowable grouts or, conversely, earth-dry.[36] Apart from the convenience of using prepackaged material, one of their main advantages is rapid curing, as just mentioned. For patches up to 1 in thick, the materials can be used as packaged (neat); for thicker patches, they can be extended with pea-gravel aggregate.

There is a price to pay for this convenience and versatility. Many of these mixes have much higher shrinkage rates than regular concrete, and high shrinkage can lead to debonding, cracking, and rebar corrosion. In an attempt to deal with the shrinkage problem, some formulations contain large amounts of expansive ingredients, namely alkalis and aluminates. While helping to reduce shrinkage, these alkalis and aluminates make repair mortars more susceptible to reactive aggregates (the alkali-silica reaction) and to sulfate attack.[62]

Some prepackaged polymer-based repair mortars produce excessive heat during hardening. A material that generates a lot of heat while it sets is said to have a high *exotherm*. Repair materials with high exotherms include those containing magnesium phosphate and methylmethacrylate, as well as the epoxy-based mortars. The coefficients of thermal expansion of these mortars tend to increase dramatically at the elevated temperatures caused by curing. As a result, these mortars tend to undergo especially large movements during setting; this can lead to cracking and debonding later. The problem is exacerbated when the high-exotherm materials are placed at elevated ambient temperatures and have to cool all the way down to normal levels. Therefore, it is best to avoid using high-exotherm prepackaged polymer-based repair mortars in exterior applications and in unheated buildings.

Many prepackaged repair mortars have extremely high compressive strengths, sometimes exceeding 10,000 psi. Why would such a strong material be needed to repair a 3000-psi concrete? The high compressive strength is an unintentional product of trying to simultaneously keep the mortar workable and achieve low rates of drying shrinkage and permeability in a formulation without any coarse aggregate.

5.8.6   Epoxy mortars

When strength and chemical resistance are important, epoxy mortars can provide an alternative to portland cement–based materials. The compressive, tensile, and bond strengths of the epoxies used in structural repairs exceed those of a typical portland cement concrete. In addition, epoxies can be applied at below-freezing temperatures. Because of these desirable properties, the phrase "Let's fix it with epoxy" is heard far too often, and some people assume that this two-part formula can fix anything, anywhere. Actually, the term *epoxy* is rather broad, and numerous materials with quite different properties can bear the name. There are some shortcomings to all of them.

First of all, epoxies not only suffer from high exotherms, but also have up to five times higher coefficients of thermal expansion than concrete. This is undesirable from the standpoint of dimensional compatibility.[13] The differences in thermal expansion between epoxy and concrete can be somewhat lessened by extending the epoxy with aggregate. The differential movement can be further mitigated by the epoxy's high elasticity and ability to stretch without cracking.

There are plenty of other limitations of epoxies in general and of epoxy mortars in particular. The structural properties of epoxies tend to weaken at elevated temperatures. When the temperature rises above a certain point—the so-called glass transition temperature, typically between 70 and 170°F—the strength, stiffness, and creep resistance of epoxy rapidly decrease. It is therefore critical to properly fireproof epoxy-repaired members if the building is to survive even a modest fire. Because epoxy mortars rely on erecting a dense protective barrier against oxygen and moisture, good compaction is critical for their successful performance.

Many epoxies creep under sustained loading, as they do at high temperatures, and Warner[50] recommends that epoxy-repaired members not be stressed above 15 percent of their ultimate strength. He also points out that:

- Epoxy material typically does not bond with concrete if the surface is wet or even moist.

- Its pot life is short, from 5 to 35 minutes.

- Its modulus of elasticity is only about one-tenth that of concrete and decreases further as the temperature increases.

- Thorough mixing is needed to allow every particle of the hardener to find its epoxy counterpart.

- The surface preparation requirements for epoxy application are especially restrictive.

Despite their shortcomings, the epoxies have their well-deserved place in many circumstances where speed of repair, strength, and chemical resistance are desired. Properly extended epoxies used in shallow patches of moderate size provide extremely strong and chemically resistant repairs. The impermeable epoxies are indispensable in cases where an abnormally small concrete cover must be maintained. In the already mentioned high-rise repair project described by Marusin,[60] low-modulus, high-viscosity epoxy mortar was chosen over many other materials to be hand-applied for patches with a maximum depth of 2 in. Still, given their poor thermal and deformational compatibility with concrete and their poor fire resistance, epoxy mortars are best used in small-scale repairs.

5.8.7 Other materials

We just said that mixes containing *magnesium phosphate cements* had high exotherm rates, a potential problem. But there are times when they should be used, such as when repairs must be conducted at low temperatures and when fast setting and rapid strength development are needed. Magnesium phosphate cements usually have good chemical resistance and can be made to withstand extremely high temperatures, up to $3632°F$ ($2000°C$).[36] Mixes containing these materials require special attention during construction; they should not be cured by moisture, for example. Also, they should be extended only with noncalcareous aggregates (typically with hard rock) to avoid harmful chemical reactions that can weaken the bond. For the same reason, all carbonated areas should be removed from existing concrete prior to placing patches made with magnesium phosphate cement.[62]

Another repair material worthy of note is *methylmethacrylate* (MMA) concrete. This acrylic resin–based repair material is prized for its extremely low permeability, good bond to concrete, excellent freeze-thaw resistance, and fast curing time. (In fact, it is used by dentists for quick-setting temporary crowns.) MMA concrete can develop a compressive strength of 5000 psi within a couple hours after placement. One study[63] that compared the effectiveness of various coating materials and overlays in blocking water entry found that an overlay with methylmethacrylate polymer concrete, while the most expensive treatment, provided the best results. It has a high exotherm, however, so that its thermal (as well as deformational) properties are different from those of concrete. Also, its fire resistance is poor. For these reasons, the use of MMA is largely confined to small patches.

Table 5.1 summarizes the properties of various materials used for repair and overlay applications.

Not included in the table is *sulfur concrete,* which many still consider an experimental material, although it was first studied more than a century ago. Sulfur concrete is made by mixing heated aggregates with molten sulfur. The reported advantages of this material include unmatched chemical and corrosion resistance, excellent durability, fast setting time, and low permeability. In addition, it can be placed in very cold or hot weather,[64] although it should be protected from high temperatures [in the range of 260 to 300°F (127 to 149°C)].[36] The sulfur-concrete floor installations of the 1970s performed satisfactorily in corrosive environments.[65]

Being less frequently used, sulfur concrete obviously costs more than regular portland cement concrete. Furthermore, sulfur concrete requires specialized equipment for manufacture and placement. But, as Crick and Whitmore[66] argue, because of its exceptional chemical resistance, the cost of sulfur concrete should be compared to that of regular concrete with a protection system, in which case the costs are comparable. As the field experience with sulfur concrete continues to accumulate, and its suppliers increase in number, the popularity of this intriguing material will probably increase.

Another material not included in the table is concrete with *high-alumina cement.* It is relatively expensive, but it offers much better chemical resistance to acids and sulfates than regular portland cement concrete. Also, it sets faster and gains strength earlier. It can be used at high temperatures—up to 572°F (300°C), or, with special refractory aggregate, up to 3272°F (1800°C).[36] On the flip side, concrete with high-alumina cement requires careful proportioning, a low water-cement ratio, and careful curing. According to Beckmann,[53] this material was used in England in the period between the two world wars to improve the sulfate resistance of concrete, but the resulting problems with strength loss have eclipsed the benefits. Neville[67] details the durability problems of high-alumina cement concrete and recommends caution in specifying it.

5.8.8 Choosing the best repair material

Space constraints preclude an expanded discussion of each repair material. A good source of additional information is ACI 546R-96, *Concrete Repair Guide.*[68] Another useful publication is the *Guide for Selecting and Specifying Materials for Repair of Concrete Surfaces* by the International Concrete Repair Institute (ICRI).[69] It lists many important properties of various repair materials, including the modulus of elasticity, creep, and rate of drying shrinkage of each.

A concise tabulation of the physical properties of various prepackaged rapid-set cementitious repair materials available on the market can be found in McGovern's article "A Guide to Rapid-Hardening Concrete."[62] The article lists the trade names and the manufacturers of various prepackaged rapid-set products, along with each material's working life, compressive strength, and rate of drying shrinkage at 28 days. The commentary emphasizes, as we did, the importance of selecting repair products with minimal rates of drying shrinkage.

So is there one "best" material for all concrete repairs? Unfortunately, no—at least, not yet. Various materials can be used with success in various applications, but no single property, formula, or test result will provide the magic bullet. A major 5-year study sponsored by the U.S. Army Corps of Engineers compared the results of laboratory testing with actual field performance of 12 various repair materials. The materials included 11 commercial prepackaged repair mortars and a regular 4000-psi concrete mix.[70] The trade names of the materials and a summary of the rankings are given in Ref. 71.

The conclusions of the study? The researchers did not find a significant correlation between material properties and field performance. However, they identified some properties, which we have already discussed, as important for good performance of repairs. Among them is tensile strength—the materials with higher tensile strength performed better. Other critical properties included modulus of elasticity, thermal expansion coefficient, and unrestrained drying shrinkage; the lower all of these were, the better the performance observed. It is these observations of the properties influencing the crack resistance of various materials, rather than the inconclusive rankings, that make the study important.

5.9 Durability of Repairs

Perhaps the most important factor in making a successful concrete repair is an accurate assessment of the underlying problems. Without

TABLE 5.1   Properties of Repair and Overlay Materials

Materials	Ingredients			Application requirements		
	Binder	Additive	Admixture	Thickness limitation, in/cm	Installation temperature, °F/°C	Curing
Portland cement mortar	Portland cement		Water reducer. Air-entr.	$\dfrac{1.5-4}{3.8-10}$	$\dfrac{40-90}{5-32}$	Wet 7 days
Portland cement concrete	Portland cement		Water reducer. Air-entr.	$\dfrac{>1.75}{4.4}$	$\dfrac{40-90}{5-32}$	Wet 7 days
Microsilica modified portland cement concrete	Portland cement	Microsilica	HRWR Air-entr.	$\dfrac{>1.25}{3.0}$	$\dfrac{40-90}{5-32}$	Wet 7 days
Latex modified portland cement concrete	Portland cement		Latex SBR	$\dfrac{>1.25}{3.0}$	$\dfrac{45-95}{7-35}$	Wet 3 days
Polymer modified portland cement mortar with non-sag filler	Portland cement	Non-sag fillers	Acrylic latex	$\dfrac{0.25-1.5}{0.6-3.8}$		
Magnesium phosphate cement concrete[a]	Magnesium phosphate cement			$\dfrac{>0.75}{1.9}$	$\dfrac{50-100}{10-40}$	Sheet 45 min– 2 days
Preplaced-aggregate concrete[b]	Portland cement	Pozzolans	Fluidifier	$\dfrac{>3.0}{7.6}$	$\dfrac{40-90}{5-32}$	Wet 7 days
Epoxy mortar[c]	Epoxy resin			$\dfrac{0.13-0.38}{0.4-1.2}$	$\dfrac{50-90}{10-32}$	4 h–2 days
Methylmethacrylate (MMA) concrete	Acrylic resin			$\dfrac{0.25-0.50}{0.6-1.3}$	$\dfrac{20-120}{-6-50}$	1–6 h
Shotcrete[d]	Portland cement	Pozzolans	Water reducer. acceler. latex	$\dfrac{>0.5}{1.3}$	$\dfrac{40-90}{5-32}$	Wet 7 days

[a] ACI 304R-23
[b] ACI 503.4
[c] Vapor may cause problems in confined area
[d] ACI 506R-90

Note. The material properties shown in this table vary from manufacturer to manufacturer and are shown for comparison purposes only.

SOURCE: Peter H. Emmons, *Concrete Repair and Maintenance Illustrated,* copyright R.S. Means Co., Inc., Kingston, Mass., (781) 585-7880, all rights reserved, 1993.

		Material properties								
Drying shrinkage*	Coefficient of thermal expansion	Compressive strength,				Elastic modulus, psi/MPa	Permeability (concrete =10)	Freeze-thaw resistance	Non-sag quality	Exo-therm
		1 h	1 day	3 days	28 days					
Moderate	Equal to substrate	0	$\frac{650}{4.5}$	$\frac{2500}{17.2}$	$\frac{5000}{34.5}$	$\frac{3.4 \times 10^6}{2.3 \times 10^4}$	9	Good	Moderate	Low
Low	Equal to substrate	0	$\frac{650}{4.5}$	$\frac{2500}{17.2}$	$\frac{5000}{34.5}$	$\frac{3.8 \times 10^6}{2.6 \times 10^4}$	9	Good	N/A	Low
Low	Equal to substrate	0	$\frac{3000}{20.7}$	$\frac{4000}{27.6}$	$\frac{7500}{51.7}$	$\frac{4 \times 10^6}{2.8 \times 10^4}$	6	Good	Good	Low
Low	Compatible w/sub-strate				$\frac{6000}{41.4}$	$\frac{2.5 \times 10^6}{1.7 \times 10^4}$	5	Excellent	N/A	Low
Moderate	Compatible w/sub-strate		$\frac{1500}{10.3}$		$\frac{5000}{34.5}$	$\frac{2.5 \times 10^6}{1.7 \times 10^4}$	5	Excellent	Excellent	Moderate
Moderate	Equal to substrate	$\frac{2000}{13.8}$	$\frac{6400}{44.1}$	$\frac{7000}{48.3}$	$\frac{8400}{57.9}$	$\frac{3.2 \times 10^6}{2.2 \times 10^4}$	9	Excellent	Low	High
Very low	Equal to substrate	0	$\frac{500}{3.4}$	$\frac{2250}{15.5}$	$\frac{4500}{31}$	$\frac{3.8 \times 10^6}{2.6 \times 10^4}$	10	Good	N/A	Low
Low	(1.5–5) × concrete			$\frac{12{,}000}{82.7}$		$\frac{2.2 \times 10^6}{1.5 \times 10^4}$	1	Excellent	Moderate	High
Moderate	(1.5–5) × concrete	$\frac{4000}{27.6}$	$\frac{12{,}000}{82.7}$		$\frac{12{,}000}{82.7}$	$\frac{3 \times 10^6}{2 \times 10^4}$	1	Excellent	N/A	High
Moderate	Equal to substrate	0	$\frac{800}{5.5}$	$\frac{3500}{24.1}$	$\frac{5000}{34.5}$	$\frac{3.8 \times 10^6}{2.6 \times 10^4}$	6	Good	N/A	Low

Note: *Drying shrinkage: Low, <0.05%; moderate, 0.05%–0.10%; high, >0.10%.

a good understanding of the reasons for ongoing deterioration, repairs can amount to a meaningless waste of money. For example, it is futile to fill cracks that leach efflorescence to the underside of slabs without investigating first whether any corrosion of reinforcement is occurring.

Even the most successful repairs will not bring the repaired structure to as-new condition. At best, the work will only extend the structure's useful life. The commonest of repairs, patching, may last 20 years or more—provided that the concrete is not heavily contaminated with chlorides. If it is, the repair may last only a few years. (The reasons for this were already stated; for further discussion, the reader is referred to Gu et al.[72]) In some cases, deteriorated concrete may be quite difficult to patch effectively. For instance, as O'Connor et al.[73] point out, most concrete in structures built prior to 1940 is not air-entrained, but present practice requires repair concrete in areas exposed to freezing and thawing to be air-entrained. This circumstance limits the repair options, because the air-entrained patches can cause damage to the surrounding areas.

Paul[74] suggests that a good first step in assuring the success of large-scale concrete repairs is to construct a repair mock-up. In this trial installation, the repair compounds can be checked for compatibility and the whole process fine-tuned and improved. Any required testing (of bond strength, for example) can also be performed at this stage. The mock-up can also serve as a demonstration of the actual surface preparation techniques, methods of material application, and acceptable final appearance.

One way to verify the adequacy of repairs is by load testing. Unfortunately, most repaired structures are not tested at all, or at best are "tested" only analytically, on paper. This is understandable; the traditional load testing described in Chap. 2 is expensive, is potentially dangerous, and often requires protective shoring. Recently, new in situ testing technologies imported from Europe have become available in the United States. One of these methods uses hydraulic load applicators, instrumentation for measuring deflections and strains, and an electronic data-acquisition system. The testing is performed only on the repaired member, not on the whole floor. The system is capable of applying a wide range of loading, from a few hundred to several million pounds.[75]

A good concrete repair project should include provisions for periodic inspection by the owner or the owner's consultants to assess the effectiveness and longevity of the repairs over a period of time.

5.10   Systematic Maintenance Programs

We complete the discussion in this chapter by emphasizing the importance of proper maintenance for the durability of concrete in building structures. What sort of maintenance is required? Neville[27] differentiates between *corrective maintenance*—fixing obviously deficient items that are in a condition below an acceptable standard—and *preventive maintenance,* which is "to intervene in the life cycle of each item immediately before it can be expected to fail." Essentially, preventive maintenance restores the structure to its proper level of performance before any failure is evident, while corrective maintenance occurs after the signs of distress are obvious. Both types of maintenance may include the concrete repairs discussed in this chapter or the strengthening described in Chap. 4. Neville observes that the repaired elements should not be expected to be as good as new; they should only meet the acceptable standard.

To be effective, a maintenance program must be systematic and conducted at periodic intervals. An integral part of any maintenance program is structural inspection, a visual examination searching for signs of trouble, as described in Chap. 2. Some government agencies in the United States have comprehensive programs for regularly inspecting their buildings and budgeting for the necessary repair and maintenance chores in a rational manner. The owners of critical facilities—nuclear power plants, for example—also have well-developed programs for upkeep. Some important structures in Europe are monitored by custom-designed corrosion sensors,[19] and systems for monitoring corrosion in concrete structures are becoming increasingly available.[76]

As with cars, it is usually less expensive to perform proper preventive maintenance than to repair something when it breaks down. The further the structure has deteriorated, the more expensive the repairs, and the costs skyrocket nonlinearly. Steen Rostam[19] quotes de Sitter's Law of Fives, which holds that $1 spent in design and construction on concrete durability issues is as effective as $5 spent during the initiation stage of corrosion, $25 spent in the early stages of deterioration, and $125 spent when serious damage has occurred.

We would add that at some point, repairs become uneconomical, and removal and replacement is the only practical option left. Unfortunately, all too often the building owners wait until some signs of failure are evident before calling in the engineers, who at that point can recommend only corrective maintenance. For many buildings, preventive maintenance is used on only two occasions: when the building is purchased and when a major renovation is undertaken.

It goes without saying that an inspection conducted by an inexperienced person who does not know what the signs of trouble look like—which crack is structural and which cosmetic, for example—does more harm than good, either because the telltale signs of distress are missed, or because unnecessary work is prescribed.

References

1. ACI 201.1R-92, *Guide for Making a Condition Survey of Concrete in Service,* American Concrete Institute, Farmington Hills, Mich., 1992.
2. Anne Balogh, "New Admixture Combats Concrete Shrinkage," *Concrete Construction,* July 1996, pp. 546–552.
3. Surendra P. Shah et al. "Shrinkage Cracking—Can It Be Prevented?" *Concrete International,* April 1998, pp. 51–55.
4. John C. Ropke, *Concrete Problems: Causes and Cures,* McGraw-Hill, New York, 1982.
5. O. C. Guedelhoefer, "Concrete Cracks Basics: Identifying the Causes," *Building Renovation,* September/October 1992, pp. 63–65.
6. A. H. Gustaferro and N. L. Scott, "Reading Structural Concrete Cracks," *Concrete Construction,* December 1990, pp. 994–1003.
7. ACI 224.1R-93, *Causes, Evaluation, and Repair of Cracks in Concrete Structures,* American Concrete Institute, Farmington Hills, Mich., 1994.
8. Rolf Eligehausen and Tamas Balogh, "Behavior of Fasteners Loaded in Tension in Cracked Reinforced Concrete," *ACI Structural Journal,* vol. 92, no. 3, May-June 1995.
9. Terry J. Fricks, "Cracking in Floor Slabs," *Concrete International,* February 1992, p. 63.
10. James Hill, "Cracks in Structures," *Concrete Construction,* March 1988, p. 316.
11. ACI 318-95, *Building Code Requirements for Structural Concrete,* American Concrete Institute, Detroit, Mich., 1995.
12. ACI 224R-90, *Control of Cracking in Concrete Structures,* American Concrete Institute, Detroit, Mich., 1990.
13. Peter H. Emmons, *Concrete Repair and Maintenance Illustrated,* R.S. Means Co., Inc., Kingston, Mass., 1993.
14. John W. Baker, "Vault Repair Provides Insights into Effective Polyurethane Grouting," *Concrete Construction,* February 1999, pp. 94–96.
15. Robert W. Gaul, "A New Generation of Epoxy Adhesives," *Structural Engineering Forum,* December/January 1995, pp. 32–37.
16. John Trout, *Epoxy Injection in Construction,* The Aberdeen Group, Addison, Ill., 1997.
17. Carolyn M. Hansson, "The Impact of Blended Cements on the Durability of Concrete Structures," in "New Technologies in Construction Materials and Building Systems," seminar notes by Trow Protze Consulting Engineers, Boston, Mass., 1990.
18. Sidney Freedman, "Why Architectural Precast Reinforcement Doesn't Corrode," *Ascent,* Summer 1997, pp. 36—40.
19. Steen Rostam, "Service Life Design—The European Approach," *Concrete International,* July 1993, pp. 24–32.
20. Ted Kay, *Assessment and Renovation of Concrete Structures,* Longman Scientific & Technical, Essex, England, 1992.
21. Ian E. Chandler, *Repair and Renovation of Modern Buildings,* McGraw-Hill, New York, 1991.
22. *Principal Types of Protective Coatings,* Koppers Company, Inc., Pittsburgh, undated.
23. Donald W. Pfeifer, "Steel Corrosion Damage on Vertical Concrete Surfaces, Parts I and II," *Concrete Construction,* February 1981.
24. Bob Tracy and Rick Klein, "Corrosion and Concrete Deterioration," *Building Renovation,* Summer 1994.
25. R. T. L. Allen et al., eds., *The Repair of Concrete Structures,* 2d ed., Blackie Academic & Professional, Chapman & Hall, Glasgow, U.K., 1993.
26. John Gregorson, "Conquering Corrosion," *Building Design & Construction,* August 1999, pp. 50—52.
27. Adam Neville, "Maintenance and Durability of Structures," *Concrete International,* November 1997, pp. 52–56.
28. Hector Gallegos and Gaby Quesada, "A Corrosion Case: Repair Procedure," *Concrete International,* June 1987, pp. 54–57.

29. ASTM STP 1276, *Techniques to Assess the Corrosion Activity of Steel Reinforced Concrete Structures,* American Society for Testing and Materials, Philadelphia, 1997.
30. Peter H. Emmons and Alexander M. Vaysburd, "Corrosion Protection in Concrete Repair: Myth and Reality," *Concrete International,* March 1997, pp. 47–56.
31. *Concrete Substrate Preparation Techniques,* Fosroc Inc., New York, undated.
32. James Warner et al., "Surface Preparation for Overlays," *Concrete International,* May 1998, pp. 43–46.
33. Remy A. Iamonaco, "Rehabilitation of Parking Structures," in "New Technologies in Construction Materials and Building Systems," seminar notes by Trow Protze Consulting Engineers, Boston, Mass., 1990.
34. "Corrosion Inhibitors," *Concrete Repair Digest,* June/July 1997, pp. 154–156.
35. *An Evaluation of Priming Systems for the Protection of Reinforcing Steel from Corrosion,* Fosroc Inc., New York, undated.
36. "Restoring Deteriorated Concrete in Pulp and Paper Mills," *Journal of Protective Coatings and Linings,* August 1995, pp. 79–90, quoted from *Practice Periodical on Structural Design and Construction,* vol. 2, no. 2, May 1997, pp. 34–42.
37. Ernest K. Schrader, "Mistakes, Misconceptions, and Controversial Issues..., Part 3," *Concrete International,* November 1992, pp. 54–59.
38. Philip Sarvinis, "Don't Use Hydrodemolition on Post-tensioned Concrete," *Concrete Construction,* August 1998, p. 708.
39. "Insights into Durable Concrete Repair," *Concrete Construction,* July 1998, pp. 619–620.
40. Charles H. Holl and Scott A. O'Connor, "Cleaning and Preparing Concrete before Repair," *Concrete International,* March 1997, pp. 60–63.
41. *Concrete Repair Methods and Procedures,* Fosroc Inc., New York, undated.
42. Paul B. Haydon, "Recasting at the Edge," *Building Renovation,* Summer 1994, pp. 24–38.
43. Ernest K. Schrader, "Mistakes, Misconceptions, and Controversial Issues..., Part 1," *Concrete International,* September 1992, pp. 52–56.
44. ACI 332R-84 *Guide to Residential Cast-in-Place Construction,* American Concrete Institute, Farmington Hills, Mich., 1994.
45. "Repair Q & A: Repairing Foundation Honeycombing," *Concrete Construction,* July 1999, p. 58.
46. Kenneth C. Hover, *Cathodic Protection for Reinforced Concrete Structures,* ACI SP-85-8, American Concrete Institute, Detroit, Mich., 1985, pp. 175–208.
47. Frank Stahl, "Bridge Rehabilitation," *Civil Engineering Practice,* Fall 1990, pp. 7–40.
48. Sam Bhuyan, "Repairing Concrete Parking Structures," *Concrete Construction,* February 1988, pp. 97–106.
49. David Whitmore et al., "Battling Concrete Corrosion," *Civil Engineering,* January 1999, pp. 46–48.
50. James Warner, "Methods for Repairing and Retrofitting (Strengthening) Existing Buildings," Workshop on Earthquake-Resistant Reinforced Concrete Building Construction, University of California, Berkeley, July 11–15, 1997.
51. Bruce A. Suprenant, "Protecting Fresh Concrete from Freezing Weather," *Concrete Construction,* February 1992, pp. 126–128.
52. George E. Ramey et al., "Structural Design Actions to Mitigate Bridge Deck Cracking," *Practice Periodical on Structural Design and Construction,* vol. 2, no. 3, August 1997, pp. 118–124.
53. Poul Beckmann, *Structural Aspects of Building Conservation,* McGraw-Hill International (UK), London, 1995.
54. Shenfu Fan and John M. Hanson, "Effect of Alkali Silica Reaction Expansion and Cracking on Structural Behavior of Reinforced Concrete Beams," *ACI Structural Journal,* September-October 1998, pp. 498–505.
55. David B. McDonald and William G. Hime, "Delayed Ettringite Formation," *PCI Journal,* March-April 1998, p. 92.

56. Terry Holland, "Using Shrinkage-Reducing Admixtures," *Concrete Construction,* March 1999, pp. 15–18.
57. Steven H. Kosmatka and William C. Panarese, *Design and Control of Concrete Mixtures,* Portland Cement Association, Skokie, Ill., 1992.
58. ACI 506R-90, *Guide to Shotcrete,* American Concrete Institute, Detroit, Mich., 1990.
59. ACI 304R-92, *Guide for the Use of Preplaced Aggregate Concrete for Structural and Mass Concrete Applications,* American Concrete Institute, Farmington Hills, Mich., 1992.
60. Stella L. Marusin, *Repairs of Concrete Columns, Spandrels, and Balconies on a High-Rise Housing Complex in Chicago,* ACI SP-85-7, American Concrete Institute, Detroit, Mich., 1985, pp. 157–163.
61. D. R. Plum, "Materials—Why They Fail," *Construction Maintenance and Repair,* September-October 1991, pp. 3–6, as quoted in Ref. 13.
62. Martin S. McGovern, "A Guide to Rapid-Hardening Concrete," *Concrete Repair Digest,* June/July 1997, pp. 164–168.
63. Donald W. Pfeifer and William F. Perenchio, "Coatings, Penetrants and Specialty Concrete Overlays for Concrete Surfaces," a paper presented at the seminar Solving Rebar Corrosion Problems in Concrete, Chicago, Ill., September 27–29, 1982.
64. Alan H. Vroom, "Sulphur Concrete Goes Global," *Concrete International,* January 1998, pp. 68–71.
65. Howard A. Okumura, "Early Sulphur Concrete Installations," *Concrete International,* January 1998, pp. 72–75.
66. Sean M. Crick and David W. Whitmore, "Using Sulphur Concrete on a Commercial Scale," *Concrete International,* February 1998, pp. 83–86.
67. Adam Neville, "A New Look at High-Alumina Cement," *Concrete International,* August 1998, pp. 51–55.
68. ACI 546R-96, *Concrete Repair Guide,* American Concrete Institute, Detroit, Mich., 1996.
69. *Guide for Selecting and Specifying Materials for Repair of Concrete Surfaces,* Guideline No. 03733, International Concrete Repair Institute, Sterling, Va., 1996.
70. "Performance Criteria for Concrete Repair Materials, Phase II Summary Report," Technical Report REMR-CS-62, U.S. Army Corps of Engineers, March 1999.
71. Martin S. McGovern, "Cracking Down on Repair Materials," *Concrete Construction,* August 1999, pp. 50–56.
72. Ping Gu et al., "Electrochemical Incompatibility of Patches in Reinforced Concrete," *Concrete International,* August 1997, pp. 68–72.
73. O'Connor et al., "Evaluation of Historic Concrete Structures," *Concrete International,* August 1997, pp. 57–61.
74. Jay H. Paul, "Extending the Life of Concrete Repairs," *Concrete International,* March 1998, pp. 62–66.
75. Matthew Phair, "Load Checks Get Easier, Cheaper," *ENR,* October 6, 1997, p. 43.
76. Peter Schiessl and Michael Raupach, "Monitoring System for the Corrosion Risk of Steel in Concrete Structures," *Concrete International,* July 1992, pp. 52–55.

Renovating Slabs on Grade

6.1 Introduction

Slabs on grade, also called floors on ground, are different from other structural members. First, they are supported directly by soil, and their success or failure may depend more on the soil qualities than on the slab construction. Second, floors on ground are among the few structural members that are open to the owner's eagle eye, and any crack or other perceived problem tends to be noticed—and sometimes magnified beyond reason. And last, they carry equipment and floor finishes, and any defect in the slab's integrity or moisture resistance affects those elements. A floor slab undergoing drying shrinkage may not only crack, but also break the brittle ceramic tile it carries.

A typical floor on ground consists of a concrete slab of uniform or varying thickness with reinforcing, joints, and finish; a subbase of gravel or crushed stone; and a compacted subgrade. A vapor barrier, with or without a sand layer, may also be present in some circumstances. (The question of whether they are needed is debated in Sec. 6.7.) The reinforcement can be deformed steel bars, post-tensioning tendons, welded wire fabric, mixed-in fibers, or nothing at all. The typical components of a slab on grade are shown in Fig. 6.1.

Contractors often question the need for a subbase—an added cost item for them. The function of a subbase is to provide drainage under the slab and to act as a cushion to help the slab span over weak spots in the subgrade. The thicker the subbase, the more effectively it can do its job. Common thicknesses range from 6 to 12 in. The subbase may not be needed if the subgrade consists of well-compacted clean

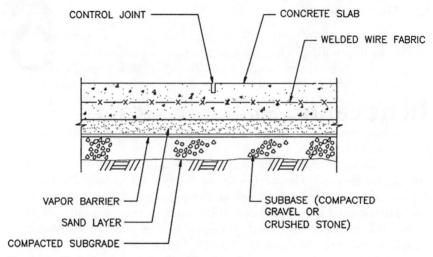

Figure 6.1 Typical components of a slab on grade.

granular material. More often than not, however, the subgrade is either poorly prepared or contaminated with fine particles that obstruct drainage and resist compaction.

Despite their apparent simplicity, slabs on grade have rather complex and frequently misunderstood structural behavior. Their rehabilitation may be required for a variety of reasons, among them age-related degradation, overload, poor design, shabby construction, or inferior materials. The goal of this chapter is to examine the common problems of slabs on grade and the best methods of solving them.

In general, slab problems can be categorized as follows:

1. Cracking

2. Surface deterioration

3. Curling

4. Settlement or heave

5. Failure of joints

6. Delamination of finishes

7. Water penetration

8. Chemical attack

9. Wearing out

These problems and the methods of their repair are examined in this chapter.

6.2 Field Investigation

Before any repair is undertaken, it is important to diagnose what exactly is happening. The general process for evaluation of concrete structures is described in Chap. 2. Here, we address only the specific issues applicable to slabs on grade. As always, the first step involves review of the existing design documents and a preliminary field visit (walk-through). If need be, any problems found during the walk-through can be documented and quantified during subsequent field visits.

A good description of the survey procedure can be found in ACI 201.3R, *Guide for Making a Condition Survey of Concrete Pavements.*[1] Although intended primarily for exterior paving slabs rather than interior slabs on grade, the guide describes and illustrates typical slab problems and includes a handy checklist for documenting them. A complementary publication, ACI 201.1R, *Guide for Making a Condition Survey of Concrete in Service,*[2] includes another checklist for items dealing with design, materials, and construction practices.

Ringo and Anderson[3] recommend that the following questions be addressed in the slab inspection:

1. *Flatness and levelness.* Is the slab generally level? Is it flat, or is it bumpy in some spots? (Checking for flatness and levelness may require specialized equipment.)

2. *General surface conditions.* Is the slab cracking, spalling, or scaling? Does it suffer from popouts or other defects?

3. *Joints.* Are the joints of the proper type (e.g., expansion vs. control joints) and spacing? Are the edges curling? (This can be checked by placing a straightedge diagonally across a four-way joint intersection.) Are the joint edges spalling, deteriorated, or otherwise damaged? What is the condition of joint fillers?

4. *Evidence of overload, heave, or settlement.* Does the slab appear to suffer from the effects of overload, such as perimeter cracks around some areas? Is the slab settling or heaving?

The general question that must be addressed by an investigation is whether the slab is adequate for the proposed use and whether it must be repaired or replaced. To address this question, field investigation can be complemented by structural analysis, especially if the new use would introduce concentrated loading. This is often the case when wheeled equipment, rack storage systems, or new columns are placed on the floor. Structural analysis for concentrated loads should follow the methods outlined in Refs. 4 and 5.

A concrete testing program, which might also be needed, involves taking concrete cores for determination of compressive strength and perhaps for petrographic examination. In finished areas, although this is admittedly inconvenient, it is the only way to determine the slab's condition. Coring is typically done with a portable water-cooled drilling machine, and some water inevitably spills on the floor. The best places to make the cores are areas where some minor water damage to finishes can be tolerated.

One way to reduce water damage in shown in Fig. 6.2. Here, while one person does the coring, another is removing water with a vacuum cleaner. (Water is pumped from the tank at right.) The picture was taken during an investigation of a slab suspected by the owner of being deteriorated because during an earlier installation of vinyl-tile flooring, the workers had reported encountering very rough, porous, and crumbly concrete. The cores showed that the weak concrete was actually a separate topping applied during construction to level the floor. The slab itself was in good condition.

Any time heave or settlement is noticed, a geotechnical investigation is advised in order to determine whether the distress is likely to continue or whether conditions have stabilized. Slab heaving and cracking is usually caused by frost, but sometimes there are other reasons. Among them are trees growing outside the building, with roots extending under the slab and pushing it up, and expansive soils that swell and shrink with changes in moisture content. Thompson[6] investigated a case of fractured slabs placed on expansive clays. He attributed the failure to the outside vegetation drawing moisture from under the edges of the slabs, causing the soil to shrink and the slabs to lose support—and crack. The result of the investigation is a report that determines whether the floor was constructed as designed, identifies and analyzes the observed problems, and recommends the repairs, if those are needed.

In this chapter, we deal primarily with plain or conventional lightly reinforced slabs on grade. Design and construction of post-tensioned slabs on ground is a specialized area covered in the PTI's design guide[7] and in ACI 360R.[4] Investigation of post-tensioned framing is discussed in Chap. 2.

6.3 Cracking

Cracking is perhaps the most common defect of concrete floors on ground. Cracking can have a variety of causes, and the underlying problem must be determined prior to undertaking any repairs. Chapter 5 discusses several possible origins of cracking. In slabs on grade, drying shrinkage and plastic shrinkage cracks are especially common.

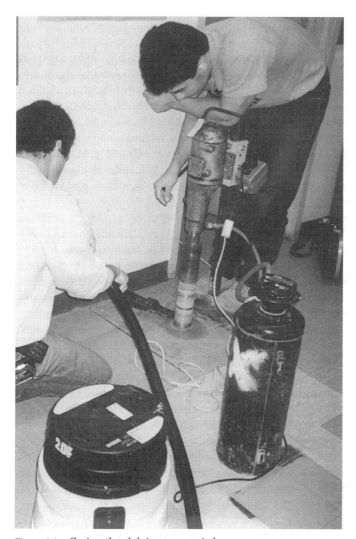

Figure 6.2   Coring the slab in an occupied space.

6.3.1 Drying shrinkage cracks

Drying shrinkage of slabs starts within the first hour after placement and continues for many months. According to one rule of thumb, about 30 percent of shrinkage occurs within the first 30 days, 60 percent within the first 6 months, and 90 percent within a year.[8] Others believe that the process takes even longer.[9] Gustaferro[10] has developed a graph showing that a 4-in-thick slab attains less than 80 percent of its ultimate drying shrinkage in 12 months and less than 90 percent in 24 months, while a 6-in-thick slab requires nearly 2 years to undergo 75 percent of its ultimate drying shrinkage.

Unrestrained slabs on grade should not crack as they shrink, but in practice some sort of restraint is always present. Among its many sources are reentrant slab corners, building columns, and embedded items—even those as seemingly insignificant as electrical conduits. An unavoidable friction against the soil also resists free shrinkage. When shrinkage is restrained, concrete is subjected to tensile stresses that it is ill equipped to resist. Concrete can stretch without cracking only about 0.002 in/ft, but the shrinkage rates are typically in the vicinity of 0.006 in/ft, so cracking in restrained concrete undergoing shrinkage is inevitable.[8]

Drying shrinkage cracks most commonly appear as jagged hairline cracks at random locations. Sometimes they have the shape of a crow-foot. Drying shrinkage cracks are rarely considered structurally important, unless their width becomes too wide. The exact mechanism of drying shrinkage cracking is not totally clear. Most experts believe that these cracks start at the top of the slab, because loss of moisture occurs faster there than at the bottom.

The most effective means of controlling the width of cracks in conventional slabs is closely spaced construction and control joints. Control joints can be saw cut or formed by cast-in removable strips. These joints cannot eliminate shrinkage cracks, but they can direct them into straight joint lines by weakening the slab there. To be effective in weakening the slab and inducing cracking, the joint depth must be substantial—at least one-quarter of the slab thickness. Otherwise, the cracking pattern will not be affected, and an embarrassing shrinkage crack might even occur next to a joint (Fig. 6.3).

Some practitioners are not afraid of cracked floors and dislike saw-cut control joints. To them, cosmetic cracking is preferable to a lifetime of maintenance chores introduced by the joints. As Fricks[8] points out, the saw cuts are usually at least 0.1 in wide—much wider than the cracks they are intended to prevent—and therefore more prone to spalling than the edges of the original cracks.

Another way of avoiding a small number of wide cracks is to control their width by using welded wire fabric (WWF). Where in the slab should it be placed? If we accept the fact that a slab cracks from the top, it is logical to place WWF as close to the top as possible. In practice, to allow for field tolerances, the mesh is usually placed 2 in below the top of the slab. But, as anyone who has ever observed the placement of a slab reinforced with WWF can testify, the task of keeping WWF at the top is extremely difficult.

Why? WWF is typically supported on concrete bricks or on special chairs. Both of these tend to sink into soil under the weight of workers. Also, the fabric is deflected between the supports when stepped on.

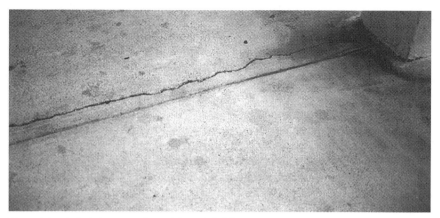

Figure 6.3   A shrinkage crack next to a slab joint.

A commonly attempted remedy is to have the depressed WWF pulled up by a hooked bar while the concrete is being placed. Unfortunately, this is easier said than done, and the scene of attempting to lift the mesh and simultaneously not step on it with heavy boots is worthy of the Three Stooges. More often than not, WWF ends up on the bottom of the slab, where it does little good.

Or does it? Cesar Kiamco[11] offers a provocative dissenting look at a cracking process. He points out that no cracking has been observed in many cases in which WWF ended up on the bottom of the slab. He then argues that drying shrinkage cracking actually starts from the bottom of concrete, not from the top. His explanation: As a slab on grade tries to shrink, its shrinkage is restrained by the friction against the soil at the bottom. Accordingly, the slab's bottom surface is where the tensile forces are introduced into the slab, and this is where the cracks develop. From this standpoint, WWF is most effective in resisting tensile forces from drying shrinkage in the location where it often ends up anyway—at the bottom—and it would be useless to put it at the top of the slab. At the present time, this alternative theory of cracking has not been widely accepted.

Other methods of controlling drying shrinkage cracking include adding steel or plastic fibers, post-tensioning, and using shrinkage-compensating (Type K) cement or special admixtures, described in Chap. 5. Like welded wire fabric, all these remedies are not 100 percent effective in preventing drying shrinkage cracking.

6.3.2   Plastic shrinkage cracks

Plastic shrinkage cracks typically appear when concrete dries out too fast—before curing begins. This may happen when curing is delayed for some reason or when the weather is very windy and hot. The cracks start at the top of the slab and progress downward, although they usually do not propagate through the whole thickness. Unlike the cracks caused by drying shrinkage, plastic shrinkage cracks are characteristically wide and short and rarely extend to the slab edges. Their length can range from several inches to several feet. Plastic shrinkage cracks typically run parallel to each other, but sometimes their pattern is irregular.

Because of their width, plastic shrinkage cracks are often considered structurally significant and are repaired. (The easiest repairs can be made when the concrete is still plastic, simply by striking the slab with a float.) Reinforcing bars cannot prevent this type of cracking, since it occurs before the bars have a chance to bond with the concrete.

6.3.3   Temperature contraction, crazing, and map cracking

Closely resembling drying shrinkage cracks are cracks caused by temperature contraction of concrete. Concrete shortens when its temperature falls, but, unlike other materials, concrete also cracks when this happens. Why? A 100-ft-long slab that was totally unrestrained against temperature contraction would shrink by approximately 0.66 in when subjected to a temperature drop of 100°F. Since in reality the slab is restrained by friction against the subgrade, it cannot fully shrink, and so it cracks instead.

Concrete is most vulnerable to thermal-contraction cracking at an early age, when it is at its weakest. The timing coincides with that of drying shrinkage cracking, and many thermal-contraction cracks are mistaken for drying shrinkage cracks. (One way to reduce temperature cracking is to place concrete at the coolest possible time—at night.[12]) Concrete made with some aggregates tends to expand more than that made with others. Quartz, for example, has a coefficient of linear expansion of about 7×10^{-6}, nearly twice that of limestone at 3.8×10^{-6}. The coefficient of linear expansion for average concrete is 5.5×10^{-6}.

Crazing and map cracking are also common in slabs on grade. They are similar in appearance—a network of shallow surface cracks resembling those of dried mud—and are differentiated mainly by size and spacing. According to ACI 302.1R,[13] crazing cracks are almost invisible and are spaced 1 to 2 in (25 to 50 mm) apart, whereas map cracking involves more visible cracks with larger spacing.

Crazing and map cracking are thought to be caused by poor finishing practices, e.g., overworking and overtroweling wet concrete, finishing concrete too early, using too much mix water, and curing improperly. The result of all these is a paste-rich top surface that shrinks faster than the rest of the slab. Crazing and map cracking seem to be more prevalent in slabs containing calcium chloride admixtures. Both these types of cracks may look disturbing, especially in moist concrete, but they are literally skin deep and do not normally affect slab strength or durability.

6.3.4 Other types of nonstructural cracks

Cracks in concrete floors on ground can be caused by soil settlement. Localized areas of settlement are often surrounded by perimeter cracks. A more ominous situation arises when settlement is widespread, as discussed in Sec. 6.5. Cracks caused by subgrade settlement can appear weeks or even months after slab placement.

Reinforced slabs on grade can crack because of local concrete settlement around the bars during construction. Similar cracks can be caused by shifting of side forms. These plastic settlement cracks appear early in the curing cycle and run directly above the bars. Sometimes they can be confused with corrosion-related cracks, which also run directly above the reinforcement but take years to develop and are typically accompanied by rust stains. One way to determine whether the observed cracking is due to plastic settlement or to corrosion is simply to ask old-time employees at the facility when the cracks appeared.

Slab cracking near the joints can be caused by misaligned joint dowels that resist the slab's normal temperature expansion and contraction movements. This problem can be overcome by using square bars with clip-on plastic sheaths.[14] Recently, another product became available for this purpose: a nailed-on plastic sheath that can accommodate regular round bars.

Some slab cracks are caused by externally transmitted forces. Albright[15] documents several occurrences of very wide cracks in residential slabs on grade that he attributes to expansion of the nearby concrete pavements. In one case, expansion of a curved pavement was transmitted by the driveway slab directly into a concrete floor of the house, where a 1-in-wide crack opened up.

Like other concrete structures, slabs on grade can suffer from alkali-silica reaction (ASR), discussed in Chap. 5. ASR cracks typically run perpendicular to the slab joints and may be related to map cracking in other slab areas. A guide to identification of these cracks can be found in Stark.[16]

A distinctive class of cracks is the so-called D cracks, which run in series roughly parallel to the joints and curving across slab corners. D cracking is caused by porous water-absorptive coarse aggregates that are susceptible to freeze-thaw deterioration. When the absorbed moisture freezes, it exerts pressure on the cement particles around the aggregate, resulting in microcracking. These tiny cracks attract more water, which freezes and expands the cracks further, until eventually, after several freeze-thaw cycles, the crescent-shaped cracks appear.

6.3.5   Structural cracks

It is not easy to produce structural failure in a slab by uniformly distributed loading alone. After all, the slab is directly supported by the subgrade. If the subgrade is perfectly compacted and leveled, a slab on grade may be considered to be just a wearing surface. When failure nevertheless occurs, it is usually the result of poor subgrade preparation or poor subgrade materials. Most structural failures in slabs on grade are caused by point loading from slab-supported mezzanines, rack systems, wheel loading of cranes and other construction equipment, and heavy stacks of material. The cracking caused by soil heaving or settlement is a separate case, discussed in Sec. 6.5.

Quite often, structural cracking can be traced to loading that was applied too early, before the concrete could gain its full strength. It is not uncommon to see heavy construction equipment placed on green concrete slabs. Unfortunately, this construction loading is usually far greater than the amount even the fully cured slab, let alone green concrete, was designed to carry.

A typical failure from structural causes manifests itself as a slightly depressed slab area surrounded by cracks. The cracks are generally the result of negative bending moments produced by a concentrated load that exceeds the slab's flexural capacity, but they can also be caused by a punching-shear overstress.

6.3.6   Which cracks to repair?

To determine whether existing cracks need repair, one should consider their number, spacing, regularity, width, and depth. The verdict also depends on whether a floor covering is planned. Do the cracks significantly reduce the structural capacity of the slab? To answer this question, we should recall that transfer of load across a crack is made possible by aggregate interlock—the ability of the protruding aggregate pieces to engage each other. Excessively wide cracks prevent aggregate interlock and, from a structural standpoint, break the slab into separate pieces.

Any crack wider than 0.04 in (about the thickness of a small paperclip wire) is presumed to make aggregate interlock ineffective. A crack like that might be a candidate for repair if it is exposed to hard-wheel traffic, which could spall its edges. A crack might also need repair if it occurs near a slab-supported post. Often even these cracks are left unrepaired if the floor appears to function normally under wheeled traffic, that is, if the wheels moving over the cracks cause no spalling and no differential vertical movements of the edges. Cracks that are not in the way of wheeled traffic and those that are located away from slab-supported columns can generally be left alone.

Does this mean that cosmetic cracks need not be repaired? Usually that's true, because most objections to floor cracking are aesthetic rather than functional. Filling these cracks with epoxy—the usual fix—might look worse than the original damage (Fig. 6.4). Repairing multiple cracks (Fig. 6.5) by injection may be too expensive.

No repair is normally needed for slabs that are scheduled to receive flexible floor coverings. In cases of brittle finishes and overlays, the cracks can be covered, rather than filled by injection, by troweling a polymer resin and embedding one or more layers of fiberglass mesh across them.

There are always exceptions, of course. Some "superflat" floors may require that all cracks wider than hairline be fixed. In food-processing plants, every crevice that can house harmful bacteria must be eliminated—and that includes floor cracks. In the "clean rooms" of electronic plants, particular attention is devoted to elimination of all dust; floor cracks there are also not welcome. Similarly, all floor cracks should probably be sealed in any process facility that generates corrosive

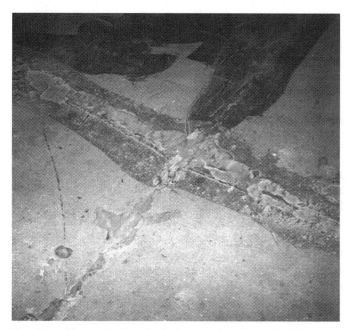

Figure 6.4   The repaired cracks look worse than the original damage.

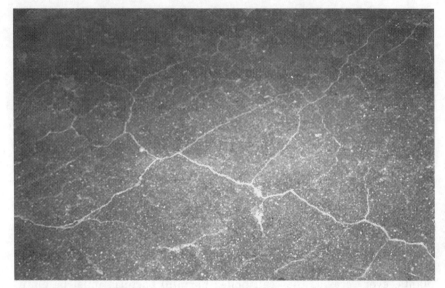

Figure 6.5   These multiple cracks are impractical to repair individually.

chemicals, to prevent concrete deterioration and contamination of the environment.

The foregoing illustrates the need to closely coordinate the proposed repair or maintenance program with the owner's requirements. There are some perfectionist owners who simply cannot bear the sight of any cracks. More often, however, the owners would rather avoid repairs that require clearing large floor areas of any equipment and severely restricting the operations.

6.3.7   Crack repair methods

The optimal method of repair depends on whether the crack is dormant or active. Dormant (static) cracks maintain their width and length over time, whereas the size of active (moving) cracks changes. Dormant cracks may be caused by shrinkage, by plastic settlement of concrete, or by a past overload. Active cracks, on the other hand, reflect continual slab movement. An argument can be made that nearly all slabs on grade are subjected to thermal movements, and therefore *all* cracks in them should be considered active. Also, drying shrinkage in floors less than 2 years old is probably not complete, and shrinkage cracks in those floors should be considered active.

Methods for structural repairs of *dormant* cracks include the epoxy injection method described in Chap. 5. Epoxy injection is best undertaken in relatively new cracks, before they have had a chance to fill with dust and debris, but after most of the drying shrinkage has taken place. As we just said, the latter could take 2 years or even longer after placement. Metzger[17] suggests that narrow shrinkage cracks—those of less than a credit card's width—can be repaired by brushed-on high-strength epoxy, provided that their edges are not spalled. The epoxy should be allowed to flow into the cracks and overlap their sides by $\frac{1}{4}$ to $\frac{1}{2}$ in.

Another method of repair involves filling dormant cracks with a rigid or semirigid epoxy or with a cementitious mix. Semirigid epoxy is ideal for slabs that must be repaired before drying shrinkage has run its course. Cracks less than ⅛ in (3 mm) wide should be "opened up" by routing or sawing, as recommended by ACI 224.1R[18] and other sources and as shown in Fig. 6.6. Some experts recommend enlarging the cracks to a minimum depth of 1 in or their full depth, whichever is less.[5] The debris resulting from the crack preparation should be removed by power washing. The epoxy can be left as placed, or the crack can be overfilled and ground down after hardening if a smooth surface is desired.

The cementitious products used for repair of dormant cracks include premixed or site-mixed mortars (a 1:3 cement-sand mix, for example) and proprietary compounds. Some experts recommend soaking the crack for 24 h and removing the water prior to mortar placement.[19] The mortar is packed into the crack, finished with a trowel, and moist-cured. Cementitious products seem to be more commonly used in Europe, while epoxy materials are more common in the United States.

If desired, a multitude of shallow cracks, such as crazing, can be repaired by grinding or by coating the floor with a polymer-resin compound.[12]

Active cracks require more flexible materials. Epoxy injection or similar "gluing" of an active crack will not be effective unless the underlying causes of movement are eliminated. If that is impossible, the only practical method of repairing active cracks is to fill them with flexible sealant. For relatively wide cracks, Gustaferro[10] suggests pouring a medium-viscosity joint filler into the crack and finishing the repair by shaving or grinding the surface flush after the filler cures. As with filling dormant cracks, the crack's mouth could be enlarged as shown in Fig. 6.6.

The ultimate repair involves replacement of the entire floor panel bounded by construction or control joints. Such replacement is a major surgery reserved for the worst of conditions; it is usually not needed for nonstructural cracks. Still, the cracking and other deterioration might be so severe that there is no alternative to replacing some of the slab. The replacement procedure is described in Sec. 6.9.

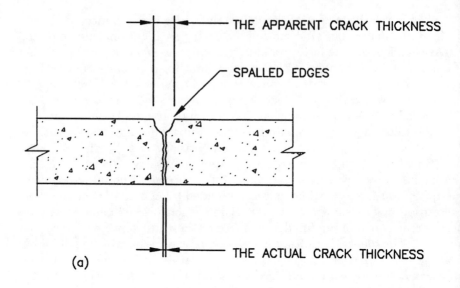

(a)

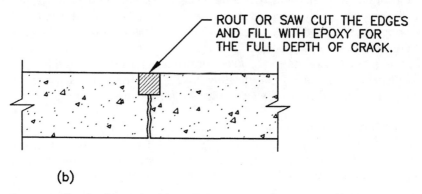

(b)

Figure 6.6   Repair of a crack with spalled edges by filling with epoxy. (a) The original condition; (b) the repaired condition.

6.4   Surface Deterioration

In addition to cracking, slabs on grade can suffer from a variety of other surface problems, such as scaling, popouts, and surface erosion.

6.4.1   Scaling, spalling, and delamination

Scaling occurs when the slab loses thin, small pieces of hardened cement from its surface by peeling or flaking. ACI 201.1[2] distinguishes among the following degrees of scaling:

- *Peeling,* or local flaking of concrete

- *Light scaling,* involving some surface loss but not exposure of coarse aggregate

- *Medium scaling,* involving surface loss of less than 5 to 10 mm in depth and exposure of coarse aggregate (Fig. 6.7)

- *Severe scaling,* involving surface loss of 5 to 10 mm in depth and not only exposure of coarse aggregate, but also 10 to 20 mm deep loss of cement around it

- *Very severe scaling,* involving loss of coarse aggregate and being at least 20 mm deep

Scaling can happen when premature finishing and curing seals the slab surface before bleeding has stopped. As a result, water and air are trapped under the surface and, in trying to escape, start to form small blisters. Eventually, these blisters burst and break the surface. Scaling can also result from overtroweling. In fact, many concrete consultants recommend totally avoiding troweling exterior slabs, and instead suggest striking the slabs off, floating if desired, and applying a broom finish.[20]

Deicing salts, especially if applied during the first winter of the slab's life in freezing conditions, can also contribute to scaling by exerting expansive pressures from within the cement paste.[21] In enclosed

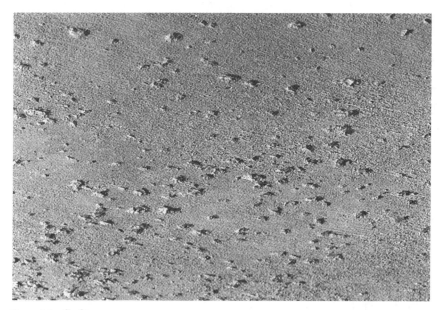

Figure 6.7   Scaling.

buildings, the slabs at the bottom level of parking garages are especially susceptible, because those areas are subjected both to freeze/thaw action and to salt dripping from cars. The repeated wetting and drying of garage slabs can result in substantial accumulation of deicing salts on the surface of concrete.

Scaling is a common problem in non-air-entrained outdoor concrete in freezing environments. Modern practice—and ACI specifications—requires air entrainment in such circumstances. However, excessive air content seems to actually promote scaling.

In new construction, scaling can be reduced not only by limiting the air content, but also by using concrete mixes with water-cement ratios less than 0.45, specifying a minimum concrete compressive strength of 4500 psi, and adding pozzolans. Also, to produce a dense and strong concrete surface that is resistant to scaling and other defects, it is essential to follow proper finishing and curing procedures. In existing construction, the probable cause of scaling can be determined if concrete does not fit these criteria. A petrographic examination can reveal the level of entrained air in existing concrete and help determine whether it contributed to scaling.

Delamination is essentially scaling of large proportions. It is caused by the same factors as scaling—excessive air content, overtroweling, sealing the surface too early—but occurs over wider areas. This problem is particularly pronounced in slabs receiving a smooth, hard, steel-troweled finish. When such a finish is required, it is difficult to avoid slab delamination even when concrete is placed by experienced contractors following good practice. Indeed, ACI 302[13] cautions that delamination is possible any time air-entrained concrete is given a smooth, hard-troweled finish. In one study described by Suprenant and Malisch,[22] a gathering of contractors placed a series of trial slab installations, using varying tools, techniques, and materials. The only slab in which delamination occurred had high levels of entrained air.

This and other studies will probably eventually lead to reassessment of the way air-entrained concrete is specified for slabs on grade.[23] Many experts already suggest the elimination of entrained air in floors, even at the price of reducing the freeze-thaw resistance. It has been suggested that entrained air may find its way into concrete as a result of cement grinding operations—without any special efforts to introduce it—especially when imported coarsely ground cements are used.[24]

Delamination has also been blamed on vapor barriers (they prevent the dissipation of excess water into the ground), on cold subgrades (they slow the rate of setting), on fly ash admixtures (they reduce the amount of bleed water and may trick the finishers into starting their operations too early), on rapid loss of moisture caused by wind, and on a host of other factors.

Spalling involves a loss of concrete fragments, as in scaling and delamination, but the broken-off areas are deeper and the damage is more severe. A small spall is defined by ACI 201.1[2] as a "roughly circular or oval depression generally not greater than 20 mm in depth nor greater than about 150 mm in any dimension, caused by the separation of a portion of the surface concrete." A large spall could equal or exceed those dimensions. A more detailed classification of spalls can be found in ACI 201.3R.[1]

6.4.2 Popouts

Popouts appear as pockmarks on the surface of the slab. Most popouts are caused by the presence of absorptive or chemically reactive aggregate particles near the surface. Simply put, when a piece of aggregate absorbs water or freezes, it expands and either breaks up or exerts pressure on the surrounding cement paste. The ensuing cracking invites more water inside the slab, and the process progresses until a thin piece of slab is ejected. (This can also cause D cracking.) Similarly, when a piece of chemically reactive aggregate is attacked by water and alkalis in cement, the resulting gel expands and pushes against the adjoining cement particles, eventually causing a popout.[20]

ACI 201.1[2] differentiates between small popouts (those leaving holes up to 10 mm in diameter), medium popouts (those leaving holes between 10 and 50 mm), and large popouts (those leaving holes over 50 mm in diameter). Logically, the most direct way to reduce the number of popouts is to use concrete that does not contain absorptive or chemically reactive aggregates.

6.4.3 Surface abrasion, erosion, dusting, and other defects

Surface abrasion is wearing out of concrete by repeated scraping and gouging, as may occur in industrial buildings containing forklifts and other hard-wheeled moving equipment. It affects the floors of many resource-recovery and recycling facilities and bulk-storage warehouses, where front-end loaders move the loose materials around, dropping them on the floor and dragging them along.

A similar term, *rutting,* refers to loss of concrete caused by abrasion from vehicle tires.[25] Rutting is rare in building slabs but is quite common in road pavements. The term *erosion* is reserved by ACI 201.1[2] for cases where surface deterioration is caused "by the abrasive action of fluids or solids in motion."

A milder surface defect is *dusting,* the appearance of a light-colored powder at the top of the slab. Dusting is usually blamed on overtroweling, on starting troweling operations too early, and on too much mixing water. Also, the use of unvented gas-powered heaters in enclosed spaces has been implicated in causing carbonation of the wet cement paste–and dusting.

An even more benign defect is *discoloration*. It can be caused by a variety of factors, among them the use of calcium chloride admixtures (which tend to cause dark patches) and the use of different types of cement in the same slab. The poor practice of placing concrete on wet subgrade can also result in discolored slabs, because rising water can produce a patch of lighter color. Another poor practice, sprinkling dry cement on the wet concrete surface to hasten the finishing process, can also be to blame, especially if the sprinkled cement is of a different type from the cement used in the mix.[26]

6.4.4 Repair methods

Should any of these surface defects be repaired? The answer depends on the extent of damage, on the future use of the slab, and on the owner's attitude toward repairs. In general, minor surface deterioration can be tolerated. If repairs are desired, Ropke[26] recommends the following measures.

Discoloration can be alleviated by scrubbing the floor with a solution of 1 pint phosphoric acid mixed with 2 gallons of water. If that doesn't help, a concrete stain can be used, but it requires periodic renewal. Dusting can be helped by one or two coats of a solution of magnesium fluosilicate and zinc fluosilicate applied with a mop or broom.

Shallow surface defects can be ground down to sound concrete or can be coated with a special liquid compound. Popouts can be filled with a self-bonding latex-modified mortar and the entire slab coated with water-repelling sealer to prevent water entry. The idea of adding a floor topping on a slab plagued by popouts is questionable, because the slab will tend to push the topping away.

Pitted, spalled, or scaled surfaces can be restored by removing all deteriorated concrete and patching the slab with an acrylic-based mortar or similar material, as described in Chap. 5. Isolated gouges can be repaired with a troweled-on low-viscosity epoxy mortar or with a polymer-modified coating specifically designed for repair of surface flaws. These materials are applied by trowel or squeegee at a maximum thickness of about $1/8$ in. A small number of large spalls or popouts can also be patched by packing epoxy mortar into a partially cored or chipped area around the popout. (Coring helps avoid feather-edging.)

Sometimes it is worthwhile to supplement patching with a surface-applied membrane. Areas subjected to high abrasion can be reinforced with steel plates or hard toppings as described in Sec. 6.11. In the worst cases, the damaged concrete can be removed and replaced.

6.5 Slab Settlement, Heaving, and Curling

When a floor on ground shows signs of heaving or settlement, its serviceability (or usability, in simpler language) clearly suffers. Building users tend to like their floors level and become alarmed at the sight of their storage shelves or equipment gradually assuming the shape of the Tower of Pisa. A settling or heaving floor generates a feeling of impending disaster, even though the slab is still firmly supported by the ground. The two main reasons for such distress are frost and soil-related problems.

One way to determine whether settlement or heave has caused slab cracking is to examine the cross sections of the cracks in the cores taken through the slab. A crack caused by heave will be wider at the top than at the bottom; a crack caused by settlement will be wider at the bottom.

6.5.1 Frost heaving and settlement

Problems with frost heaving are common in outdoor slabs and in unheated buildings located in northern climates. A typical symptom of frost heaving is a bulging and cracked slab. Bulging is caused by ice lenses forming in the soil and pushing up on the slab. The cracks occur when the ice lenses disappear in the spring, leaving behind voids in the ground that eventually collapse. There is also another spring problem: The topmost soil has thawed and is saturated with water, but this water cannot drain away because the soil below is still frozen. As a result, the saturated soil turns into mud and loses its loadbearing capacity. A slab founded on such soil is likely to break up and settle under load.

Frost heave is most common in soils that expand dramatically during freezing and thawing, such as silts, clays, and fine sands. Of these, silts with high permeability and strong capillary action are the most prone to frost heave. Essentially, such silts soak up moisture and hold it like a sponge.[27] Coarse sands and gravels are the least susceptible to frost heave.

As ACI 360[4] points out, three conditions must be met for damage to occur: The soil must be at the freezing temperature, the water table must be close enough to the surface to form ice lenses, and the soil must exhibit capillary action to act as a wick carrying groundwater into the freezing zone. If any one of these factors is absent, frost heave or settlement will not occur. Frost heave can also be prevented if the building is at least minimally heated and if drainage under the slab can remove water from the freezing zone. The drainage can be provided by a thick (at least 12 in) layer of coarse sand, gravel, or crushed stone, all of which are capable of interrupting the capillary action of soil.

6.5.2 Settlement caused by inadequate soils

Settling soil usually leaves the slab partly depressed and broken, with some of it staying at the original elevation and the rest sinking. This partial settlement typically occurs at the slab joints. When located in the path of moving equipment, the resulting step quickly breaks down, and the slab begins to disintegrate. When it is located in a path of people's movement, it becomes a tripping hazard (Fig. 6.8).

The most common reason for settlement is inadequate soil compaction. One example of this problem can be seen in shopping-mall parking lots built on loosely dumped fill—some of those pavements resemble ocean waves. Settlement occurs in slabs placed on poorly compacted subgrade materials or in those on materials whose poor quality cannot be improved by compaction. Even if a slab placed on such poor soil does not settle, its thickness will typically be uneven, inviting cracking in thin spots. Figure 6.9 shows a cut-out piece of a

Figure 6.8 A settled slab located in the path of people's movement becomes a tripping hazard.

slab on grade (placed on end) that was undergoing excessive cracking. As can be seen, the thinnest part of the slab near the ruler is less than 3 in thick, while the thickest is almost twice that—all within a space of about 1 ft!

Slab settlement can sometimes be traced to soil washout by movement of rain water under the slab. This movement can sometimes be detected as *pumping*—the appearance of water, with or without soil fines, on top of the slab. Pumping typically occurs along slab cracks, joints, and edges.

In some areas of the United States, slab settlement can also be caused by expansive soils. Typically, any soil with a plasticity index of

Figure 6.9   A cut-out piece of slab placed on end. Note the variability of its section. *(Maguire Group Inc.)*

over 20 is considered expansive, that is, capable of drastically swelling and shrinking with changes in moisture. In dry conditions, these unstable soils may have deep furrows caused by shrinkage.

Because the possible causes of slab settlement are many, it is important to determine the exact one prior to attempting any repairs. The guidance of a geotechnical engineer should be sought whenever soil problems are suspected. This includes most cases other than obvious overload.

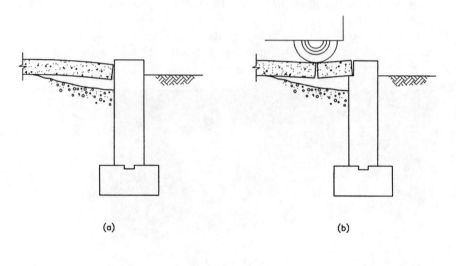

(a)   (b)

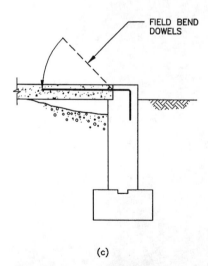

(c)

Figure 6.10   Details of slab on grade near foundation walls, the area of potentially poor compaction. (*a*) Placing premolded compressible joint material between the slab and the wall; (*b*) supporting the edge of the slab by a foundation seat; (*c*) using steel dowels hooked into the wall.

A special case of floor settlement involves slabs on grade placed near foundation walls. The usual detail, intended to thermally isolate the slab from the wall and to eliminate restraint on shrinkage, involves placing a premolded compressible joint material between the slab edge and the wall. These are laudable objectives, but there is a hitch: The slab tends to settle near the wall because soil compaction there is difficult to achieve (Fig. 6.10a). Unfortunately, this happens quite often.

One solution is to support the edge of the slab on a foundation seat 1.5 to 2 in wide. This prevents slab settlement by making the slab span over the poorly compacted area. However, the unreinforced slab can break if a concentrated load is placed near the wall (Fig. 6.10b).

The next logical step is to add a steel dowel hooked into the wall to reinforce the slab over the area of potentially poor compaction (Fig. 6.10c). This detail seems to solve the problems of settlement and slab breaking, but introduces other problems. First, tying the slab at the perimeter provides an unwanted restraint and increases the chances of drying shrinkage cracking. Second, the dowels require field bending, because it is impossible to compact the slab with them sticking out from the wall, and represent a certain safety hazard. The typical contractor's solution is to bend the dowels out of the way, parallel to the wall, and then bend them back, potentially weakening the steel. Despite its shortcomings, however, the supported-edge solution, with or without the dowels, is preferable to simply letting the slab float and deflect at the edges.

6.5.3 Curling

Curling of slabs on grade (Fig. 6.11) has causes entirely different from those that produce settlement or heave. Theoretically, a discussion of curling should follow that of slab shrinkage, but, regardless of causes, repairs of curled slabs are more in line with those for correcting out-of-level floors than with fixing shrinkage cracks.

The classic explanation of why curling occurs is based on the principle of nonuniform drying—the top of a new slab on grade dries out faster than the bottom. With nonuniform drying comes nonuniform shrinkage: The top shrinks, and therefore shortens, before the bottom does. Curling intensifies when a vapor barrier is placed directly under the slab, because it prevents the water in the concrete from dissipating into the subgrade soil and keeps the excess moisture at the bottom.

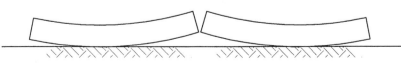

Figure 6.11 Curling.

The longer the bottom of the slab is kept moist while the top dries out, the more pronounced the curling becomes.

It is not uncommon to see curling of $\frac{1}{8}$ in (3 mm) or more at slab edges and construction joints, with perhaps twice that at the corners. Curling of up to 1 in has been observed. Curling can also occur at control joints, but to a much smaller degree. According to one rule, the amount of curling is roughly proportional to the square root of the joint spacing.[28] Curling not only lifts up the slab edges, but also induces tensile stresses in the slab. If these stresses accumulate beyond a certain point, the slab can crack. The slab's own weight counteracts the effects of curling and tends to keep the slab level, so curling is less pronounced in thick slabs. One other effect of curling is that when the slab edges are lifted, the slab partially loses contact with the soil. This can lead to joint spalling by wheeled traffic, as illustrated in the next section.

The amount of curling depends on some other variables as well, among them the type of coarse aggregate. Some aggregates produce concrete with high rates of shrinkage and curling. Also, curling can be worsened by the size and position of steel reinforcement, by load-transfer devices at the joints, and by the amount of water in the mix and in the subgrade.

One way to minimize the harmful effects of vapor barriers is to avoid them altogether, if no moisture-sensitive finishes are to be used, as recommended by ACI 302.1,[22] for example. Another is to add a 2- or 3-in-thick "blotting layer" of sand on top of the vapor barrier, to soak up the excess water from the bottom of the concrete and to equalize moisture content across the slab. Some experts question the effectiveness of both of these approaches, and even the whole theory of curling presented here. In any case, we are interested not so much in the theory of curling as in the methods of correcting it.

6.5.4 Repairing settled, heaving, and curled slabs

Depending on the magnitude and cause of damage, heaving, settled, or curled slabs can be left alone, repaired, or removed and replaced. When damage is slight, the repair can consist of a simple slab leveling, done by grinding down the high areas (assuming that the reduced slab thickness is still acceptable). The partly settled slab in Fig. 6.8 could be repaired by patching with an epoxy mortar or similar material, as described in Chap. 5. Following proper patching procedures is essential to success. (For example, the temptation to feather the edges with the mortar should be avoided.)

For major settlement, slabjacking can be very cost-effective. In a slabjacking operation, liquid grout is pumped through holes drilled in the slab; the grout pressure lifts the slab to a desired elevation. Meyers[29] describes the procedure in detail; it is summarized here.

A series of 2- to 2.5-in-diameter holes from 3 to 8 ft apart and at least 1.5 ft from the edges, with larger spacing used for thicker slabs, is drilled. Grout for slabjacking typically consists of water, cement, and either limestone dust or fly ash. Some contractors attempt to use sand or even soil instead of limestone dust or fly ash, a practice that may result in a grout that is weak, difficult to pump, and vulnerable to breakdown. Such a weak grout can cause the slab to settle again in a short period of time. The grout's setting time can be controlled by varying the cement content, which typically ranges from 5 to 20 percent by dry weight. The higher the cement content, the shorter the setting time and the stronger the grout.

The pumping starts at the low side of the slab. When the slab rises 0.25 to 0.5 in, pumping is moved to an adjacent hole. The consistency of the grout can be varied, with stiffer grout used to lift the slab and thinner grout then pumped into the separate support holes to fill the voids under the slab after it has risen. Slabjacking is as much art as science, according to Meyers, and many things can go wrong during the process, among them cracking, grout leaking, and lifting of the slab in the wrong place. The contractor's experience is crucial in obtaining successful results.

A process somewhat similar to slabjacking is slab stabilization or subsealing (undersealing). Subsealing involves filling the voids underneath the slab without lifting it. It is typically specified to stop slab pumping and subgrade erosion or to fill the areas under curled slabs. The procedure is described in Ref. 30.

Another promising variation on the slabjacking theme is polyurethane injection. The process works on the same principle that has made polyurethane grouts successful in stopping leaks in concrete structures: When mixed with water, polyurethane foams up to several times its original volume. But, instead of plugging a leak, the pressure of expanding foam can be employed to raise the slab. One specialized contractor uses a polyurethane mix that produces foam with an upward pressure of up to 8000 psf when injected in $^5/_8$-in holes spaced 6 to 8 ft on centers. The advantages of polyurethane injection are speed (the foam reaches 90 percent of its design strength in 15 min) and a need for much smaller holes. Polyurethane injection is more expensive than slabjacking.[31]

Slabs often settle in a nonuniform fashion. Sometimes, slabjacking lifts the sunk side of the slab but does not depress the raised side. What can be done in this situation? One solution is to raise the adjacent slabs to the elevation of the high side, without depressing that side. When this is not desirable because the size of the resulting bump would be too large, the jacked slab probably has to be replaced. Another solution is possible for a small slab: It could be pried up at the raised side, some soil under it removed, and the slab lowered into position.

Repair of curled slabs involves two operations: sealing the voids under the lifted slab areas, to restore contact with soil, and removing the ridges caused by curling. The voids under the floor can be filled with subsealing. Another approach would be to make small holes through the slab at the lifted corners and fill the voids with flowable nonshrink grout. Some joint grinding is usually sufficient to remove the curled areas, but grinding is expensive and may not be needed if sealing or grouting solves the problem of joint movement (see Sec. 6.6.3). If used, slab grinding may be confined to the areas subjected to wheeled traffic that might damage the ridges resulting from curling. Areas to be covered with ceramic tile can also be leveled by grinding. In lieu of grinding, it might be cost-effective to make a partial-depth patch by removing the top area of concrete and replacing it with epoxy mortar or other patch material. The patched areas would be recut at the joints and receive new joint sealants.[32] Sometimes, it is suggested that the slab be cut near the curled corners, to allow the curled areas to lie flat. This does not seem like a good solution, because it introduces many more joints to maintain—all of them without any means of load transfer across the joint, such as keys or dowels.

6.6   Joint Failures

6.6.1   Joint fillers

Slabs on grade contain three types of joints: *expansion joints* intended to accommodate thermal expansion of concrete; *construction joints* separating the pours; and *control joints,* whose function is to induce the slab to crack at the joint locations rather than elsewhere. The closer the joints are spaced, the better they control cracking. On the other hand, the more joints there are, the more potential problems there are with their performance and periodic maintenance. Indeed, the largest part of a typical floor-maintenance budget goes to joint repair.[5]

A typical expansion joint contains joint filler, which extends through the thickness of the slab. The main task of joint filler is to keep debris and water from entering the joint, without impeding slab temperature and shrinkage movements. The material must be flexible, dimensionally stable, and able to maintain a tight seal under a variety of conditions.

There are two types of joint fillers: preformed and field-molded. Preformed (premolded) compressible materials are made of spongelike elastomers or asphalt-impregnated fibers. Preformed fillers can be installed quickly and without curing, but a separate sealant must be applied on top of the joint. Another type of joint fillers is field-molded materials, among them the rubberlike elastomeric sealants made of polyurethane or polysulfide. Field-molded fillers are poured into joints and cure in place; they do not need a separate sealant. A typical construction or control joint contains only a field-molded filler or sealant applied after the slab is saw cut or indented by removable inserts.

In buildings *without* moving equipment, the most important property of joint filler is great flexibility: the thin joints–they are rarely wider than $\frac{1}{4}$ in—must be capable of undergoing extreme movements, as the rate of drying shrinkage is about $\frac{3}{16}$ in for a typical 25-ft slab panel. In industrial facilities *with* moving equipment, the function of construction or control joint fillers is more complicated. There, the fillers must not only be capable of large elongation and contraction, but also provide good support for the joint edges, to prevent their unraveling under wheel traffic.

Unfortunately, there is no ideal material to meet these two conflicting requirements: The ease of movement is best satisfied by elastomeric sealants, while proper edge support is best provided by rigid mortars or epoxies. The compromise usually reached is to specify semirigid epoxy joint fillers, even though these materials do not excel at either task. Another class of materials that was recently developed for industrial applications is polyurea joint fillers. Polyureas have physical properties similar to those of semirigid epoxies, but unlike epoxies, which typically require a minimum application temperature of 50°F, they can be placed at low temperatures. Polyureas also cure faster than epoxies, but are more flexible and deflect more under load.

6.6.2   Failures of joint fillers and sealants

Joint fillers need periodic replacement when they show evidence of splitting or loss of bond. The most common source of joint deterioration is aging. Joint failures can also be caused by overload, poor subgrade preparation, or improper joint design, construction, or spacing. The symptoms of these problems range from squeezed-out sealants to unraveled joint edges.

The simplest type of joint repair—maintenance, really—is replacing the joint filler materials. This operation typically involves removing the old joint fillers and/or sealants to a depth needed for installation of new materials. The job is done with the help of a hooked bar that is maneuvered under the old sealant and pulled up, after which the remnants are cleaned off with an electric wire brush. If the existing joint edges are in good shape and the joint is deep and wide enough, new field-molded elastomeric sealant can be poured directly into the cleaned-out space. Allen et al.[19] recommend placing masking tape along the joint edges, so that after the filler is dispersed and troweled flush with the concrete, any surplus filler material can be easily removed from the surface.

If the existing joint filler is being squeezed out or pulled out of position, this probably means that the joint size and spacing are insufficient. Under these circumstances, instead of simply replacing the filler, it is wise to evaluate the adequacy of the joints and perhaps to add some new joints or enlarge the existing ones. Otherwise, the repair may be short-lived. The joint can be made wider and deeper by saw cutting; it can also be opened up in a procedure similar to that used for cracks (Fig. 6.6).

Introducing new or larger joints can cause new problems: The new or enlarged joints will break the slab into structurally separated areas. Unlike construction joints, where keys or dowels help transfer the load across them, and control joints, which rely on aggregate interlock for this purpose, the new expansion joints (that's what they will be) have no such means of load transfer. This may present a problem in buildings with moving equipment or slab-supported racks.

The worst thing that can happen to a joint when its flexible filler fails is to be filled with a rigid epoxy. The epoxy essentially welds the slab together and destroys any ability of the joint to accommodate thermal or drying shrinkage movements. As a result, new cracks can be expected to appear nearby.

6.6.3   Repairing spalls in joint edges

What causes the common problem of spalled joint edges? In nonindustrial buildings, joint edges have been known to unravel from saw cutting them too early and from overtroweling. Spalls can also occur if the joint was jammed with stones and other debris that prevent slab movement. In exterior slabs, spalls that occur at four-way joint intersections are typically blamed on freeze-thaw damage (Fig. 6.12).

In industrial buildings containing hard-wheeled moving equipment, joint spalls are most commonly caused by wheel impact. The spalls are most likely to appear in slabs with edges that deflect under wheel loads and in curled slabs. Figure 6.13 illustrates what happens: As the wheel rides over one edge of the joint, that edge moves downward, exposing the opposite edge to impact and eventual spalling.[33]

Curling of the slab edges is easy to detect by placing a level across the joint, but some investigative work is necessary to verify whether the joint indeed moves under load. The simplest way to find out whether that occurs is to observe the joint as the equipment passes over it. Deflection of the joint edges under a moving wheel is easily noticed by the naked eye; a more precise measurement can be made

Figure 6.12   Spall at a four-way joint intersection.

with the aid of a laser level. However it is detected, joint instability indicates poor condition of the subgrade. (The subgrade can be repaired as described in Sec. 6.5.4.)

How much movement would justify repairs? The precise amount is a matter of debate. While some sources recommend repairs when joint movements are only 0.02 to 0.03 in, others consider this overly conservative. In any event, the floor's age is important. As Suprenant and Malisch[32] observe, joint movements of 0.03 in in a 10-year-old slab are not likely to worsen, and the slab probably need not be repaired, but in a new slab, this amount of movement is a cause for concern, because it will probably lead to deterioration.

Metzger[17,34] identifies several other causes of joint spalls. The first is insufficient edge support provided by the existing joint filler. Elastomeric fillers can absorb temperature movements well, but are too flexible to protect the joint edges from wheel impact (Fig. 6.14a). Conversely, the edges of joints filled with high-strength epoxy materials are essentially glued together and may spall or crack when slab shrinkage stresses pull them apart (Fig. 6.14b).

The improper use of metal keys at construction joints can also contribute to edge spalling. Good practice dictates that a joint with a metal key be saw cut and filled after concrete placement. The intent is to prevent shrinkage stresses from pulling the edges apart and creating a gap filled with debris. If saw cutting and filling is not done, the unsupported overhanging concrete portion can break under wheel

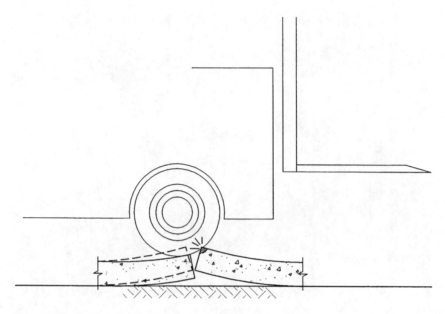

Figure 6.13 Failure of curled joint edge caused by a hard wheel.

loading (Fig. 6.14c). The plastic crack-inducing strips sometimes used to form control joints can also create a problem: If placed out of plumb, they create a similar overhanging area of concrete that is vulnerable to spalling (Fig. 6.14d).

Regardless of the actual cause, joint edges should be repaired to stop further deterioration. Figure 6.15 illustrates a method of repairing badly deteriorated joint edges in spalls wider than 1.25 in, as recommended by Metzger.[35] This method involves removing the damaged areas by making saw cuts at least $\frac{1}{2}$ in deep, and preferably 1 in deep, cleaning out the debris, and inserting a temporary plastic strip covered with a bond breaker in the middle of the joint. The exposed edges and bottom of the joint are coated with a primer consisting of neat epoxy. Then the joint is filled with high-strength epoxy mortar made by mixing epoxy and sand, and the reconstructed area is finished with a trowel or left slightly protruding. Once the mortar sets, the temporary plastic strip is removed, the joint cleaned, and the remaining space filled with a semirigid epoxy joint filler. After the filler is cured, the top of the joint is ground flush with the floor surface.

A variation on this method involves saw cutting the newly formed area instead of using the removable strip. The cut should be made as soon as possible after placement, be as wide as the original joint, and extend the full depth of the patched area.[5]

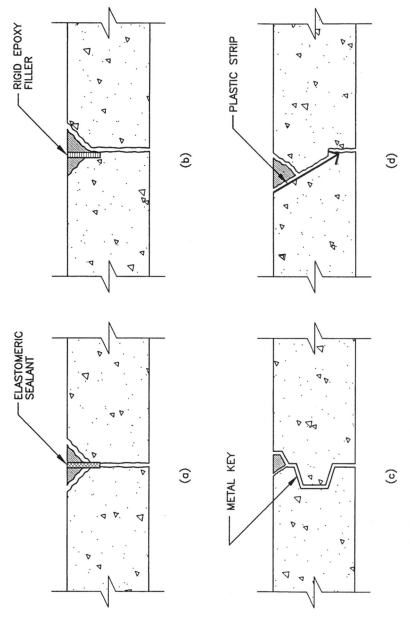

Figure 6.14 Spalling of joint edges caused by (*a*) too-soft elastomeric sealant; (*b*) too-hard epoxy filler; (*c*) metal key; (*d*) improper placement of plastic crack-inducing strip. (*After Ref. 34.*)

347

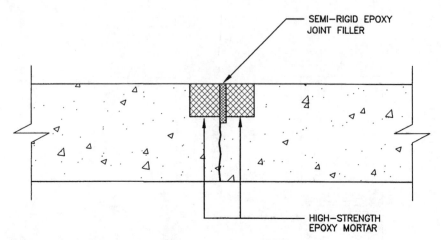

Figure 6.15 Repair of badly deteriorated joint edges. *(After Ref. 35.)*

For spalls less than $1^3/_{16}$ in wide, the suggested repair is similar to that shown in Fig. 6.6, except that the filler material is semirigid epoxy. The typical depth of this repair is 0.5 to 1 in. Metzger recommends using a semirigid epoxy with a minimum Shore hardness of A80 or D50 as specified in ASTM D2240. For areas larger than 0.5 in, he suggests adding aggregate to the epoxy to increase the filler's ductility. A sample formula involves 2.5 parts of silica sand and 1 part epoxy.

Similarly, ACI 302.1R[22] recommends that epoxy resins used as joint fillers consist of 100 percent solids and have a minimum Shore D hardness of 50 and elongation of 6 percent. The minimum specified hardness level allows the filler to protect the slab edges from spalling under the impact of hard wheels, but stay flexible enough to deform under joint movements. In buildings without moving equipment, there is no danger of unraveling the joint edges, and elastomeric sealants can be safely used instead of semirigid epoxy.

6.7   Water Penetration or Emission

6.7.1   The signs of problems: Damage to floor finishes

Water penetration through slabs on grade concerns owners a great deal and is frequently on the wish list of items to repair. Excessive moisture is often accompanied by growth of mold and mildew and a concurrent rise in the occupants' allergies. The most common problem is damage to floor finishes and coatings, as no finish in existence is totally immune to water. The signs of water damage are unmistakable: Wood flooring becomes warped, discolored, and decayed; carpet bubbles up and gets covered with water stains and leached salt deposits; industrial sealers blister and delaminate; and vinyl tiles loosen, curl, and spall off.

Moisture can get through the slab in a variety of ways. One avenue for steady water intrusion is capillary action, when groundwater rises through the minute channels in the soil and slab. The second is movement of water vapor from areas of high concentrations to those that are relatively dry. The third is leakage of surface water or leakage from broken pipes. And finally, the most significant amounts of penetration could be caused by hydrostatic pressure of groundwater.

And then there is water contained in concrete pores. Concrete does not cure overnight: As discussed in Sec. 6.3.1, it can take nearly 2 years for a slab to undergo 75 percent of its ultimate drying shrinkage and to lose three-quarters of its free pore water. Few owners are willing to wait that long to apply floor finishes. A potential for problems is obvious.

How does the water contained in concrete actually damage finishes? Some experts believe that blistering and delamination of sealants, hardeners, and other coatings is caused by vapor pressure. However, Gaul[36] demonstrates that it is quite improbable that that is the case, as the magnitude of capillary pressure never approaches the bond strength of a typical coating. Instead, he proposes that moisture damage to floor coatings occurs during their curing (or curing of the adhesives), because moist concrete can either prevent proper cure of the coatings or prevent their bond to the surface.

Improperly cured coatings are typically soft, sticky at the bottom, and easily removed, and can retain the smell of one of the components. Poorly bonded coatings can also be easily removed, but their bottom surface is hard; after their removal, the slab appears damp. In both cases, blisters may be present. In improperly cured coatings, blisters may be small (less than 0.25 in), closely spaced, and numerous. In poorly bonded coatings, blisters are typically large and are sometimes filled with water.

To be sure, sealers and coatings can delaminate not only because of excessive moisture in the slab, but also because of poor finishing practices. Chief among these is hard troweling, which tends to bring excessive laitance to the top. The life of a sealer applied on top of the laitance will be short. Also, failures of brittle floor finishes, such as breakage or delamination of ceramic tile, can result from slab shrinkage. As already noted, shrinkage is likely to continue well after these brittle finishes are applied. Failures of floor finishes can also be caused by slab curling and by expansion and contraction from moisture and temperature changes.

The speed of water entry is greatly facilitated by cracks, but some moisture will migrate even through a perfectly dense slab, unless it is protected by waterproofing. Regardless of how water gets through the slab, most of it evaporates inside the building. The level of evaporation depends on the relative humidity within the building. Typically, moisture migration continues undetected, unless floor finishes are present.

There is a big difference between the amount of infiltration that produces only minor mustiness and the amount that causes damage to finishes. To assess the severity of water penetration and to develop a rational repair strategy, the actual amount of water entering through the slab must be measured first.

6.7.2 Measuring water infiltration

The first step in planning repairs of water-related problems is to determine whether significant moisture emission through the floor is actually taking place. A simple check involves taping an 18-in-square sheet of 4-mil-thick polyethylene to the slab, waiting at least 16 h, and removing the sheet. If the slab is damp or wet, or if water has condensed on the plastic, excess moisture is indeed present. This qualitative test is described in ASTM D4263-83, *Standard Test Method for Indicating Moisture in Concrete by the Plastic Sheet Method.* The test does not tell whether this moisture is the result of water coming in from outside or of emission of the pore water from concrete.

When the signs of moisture emission are unmistakable, the next step is to quantify the amount. The *moisture dome test* commonly used for this purpose was developed by the Rubber Manufacturers Association Inc. and is sometimes called the RMA test. It is also described in the new ASTM F1869-98, *Standard Test Method for Measuring Moisture Vapor Emission Rate of Concrete Subfloor Using Anhydrous Calcium Chloride.* The test involves putting a dish on the slab, placing a premeasured amount of calcium chloride into it, covering the dish with transparent plastic, and sealing the edges of the cover to the floor (typically with duct tape). Any moisture transmitted through the slab will be absorbed by the calcium chloride.

The amount of accumulated water is the difference between the original and final weights of the calcium chloride. The test determines the *vapor flow rate V*—the emission rate per 1000 ft^2 of slab. The vapor flow rate is equal to the weight of water Q multiplied by 1000 and divided by the sealed area A and by the number of days the test was performed T:

$$V = 1000 \, \frac{Q}{AT} \text{ lb/day}$$

The moisture dome test should run for at least 60 h. The test is not very accurate—it can underestimate the emission rates when major amounts of moisture penetration are present. Still, it gives a simple and inexpensive means of assessing the severity of the problem.[37]

What amount of moisture penetration is acceptable? No definite "safe" number exists, but a maximum emission rate of 3 lb/day is a common ceiling recommended by various flooring manufacturers for safe installation of their products. According to the trade publications quoted by Day,[37] this rate is acceptable for rubber, vinyl, most carpet, and perhaps wood. Carpet with porous backing and some types of vinyl can usually tolerate a vapor flow of up to 5 lb/day. Suprenant and Malisch[38] have found that the bond of the adhesives used to attach the flooring decreases dramatically when the water vapor emission rate exceeds 5 lb/day.

However, concrete dries slowly, and it typically takes at least 3 months—and perhaps much longer if the concrete has a high water-cement ratio—to reach an emission rate of 3 lb/day.[10] Malisch[39] describes a case in which the emission rates of a 3-month-old slab on grade ranged between 7 and 10 lb/day. It is quite probable, therefore, that concrete will have a much higher vapor emission rate when the floor finishes are installed than is allowed by the manufacturers of the finishes.

Other methods of measuring moisture content in concrete include using embedded meters and surface meters. It is also possible to remove a concrete core and determine the amount of moisture it contains by drying it in an oven.

6.7.3 Methods of repair

As we have seen, it is relatively easy to determine whether damage to floor finishes is caused by excessive levels of moisture penetration. To pinpoint the exact source of moisture may not be that easy, and to fix the problem is even more difficult.

The commonly prescribed means of protecting *new* slabs from excessive levels of moisture penetration include placing a vapor barrier under the slab, with a 4- to 12-in-thick gravel course under the vapor barrier. Gravel is needed to prevent water rise by capillary action; to be effective, it should be "clean"—without void-clogging fines and silt.

Adhering to these rules does not guarantee success, as demonstrated by the case studies presented by Amundson et al.[40] (The case studies also suggest that sometimes it is nearly impossible to find the exact source of water penetration.) Also, as noted already, many engineers prefer to avoid vapor barriers altogether, because they can aggravate shrinkage, curling, and delamination problems. To reduce these problems caused by vapor barriers, some sources suggest placing a blotting sand layer on top of them. Other experts argue that the sand is unnecessary, because it tends to soak up water and simply hold it under the slab. Still others advocate replacing the blotting sand layer with a minimum of 4 in of clean gravel.[41] The cases for and against the granular layer were debated in some depth by Suprenant and Malisch.[42]

But what can be done with an *existing* slab on grade that is leaking moisture from below? The radical solution would be to remove the existing floor and replace it with a slab protected at the bottom by an impermeable waterproofing membrane. This approach is frequently pursued in cases of massive amounts of water infiltration caused by hydrostatic pressure of groundwater. One difficulty of implementing it concerns the edge terminations of the waterproofing membrane. In new construction, the whole building envelope can be protected by splicing the underslab waterproofing with that at the exterior of the walls into a single impermeable barrier. When only the slab is replaced, the edges are very difficult to seal, and coating failures after these repairs are frequent.

A light-duty variant of this solution involves using a polyethylene vapor barrier covered with a blotting sand layer in lieu of the waterproofing membrane under the replaced slab. Here, the problem of sealing the membrane edges is compounded by the difficulty of joining separate polyethylene sheets and maintaining the vapor barrier free of punctures. The practical challenges make this repair method difficult to recommend. Another possible solution is a program of removing groundwater by drainage or permanent dewatering. One potential problem with long-term lowering of the water table is that it could cause the surrounding building foundations to settle.

The least expensive repair, appropriate for modest rates of water infiltration, is to seal the cracks in the slab and replace the existing floor finishes with moisture-tolerant ones. These finishes include breathable coatings and membranes that do not trap moisture, even though they may not protect concrete as well as impermeable floor coverings. An example of a breathable membrane is a latex-modified cementitious coating $1/16$ to $1/8$ in thick covered with a fast-setting primer.[36] For office settings, synthetic carpet with porous backing can be used. The next logical step is to convince the owner not to use any floor finishes at all, if possible.

Another avenue is to seek out quality floor-covering products. An interesting point is made by Cox,[41] who argues that the recent increase in flooring (mostly vinyl tile) failures should be blamed more on poor adhesive formulations than on excessive concrete moisture. He explains that, while concrete design has not changed much in many years, tile adhesives have changed from solvent-based to water-based to comply with EPA requirements.

The failure of finishes placed on still-curing concrete with excess moisture can be remedied relatively easily, at least conceptually: Remove the finishes, let the slab drying run its course, and reinstall new finishes when the vapor emission rate is within the allowable limit. Unfortunately, this theoretically correct approach ties up building operations for an indeterminate amount of time and is not popular with owners. Recent research[43] confirms that it is difficult to accurately predict the drying time required to reach a specified emission rate.

Can the slab be dried out faster? Malisch[39] suggests one solution: Retain a firm specializing in dehumidification services. The workers will typically enclose the floor with plastic barriers and dehumidify the resulting plenum by circulating dry air. A less expensive alternative is to use portable dehumidifiers or large exhaust fans to remove moist air. It still takes some time—and money—to realize the benefits of either approach.

6.7.4 The case of bulging joints in floor coverings

Our discussion about covering the still-curing slabs with impermeable coatings should include a particular problem. It concerns floors placed on vapor barriers and covered with vinyl composition tile (VCT) or with high-build coatings. According to Metzger,[44] some of these floors experience bulging of semirigid epoxy joint fillers and loosening and cracking of floor coverings. The problem is relatively rare; it affects less than 2 percent of installations involving these materials.

Owners tend to assume that this problem is a result of some chemical reaction between the filler and the floor covering. However, the most sensible explanation is this: By the time floor covering is applied, the concrete at the top of the slab has lost much more moisture than the concrete at the bottom. Therefore, some shrinkage of the top surface will already have occurred prior to placement of finishes (see the discussion on curling in Sec. 6.5.3). Once the covering is applied, evaporation slows down, and the moisture level at the top of the concrete increases again. The top surface of the concrete reexpands, squeezing the joint fillers in the process.

The simple method of repair offered by Metzger involves removing the floor covering, cutting the joint-filler bulge down to the top of the concrete, and reapplying the finish. This procedure may have to be repeated. To prevent this problem in new construction, he recommends filling the joints that receive floor coverings earlier than normal—15 to 30 days after slab placement, or 15 days before the covering is applied—rather than following the usual rule of filling the exposed slabs as late as possible. The early application of the filler gives the filled joint a chance to open before the reexpansion takes place. Other recommendations include reducing the joint spacing and allowing for compression space by filling the joints to only about 75 percent of their depth.

6.8 Chemical Attack

6.8.1 Investigation and repair choices

Aggressive chemicals are used in many industrial buildings for cleaning or manufacturing processes. These chemicals can cause damage to concrete. The general issues of dealing with chemical attack are discussed in Chap. 5; here, our focus is on floors.

When an existing slab on grade is damaged by chemicals, the fundamental decision that must be made is whether the floor can remain in service or should be replaced. The answer may appear to be obvious if the slab is visibly disintegrated, but the visual clues could be misleading. For example, as mentioned in Chap. 5, one study concluded that beams suffering from an advanced alkali-silica reaction actually did not behave much worse than control specimens. The corrosive effects of some chemicals on concrete are not entirely clear, and minor surface deterioration may appear worse than it is. A definitive method of strength investigation involves slab coring and compressive testing.

One example: Are greasy stains from lubricating and fuel oils that are frequently spilled on the floors harmful? In theory, these oils should not attack well-cured concrete. However, some additives in them, such as animal or vegetable fats, may be corrosive. Also, some researchers point out that used lubricating oils become more acidic than new oils, because of oxidation. Is the increased acidity harmful? There is some evidence[45] that prolonged exposure to petroleum products does in fact cause deterioration of concrete. A good source of information about the effects of various chemicals on concrete can be found in Ref. 46.

Even if they are structurally adequate, rough and partly deteriorated slabs may interfere with cleaning or with production operations. What should be done with these slabs? Should they be patched? Should they be replaced? There are circumstances in which patching is appropriate, but in most cases it is counterproductive. The chemicals are likely to migrate into the patches, and deterioration will continue. A better course of action is to patch the damaged areas and cover the floor with a membrane that is impervious to the chemicals. If the existing floor does not yet exhibit significant deterioration, it may be possible to use a variation of this theme by simply covering the slab with the protective material.

The most radical solution is to remove and replace the damaged concrete, as discussed in Sec. 6.9. The cost of replacement may only slightly exceed that of a thorough patching, but the new concrete can be made more chemically resistant and survive longer. In either case, the slab can be protected with a coating or overlay that is unaffected by the chemicals. Emery and metallic floor toppings installed primarily for abrasion resistance (and discussed in Sec. 6.11) can also improve the chemical resistance of concrete.

At the other extreme, if the existing concrete is in good condition and the chemical attack is extremely mild, the least expensive solution is simply to coat the slab with a good sealer. The popular silane and siloxane sealers not only are water repellants, but also can prevent absorption of water-soluble chemicals. These sealers need to be reapplied every few years. Silanes and siloxanes are not very effective against absorption of oils, but low-solids acrylic and urethane sealers are.

6.8.2 Using repair materials with moderate chemical resistance

For an intelligent selection of the repair method, one must learn about the chemical resistance of various repair materials. The discussion in this section is based largely on McGovern.[47]

Most of the materials with moderate chemical resistance are polymer resins applied as bonded coatings in thicknesses ranging from a surface film to $3/_8$ in. The thicker coats may be reinforced with fabric to bridge over cracks. Most of these materials require a dry substrate for successful application. Our discussion deals with generic materials, and there is some variation in properties within each category. It is therefore important to investigate the actual formulation and chemical resistance of the proposed product before specifying it.

Epoxies are perhaps the most commonly used chemical-resistant coatings. The properties of these two-component materials can vary, but most epoxies are impervious to moisture, bond well to concrete, and are inexpensive in relation to some of the other materials discussed here. Among their shortcomings is sensitivity to ultraviolet light, a tendency to soften and creep at high temperatures, and slow curing in cold environments. In general, epoxies are resistant to alkalies and to weak and moderate acids. *Epoxy novolacs,* more dense than regular epoxies, are resistant to many solvents and can withstand higher temperatures.

Polyester coatings can be used at higher temperatures than either epoxies or epoxy novolacs. Polyesters are frequently specified for their bleach resistance. *Vinyl esters,* with a slightly different composition, have a better chemical resistance to weak and moderate acids, weak alkalies, and solvents. Both polyesters and vinyl esters require a minimum application temperature of 45°F, and both can accept foot traffic in a few hours. Both have a strong odor. Among their weak points are a high coefficient of thermal expansion and a high shrinkage rate, meaning that the coating and the concrete substrate will move at slightly different rates and could eventually lose bond with each other.

The situation can be improved by extending these materials with silica sand or other fillers and by reinforcing them with fiberglass fabric.

Potassium silicates are known for their excellent acid resistance (except to hydrofluoric acid, HF). They are ideal for floors subjected to hot acid spills. Unfortunately, potassium silicates have many disadvantages, among them poor resistance to alkalies, very high shrinkage rates, and susceptibility to cracking. To some degree, these shortcomings can be mitigated by reinforcing the material with stainless steel fibers and by placing it on a membrane that is resistant to the chemicals used. Potassium silicate toppings must be applied in a minimum thickness of 1.5 in.

Urethane coatings are widely used because they resist both acids (in weak and moderate concentrations) and alkalies. Owing to their high cost, urethanes are often applied as topcoats over other materials, usually epoxies. The desirable properties of urethanes include fast curing time (about 12 h), good abrasion resistance, and toughness. These materials can be classified as either two-component (those requiring the mixing of two ingredients for curing) or single-component urethanes (those cured by moisture). Two-component urethanes are normally superior to the moisture-cured variety, although the latter remain popular because of their ease of application. Another classification based on slight chemical differences divides urethanes into aliphatic and aromatic. The main advantage of aliphatic urethanes is their UV resistance.

When a repair material must have a fast curing time or must be placed at low temperatures, *methylmethacrylate* (MMA) can be the answer. This two-component material cures in as little as 1 to 2 h. The twin advantages come at a steep price: MMA can withstand only weak acids and alkalies, plus some solvents, and requires specially graded aggregate fillers. Although MMA is UV resistant, which makes it suitable for exterior slabs, it should not be used at temperatures exceeding 140°F. Some additional discussion of MMA can be found in Chap. 5.

The chemical resistance of various materials is summarized in Table 6.1. A note of caution: The degree of chemical resistance of all these coatings depends on their integrity, since no coating can fully protect against chemical attack through cracks or chipped and abraded areas. And certainly no coating can help against chemicals that seep under the slab from outside and degrade the concrete from below.

6.8.3 Using repair materials with excellent chemical resistance

A long-term exposure to strong concentrations of aggressive chemicals typically requires better protection of concrete than simply coating it with one of the materials listed in the previous section.

TABLE 6.1   Chemical Resistance of Various Materials

Type of surface protection	Maximum service temperature, °F/°C	Alkalies		Acids				Solvents		Bleach
		Strong	Weak	Inorganic	Weak	Organic		Organic	Inorganic	
						Moderate	Strong			
Epoxies	150/66	✓	✓		✓	✓		[a]	[a]	✓
Epoxy—Novolac	180/80	✓	✓	✓	✓	✓		✓[b]	✓[b]	
Furans	360/180	✓	✓	✓	✓	✓	✓	✓	✓	✓
Methacrylates (MMA)			✓		✓	✓		✓[b]	✓[b]	
Polyesters	230/107		✓	✓[c]	✓	✓	✓	✓[a]	[a]	✓
Potassium silicates	2000/1093				✓	✓				
Sulfur cement	190/88			✓[d]	✓	✓				
Urethanes	150–250 / 66—122	✓	✓	✓[d]	✓	✓				
Vinyl esters	220–250 / 104–127		✓	✓	✓	✓		✓	✓	✓
PVC	High	✓	✓	✓	✓	✓	✓			
Acid brick		✓	✓	✓[c]	✓	✓	✓			
Carbon brick		✓	✓	✓	✓	✓	✓			

[a]Resistant only to some solvents.
[b]Moderate resistance to solvents.
[c]Not resistant to hydrofluoric acid (HF).
[d]Moderate resistance to acids.

Note: This table should be used as a guide only. Actual performance may differ depending on formulation.

SOURCE: Peter H. Emmons, *Concrete Repair and Maintenance Illustrated*, copyright R.S. Means Co., Inc., Kingston, Mass., (781) 585-7880, all rights reserved, 1993.

Of particular concern for slabs on grade and foundations are sulfates (SO_4), which are sometimes present in soil or groundwater. Sulfate attack can necessitate the use of sulfate-resisting concrete. Unlike the coatings, sulfate-resisting concrete is chemically resistant throughout its thickness. The recommended proportions of this material—the type of cement, the maximum water-cementitious materials (w/c) ratio, and the minimum concrete strength—are described in Table 4.3.1 of ACI 318[48] for various degrees of sulfate exposure. For the most severe exposures, the table recommends using Type V cement with added pozzolans, a maximum w/c ratio of 0.45, and a minimum concrete strength of 4500 psi. ACI 318 also reminds us that adequate air entrainment, low slump, and proper consolidation and curing practices are essential for achieving durable sulfate-resisting concrete. We should add that these measures will also make concrete resistant to moderately acidic water. Another material with excellent sulfate resistance is sulfur concrete, discussed in Chap. 5.

When strong acids are involved, the ultimate solution is paving the slab with *acid brick* joined by chemical-resisting mortar. Acid brick can withstand all acids—some in very high concentrations—except hydrofluoric acid. It offers poor defense against chemicals other than acids. In contrast, *carbon brick* is resistant to both acids and alkalies, but it is more expensive and more absorptive than acid brick.

Acid brick is similar to regular face brick, but is fired longer to make it less porous. Still, porosity is in the nature of the material, and no brick, acid or not, will stop chemicals from eventually penetrating it and attacking the concrete below. For this reason, a chemically impervious membrane, typically made of asphalt or modified urethane, is added under the brick. Why, then, use acid brick at all? Mainly because of its high durability and abrasion resistance; the membrane alone will not last under foot and forklift traffic. The thickness of the brick can range from $1^3/_{16}$ in for relatively light traffic to $2^1/_2$ in for buildings with heavy trucks. Even thicker bricks, up to $4^1/_2$ in, can be used for especially severe applications.[47]

Acid brick is expensive and is reserved for the most demanding environments, such as floors of chemical factories. The brick floors should be sloped for drainage, so a separate underlayment course often needs to be placed on the existing slab. Expansion joints in the brick are required over the slab expansion joints, around the perimeter, and at all drains and similar penetrations.[49]

To be effective, acid and carbon bricks must be joined with one of the several available chemical-resisting mortars. A good overview of the mortars, which we used as a reference, is given by Wallace.[49] As with overlay materials, there is no single formulation that can resist all aggressive chemicals. One of the best and most popular mortars is made with *furan,* a two-part material based on organically derived alcohol combined in the field with catalyst and filler. As Table 6.1 shows, furans possess excellent resistance to most types of chemicals except bleach, and are specified for a wide variety of uses. Furans made with 100 percent carbon fillers are especially durable and have low absorption, but even they are not infallible to strong concentrations of nitric and sulfuric acids.

Other chemical-resisting mortars are made with the same materials listed previously as coatings. For instance, epoxy, truly a versatile formulation, can produce the strongest mortars. Epoxies based on bisphenol-F resin are especially acid-resistant. For applications involving accidental spilling of bleach, the best mortar to use is polyester-based. Vinyl esters are well suited to withstand the chemicals found in pulp and paper plants. If the floor must act as an electrical insulator, phenolic mortars with 100 percent silica filler are useful. The minimum requirements for organic mortars—those made with furans, epoxies, vinyls, polyesters, and phenolic resins—are contained in ASTM C395, *Standard Specification for Chemical-Resistant Resin Mortars.*

When the conditions of use require acid resistance combined with very high temperatures—up to 2000°F—(as in chemical plants making sulfuric acid, for example), a mortar made of potassium silicate would be a good choice. Sulfur mortars can be used if high early strength, economy, and resistance to oxidizing acids and salts are desired.

6.9   Slab Replacement

Concrete removal and replacement was mentioned earlier as the ultimate method of slab repair. It is unavoidable during modification or relocation of existing utilities embedded in the floor or placed under it. There is no alternative to replacement of badly deteriorated slabs, such as that shown in Fig. 6.16. The new slab can have its thickness, protection membrane, shake-on hardeners, and joints custom tailored to the specific requirements of the facility. These obvious benefits are tempered by the need to interrupt operations and by the cost of equipment removal and of slab demolition and disposal. Also, the color of the replacement slab will be slightly different from that of the rest of the floor, and therefore it will be easily noticeable.

From a structural standpoint, slab replacement is likely to destroy any tension-tie action of the floor. In some structures, such as metal building systems, slabs on grade are occasionally designed as tension ties between the opposite lines of columns. As the author demonstrates elsewhere,[50] cutting and replacing the slab in such circumstances can totally destroy the system of lateral load transfer. In these cases, and whenever the structural continuity of the slab is essential in the

Figure 6.16   This badly deteriorated slab should be replaced.

design, the new and old slabs can be tied together by drilling holes into the sides of the opening and placing new reinforcing dowels into chemical anchors (Fig. 6.17).

Existing slabs can be removed by saw cutting, jackhammering, or hydrodemolition. Saw cutting is often preferred for removal of slabs of moderate size and thickness, although it requires care to avoid damage to the surrounding concrete. In particular, the workers should be careful not to overcut the corners of the proposed opening. The best cutting technique is to start by coring out each corner and then saw cut from core to core.

When a new slab segment is placed within the existing floor, and the original soil compaction was poor, measures must be taken to improve the existing subbase or subgrade. The soil may be densified by portable plate compactors or, for small areas, simply removed and replaced with crushed stone or flowable fill, both of which require no compaction. During slab removal, some soil can be lost from under the edges of the existing slab. One way of dealing with this problem is to specify that the new concrete extend under the old slab and fill any voids (Fig. 6.18).

Saw cutting leaves the edges of the slab smooth. Therefore, the aggregate interlock between two sides of the joint is lost. To transfer loads across the joint, bar dowels placed in chemical anchors are typically used. In some industrial slabs on grade with extremely heavy traffic, dowels alone may not suffice. In such cases, one can use a technique found in replacement of highway pavements that combines dow-

Figure 6.17   Drilled-in dowels for partial replacement of slab.

els with cut-in keys and aggregate interlock. To make a key, the slab is cut twice. The first cut is full-depth and the second of partial depth, to about one-third of the slab thickness. The cuts are parallel to each other and spaced $1\frac{1}{2}$ in apart. A key is made by chipping concrete from the area between the two cuts. Load transfer across the joint can be increased by roughening the smooth surface of the cut slab with a lightweight pneumatic chipping hammer.[51]

After the replacement slab is cured, the interface with the surrounding concrete is checked to make certain that both surfaces are aligned. If they are not, some grinding may be required. The joint can be saw cut and filled with an elastomeric joint filler sized to accommodate the anticipated slab shrinkage rate.

How soon after a slab replacement can floor finishes be reapplied? This is a thorny question, because the answer is not likely to please the owner. As discussed in Sec. 6.7, successful installation of floor finishes demands that the slab's rate of moisture emission be below a certain maximum. It is difficult to accurately estimate the drying time required to reach a specified emission rate, but this time is measured in many months, and perhaps years. A delay of such magnitude is difficult to accept even during new construction, when the building is not scheduled to open immediately after the slab placement. It can be infuriating to the owners of buildings undergoing renovations when the new slab must stay out of service because the finishes cannot be

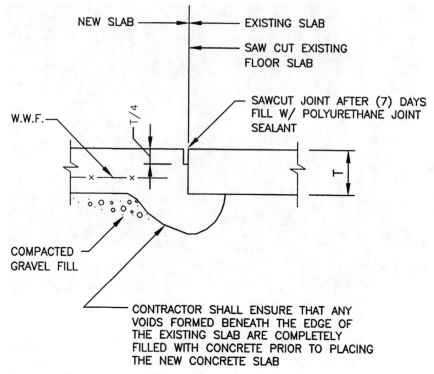

NEW SLAB ————————— EXISTING SLAB

SAW CUT EXISTING
FLOOR SLAB

SAWCUT JOINT AFTER (7) DAYS
FILL W/ POLYURETHANE JOINT
SEALANT

W.W.F.

$T/4$

T

COMPACTED
GRAVEL FILL

CONTRACTOR SHALL ENSURE THAT ANY
VOIDS FORMED BENEATH THE EDGE OF
THE EXISTING SLAB ARE COMPLETELY
FILLED WITH CONCRETE PRIOR TO PLACING
THE NEW CONCRETE SLAB

Figure 6.18 New concrete extends under the existing slab to fill any voids caused by slab removal.

replaced. A step that can speed up the drying process is dehumidification, either by a specialized service or by portable dehumidifiers; a few large exhaust fans can also help. Another approach could be to use special floor sealers designed to limit moisture emission rates; however, the approval of the floor-finish manufacturer should be obtained first. Instead of using tile or stone, one could choose a look-alike made of stamped concrete coated with a permanent concrete stain. And, of course, there is always the possibility of using a breathable floor covering—or no covering at all.

6.10   Slab Overlays

6.10.1   Overlays vs. resurfacing

When only the surface of the slab is damaged, floor replacement may be too drastic a solution. Unless the existing concrete is totally disintegrated or covered with many deep cracks, it is usually more economical to overlay it with a new material that has the desired chemical or abrasion resistance.

The overlay materials include many of the repair products discussed in Chap. 5, among them portland cement concrete of either normal or high-early-strength varieties. Overlay concrete can be modified with latex or polyvinyl acetate (PVA) or can contain high-alumina cement. The topping material can be also based on two-component polymer or resin products—epoxies or polyesters.[5] Of all these, concrete with high-early-strength cement and latex-modified concrete are the most commonly used.

Slab overlays can be applied as patches, thin bonded toppings, or unbonded toppings. All of these can be sloped for drainage and may include the abrasion-resisting aggregates described in Sec. 6.11. The minimum depth of an overlay is typically 1 to 1.5 in, allowing for addition of pea-gravel aggregate to reduce shrinkage and cost. Patching is useful for partial overlays from 0.75 to 2 in thick. The suggested procedures for patching and the available materials are described in Chap. 5.

Epoxy-modified concrete toppings require great care in installation and curing. Murray[52] investigated several cases involving applications of these toppings. He found that epoxy-modified concrete toppings performed extremely well but were marred by widespread superficial shrinkage cracking. He attributed the cracking to the very low water-cement ratios of these materials and suggested that there could be insufficient water in the mix to fully hydrate the cement, so that an early moisture curing was critical.

Slab resurfacing is a largely cosmetic treatment intended to revitalize cracked and tired-looking concrete slabs by giving them interesting colors and textures. The finish surface may look like a brick pavement, stone, or terrazzo. Resurfacing is essentially the application of a thin overlay of polymer-modified concrete with a patterned template. The ingredients—portland cement, fine aggregate, liquid polymer, and pigment—are mixed with water at the site and applied in one or two coats.[53] Uneven floors can be leveled with a cementitious grout prior to resurfacing. Like any overlay, this procedure requires a thorough surface preparation. A typical resurfacing operation involves placing a base coat, installing the template, covering it with the finish coat, and sealing the surface. Some additional information about slab resurfacing can be found in Ref. 53 and in the manufacturers' literature.

6.10.2 Surface preparation, placement, and curing for overlays and toppings

A successful overlay project requires proper surface preparation. Depending on the condition of the existing concrete, surface preparation can range from a thorough cleaning to removal of several inches of the slab's thickness. Cleaning is usually sufficient for undamaged floors with few cracks; some of the cleaning methods are listed in Ref. 5. A partial slab removal may be needed when the concrete has been damaged by deicing salts, has been penetrated by harmful chemicals, or suffers from widespread cracking and other surface deterioration. Also, if only a part of the floor receives an overlay, the top of the new material must match that of the surrounding concrete, and some sound concrete must be removed to make space.

The typical methods of removing sealers, curing compounds, and similar finishes include sandblasting, captive shotblasting, and water blasting. Of these, captive blasting by steel shot is perhaps the most suitable for use inside existing occupied buildings, because it is the cleanest and fastest. The operation involves attacking the floor with steel shot and immediately vacuuming it back. Captive blasting should be followed by pressure washing or by brushing and vacuuming to remove dust and steel remnants. Sandblasting generates a lot of dust and requires too much elaborate preparation to be cost-effective inside buildings. A new method of cleaning concrete in a gentle manner, without scratching, is by blasting with baking soda. Water blasting is similar in principle to hydrodemolition (see Chap. 5) but uses lower water pressures.

Ironically, sealers and other flexible coatings that were properly applied are the most difficult to remove, while those placed without much thought to surface preparation are the easiest to take out.[54] Existing dry-shake hardeners are so tough that their removal presents particular difficulties; several passes of blasting may be required to achieve the needed profile.

Mechanical tools for concrete removal include wheeled scabblers, scarifiers, planers, and scalers. When used properly, all of these can effectively remove concrete to a desired depth and produce level surfaces suitable for overlays. Most larger machines incorporate vacuum cleaners and require no further surface preparation. Smaller hand tools—needle guns, long-reach scalers, and bush hammers—are useful for difficult locations, but are slow and generate a lot of dust and debris. And, as discussed in Chap 5, scabblers, heavy jackhammers, bush hammers, and similar instruments have been blamed for producing microcracking in the parent concrete. Microcracking, also called bruising, can lead to debonding just below the prepared surface.

Warner et al.[55] document a variety of cases in which overlays failed because of bruising of concrete by mechanical demolition methods. Their conclusions state that scabblers and bush hammers are the worst offenders, whereas blasting, including water blasting, tends to produce the least amount of bruising. Because abrasive blasting, shotblasting, or hydroblasting cannot be used in every application, the authors suggest using them as secondary methods of surface preparation to reduce the bruising effects of the other concrete-removal means.

Another method sometimes used for surface preparation is acid etching. In this method, muriatic acid (a 10 percent solution of hydrochloric acid) is applied to the floor to remove adhered laitance and mortar droppings. Acid etching is inexpensive, but it cannot remove anything more than a thin surface layer and requires great care in application. It must be followed by an alkali wash to neutralize the acid and by a subsequent power washing. And, as stated in Chap. 5, this operation can precipitate corrosion of embedded steel items. Given the disadvantages and the limited effectiveness of acid etching, it should never be the first choice for floor surface preparation. Additional information about surface preparation methods can be found in Refs. 56 and 57.

Once the surface has been prepared, the concrete is checked to verify that all dust and oil have indeed been removed. A simple check is to wipe the surface with a dark cloth; an excessive amount of dust indicates that another cleaning is required. A check for oil residue can be made by sprinkling some water on concrete: An oil-free floor allows water to spread out rather than bead up in droplets.[57,58]

Following surface preparation, concrete is normally presoaked for a minimum of 24 h prior to placement of the topping. As with any patching, wetting of the concrete substrate is done to prevent it from absorbing too much water from the topping. Immediately prior to topping placement, standing water is removed and the bonding agent recommended by the manufacturer is applied. After the topping has been placed and troweled to a proper elevation, any required floor hardener can be installed. For cementitious materials, standard concrete curing practices apply, such as using wet burlap, plastic sheets, or at least a good curing compound. If the floor topping is a proprietary mix, the manufacturer's recommended procedures for placement and curing are followed.

Construction and control joints in both bonded and unbonded toppings should coincide with those of the base slab. Some additional joints may be needed to keep the bay size less than 215 ft^2 (20 m^2) for bonded toppings and less than 161 ft^2 (15 m^2) for unbonded toppings.[19]

6.10.3 Protection by sealers and membranes

The new overlay may need the same kind of surface protection as a new slab: a sealer, a special coating, or a waterproofing membrane. (Floor sealers are coatings and should not be confused with joint sealants, the caulking-type materials.) The choice of surface protection depends on the use of the building. An in-depth discussion of these materials is beyond the scope of this book, and we can afford only a brief note. There are several good sources of information on the subject, among them Chap. 4 of *Concrete Repair Guide* by the American Concrete Institute.[59] According to this source, surface treatments can be classified according to thickness as

- *Penetrating sealers* of negligible thickness that get absorbed into concrete

- *Surface sealers* up to 10 mils (0.25 mm) thick

- *High-build coatings* from 10 to 30 mils (0.25 to 0.76 mm) thick

- *Membranes* from 30 to 250 mils (0.76 to 6.25 mm) thick

- *Overlays* more than 250 mils (6.25 mm) thick

Penetrating sealers are the least expensive and the easiest to apply, but, not surprisingly, provide the least amount of protection and have to be renewed every 3 to 5 years. Their main function is to decrease, but not totally eliminate, water entry into the slab. This, in turn, limits the penetration of water-borne chemicals and thus helps increase the chemical resistance of the floor. These sealers also reduce dusting and somewhat improve surface hardness. Most sealers penetrate only about 2 mm into the slab and allow for some water entry, especially through cracks. Penetrating sealers are commonly specified for parking garage slabs.

The favorite penetrating sealer of yesterday was boiled linseed oil. It was effective in protecting exterior slabs from deicing chemicals and freeze-thaw damage. Among the newer sealer materials is sodium silicate, also known as water glass. Sodium silicate reacts chemically with calcium compounds in hardened concrete, filling the concrete's surface pores with a hard and glassy material. It is typically applied in three or four coats over several days. Another relatively new class of sealers is silicofluorides; its application is similar to that of sodium silicate.[5] The most common sealers of today are silanes and siloxanes. Although perhaps 10 times more expensive than boiled linseed oil, silanes and siloxanes provide better protection.

Surface sealers and high-build coatings are similar materials, differentiated mainly by thickness. They provide better protection and are more expensive than penetrating sealers. These coating materials include epoxy, polyurethane, chlorinated rubber, polyester, and paints. Some of them are breathable, some are not; most are susceptible to wear and to UV damage. Surface sealers and high-build coatings have reasonably good waterproofing qualities, but neither can bridge moving cracks or hide prominent surface defects.[60]

Waterproofing membranes provide superior protection from chemicals and moisture. In addition, they can bridge minor moving cracks and hide some surface defects. There are two basic types of waterproofing systems: thick and thin. A thick system typically consists of a rubberized asphaltic membrane covered with asphalt in the form of mastic or bituminous concrete. A thin system usually has two parts: a field-applied membrane and a wearing surface. The membrane consists of a primer and a main coat of urethane or neoprene; the wearing surface has an epoxy or urethane base with embedded mineral aggregate.[61] The biggest problem with specifying membranes for slabs on grade is the fact that they can trap moisture, blister, and delaminate. In contrast, cementitious overlays are permeable to vapor and therefore do not suffer from this problem.

One study[62] compared the effectiveness of various coating materials, from linseed oil to several overlays, in blocking water entry. The testing was done by soaking the materials in salt water for 25 days and recording the weight gain. Not surprisingly, the study found that the most expensive treatment—an overlay with methylmethacrylate polymer concrete—provided the best results: zero weight gain. Overlays with epoxy and latex-modified concrete also fared well. Low-slump concrete, while not in the same league, provided a very economical added protection.

6.11   Improving Abrasion Resistance

6.11.1   Dense concrete

The floors of some industrial buildings undergo a great deal of damage caused by heavy steel-wheeled traffic, repeated scraping and abrasion, and chemical attack from spilled oils and solvents. Under these conditions, regular concrete floors may deteriorate rather quickly. Simply replacing the floors with the same materials when this occurs is not the most cost-effective solution. But what is?

In most industrial buildings—auto assembly plants, production facilities for high-technology, medical and plastics manufacturing— relatively little floor scraping takes place. Here, the two biggest causes of floor deterioration are wearing out by hard-wheeled equipment and impact caused by dropped objects, and the answer is to make the replacement slabs more abrasion resistant. To accomplish this, Halvorsen[21] suggests the following. First, specify higher-strength concrete; it is usually more resistant to wear than regular concrete. Second, pay particular attention to the gradation of the aggregate, since denser concrete is also more abrasion-resistant. Third, use coarse aggregate with tough wearing properties, such as traprock, emery, and quartz. Fourth, avoid using more than 3 percent entrained air (unless freeze-thaw resistance is also required). And finally, cure the concrete carefully: The surface abrasion resistance can be reportedly doubled by extending the wet curing time from 3 to 7 days.

6.11.2 Dry-shake hardeners

For the more demanding environments, such as those found in food-processing facilities or machine shops, dry-shake hardeners (also known as surface hardeners) can be used. Dry-shake hardeners are broadcast into plastic bull-floated concrete by hand (by shovel, that is) or by mechanical spreaders. Spreaders allow for a much better distribution of the material, but this obviously costs more than hand application and might not be practical for large areas. Dry-shake hardeners may require application in two passes. After the hardener is applied, the surface is bull-floated again to embed the aggregate, then power-floated and troweled as usual.

Dry-shake hardeners are mixtures of portland cement with crushed mineral aggregates (emery, traprock), metallic aggregate, or ferrosilicon. There is a good deal of rivalry between the manufacturers of emery and the manufacturers of metallic products, and some conflicting claims have been made about the actual performances of those materials. What is clear is that both products provide significantly higher abrasion resistance—four to eight times that of plain concrete—and improve the chemical resistance of slabs. (Occasionally, they are used not for strength enhancement but as a powder covering the pinholes caused by steel-fiber reinforcement.)

Among the most commonly used dry-shake hardeners are those with emery aggregate. Emery is an extremely dense natural abrasive material consisting of corundum, a form of aluminum oxide (found in premium sandpaper, among other uses), and other ingredients. Emery's hardness is eclipsed only by those of diamond, silicon carbide, and fused aluminum oxide.[63]

The advantages of metallic aggregates include their toughness, ductility, and electrical conductance. A metallic aggregate can deform slightly under impact and thus can act as a shock absorber of sorts, preventing local damage to concrete. The aggregate typically consists of ground cast-iron powder, kiln-fired to remove impurities. Lately, a new line of products incorporating ground steel has been introduced; it is claimed that these have nearly twice the abrasion resistance of iron. The aggregates can be premixed and prepackaged or can be mixed with portland cement at the site. Dry-shake hardeners with metallic aggregate can be customized to provide light reflectance, color, static conductance, or even resistance to moisture. The rate of application ranges from 1 to 3 lb/ft^2. However, the hardener's penetration is only skin deep: At a rate of 1.5 lb/ft^2, only the top $^1/_{16}$ in of the slab is reinforced.[64]

Ferrosilicon is a relatively new material that bridges the gap between emery and metallic aggregates. It is essentially a synthetic metallic aggregate containing iron with 15 to 18 percent silicon and 5 to 8 percent titanium. Like emery, it has excellent durability and chemical resistance, but it can withstand impact better. Like iron, it has ductile properties and can conduct heat and electricity, the properties important for applications requiring spark resistance and heat dissipation. Unlike iron, ferrosilicon does not rust. This material can provide a light-reflective surface.[63]

Dry-shake hardeners offer some improvement over plain concrete floors, mostly because they result in slabs with high aggregate concentrations near the top, but their contribution is incremental and should not be overestimated.

6.11.3 Floor toppings: General types

The idea of a floor topping is far from new: Many heavy-duty building floors of yesterday were covered with what were called "granolithic toppings" that used granite aggregate for wear resistance. There are two basic types of floor toppings: monolithic and separate.

Monolithic toppings are placed on plastic concrete and compacted and finished along with it; they are included in the design thickness of concrete for structural purposes. Monolithic toppings can be placed in thicknesses smaller than separate toppings, from $^1/_2$ to 1 in. Separate toppings are applied after the slab has hardened. There are two types of separate toppings: bonded and unbonded.

As the name implies, a bonded topping requires complete adhesion to the substrate, while an unbonded topping is essentially a separate slab laid on top of the existing floor. Bonded toppings adhere to concrete with the help of bonding agents—cementitious slurries, epoxies, and polymers. Of those, the pure two-component epoxy provides the best bond strength. The bonding agents are applied immediately prior to the topping installation to prevent their drying up and becoming bond breakers instead (see the discussion in Chap. 5). Like monolithic toppings, bonded toppings are included in the design slab thickness. A common thickness of bonded toppings is between 0.75 and 1.5 in.[5]

A separate unbonded topping may be used when the existing slab is in poor condition and cannot serve as a base for the bonded topping. Separate unbonded toppings with a minimum thickness of 3 in (thicker toppings may be needed for heavy-duty industrial applications) are placed on polyethylene sheets laid on top of the substrate.

Floor toppings are specified for demanding applications, where dry-shake hardeners would not provide adequate protection. Perhaps the most punishing floor use occurs in the so-called tipping floors of resource-recovery and materials-recycling facilities, where solid waste is dumped from garbage trucks onto the slab and then moved around by front-end loaders or bulldozers (Fig. 6.19). Here, the slab is scraped, gouged, abraded, abused, and impacted by equipment and by dropped refuse. Harsh treatment from overzealous operators of the front-end loaders and similar equipment can destroy concrete floors in only a few months, or at most a few years.

The most practical method of protecting such floors is by abrasion-resistant floor toppings. As with dry-shake hardeners, the two most commonly used kinds are emery and metallic. The superiority of floor toppings over dry-shake hardeners that have a similar rate of wear is easily explained: The toppings are much thicker. A typical topping thickness is 1 to 2 in, while the thickness of surface hardeners is only about $\frac{1}{8}$ to $\frac{1}{4}$ in. A typical emery dry-shake hardener is applied at a rate of 1 to 1.5 lb/ft^2, while a 1-in-thick emery topping weighs 14 lb/ft^2.[65] Emery and metallic toppings are described separately in the next sections.

6.11.4   Emery and other natural-aggregate toppings

The best emery toppings consist of 100 percent pure emery aggregate and contain at least 58 percent aluminum oxide, the ingredient responsible for emery's hardness.[65] By comparison, traprock usually contains less than 10 percent aluminum oxide.[63] Floors protected with emery and other natural aggregates can be easily cleaned by washing and can resist mild concentrations of acids and alkalies, but the natural aggregates are brittle and may break down under repeated abrasion and impact.

Emery (and also metallic) toppings perform best when they are applied to new slabs as monolithic or separate bonded overlays. One of the most effective application methods involves a dry-tamp process, in which emery aggregate mixed with cement, a special densifying admixture, and water is placed at a zero-slump consistency. The mixture is then well compacted by a heavy power roller and finished as usual. Obviously, this procedure cannot be used on existing slabs, and for this reason alone it might be worthwhile to remove and replace the slab in critical installations. Keep in mind that the cost of the emery or metallic topping often vastly exceeds that of the slab itself.

If slab replacement is impractical and the topping absolutely must go on top of an existing slab, careful surface preparation is required. (The specific methods are discussed in Sec. 6.10.) For bonded toppings usually specified for existing buildings, the requirements include removing at least $^3/_4$ in of old concrete (or more if the topping must be flush with the existing floor), roughening the surface to an amplitude of $^1/_4$ in, and applying a bonding agent, such as epoxy. The success of this installation depends greatly on the quality of workmanship, and it is wise to request the presence of the topping manufacturer's representative on the job to assist with planning and inspection.

How can one verify that the emery topping has been properly installed? The material can be tested. A commonly performed compressive-strength test is of limited value for this purpose, although a measured strength of only 5000 to 6000 psi, instead of the required 9000 to 10,000 psi, would be alarming. (The tested specimen should be taken from the mixer to avoid contamination with concrete, a much weaker material.) What really needs to be tested is the abrasion resistance of the topping.

The test for abrasion resistance conforming to ASTM C944 is purely comparative and cannot be used prescriptively. The test involves applying the topping to a 4000-psi concrete slab, curing it fully, and abrading it with a dressing wheel for a specified period of time. Then, a 5-in-square area of the slab is removed and the resulting weight loss is measured against that sustained by a control specimen of the same slab without topping. A properly placed topping should lose from 6 to 9 times

Figure 6.19 The floors of resource-recovery and materials-recycling facilities require abrasion-resistant surfaces to reduce damage by front-end loaders.

less weight than concrete alone. To compare apples to apples, it is important to make certain that the control slab is made with the same aggregate as the tested slab.

6.11.5 Metallic toppings

Metallic-aggregate toppings have been used in waste-processing applications since 1959. The benefits of metallic aggregates include their excellent durability and their near impermeability to oil, grease, and other elements.[66] Metallic toppings are designed to attain a 28-day compressive strength of from 10,000 to over 12,000 psi. There are two types of metallic toppings: cementitious and epoxy-based. Epoxy metallic toppings are twice as expensive as cementitious, but they are nearly 30 percent more abrasion-resistant. In addition, epoxy toppings cure much faster than their cementitious brethren—in about 36 h vs. 96 h.[67] Metallic toppings are impact-resistant and conduct heat and electricity.

One obvious disadvantage of all metallic toppings is their poor resistance to moisture, which can impart a rusty appearance to the floor. Also, these toppings are heavy—a 1-in-thick material weighs about 19.2 psf—and are more difficult to place than concrete.

Metallic toppings can be applied to both new and existing slabs. In both cases, the surface should be roughened, but the bonding agent may be different: latex-based for new slabs and epoxy for existing slabs.[67] Other construction requirements are similar to those for emery topping.

Despite their high abrasion resistance, both emery and metallic toppings will eventually wear out and may require replacement in a few years. The floor should be inspected at periodic intervals, so that slab deterioration is not allowed to proceed too far, beyond repair.

6.11.6 Steel plates

One other method of surface protection is the use of steel plates. The benefits here are the speed of application and the absence of surface preparation and curing requirements—a critical consideration for waste-processing operations. In addition, 1 in-thick plates are widely available and are easy to replace when they wear out.

Steel plates need to be fastened to the floor; if they are not, they will be caught by the scoop of a front-end loader or by other machinery and moved around. The best method of attachment is by countersunk bolts, either expansion-type or placed in chemical anchors. While this method avoids having protruding fastener heads at the beginning, after a while repeated abrasion will wear out the heads, and they will eventually break. Then, either the bolts or the whole plate will need replacement. Another option is to anchor the plates with adhesives or grout, but that is also fraught with problems, because plate movement under load tends to break the bond and fracture the grout.

Some other disadvantages of steel plates include their smooth and slippery surfaces and the fact that the plate joints tend to curl after prolonged use, presenting a potential tripping hazard. A floor covered with steel plates is also noisier than a floor with toppings. The best way to use steel plates is not to cover the whole floor but to reinforce the limited areas that suffer the worst abuse. For example, the edges of the floor pits in resource-recovery and materials-recycling facilities are often lined with embedded plates. These areas experience some of the worst wear imaginable.

6.12   Repair of Deteriorated Overlays, Toppings, and Hardeners

Floor toppings and overlays, especially metallic or emery toppings, rarely fail in uniform fashion. Typically, the most heavily used areas are worn out, while the rest of the floor looks good. What to do with these partly deteriorated toppings and overlays? Another common problem arises during a change in occupancy when a change in the floor finish or chemical resistance is necessary. If the existing floor already has a surface hardener or topping, it must be removed. But how can one deal with extremely strong and tough existing coatings that are resistant to the usual methods of materials removal?

The repairs are difficult but possible. Existing metallic floor toppings can be repaired with compatible patching materials, such as iron-modified epoxy mortars. These mortars offer excellent wear resistance, bonding capability, and strength characteristics, matching those of the original topping. It is also possible to place a new metallic topping or another overlay material over the old one.

The obvious difficulty lies not only in removal, but even in roughening the tough surface to accept the new overlay. One method that can adequately roughen the metallic surface is very aggressive shotblasting, perhaps using steel grit instead of steel shot. Another approach is to coat the surface with a 20 percent solution of phosphoric acid.[68] The adequacy of these procedures can be verified by conducting an adhesion test in a trial installation, as described in Appendix A of ACI 503R-93.[69]

Occasionally, improperly installed overlays delaminate. What can be done about this? The most obvious solution is to determine what caused the problem, remove the overlay, properly prepare the surface (or, if applicable, fix some other cause of the delamination), and try again. This is not always practical, and the owner may want to have the overlay rebonded rather than replaced. Rebonding can be accomplished by epoxy injection of the delaminated interface or by placing steel dowels into epoxy-filled holes drilled at regular intervals in the area of delamination.

The procedure utilizing epoxy injection is more common in roads and bridges than in buildings. In one example, described in Ref. 70, an 18-ft-wide by $\frac{1}{4}$-mile-long section of pavement in Texas was repaired by rebonding. A 7.5-in-thick delaminated overlay was rebonded by injecting low-viscosity (200 to 300 cP) epoxy resin into ⅝-in holes spaced 2 ft apart in each direction. Since the drilling terminated in the delaminated area, there was a potential for filling it with dust, which could interfere with bonding. To avoid introducing dust into the void, vacuum drills with hollow-core bits were used. The rebonding was a success, verified by impact-echo testing that detected no voids in the repaired areas.

This type of repair requires excellent workmanship, and it remains to be seen whether it will receive widespread use in rebonding slabs on grade. It is easy to visualize the pressure of the injected epoxy pushing the delaminated area upward rather than gluing it in place. As Ref. 71 points out, even a low injection pressure of 5 psi exerts a force of over 700 lb/ft² against the underside of the repaired overlay. Also, if the delamination occurred because the slab surface was poorly prepared, the epoxy may not adhere well either.

Placing steel dowels into holes filled with low-viscosity epoxy provides a mechanical connection and seems more reliable. Trout[72] states that the design engineer should determine the size and spacing of the pins. For example, steel pins with a $\frac{3}{8}$-in diameter placed in $\frac{1}{2}$-in holes drilled 2.5 in into the slab might be used. To contain the epoxy, he recommends sealing the voids between the bottom of the delaminated overlay and the slab with thin resin grout. Once the overlay is pinned in some locations, additional holes can be drilled in between, with a maximum spacing of 2 ft.

Being essentially a thin version of abrasive toppings, dry-shake hardeners suffer from the same problems of local wearing out and delaminations. How can they be repaired? The answer depends on the extent and probable cause of the failure. If the hardener failed throughout the whole floor, and the slab is deficient in some way (has an air content of more than 3 percent, for example), removal and replacement of the slab may be the best option. Presumably, when the new slab is placed, extra attention can be paid to the details of construction, air content, curing, and hardener application. This option offers the most predictable outcome, but is also the most expensive.

Another possibility is to remove the hardener from the whole area and apply a floor topping as described previously, since dry-shake hardeners cannot be applied to existing concrete; this has the minor disadvantage of increasing the height of the floor.

If the area of delamination is relatively large, but does not cover the whole floor, surface patching can be attempted. As in the classic patching procedure of Chap. 5, the repaired area is saw cut to the minimum depth required by the repair material, typically 1 in. The existing concrete is then removed to this depth with light chipping hammers, followed by sandblasting or shotblasting. After the dust has been vacuumed and the surface dried out, an epoxy bonding-agent paste is spread out, the area is filled with pea-gravel concrete, and a new dry-shake hardener is applied. The hardener may have to be introduced in three passes instead of the usual two. The repair concrete mixture requires plenty of water to saturate the dry-shake hardener, and shrinkage cracking is an obvious concern. The cracking can be reduced, but not totally eliminated, with careful curing. Isolated spots of hardener delamination can be repaired by coring them out to a depth of about $^1/_4$ in, removing the dust, applying the same epoxy-paste bonding agent, and packing the area with moistened dry-shake hardener. The repair should then be cured. Small surface blisters can be rebonded with epoxy injection.[73]

References

1. ACI 201.3R-86, *Guide for Making a Condition Survey of Concrete Pavements,* American Concrete Institute, Farmington Hills, Mich., 1986.
2. ACI 201.1R-92, *Guide for Making a Condition Survey of Concrete in Service,* American Concrete Institute, Farmington Hills, Mich., 1992.
3. Boyd C. Ringo and Robert B. Anderson, *Designing Floor Slabs on Grade,* 2d ed., American Concrete Institute, Farmington Hills, Mich., 1996.
4. ACI 360-92, *Design of Slabs on Grade,* American Concrete Institute, Farmington Hills, Mich., 1992.
5. *Concrete Floors on Ground,* 2d ed., Portland Cement Association, Skokie, Ill., 1983.
6. H. Platt Thompson, "Supplementary Considerations for Slab-on-Grade Design," *Concrete International,* June 1989, pp. 20–25.
7. *Design and Construction of Post-Tensioned Slabs-on-Ground,* Post-Tensioning Institute, Phoenix, Ariz., 1991.
8. Terry J. Fricks, "Cracking in Floor Slabs," *Concrete International,* February 1992, pp. 59–63.
9. William F. Perenchio, "Designers Must Consider the Characteristics of Materials," *Concrete Construction,* September 1998, p. 754.
10. Armand H. Gustaferro, "Owner's Manual for Concrete Floors," *Concrete Construction,* November 1998, pp. 951–957.
11. Cesar Kiamco, "A Structural Look at Slabs on Grade," *Concrete International,* July 1997, pp. 45–49.
12. George Garber, *Design and Construction of Concrete Floors,* Edward Arnold, London, 1991.
13. ACI 302.1R-96, *Guide for Concrete Floor and Slab Construction,* American Concrete Institute, Farmington Hills, Mich., 1996.

14. Ernest K. Schrader, "A Solution to Cracking and Stresses Caused by Dowels and Tie Bars," *Concrete International*, July 1991, pp. 40–45.
15. Richard O. Albright, "Beware of Unrestrained Expansion," *Concrete Construction*, April 1998, pp. 371–374.
16. David Stark, *Handbook for the Identification of Alkali-Silica Reactivity in Highway Structures*, Strategic Highway Research Program, Portland Cement Association publication LT165, Skokie, Ill.
17. Steven N. Metzger, "How to Prevent Failures of Industrial Floors," *Plant Engineering*, April 14, 1983, pp. 111–114.
18. ACI 224.1R-93, *Causes, Evaluation, and Repair of Cracks in Concrete Structures*, American Concrete Institute, Farmington Hills, Mich., 1994.
19. R. T. L. Allen et al., eds., *The Repair of Concrete Structures*, 2d ed., Blackie Academic & Professional, Chapman & Hall, Glasgow, U.K., 1993.
20. Ward R. Malish, "Avoiding Common Outdoor Flatwork Problems," *Concrete Construction*, July 1990, pp. 632–638.
21. Grant T. Halvorsen, "Durable Concrete," *Concrete Construction*, August 1993, pp. 542–549.
22. Bruce A. Suprenant and Ward R. Malisch, "Diagnosing Slab Delaminations," *Concrete Construction*, January 1998, pp. 29–35.
23. Carl Bimel, "Is Delamination Really a Mystery?" *Concrete International*, January 1998, pp. 29–34.
24. Shondeep L. Sarkar, "Imported Cement May Be Cause of High Air Contents," *Concrete Construction*, February 1999, pp. 13—14.
25. ACI 546.1R-80, *Guide for Repair of Concrete Bridge Superstructures*, American Concrete Institute, Farmington Hills, Mich., 1980.
26. John C. Ropke, *Concrete Problems: Causes and Cures*, McGraw-Hill, 1982.
27. Mark Wallace, "How to Prevent Frost Heave," *Concrete Construction*, April 1987, pp. 369–372.
28. Armand H. Gustaferro, "What Every Engineer Should Know about Concrete," *Concrete Construction*, February 1981, pp. 89–90.
29. John G. Meyers, "Slabjacking Solutions for Settled Slabs," *Concrete Construction*, July 1988, pp. 651–657.
30. John G. Meyers, "Stabilizing Slab Deflection in Industrial Floors," *Concrete Repair Digest*, April/May 1992, pp. 56–60.
31. "Repair Roundup: Polyurethane Injection Raises Sunken Slabs," *Concrete Construction*, October 1998, pp. 899–900.
32. Bruce A. Suprenant and Ward R. Malisch, "Repairing Curled Slabs," *Concrete Construction*, May 1999, pp. 58–65.
33. "Repair Q & A: Repairing Industrial Floor Joints," *Concrete Repair Digest*, February/March 1998, pp. 6–7.
34. Steven N. Metzger, "Repairing Joints in Industrial Floors," *Concrete Construction*, June 1989, pp. 548–551.
35. Steven N. Metzger, "A Closer Look at Industrial Floor Joints," *Concrete Repair Digest*, February/March 1996, pp. 9–14.
36. Robert W. Gaul, "Moisture-caused Coating Failures: Fact and Fiction," *Concrete Repair Digest*, October/November 1996, pp. 255–258.
37. Robert W. Day, "Moisture Penetration of Concrete Floor Slabs, Basement Walls, and Flat Slab Ceilings," *Practice Periodical on Structural Design and Construction*, vol. 1, no. 4, November 1996, pp. 104—107.
38. Bruce A. Suprenant and Ward R. Malisch, "Effect of Water-Vapor Emissions on Floor-Covering Adhesives," *Concrete Construction*, January 1999, pp. 27–33.
39. Ward R. Malisch, "Concrete Floors and Fondue Tables" (editorial comment), *Concrete Construction*, November 1997, p. 867.
40. John A. Amundson et al., "Analyzing Moisture Problems in Concrete Slabs," *Concrete Construction*, March 1997, pp. 306–311.
41. Robert G. Cox, "Wet Slabs...So...???" *Concrete International*, September 1997, pp. 68—69.

42. Bruce A. Suprenant and Ward R. Malisch, "Where to Place the Vapor Retarder," *Concrete Construction,* May 1998, pp. 427—433.
43. Bruce A. Suprenant and Ward R. Malisch, "Are Your Slabs Dry Enough for Floor Coverings?" *Concrete Construction,* August 1998, pp. 671–676.
44. Steven N. Metzger, "Preventing Bulging Joints under Floor Coverings," *Concrete Construction,* November 1998, pp. 941—943.
45. G. W. Spratt, "Problem Clinic Q&A: Petroleum Products Cause Concrete Deterioration," *Concrete Construction,* June 1997, p. 529.
46. *Effects of Substances on Concrete and Guide to Protective Treatments,* IS001.04T, Portland Cement Association, Skokie, Ill., 1981.
47. Martin S. McGovern, "Protecting Floors from Chemical Attack," *Concrete Repair Digest,* August/September 1997, pp. 198–203.
48. ACI 318-95, *Building Code Requirements for Structural Concrete,* American Concrete Institute, Detroit, Mich., 1995.
49. Mark Wallace, "For Chemical-proof Floors, the Choice Is Acid Brick," *Masonry Construction,* October 1996, pp. 464–468.
50. Alexander Newman, *Metal Building Systems: Design and Specifications,* McGraw-Hill, New York, 1997.
51. Hank Brown, "Tips for Concrete Pavement Repair," *Concrete Construction,* January 1992, pp. 7–11.
52. Myles A. Murray, "Applications of Epoxy-Modified Concrete Toppings," *Concrete International,* December 1987, pp. 36–38.
53. Bruce A. Suprenant, "Resurfacing Systems Revitalize Old Concrete," *Concrete Construction,* August 1997, pp. 632–636.
54. "Concrete Substrate Preparation Techniques," Fosroc Inc., New York, undated.
55. James Warner et al., "Surface Protection for Overlays," *Concrete International,* May 1998, pp. 43–46.
56. "Selecting and Specifying Concrete Surface Preparation for Sealers, Coatings and Polymer Overlays," Guideline No. 03732, International Concrete Repair Institute, Sterling, Va., 1997.
57. Charles H. Holl and Scott A. O'Connor, "Cleaning and Preparing Concrete before Repair," *Concrete International,* March 1997, pp. 60–63.
58. Kim Basham, "Preparing Surfaces for Coatings," *Concrete Construction,* June 1998, pp. 525–531.
59. ACI 546R-96, *Concrete Repair Guide,* American Concrete Institute, Farmington Hills, Mich., 1996.
60. Jay H. Paul, "Extending the Life of Concrete Repairs," *Concrete International,* March 1998, pp. 62–66.
61. Remy A. Iamonaco, "Rehabilitation of Parking Structures," in "New Technologies in Construction Materials and Building Systems," seminar notes by Trow Protze Consulting Engineers, Boston, Mass., 1990.
62. Donald W. Pfeifer and William F. Perenchio, "Coatings, Penetrants and Specialty Concrete Overlays for Concrete Surfaces," paper presented at the seminar Solving Rebar Corrosion Problems in Concrete, Chicago, Sept. 27–29, 1982.
63. James A. Knight, "Properties and Uses for Emery and Ferro-silicon Aggregate in Industrial Floors," *Concrete Construction,* April 1992, pp. 308–311.
64. Anne Balogh, "Armor Floor Surfaces with Metallic Aggregates," *Concrete Construction,* July 1994, pp. 551–555.
65. *EMERY TOP,* L&M Construction Chemicals, Inc., Omaha, Neb., November 1990.
66. Barry Tyo, "Surface Treatments for Industrial Floors," *Concrete International,* April 1991, pp. 36–38.
67. William S. Phelan and Moorman L. Scott, "Tipping Floors Require Tough Toppings," *Concrete Construction,* May 1991, pp. 407–410.
68. "Repair Q&A: Coating a Metallic Dry-shake Hardened Floor Surface," *Concrete Repair Digest,* April/May 1997, p. 107.
69. ACI 503.1R-93, *Use of Epoxy Compounds with Concrete,* American Concrete Institute, Farmington Hills, Mich., 1993.

70. "Rebonding a Delaminated Pavement Overlay," *Concrete Construction,* July 1998, p. 619.
71. "Rebonding a Delaminated Overlay," *Concrete Construction,* February 1999, p. 98.
72. John Trout, *Epoxy Injection in Construction,* The Aberdeen Group, 1997 (quoted in Ref. 71).
73. "Repair Q&A: Repairing Dry-shake Hardener Delaminations," *Concrete Construction,* July 1998, pp. 623–624.

7

Renovating Post-Tensioned Concrete*

7.1 System Overview

7.1.1 Introduction and historical perspective

Deterioration of post-tensioned concrete buildings is difficult to diagnose and repair. These proprietary structures often reveal very little about their true condition until failure is evident. In this chapter, we examine the issues involved in the study and evaluation of distressed concrete reinforced with post-tensioned tendons and the repair methods available. Expertise in this type of renovation is quite valuable—and rare. We hope that this brief discussion of the history of post-tensioned structures, evaluation techniques, and basic procedures necessary to perform repairs to these structures will benefit both the experienced restoration engineer and the novice.

In essence, post-tensioning of concrete improves its flexural behavior by imposing additional stresses that counteract those from service loads. It involves applying large tension stresses to post-tensioning steel and anchoring the steel in a stretched condition at the ends of the member. A similar idea is used in prestressed concrete, except that there the tendons are tensioned before the concrete is cast.

Tendons can be tensioned from one end (the "live" end) or from both ends. Stressing anchors are placed in pockets and grouted for corrosion

*The author is grateful to Mr. George Tapas, P.E., Chicago, Ill., for providing most of the material for this chapter and supplying the illustrations for Figs. 7.2, 7.3, and 7.7 through 7.14.

protection. The stressed post-tensioning steel transmits compressive forces to concrete by bearing of mechanical anchors. If the steel is properly placed, this compression reduces the applied stresses in the tension zones of the structural member. As a result, post-tensioned structures can be thinner and lighter, and can span longer distances; their deflections and porosity are reduced. Concrete cracking in them is also lessened. Figure 7.1 illustrates the typical shape of post-tensioning tendons.

Tendons are placed in ducts, both for corrosion protection and for ease of stressing. Depending on the material placed in the duct around the tendons, post-tensioned systems can be of two main kinds: bonded and unbonded. Bonded tendons are placed into the duct that is later filled with grout; they transfer stresses to the concrete by bond. Unbonded tendons are enclosed with grease and are free to move within the duct; they are isolated from the concrete, transferring stresses to it only at the points of anchorage.

The first use of unbonded tendons is attributed to R. E. Dill in 1925. In the 1930s, unbonded rods stressed by large turnbuckles were used as prestressing for cylindrical tanks. To protect these rods from exposure to the elements, they were either galvanized or painted. As experimentation continued, unbonded tendons began to find their way into buildings, starting with lift-slab construction, in the mid-1950s.[1]

By the mid-1960s, post-tensioning of building structures was becoming popular in North America, after many years of use in Europe. A virtual explosion in post-tensioned structures followed as the advantages of this type of construction were recognized and the possibilities of building more economical and durable concrete structures beckoned. From 1965 to 1989, approximately one million tons of bonded and unbonded post-tensioning tendons were installed; of these, about 730,000 tons were unbonded tendons.[2] It is estimated that by the 1990s, there were more than 2 billion ft^2 of unbonded post-tensioned construction, approximately 40 percent of that amount in parking

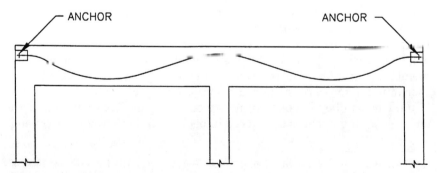

Figure 7.1 Typical shape of post-tensioning tendons.

structures.[3] As the post-tensioned structures of the first generations begin to age, the need to renovate and repair them is becoming urgent.

7.1.2 Types of post-tensioning systems

The main types of post-tensioning systems that have been used in the United States can be divided into three main groups:

1. Unbonded (greased and wrapped)

2. Bonded (grouted)

3. External (used primarily for rehabilitation)

The differences between bonded and unbonded systems have already been explained. Which is better? As Schupack[4] observes, unbonded tendons can be used in relatively shallow construction; they also provide excellent deflection and crack control for long-span members. On the other hand, unbonded tendons lack the redundancy of bonded systems. If cut, bonded tendons still provide prestress for at least a part of the structure, but a severed unbonded tendon is useless. Also, the ultimate strength of a simple beam with unbonded tendons is nearly 30 percent less than that of a similar beam with bonded tendons; for a continuous beam or slab, the strength is about 20 percent less. And, unlike bonded tendons, unbonded tendons offer little help in controlling crack sizes.

Today, bonded tendons tend to be specified for main structural members, such as beams and joists, while unbonded tendons are typically used in slabs and as reinforcement for temperature and crack control. As the technology of post-tensioned systems continues to improve, and their quality and effectiveness increase, new possibilities arise. Some structures are already being built with multiple systems of bonded and unbonded tendons.

Steel tendons used for post-tensioning systems can be of the three following types:

1. Wire

2. Strand (either monostrand or multistrand)

3. Bar (threaded rod)

Of the three, the unbonded wire system, commonly called the "buttonhead" system, has been used most commonly (Fig. 7.2). The term *button* refers to the shape of the wire's end after it is anchored. This system is no longer specified in the United States but lives on in numerous existing structures, all potential candidates for future repairs. Accordingly, our main focus will be on repairing unbonded

Figure 7.2 Post-tensioning wire system with externally threaded anchor head, bearing plate, and wires.

post-tensioning wire tendons. The three tendon systems are further described in Sec. 7.2.2.

7.2 Evolution of Post-Tensioned Structures

In order to assist later discussions, here is a brief review of design standards and of the components of post-tensioning systems likely to be encountered in existing buildings.

7.2.1 Design and construction standards

Post-tensioned structures were being built in the United States as far back as the mid-1950s, but the first comprehensive document dealing with their design appeared only in February 1969. The ACI-ASCE Committee 423 report "Tentative Recommendations for Concrete Members Prestressed with Unbonded Tendons" was published then, with minor modifications in 1983. In 1985 the Post-Tensioning Institute (PTI) published its *Specification for Unbonded Single Strand Tendons,* which increased the requirements for corrosion protection of tendons. The PTI specification's provisions were referenced in the revised 1989 *Recommendations for Concrete Members Prestressed with Unbonded Tendons* (no longer "tentative").[1]

The PTI specification included the recommended details of tendon fabrication for normal (noncorrosive) and aggressive (corrosive) environments. In corrosive environments, watertight enclosure for tendons along their whole length was required. The enclosure could be provided by a heavy-gage high-density polyethylene or polypropylene sheath completely filled with grease containing anticorrosive additives. Special attention was directed to patching of stressing pockets.[3]

Up until the 1980s, engineers involved in specifying post-tensioning systems depended almost entirely upon suppliers of tendons and contractors for design information. A wide variety of systems and components were on the market at that time, and design uniformity was sorely lacking. Even buildings constructed after the PTI specifications were adopted and the strand system became predominant exhibited major differences in design. To bring a measure of uniformity to construction practices, in 1989 the PTI published its *Field Procedures Manual for Unbonded Single Strand Tendons.*

Concrete codes were slow to address some of the major deficiencies in the early designs of post-tensioned structures. For example, requirements mandating minimum amounts of mild reinforcing steel were not introduced until the 1989 edition of *Building Code Requirements for Structural Concrete* (ACI 318). While some design engineers had already been meeting these minimum reinforcement requirements for years, others had not. As a result, many structures were built without this extra measure of ductility that could prevent collapse if tendons were damaged. Also, many designs of the 1960s and 1970s were based on assumed zero tension in concrete or on limiting the concrete tensile stress at working loads to $3 \sqrt{f'_c}$ or less.[5]

Those early designs followed the allowable-stress approach rather than the ultimate-strength design that is standard today.

7.2.2   Early post-tensioning systems

The PTI's *Post-Tensioning Manual*[6] provides an in-depth discussion of the details of the various systems described below. This reference, and its earlier editions, provide invaluable information for both the engineer and the contractor.

Wire system. In this old system, tendons typically consisted of $1/4$-in- (and later up to 0.6-in) diameter wires made of 240,000-psi high-strength steel. They were assembled and greased by hand and were either spirally hand-wrapped in kraft paper or installed in flexible conduits of galvanized steel. It is the paper wrapping, which often produced tendon cross sections of irregular shapes (typically oval), that complicates repairs of this system today. In a typical installation, all the wires were cut to the same length and then passed through predrilled holes in the anchor head. To secure and seat the wires against the anchor head, a small buttonlike configuration was cold-formed ("button-headed") onto the end of each wire by a special machine. Dead-end anchors may have been cast into the structure during concrete placement at the opposite end of the member. The dead-end anchors were plates with predrilled holes that allowed the wires to pass through and be seated (Fig. 7.3a). After the tendons were stressed, they elongated, causing a gap to appear between the "live-end" anchor head and the embedded bearing plate. In order to maintain the tension force in the tendon and transfer it to the bearing plate, shim plates or threaded bushings were installed. Three main live-end anchor types were commonly employed: split-shim, threaded-bushing, and stand-up, all indidcated in Fig. 7.3(b–d). As stated previously, this system is now mainly defunct in the United States, but it was used in a large percentage of the post-tensioned structures constructed before the late 1970s. The reasons why this system has not performed as initially planned included the high labor-intensiveness of wrapping tendons by hand and poor corrosion protection of tendons in harsh environments. The wire system was soon displaced by strand system.

Strand system. This system was originally developed for concrete prestressing, not post-tensioning, but the desire for material uniformity among the two types of concrete work made it popular. With further advances in the technology of tendon sheathing and protection, the post-tensioning industry adopted the strand system as standard. The strands are usually $^3/_8$, $^7/_{16}$, $^1/_2$, or 0.600 in in overall diameter and are made of 270,000-psi high-strength, low-relaxation or stress-relieved steel. Each strand consists of individual wires of smaller diameters mechanically twisted together, usually into a seven-wire configuration. The tendons are either monostrand or multistrand. The single-strand design came first, in the early 1960s, when it started to replace wire tendons. Prior to 1969, tendons were used as uniformly spaced single elements; after that date, tendons increasingly became combined in groups (bands) of two or more tendons. The banded tendons are placed in one direction, while in the other direction tendons are spaced uniformly. This approach has been shown to be effective and practical and is now provalent.[1] In flat-plate slab construction, multiple tendons or strands are often banded together to create a system of beams and girders located at column lines within the slab. Each strand is anchored by a two- or three-piece steel wedge clamped onto the strand and held by a prefabricated steel wedge plate at its bearing point at the edge of the structure.

Bar system. Here, steel bars are typically $^5/_8$ to $1^3/_8$ in in diameter, made with threads or uniform deformations, and fabricated from steel with an ultimate tensile strength of 160,000 psi. Bar reinforcement

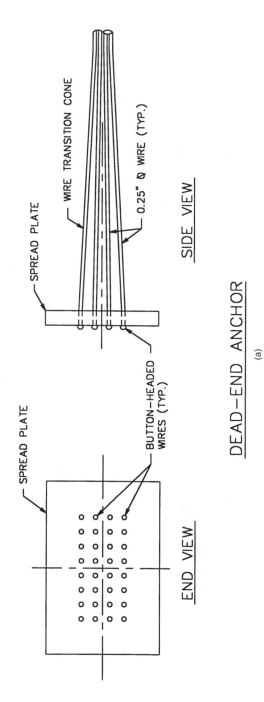

Figure 7.3 Typical stressing-end anchor-head configurations for various wire systems.

385

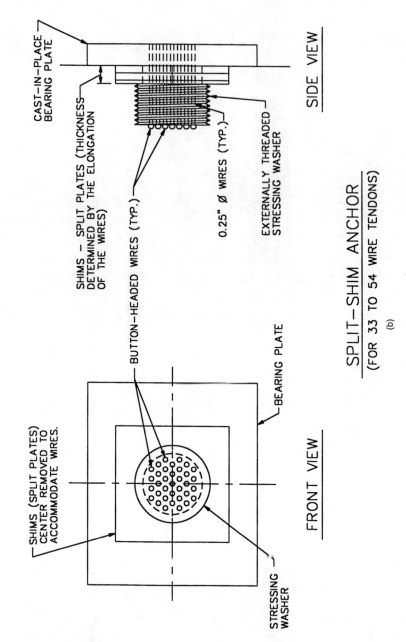

CAST-IN-PLACE
BEARING PLATE

SHIMS — SPLIT PLATES (THICKNESS
DETERMINED BY THE ELONGATION
OF THE WIRES)

BUTTON-HEADED WIRES (TYP.)

0.25" Ø WIRES (TYP.)

EXTERNALLY THREADED
STRESSING WASHER

SIDE VIEW

SHIMS (SPLIT PLATES)
CENTER REMOVED TO
ACCOMMODATE WIRES.

BEARING PLATE

STRESSING
WASHER

FRONT VIEW

SPLIT-SHIM ANCHOR
(FOR 33 TO 54 WIRE TENDONS)

(b)

Figure 7.3 (Continued)

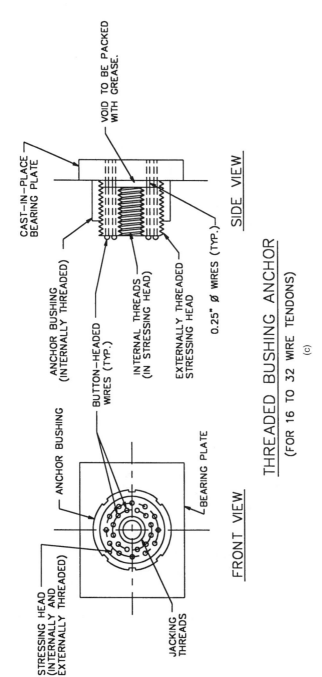

STRESSING HEAD
(INTERNALLY AND
EXTERNALLY THREADED)

ANCHOR BUSHING

BEARING PLATE

JACKING
THREADS

FRONT VIEW

CAST-IN-PLACE
BEARING PLATE

VOID TO BE PACKED
WITH GREASE.

ANCHOR BUSHING
(INTERNALLY THREADED)

BUTTON-HEADED
WIRES (TYP.)

INTERNAL THREADS
(IN STRESSING HEAD)

EXTERNALLY THREADED
STRESSING HEAD

0.25" Ø WIRES (TYP.)

SIDE VIEW

THREADED BUSHING ANCHOR
(FOR 16 TO 32 WIRE TENDONS)

(c)

Figure 7.3 (Continued)

387

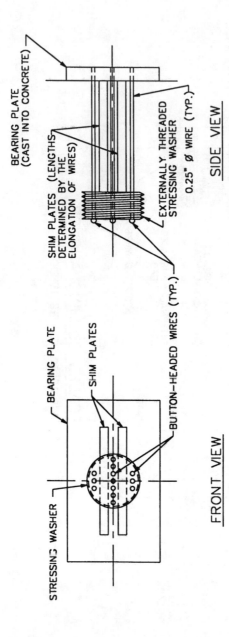

BEARING PLATE
(CAST INTO CONCRETE)

SHIM PLATES (LENGTHS
DETERMINED BY THE
ELONGATION OF WIRES)

EXTERNALLY THREADED
STRESSING WASHER

0.25" ∅ WIRE (TYP.)

SIDE VIEW

BEARING PLATE

SHIM PLATES

BUTTON-HEADED WIRES (TYP.)

STRESSING WASHER

FRONT VIEW

STAND-UP SHIM ANCHOR
(FOR 2 TO 15 WIRE TENDONS)

(d)

Figure 7.3 (Continued)

has been largely confined to bonded post-tensioning systems and, despite its simplicity and the ease of construction, underutilized. At this time, the most common use of bar system is in renovation work and for external post-tensioning (see Example 4.5 in Chap. 4).

7.2.3 Coatings and sheathing systems for unbonded tendons

Early unbonded tendons were coated with grease and mastic. The primary function of the coating was not to protect the tendons from corrosion but to lubricate the wires in order to reduce friction losses during stressing operations. As understanding of the effects of chloride contamination on steel corrosion grew, high-quality corrosion-inhibiting specialty tendon greases were developed. (Much of the research came from the nuclear field.) The importance of coatings for workability during post-tensioning operations was demonstrated by Tapas,[7] who describes the restoration of post-tensioning tendons in an elevated parking garage at O'Hare International Airport in Chicago, the world's largest stand-alone parking structure. During the repair of tendons coated with mastic and grease, at a slab temperature below 50°F, the stressing operation encountered significant resistance (drag effects) and large friction losses. The reason for the difficulty? The mastic-and-grease coating had solidified, becoming an obstacle rather than a helper.

The earliest sheathing system—wrapping tendons (either wire or strands) in kraft paper—offered poor corrosion protection. The paper could easily tear, absorb water, or allow grease to leak; all these factors reduced its effectiveness. In the mid- to late 1960s, polyethylene and polypropylene materials were developed, and plastic sheathing replaced paper. The split-sheathing system, in which the sheathing had a continuous longitudinal split, was tried first. But, while an improvement over paper wrapping, split sheathing still allowed water to enter.[8] Today, the three common types of plastic sheathing systems are push-through preformed (introduced in the early 1970s), heat-sealed, and extruded.[9] These are shown at the bottom of Fig. 7.4.

Push-through preformed tubes, in which strands are inserted (pushed through) simultaneously with grease application, and heat-sealed sheathing, formed from flat strips as grease is placed, leave a lot of space around post-tensioning steel. Any water that penetrates the outer shell can collect in this space and cause corrosion. Extruded sheathing, formed by continual extrusion of plastic over the strand as grease is applied, offers a tighter fit around the tendons.

7.3 Typical Reasons for Repair of Post-Tensioned Buildings

Repair of post-tensioned buildings may be needed for a number of reasons. The tendons might have been accidentally damaged by care-

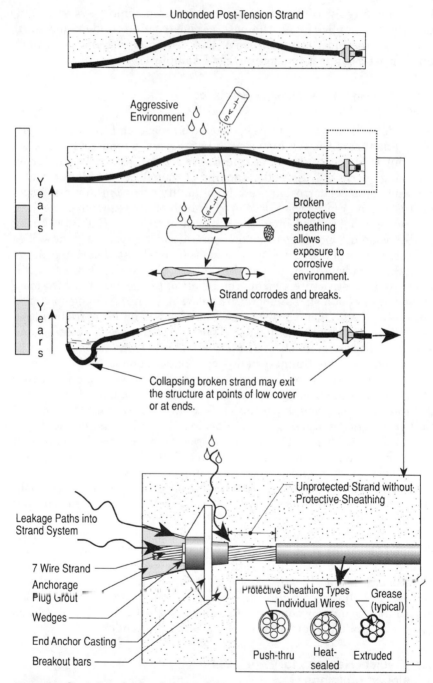

Unbonded Post-Tension Strand

Aggressive Environment

Years

Years

Broken protective sheathing allows exposure to corrosive environment.

Strand corrodes and breaks.

Collapsing broken strand may exit the structure at points of low cover or at ends.

Unprotected Strand without Protective Sheathing

Leakage Paths into Strand System

7 Wire Strand

Anchorage Plug Grout

Wedges

End Anchor Casting

Breakout bars

Protective Sheathing Types
Individual Wires

Grease (typical)

Push-thru   Heat-sealed   Extruded

Figure 7.4 Corrosion of unbonded post-tensioning strands. [*Reprinted with permission from Peter H. Emmons*, Concrete Repair and Maintenance Illustrated, *copyright R.S. Means Co., Inc., Kingston, Mass., (800) 334-3509, 1993, All rights reserved.*]

less floor coring during utility installation or maintenance, or perhaps by mechanical impact. The loading requirements for the building could have changed, or the structure might have been improperly designed and constructed. But the most common reason for repairs is deterioration.

Several factors affect the performance of post-tensioned structures, among them the age, the use, and the degree of exposure to aggressive chemicals. Post-tensioning tendons carry large forces from stressing operations and are extremely sensitive to section weakening brought about by corrosion and the resulting increase in stresses in the remaining section. Wires and strands are typically stressed to approximately 65 percent of the guaranteed (that is, minimum) ultimate tensile strength (G.U.T.S.) of the steel. With these extremely high levels of stress, even minor section losses can increase the stresses in individual wires to 80 to 90 percent of G.U.T.S. If corrosion continues undetected, it will ultimately conclude the process of reducing the section and increasing stresses in it by breaking the tendon.

Depending on the rate of corrosion, stress levels can exceed the ultimate capacity of tendons rather quickly. If water, air, and chlorides are permitted to attack the tendons, the resulting corrosion will occur at a much faster rate than in mild reinforcement. (Kesner and Poston[10] point out that tendon corrosion can take place even without the presence of chlorides.) When the remaining section becomes too small to carry the load, it breaks. According to Emmons,[11] the strands might be stretched as much as 8 in per 100 ft of length before breaking. The eventual failure can be quite violent, resulting in the concrete cover being spalled or broken away from the tendon. When certain combinations of tendon length, rate of failure, and location of the break are present, the end cover of the anchorage can crack and even fall off.

Tendon breakage can create forces in excess of 30,000 lb and can cause damage not only to the structure, but also to its occupants.[12] Unfortunately, until that happens, deterioration of tendons is difficult to detect. There have been cases in which the entire tendon length was consumed by corrosion without any signs of distress being visible.[8]

Corrosion of post-tensioned systems is especially prevalent in parking garages. Most of them, it is said, will eventually require repair. Also vulnerable are waterfront buildings, especially those with thin balcony cantilevers and where stressing pockets are located in the exposed slab edges, and other structures subjected to application of deicing salts or saltwater spray.

Deterioration often starts at the end anchorages—perhaps the most vulnerable area of the structure—where the steel may be protected only by a grout plug. Also, chlorides and other chemicals can attack and eventually penetrate the protective sheathing at locations where the concrete cover is eroded or broken, where the sheathing itself is damaged, and, near the ends, where the sheathing must be terminated to make the anchorage connection possible. The process of corrosion in post-tensioning strands is illustrated in Fig. 7.4.

Deterioration of post-tensioned buildings can be caused by factors other than corrosion. Prominent among them is concrete cracking caused by incorrect draping of post-tensioning cables. In a proper position, the cables exert forces that counteract the effects of the applied loads. Conversely, improper placement of cables can damage the member by magnifying the stresses from service loads. Figure 7.5 illustrates a case in which incorrect draping of the tendon results in its being placed near the end that is too high. This produces a downward, instead of an upward, force on the member. If the thickness of the concrete on which this force acts is insufficient, an internal crack develops.

7.4 Planning for Repairs

7.4.1 Starting with a clear goal

The preceding discussion should set the tone for the serious and difficult nature of repairs in post-tensioned structures. Locating and effectively repairing a deteriorated post-tensioned tendon is often akin to finding the proverbial needle in a haystack. Monitoring, investigating, inspecting, evaluating, and maintaining post-tensioned structures should be performed only by qualified engineers and contractors. The following sections should assist them in those activities.

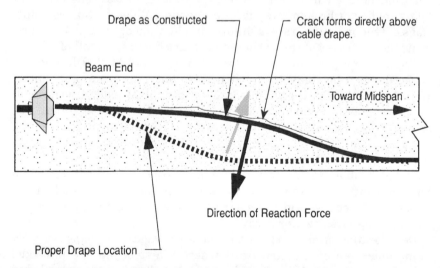

Figure 7.5 Concrete cracking caused by improper draping of post-tensioning cables. [*Reprinted with permission from Peter H. Emmons,* Concrete Repair and Maintenance Illustrated, *copyright R.S. Means Co., Inc., Kingston, Mass., (800) 334-3509, 1993, all rights reserved.*]

As with other materials, proper inspection methods and experience are both essential in uncovering defects in post-tensioned concrete. The process of field investigation is described in detail in Chap. 2. Here, we highlight some specific items that should be investigated prior to performing post-tensioning repairs.

The first step is to develop a clear plan of investigation. Successful repairs are preceded by well-defined and effective evaluation strategies. The engineer's plans must be coordinated with the client's expectations. To determine a comprehensive scope of work, a site visit should be arranged by the key engineering personnel and the client's representative. The outcome of the site visit should be an understanding of what the investigation phase will include and what the discovery of any deteriorated tendons will entail. The extent of the investigation depends on the location of the structure, its general condition and age, construction and budgetary constraints, and the efforts required to assure the safety of users.

7.4.2 Finding the available information

Finding information about the original design is time-consuming. If the client is a government agency or a municipality, this information is usually available, but may take time to find; with private clients, the information may be difficult to obtain. Some of the possible sources are mentioned in Chap. 2. Ideally, all of the following should be obtained:

1. Original as-built design drawings, specifications, and calculations

2. Shop drawings

3. Stressing records and other information from the field inspection team

4. Design documents from any previous alterations and repairs

If found, the original design documents (tendon shop drawings are most helpful) can be compared to the age of the system and to the applicable design code. They should be examined to see whether the calculations were made using the allowable-stress or the ultimate-strength method; the difference in the required amount of mild reinforcement could be dramatic. It may even be necessary to talk to the original designers to clarify some issues.

If the original drawings are not available, the sizes and general layout of the tendons can be determined by field exploration, an expensive but in this case unavoidable effort. It is possible to determine the size and spacing of tendons with rebar locators, in combination with making some exploratory openings in concrete and with a search for tendon anchors at the edges.[13]

The main areas of interest in the information-gathering phase, and prior to beginning any field investigation work, are

- The maintenance history of the structure and locations of prior repairs and modifications
- Drainage profiles and water-flow patterns
- Environmental aspects (proximity to salt water, location of nearby buildings with large carbon output, etc.)
- Traffic patterns and the intended use of the structure
- Locations of inadequate concrete cover
- Types of tendon sheathing and protection system
- Type of concrete (e.g., high-strength, lightweight, or special-aggregate)

7.4.3 Performing field investigation

If the theoretical weaknesses of the structural and post-tensioning building systems are well understood in advance of the investigation, and if a clear goal is established, a systematic field inspection will reveal many of the deficiencies. Since field inspection is often performed by specialized personnel, it is essential to share the findings of the information-gathering phase with them. The inspectors should know what to look for and should develop and maintain a detailed set of plans, to be used in analyzing and processing the field data. The inspection should follow the same approach for all areas of the building.

During inspection, typical problems and locations of suspected deficiencies are identified. These areas are marked for further investigation and physical testing. The goal of this activity is not simply to catalog the defects but also to look for their causes, in order to transform the raw field data into the repair plans. By finding the cause of the observed distress, the engineers can find the areas where similar defects can be prevented in the future.

The easiest defect to identify is an exposed tendon; the others are much more difficult to find. The focus of field investigation should be on detecting subtle clues pointing to a latent structural distress. Some items that require particular attention during field inspection include

- The condition and quality of the existing concrete at the locations of shallow cover (it should be clear from Fig. 7.1 where they might be), at construction and expansion joints, at anchorage points, and at grouting ports. Evidence of heavy leakage, combined with rust

Figure 7.6 Heavy leakage and rust stains at a construction joint indicate likely corrosion.

stains, indicates probable corrosion of tendons, hardware, or mild reinforcement (Fig. 7.6).

- Signs of water intrusion, spalls, and cracks at locations of shallow concrete cover.

- Cracking, both parallel and perpendicular to the tendons. A note should be made of the crack size, relation to the tendon, and pattern.

- Rust staining, in particular large concentrations of rust stains.

- Grease staining on the underside of the structure. These stains are dark, pasty, and often dirty. When this condition is located, the top surface of the structure over the tendon centerline should be closely examined for evidence of water infiltration points.

- Indications of large deflections, as might be suggested by large amounts of ponding water over some structural members.

- Efflorescence on the underside of the structure.

Some engineers believe that a close observation of the anchorage regions is sufficient to detect signs of corrosion. These places are certainly among the most troublesome, because the ends of tendons are typically protected only by porous grout plugs. However, Kesner and Poston[10] argue that neither the condition of the anchors nor that of the strands near the anchors is a reliable predictor of corrosion in other parts of the tendon. They suggest that only a selective removal and examination of a statistically significant number of tendons can meaningfully testify to their true state of pitting and corrosion.

7.5 Nondestructive Testing

7.5.1 Types of nondestructive tests

The most definitive method of determining and verifying the severity, location, and extent of problems in post-tensioned buildings is by testing. By making tests at a few key locations, the investigators can detect some general trends that can apply to the whole structure. The most common methods of nondestructive testing are discussed in Chap. 2. Here, we outline some other available testing methods and procedures used for evaluation of post-tensioned concrete structures.

Nondestructive testing methods are the easiest to use—and therefore are most frequently used—even though their effectiveness at finding locations of tendon breaks is questionable. At present, there simply is no reliable nondestructive method of locating corroded tendons or finding the extent of corrosion.[8,10] The term *nondestructive* does not mean that no concrete removal at all takes place; it means that the integrity of the post-tensioning system is not compromised by testing. Nondestructive testing can be subdivided into external (nonphysical) and internal (physical) methods. These are listed separately in the next sections.

7.5.2 External testing

This testing of post-tensioned concrete is performed without concrete removal or excavation and in a sense is made "externally" to the tendon. The methods of investigation are essentially the same as those listed for nondestructive testing of concrete in Chap. 2. Among them are pachometers, rebar locators, and cover meters, the least expensive but, not surprisingly, also the least useful testing instruments. These devices are typically used to locate tendons and mild steel prior to removing any concrete for physical testing; they cannot find tendon breaks or deteriorated areas.

Other testing methods include x-ray, electric potential, ultrasonic pulse velocity, and ground-penetrating radar, all discussed in Chap. 2. These tests are useful if more definitive physical tests cannot be made, or if a large number of tendons must be examined quickly. Most of these methods provide only two-dimensional information and require experienced operators for their interpretation.

7.5.3 Internal testing

Internal testing methods allow examination of the actual condition of a tendon by gaining physical access to it. The procedures require removal of the surrounding concrete, but without altering or compromising the tendon's integrity.

Making test wells (inspection pits). Making test wells (also called inspection pits or inspection openings) is the most direct and reliable method of determining the tendon's condition. Inspection openings are typically made in places where a good probability of a tendon break exists, as can be suggested by external testing methods.

The process involves careful saw cutting at the perimeter of a planned opening and removal of the concrete until the tendon covering is visible. The tendon sheathing is then examined and removed, and the tendon wires or strands exposed for inspection. If the extent of deterioration happens to be confined within the opening, the size and location of the test well was well chosen. If damage extends beyond the borders of the opening, additional inspection pits or testing will be required.

It is essential to check the tendon's condition along its entire length and to make repairs beyond the strict limits of deterioration. When a break in the tendon is suspected but not yet located, additional test wells are opened, a lift-off test is performed, or a destructive test is made.

Screwdriver test. This simple test is often performed to quickly determine if the tendon's wires are under tension. It requires making a test well to uncover the wires. The inspector attempts to wiggle or pry the wires away from each other by forcing a screwdriver between them. If the wires are still stressed, they will not move appreciably; a movement usually indicates that the tendon has lost at least some of its tension.

Lift-off test. A lift-off test can yield information about the current load capacity of the tendon. During the test, a stressing-end anchor is excavated, exposed, and cleaned, and a force is applied to it by a hydraulic ram (jack). The magnitude of the force that causes the anchor to lift off—to lose contact with the shim and bearing plates—is noted and recorded. This lift-off force is then compared with the design loading on the tendon, allowing for some post-tensioning losses, creep, and shrinkage. In a properly installed and well-maintained post-tensioning system, there should be a direct relationship between the two forces.

Occasionally, despite the fact that a substantial force is applied by the jack, separation between the anchor and the plates does not occur, even at stresses as large as 65 percent of G.U.T.S. The standard procedure in this situation is to release the force and repeat the test after a few minutes, a technique called rebounding the tendon. Its purpose is to allow the lift-off testing force to "rebound" (get back into the tendon), in the hope that any binding points that might have interfered with the first try were released. The testing force is then reapplied. If, again, no separation occurs at the force level producing 65 percent of G.U.T.S., the most likely scenario is that an obstruction is present somewhere along the tendon length. One culprit could be excessive rust buildup, causing what is known as a rust bind. A tendon that is heavily rusted along most of its length could develop over 55 percent of G.U.T.S. in a lift-off test without separation.

A lift-off test can also help in finding broken wires. When an anchor separates from bearing and shim plates in a properly performed test, the wires elongate. Since the stresses do not exceed the elastic limit of the steel, the intact wires snap back (recover) after the load is removed, but the broken wires do not. The button heads of the broken wires may become unseated and remain popped out (Fig. 7.7).

The wires with unseated heads can be extracted and examined. To locate the break, an extracted wire is placed on top of the structural member it came from; it will end at the break location. The point of a probable break is marked on the surface of the concrete, to be explored by a subsequent test well that will confirm it. This procedure is helpful when a wire break occurs behind the bearing plate, within the transition cone.

Like the other tests discussed here, the lift-off test should be performed by a qualified post-tensioning contractor in the presence of an experienced engineer. The lift-off test is moderately expensive, but it can provide definitive information about the tendon's condition and the location of any breaks in it.

Figure 7.7 Unseated wire button heads at deteriorated stressing-end anchor.

Vacuum test of anchors. Post-tensioned structures tend to deteriorate near the anchors, for several reasons. The anchors are typically located at the edges of the structure, where the largest amounts of weather, temperature, and environmental changes take place. Also, the anchor heads are located in concrete pockets, where they are typically protected only by grout plugs placed after stressing. As the grout shrinks, a crack can open between the grout and the surrounding concrete—a prime infiltration point for water and chlorides. The resulting anchor corrosion can proceed quite rapidly.

To reduce the problem of corrosion and lessen deterioration in new construction, high-quality nonmetallic, nonshrink grouts should be used. Corrosion-inhibiting concrete admixtures can also help. Among the emerging technologies is the electrically isolated tendon, a system in which tendons and their entire anchor head assemblies are coated and encased in plastic, removing them from the chain of corrosion.

The effectiveness of corrosion protection at an existing end anchorage can be determined by a vacuum test, also called the tightness test.[14] The test measures the tightness of the joint between the parent concrete and the grout plug, as well as their relative porosities. The test is performed by installing a transparent testing head over the anchor pocket and sealing it with a gasket. The space under the testing head is placed in a vacuum provided by a battery-operated tester. The level of maximum negative pressure that can be sustained, or its rate of change, can indicate how easily aggressive materials could enter at the anchor location.

7.6   Destructive Testing

Testing methods in this category are used when one or more tendons have a confirmed break and require repairing. Destructive tests must be conducted by experienced qualified restoration contractors and engineers. The tests can be quite costly and risky, because of the ever-present possibility of collapse. They should be undertaken only under the safest of conditions and include provisions for adequate shoring.

7.6.1   Cut and recoil

As discussed previously, tendons are typically stressed to between 60 and 65 percent of the G.U.T.S., still well within the elastic region of the steel. If a wire is cut, the load on it is removed, and the wire shortens as it undergoes elastic rebound. (This is a phenomenon quite familiar to guitar players.) The distance that the wire snaps back (recoil) can be measured and correlated with the previous stress level in the wire. This is the principle behind the cut-and-recoil testing procedure used to find broken tendon wires.

The test involves making an opening in the concrete to gain access to the tendon, uncovering the wire or strand, clearly marking the proposed cut location on the tendon and on the adjacent concrete, cutting one wire in the tendon, and measuring the amount of recoil. If no recoil occurs, the tendon was unstressed—that is, it was already broken; if recoil does occur, its length is measured. If the type and locations of existing anchors are known, the nature of the tendon break can be gleaned from the magnitude and direction of the recoil.

The expected magnitude of the recoil, computed prior to cutting the wires, can be compared to the actual recoil length on either side of the cut. Essentially, the wire will shorten to its original unstressed length minus an allowance for long-term creep. The percentage of recoil in each direction should be directly proportional to the elastic elongation and the distance from the cut to the anchors. If the cut wire or strand was not damaged before the test, the measured recoil distance should equal at least 90 percent of the theoretical elongation. If it does not, one or more breaks in the tendon are likely. Some other clues, such as signs of concrete damage caused by the recoil, can help pinpoint the break locations.

When deterioration destroys some of the wire section, the tension stresses in the remaining steel continue to increase until the tendon breaks. A broken and recoiled tendon may pop out of concrete in regions of low cover or those with a large amount of spalling; it can also produce cracking near the anchors. The cracks can help determine the location of the break—for example, large cracks at the anchor ends may indicate that the tendon is broken somewhere within the middle half of its length—but their absence does not guarantee that no tendon breakage has occurred. Quite often, wire deterioration at anchor heads simply cannot produce the forces necessary to crack concrete. Conversely, observed cracks could result from factors other than tendon breakage.

Despite being rather common, cut-and-recoil testing requires care and experience; like other destructive tests, it can be dangerous if not handled properly. It is wise to begin the investigation only after a thorough study of the original plans for such items as the amount of concrete cover and location of the end, dead, and intermediate anchors. It is helpful to mark these items on a field investigation sheet prior to starting field work, so that defects can be identified quickly and easily.

7.6.2   Wire extraction

This method may be considered an extension of the cut-and-recoil procedure. Wire extraction is an expensive, dangerous, and destructive test, since the tendon must be physically cut and removed and two or more openings created (one typically at an anchor), severely weakening the member. This invasive procedure is generally performed only when it is obvious that the member must be repaired anyway. Wire extraction often requires shoring of the structural member until the repair is completed.

The main purpose of this test is to determine the condition of the tendon along its length by physically removing one or more wires from it. The wires are removed at an anchor using a small "ram" (hydraulic jack) or other mechanical means. After removal, the wire is laid down on top of the member. An examination of the wire and its damage may reveal the probable locations of infiltration points and other areas of deterioration. Examination of an extracted wire can yield two possible results:

1. A tendon break is found in the extracted wire, meaning that the problem area has been identified and repairs can proceed.

2. The tendon is break-free and in good condition. This indicates that the break, if it exists, is in the wire section remaining in the concrete, and so the rest of the wire must now be extracted and examined. To avoid repeated testing, the engineer should carefully review the results of the cut-and-recoil test prior to extracting any additional wire.

7.7   Repair Methods

7.7.1   Maintain, repair, or replace?

The results of field investigation and testing will place the structure in one of three categories:

1. *Deterioration is minor, and tendon integrity is not yet affected.* It is too early to repair. Floor maintenance—repairing minor spalls, filling cracks, replacing joint fillers, and sealing the surface—should be sufficient to restore the structure.

2. *Some tendon damage is detected, but no significant strength reduction has yet occurred.* In this case, the situation is more complicated. The most economical approach is probably to perform the just-mentioned maintenance and carefully watch the condition of the structure for any signs of further deterioration. However, as discussed in Chap. 1, this is not always practical if the owner has the funds for repair this year but not next. Another option is to undertake an expanded maintenance program, including excavating and sandblasting all the exposed anchor hardware, rewrapping the transition cone and pumping it with grease, and coating all exposed steel items with epoxy.[15]

3. *The structure is seriously damaged at least in some places.* In this case, repair or replacement is necessary. If the damage is caused by corrosion, the adjacent concrete members will probably need attention as well.

Some issues relating to maintenance of concrete floors are examined in Chap. 6, and the discussion in this chapter is primarily concerned with the last case. When widespread serious deterioration is present, and the structure requires either large-scale strengthening or replacement, there are several options, some of them described by Nehil[15]:

1. Repair the existing tendons and anchors

2. Internally replace the post-tensioning system in kind

3. Externally replace the existing post-tensioning system with similar construction or by means of external mild reinforcement

4. Totally replace the deteriorated slab, the floor structure, or even the whole building

The last option is self-explanatory. The other three are addressed in the remainder of this chapter in various degrees of detail, with the main emphasis being on repairs. The selected course of action depends on many variables, including the overall condition of the structure, the extent of deterioration, and the available budgets for present and future repairs. If the damage is severe but localized and affects relatively few tendons, repair might be considered. At the other extreme, damage that is severe *and* widespread, perhaps even affecting most tendons, requires replacement of the system or of the whole structure.

7.7.2 Making repairs

Moderately corroded tendons can be repaired or partly replaced. Tendons suffering from accidental mechanical damage are also candidates for repair. The repair should include not only tendons, but also patching deteriorated concrete and replacing the means of corrosion protection. The latter might involve removing water from inside the sheaths (by dry gas pressurization, for example), cleaning off rust, and reapplying grease with corrosion inhibitors along the whole length of the tendons. The typical types of repair are

- Partial replacement of a tendon section
- Replacement of dead-end anchorage
- Replacement of intermediate stressing anchorage

- Replacement of live or stressing anchors
- Greasing and rewrapping the tendon

Among these repairs, replacement of a live or stressing anchor is perhaps the most common. Accordingly, this procedure is described in detail at the end of this section. Some other repair procedures are illustrated in Ref. 16. Quite often, the same tendon requires several types of repairs, and planning the operations in the proper sequence is critical for a successful outcome.

Partial replacement of a tendon requires splicing of the new and existing tendon sections. The draped tendon profile shown in Fig. 7.1 makes installation of any splicers in the areas of low cover difficult. The most practical splice locations are those places where tendons happen to be at middepth of the concrete member. When many tendons are replaced in this manner, they are often detensioned in alternating locations.

If it is overly difficult to simply pull the tendons out of the concrete, they can be removed by using the patented *trenching method,* in which concrete is partly excavated both at the top and at the bottom of the member in the areas of low cover. If tendons in a beam-supported slab are being replaced, full-depth openings are also required at approximately quarter points of the slab's span. New tendons are inserted through the opened areas (when this is done, it is recommended that a 40-mil-thick tendon sheathing be used) and restressed after the anchors are set and the slabs are patched and cured.[15]

A decision as to whether to replace only the damaged part of the tendon or the whole length requires careful study, especially if the floor is contaminated with chlorides. As discussed in Chap. 5, leaving some existing steel in place in such circumstances can lead to accelerated corrosion at the edges of the repair. Granted, the existing tendons are protected from the concrete by grease and sheathing, but the effectiveness of these elements may have been compromised over the years. If parts of the existing tendons are left in place, it might be wise to install an external corrosion-protection system, such as a traffic-bearing membrane or latex concrete overlay, to prevent further deterioration.

To develop a budget for repairs, the following prices are suggested (in 1998 dollars): $4000 to $7000 per splice and per anchorage requiring reconstruction; $1250 to detension and retension a tendon at each anchor, or to expose and recondition an anchor that requires no repairs. Tendon replacement might cost from $250 to $350 per linear foot. These costs should be somewhat increased for a small number of tendons and reduced for large quantities.

7.7.3 Internal replacement of several post-tensioning tendons

When there is a relatively high level of certainty that damage, even if severe, is localized, replacement can include only the affected tendons. The basic technique for replacing tendons, few or many, is the same: The damaged member is shored, the tendons detensioned, the anchors removed and replaced, the concrete and steel around the anchors repaired and given additional protection from corrosion, and new tendons installed and stressed. The actual tendon removal can be done from anchor locations, from the middle of the member, or from some other point. These steps are further described in Sec. 7.8.

When single-strand tendons are replaced, the new ones can be threaded through the existing plastic sheathing, if its shape is approximately round. Since the sheathing may not survive the operation intact, or could have already been damaged, some experts recommend giving the replacement tendons a 30-mil epoxy coating for corrosion protection.17 Replacing tendons in a paper-wrapped wire system is much more difficult because it involves inserting new tendons into existing ducts of irregular cross section. In that case, the trenching method may be the most practical.

Kesner and Poston10 suggest that when replacement of a post-tensioning system is required, the new system need not necessarily mimic the existing one. It could be beneficial to change the tendon layout to a more efficient one. For example, a floor that is post-tensioned in two directions could be replaced with a one-way banded post-tensioned system, a simple and practical solution. Moreover, repairs to modestly deteriorated post-tensioned structures may not be even needed in some cases. Kesner and Poston point out that the ACI Building Code (ACI 318-9518) Sec. 18.18.4 allows for up to a 2 percent loss of prestress force "due to unreplaced broken tendons." This provision can presumably be applied to repair work. Thus, if only one or two of the hundreds of tendons are damaged, the structure might be considered adequate without strengthening. Also, higher effective steel stresses could probably be assumed for the tendons used in repairs. The reason: Most post-tensioning losses typically result from concrete shrinkage and creep, which would have largely taken place by the time of the repair work. It is also possible that the design live load on the building has become smaller, so the required post-tensioning forces might be less than those originally provided. And finally, the existing tendons might have been tensioned to a larger degree than necessary. All these factors could help reduce the number of repaired or replacement tendons, or provide justification for leaving the existing tendons in place. Or course, careful investigation and analysis by qualified engineers is required to determine if any of these factors apply.

This approach is more appropriate when easy-to-see mechanical damage has occurred, rather than hidden deterioration resulting from corrosion, for the simple reason that in the case of corrosion, no one can be certain how many tendons are damaged. Also, some tendons are located in more critical areas than others.[19]

Sometimes internal tendon replacement is not practical because of budgetary constraints, drastic increases in loading, or the practical impossibility of installing new tendons into existing tendon ducts. When the traditional post-tensioning repair methods cannot be used, external post-tensioning—attaching new tendons outside of the existing structural member—can often help.

7.7.4 Using external post-tensioning

The new tendons used for strengthening an existing post-tensioned member need not be located where the original tendons were. They can be placed at the exterior—along the sides or underneath the member. For tendons of straight lengths, high-strength threaded rods are typically used; for tendons with draped configurations, wire strands. Hardware for external post-tensioning can be connected to the existing member either by installing an end-bearing assembly or by means of a shear transfer mechanism. The latter may be mounted on the sides or underneath the member and bolted through it.[11]

Regardless of the details and techniques, the first step of an external post-tensioning job usually involves inspection and repair of the existing concrete, including all flexural cracks. If end-bearing assemblies are used, concrete at both ends of the member is roughened and prepared to accept them. Then, tendons are installed and draped to the profile required by design.

If the new tendons require protection from fire and corrosion, a new concrete enclosure are attached to the existing member with drilled and grouted mild reinforcement stirrups. Longitudinal reinforcement and other required bars are then placed, along with new tendon hardware. After the tendons are stressed, they are encapsulated in concrete.

External post-tensioning can also be an effective means of repairing construction that is not post-tensioned. This repair method has been used to increase the shear strength and flexural capacity of concrete members; it is particularly effective for strengthening long-span beams. It is also appropriate for reinforcing sagging cantilevered slabs. In fact, according to published reports, external post-tensioning was chosen to stop the sagging of the distressed concrete cantilevers in Frank Lloyd Wright's "Fallingwater."[20]

When should external post-tensioning not be used? Problems might arise if other members intersect the one being repaired; this can raise the cost of this method beyond reasonable limits. Some methods of external post-tensioning are discussed and illustrated in Chap. 4.

7.7.5 Repairs other than post-tensioning

When neither internal nor external post-tensioning methods of repair are practical, other approaches can be tried. In one scenario that may be useful when the existing post-tensioning tendons in slabs are badly deteriorated, but the beams and columns show no sign of distress, the existing slab tendons can be assumed to be ineffective and simply ignored, and the slab considered as unreinforced or reinforced only with mild temperature steel. The slab can then be strengthened by some of the methods discussed in Chap. 4, such as shortening the span by adding steel beams underneath, with the beam spacing being determined by the slab's spanning capacity; adding steel or plastic-fiber-composite reinforcing plates or strips; or placing overlays containing new reinforcing bars.

Another approach is to remove and replace the slabs while keeping the beams and columns intact. If any load on those members remains, they should be checked for temporary loading conditions. A building that is partially occupied during construction may require lateral bracing of columns and shoring of beams.

When a slab is removed, great care should be taken to preserve the beam and column bars and stirrups contained in the slab. If this is not done, the supporting framing will be severely compromised, and the most practical solution will be to remove and replace this framing as well—or perhaps even rebuild the whole structure. The idea of tearing down and rebuilding may not please an owner who is expecting basic repairs, but it could prove to be the most economical long-term solution, and deserves consideration. Indeed, a total building replacement is the most definitive way of repairing damaged floors, because any uncertainties about deterioration in other members are sidestepped. The owner may have to be educated on the basic notion of limits on the useful life of buildings. For example, parking garages in aggressive environments should not be assumed to last more than about 40 years.

Quite often, even when the owner is convinced that a total replacement is the best option to pursue, the useful life of the building can be extended for a few years by making some emergency repairs. This temporary fix is beneficial to everyone: It allows the owner to obtain funding, the users to find alternative locations, and the designers to prepare plans for a new structure.[15]

7.8   A Step-by-Step Example: Replacing Post-Tensioned Stressing-End Anchorage

The method illustrated in Fig. 7.8 represents one common approach to replacing post-tensioned stressing-end anchorages. The actual repair procedure may vary from project to project and from contractor to contractor, but the following steps are typically involved:

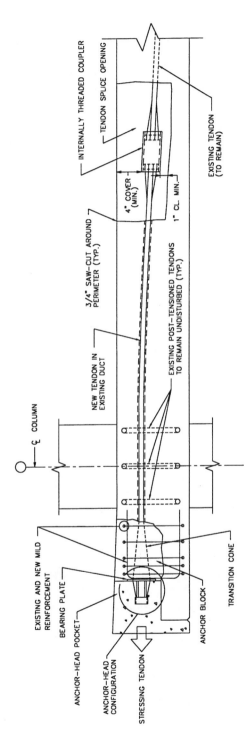

Figure 7.8 Schematic sketch of a typical stressing end-anchorage replacement.

Labels in figure:

INTERNALLY THREADED COUPLER

TENDON SPLICE OPENING

EXISTING TENDON (TO REMAIN)

4" COVER (MIN.)

1" CL. MIN.

3/4" SAW-CUT AROUND PERIMETER (TYP.)

NEW TENDON IN EXISTING DUCT

EXISTING POST-TENSIONED TENDONS TO REMAIN UNDISTURBED (TYP.)

℄ COLUMN

EXISTING AND NEW MILD REINFORCEMENT

BEARING PLATE

ANCHOR-HEAD POCKET

ANCHOR-HEAD CONFIGURATION

STRESSING TENDON

ANCHOR BLOCK

TRANSITION CONE

1. *Shore and properly support the member.* This step is among the most critical aspects of repair work in post-tensioned concrete; it is essential for protection of the structure and its occupants. Determination of proper shoring and support locations requires a careful study of the existing framing. It is a good idea to provide shoring not only for the members under repair, but also for the members supported by them. In parking facilities, the location of supports must take into consideration traffic patterns, and the shoring system must be protected from vehicle impact. Other factors that need to be considered in shoring design include provisions for temporary loading, including the weight of construction equipment, and the forces generated during restressing operations and replacement of multiple tendons.

Shoring assures safety, but it is expensive. In some cases it is possible to carefully analyze the existing structure and find a way to avoid shoring by detensioning and reanchoring tendons in several phases. Popovic and Donnelly[13] describe a parking garage renovation project in which bundled tendons were detensioned and reanchored in three phases. In each phase, every third tendon bundle was detensioned, new end anchors were installed, and new concrete in the anchor zone was allowed to reach sufficient strength. Then, the work in the next phase was done in the same manner. (The affected floors carried no live load during the repairs.)

2. *Locate and expose deteriorated anchor heads.* Concrete is removed back to the existing bearing plates, so that shim plates can be replaced if needed. When a post-tensioned structural member contains multiple tendons and anchors, the work should proceed in a fashion that protects the adjacent elements from damage. To protect anchor heads, this work should be done with extreme caution and only by skilled and experienced personnel. Hasty concrete removal may cause damage to threads of externally or internally threaded anchors. Figure 7.9 shows an exposed intermediate anchor head that has deteriorated.

3. *Detension the tendon.* A tendon can be detensioned anywhere along its length, provided that the break is located between the accessible stressing anchors. A tendon slated for repair could be cut at the test-well opening, although in operations involving anchor replacement the tendon is usually detensioned at the anchor head. A stressing ram could be used to lift the anchor head and remove the shim plates, but quite often the existing button heads are simply ground off at the face of the anchor. The detensioning operation should proceed in a slow and systematic manner, so that the stress in the tendon is released evenly. Quick or uneven detensioning may cause significant damage to the existing structure and to the adjacent anchors. The work should not begin until shoring is installed and all personnel in

Figure 7.9   Exposed deteriorated intermediate anchor configuration with numerous broken and unseated wires.

the area are notified about it. The member being detensioned should be closely monitored during the operation and carefully inspected immediately after its completion, as new cracks and delaminations might appear.

4. *Remove the existing anchor and bearing plates, if needed.* Corroded anchors and bearing plates must be removed, but the work may not end there. Wires in a transition zone directly behind the bearing plate are typically unprotected and vulnerable to corrosion. If the anchor head or the wires between the anchor head and the bearing plate are deteriorated, the condition of the wires behind the bearing plate must also be checked. Very seldom will the damage stop at an anchor head.

5. *Remove all deteriorated concrete, wire, and reinforcement in the transition zone.* To expose the wires behind the bearing plate, concrete in this area is usually removed. If the anchor and bearing plates must be removed, some tendon length must also be replaced, because it may be practically impossible to feed new tendon through the transition zone unless the tendons are sheathed in that area. All other deteriorated concrete and reinforcement in the transition zone should be replaced to extend the life of the repair.

6. *Install a new bearing plate and reinforcement.* The high concentrated loading imposed by the bearing plate is resisted by the concrete behind the plate. This area is typically reinforced with mild rebars

referred to as the anchor-block, or backing, steel. The arrangement of the bars in this area can be quite complex. It can be determined from the original design drawings, or it may have been prescribed in the governing building code in effect at the time of construction. If the steel reinforcement behind the bearing plate is deteriorated, it should be replaced with the same or a greater amount of steel.

7. *Clean the void left after removal of the existing duct.* When an older-type unbonded tendon wrapped in kraft paper is removed, most of the wrapping tends to remain in the resulting void. This wrapping should be removed, because it is often saturated with water and contaminated with chlorides and rust. The removal can be done by pulling a rag or a special wire brush through the duct space, then cleaning out the remnants by compressed air or vacuuming.

If the void is filled with a lot of water that had seeped through the cracks, simply cleaning the void will be insufficient—it must be dried out. Some recently introduced patented drying techniques utilize air or nitrogen injection. The air or nitrogen is injected into the duct at one end, while the humidity and temperature are monitored at the other end. The injection stops when the duct is dried out and the desired results have been achieved.[19]

8. *Install new post-tensioning tendons.* At this point, new tendons can at last be installed, often the most difficult and time-consuming part of the work. As already noted, early unbonded wire tendons were often paper-wrapped, resulting in oval or egg-shaped tendon void cross sections. As can be imagined, it is quite difficult to feed new round tendons through existing holes with those configurations. The two most commonly used methods of feeding new wires through existing voids are trenching and broaching the void.

The trenching method was already mentioned. It uses a series of strategically placed slots or openings along the length of the member to facilitate easier insertion of tendons. The trench openings, which divide the existing duct voids into several short segmented lengths, are located at the high and low points of the tendon's profile (Fig.7.10).

The broaching method is similar to cleaning out a clogged drain pipe. Broaching (routing) heads are inserted into the duct void and are slowly fed through to the other end of the duct. The heads are then increased in size and the process is repeated, leaving a progressively larger hole size each time, until a clean, uniform, and smooth duct void is attained. This method has been used with much success, but it is more costly and time-consuming than trenching.

9. *Repair all other areas of deterioration.* A deteriorated post-tensioned member often requires several different types of repair. Replacement of stressing-end anchors often goes hand in hand with replacement of a certain length of tendons and sometimes intermediate anchor assemblies and dead-end anchors. All these types of repairs

Figure 7.10 A trench opening around the coupler in a concrete girder.

are best performed concurrently, while the member is shored and detensioned.

10. *Connect new and existing tendons.* The next step in the repair process is to physically connect the new and existing tendons, typically by internal threaded couplers. The cross section and a photograph of an internally threaded coupler are shown in Figs. 7.11 and 7.12.

The coupling process involves several steps. First, the wires are seated against an existing anchor head. The tendon wires are then cut evenly, so that they sit flush with the new anchor head and are of the same length. If the wire lengths were different, this would result in

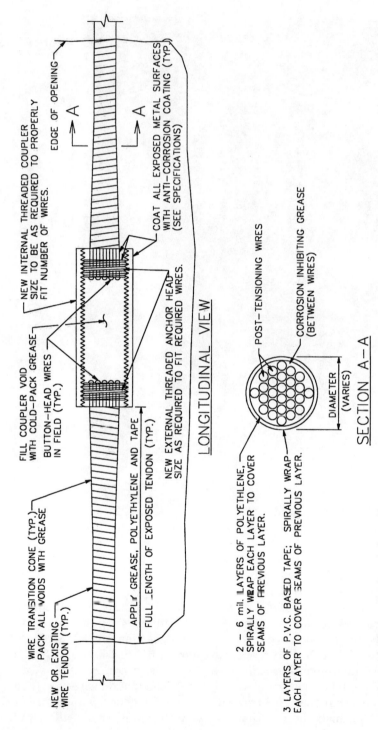

Figure 7.11 An internally threaded coupler.

Figure 7.12   Photograph of an internally threaded coupler.

Figure 7.13   Field cold-formed button-heading machine for tendon field splice.

unequal wire tension; some wires would be overstressed and some not stressed enough. Occasionally, the anchors themselves rotate or twist, further compounding the effects of uneven stressing. Next, the wires are placed through the anchor head and their ends are deformed by a button-heading machine (Fig. 7.13).

To be effective, the couplers need ample concrete coverage around them. They must also be allowed to move during stressing operations. Why is this important? As previously discussed, a tendon being stressed within the elastic range elongates. As the stretched wires move, so should their couplers. If the couplers are restrained and are not allowed to freely move during tensioning, they become intermediate dead-end anchors. The tension loads are then concentrated between the stressing end and the coupler, while the part of the tendon beyond the coupler is left unstressed—and incapable of supporting design loads.

The most common solution that allows couplers to move during tensioning is to leave openings in concrete around them. To avoid severely weakening the member, the depth of the openings should not exceed 30 percent of the member thickness. Another approach is to install the couplers in plastic pipe sleeves filled with grease. This allows the couplers to move, eliminates the need to patch the pockets at a later date, and provides maximum protection for the couplers.

11. *Cast new concrete at the anchor block and other locations.* When all of the new tendon segments have been inserted and properly draped, and the anchors have been properly installed, concrete can be placed. As with any concrete repair, proper surface preparation, avoidance of unnecessary cold joints, and high quality of workmanship are essential for success. To speed up the process, high-early-strength concrete can be used, as was done in the above-mentioned parking garage renovation project described by Popovic and Donnelly. (This allowed the next operation, stressing of the tendons, to begin only 2 or 3 days after the concrete was placed.)

12. *Stress the tendons.* Tendon stressing should start only after the concrete is strong enough to resist the imposed forces. Concrete strength is best verified by making concrete cylinders and having them tested by a certified testing laboratory. The first set of cylinders, tested 3 days after placement, should indicate that the concrete has attained the specified desired strength prior to stressing—3500 psi, for example. If it has not, the concrete should be allowed to harden for an additional period, until the proper strength required for stressing is reached.

When the proper strength has been reached, the tendon can be stressed, but to what level? In new construction, tendons are typically tensioned to 65 percent of G.U.T.S., held at that level for a short period to allow stresses to equalize along the tendon length, and then stressed to 70 to 75 percent of G.U.T.S. Next, the tensioning force is slightly reduced, and the stress is returned to 65 percent of G.U.T.S.; the tendon is then anchored (locked off). Since approxiamately 5 percent of the stress level is lost during the lock-off procedure, the design stress in the tendon often ends up being about 60 percent of G.U.T.S. This is the level of stress that many engineers strive to achieve in both repair applications and new construction.

Tendons are typically stressed by means of stressing rams, one of which is shown in Fig. 7.14. Monitoring tendon elongation is as important as assuring the desired level of stressing force. Tendons typically stretch at a rate of $^5/_8$ in per 10 ft of their length, or approximately 6 in for every 100 ft. The expected elongation is calculated prior to starting the work and verified during the stressing operation by the contractor and the engineer. If the measured elongation is significantly different from the expected value, the tendon may be binding, and stress distribution along its length is likely to be uneven—a highly undesirable outcome.

There could be other reasons for an observed difference between the expected and the actual elongations, in particular the fact that existing tendons are spliced with new. Since the existing tendons have been stressed for a long period of time, creep and shrinkage in them have probably stabilized, but the opposite is true for the new tendon segments. Thus the difference between the actual field-measured elongations and the theoretical values can vary according to the age of the

Figure 7.14 A 200-ton stressing ram.

existing tendons and the ratio of their length to the length of the new tendons. The elongation calculations should attempt to account for these factors, to avoid confusion in the field.

13. *Restress the tendons.* The stressing operation often fails to provide the anticipated levels of stress and elongation, especially if older tendons are involved. The main reason for falling short of the desired results is high friction losses owing to old and ineffective original grease. Many problems were reported with old mastic-type lubricants; low ambient temperatures are also among the detrimental factors. To check whether the first stressing operation was successful, the level of transfer stresses should be verified 48 h after completion. If insufficient levels of stress are found, the tendons may have to be restressed, followed again by verification of the actual stress and elongation levels. After restressing, the measured and the anticipated tension values will often agree or will at least be close enough.

14. *Apply a protective coating to the anchor head.* After the tendons are tensioned, and restressed if necessary, the anchor-head assemblies should be given some sort of corrosion protection. There are custom-made protection covers available for this purpose, but moisture-tolerant cementitious corrosion-protective coatings work best. If the selected coating can double as a bonding agent, it may help eliminate shrinkage cracking around the assemblies.

15. *Fill anchor-head pockets and other locations along the tendon.* Properly sealing anchor-head pockets is essential for the durability of the repair. After all, a problem in these areas has already occurred once, and the owner rightfully expects that it will not recur for a long time. This demanding application deserves a high-quality repair material, such as a nonmetallic, nonshrink grout with corrosion inhibitors or other formulations discussed in Chap. 5. Of course, the material must be properly applied, to reduce shrinkage of the patch and to ensure a good seal at the edges.

16. *Remove shoring.* Prior to removing the shoring, all stressing and elongation records should be reviewed and approved, and the member inspected. The member should be inspected again after shoring removal, to make certain that all defects have been corrected and no unanticipated movements have occurred.

The repair procedure illustrated here applies to an unbonded-wire tendon system; repairs to other unbonded systems follow a similar path. The bonded-tendon systems have their own specific problems, such as air and moisture left in the grout duct during the original construction, which could precipitate corrosion-related failures with little notice. Bonded-tendon systems were banned in bridge construction in the U.K. because of these corrosion concerns. Bonded-tendon systems are sound in principle, but tend to suffer from poor grouting practices, according to Schupack.[21]

Regardless of these shortcomings, the use of bonded-tendon systems has dramatically increased in the past decades, and they are being introduced in locations with harsh environments. As these systems age, cost-effective methods for their diagnosis and repair will probably be developed—a subject for another time.

References

1. Morris Schupack, "Evaluating Buildings with Unbonded Tendons," *Concrete International,* October 1991, pp. 52–57.
2. "Summaries of Post-Tensioning Industry Tonnage Statistics from 1965 through 1989," Post-Tensioning Institute, Phoenix, Ariz., May 1, 1990.
3. Clifford L. Freyermuth, "Durability of Post-Tensioned Prestressed Concrete Structures," *Concrete International,* October 1991, pp. 58–65.
4. Morris Schupack, "Unbonded Performance," *Civil Engineering,* October 1989, pp. 75–77.
5. Thomas E. Nehil, "Rehabilitating Parking Structures with Corrosion-Damaged Button-Headed Post-Tensioning Tendons, Part I," *Concrete International,* October 1991, pp. 66–73.
6. *Post-Tensioning Manual,* 3d ed., Post-Tensioning Institute, Phoenix, Ariz., 1982.
7. George A. Tapas, "Repair at O'Hare," *Concrete Construction,* July 1998, pp. 625–629.
8. Terry A. Campion and David H. Nicastro, "Evaluation and Repair of Post-Tensioned Concrete Structures," *The Construction Specifier,* November 1999, pp. 25–29.
9. Morris Schupack, "Unbonded Tendons—Evolution and Performance," *Concrete International,* December 1994, pp. 32–35.
10. Keith Kesner and Randall W. Poston, "Unbonded Post-Tensioned Concrete Corrosion: Myths, Misconceptions and Truths," *Concrete International,* July 1996, pp. 27–32.
11. Peter H. Emmons, *Concrete Repair and Maintenance Illustrated,* R.S. Means Co., Inc., Kingston, Mass., 1993.
12. Scott Greenhaus, "Special Concerns with Post-Tensioned Parking Structures," *Skylines—News of the Office Building Industry,* September 1992, pp. 30–34.
13. Predrag L. Popovic and James P. Donnelly, "Renovation Adds Space and Value to Parking Garage," *Concrete Construction,* February 1999, pp. 29–33.
14. Morris Schupack, "Corrosion Protection for Unbonded Tendons," *Concrete International,* February 1991, pp. 51–57.
15. Thomas E. Nehil, "Rehabilitating Parking Structures with Corrosion-Damaged Button-Headed Post-Tensioning Tendons, Part II," *Concrete International,* March 1992, pp. 24–30.
16. *Repair Guide for Unbonded Post-Tensioned Structures,* Structural Preservation Systems, Inc., Baltimore, 1997.
17. Bijan Aalami and David Swanson, "Innovative Rehabilitation of a Parking Structure," *Concrete International,* February 1988, pp. 30–35.
18. *Building Code Requirements for Structural Concrete* (ACI 318-95) *and Commentary* (ACI 318R-95), American Concrete Institute, Farmington Hills, Mich., 1995.
19. Martin S. McGovern, "Post-Tensioning Primer," *Concrete Construction,* February 1999, pp. 90–92.
20. "Post-Tensioning to Settle Tensions over Fallingwater," *ENR,* April 19, 1999, p. 20.
21. Morris Schupack, "Bonded Tendon Debate," *Civil Engineering,* August 1993, pp. 64–66.

Renovating Wood Structures*

8.1 Historical Background

8.1.1 Introduction

Timber is among the oldest and most widely available construction materials. Unlike steel, concrete, and masonry, timber grows naturally—with natural defects—and is renewable. It is used in a variety of building structural elements: beams, columns, girders, trusses, wall and floor panels, pilings, poles, and temporary forms in concrete construction. Most light structures built in North America, including schools, single-family homes, and two- to three-story commercial and apartment buildings, are built with wood and wood products. Many large existing institutional and industrial buildings, not to mention transportation and marine structures, have timber framing as well.

A mention of wood's anatomy and terminology is in order. A piece of lumber is essentially an assembly of fibers held together by lignin, a natural binder of cellulose and hemicellulose. The "structural" part of a tree that is responsible for its support is called xylem. Xylem carries water and nutrients (sap) from the roots to the branches and leaves; it is enclosed by bark (cambrium) outside, and contains pith in the center. The younger cells carrying sap are porous and light; they are called sapwood. As the cells get older, they gradually become darker and clogged with various deposits. These older cells form the center of the tree (heartwood) and are enveloped by new sapwood cells. Heartwood

---

*The author is grateful to Wood Science & Technology Institute (N.S.), Ltd., for providing much material for this chapter, including Examples 8.1 through 8.4, and for supplying the illustrations for Figs. 8.5, 8.12, 8.13, 8.19, 8.23, 8.24, and 8.29 through 8.32.

and sapwood are similar in strength, but heartwood is more resistant to decay. The tree's annual rings reflect different rates of growth: fast in the spring, slow in the summer, and almost zero during the rest of the year. Lumber contains natural defects, such as knots, splits, and shakes, some of which affect strength.

There are three main characteristics of wood that make it a unique construction material. One of them is its anisotropic nature: the properties of wood in different directions are not the same. Since the fibers run only one way, wood strength in tension and compression parallel to the grain vastly exceeds its compressive strength perpendicular to the grain. Tension perpendicular to the grain tends to tear the fibers apart from one another; it should be avoided at all costs. Also, the allowable shear stresses in the plane parallel to the fibers are much smaller than those across the grain. Unlike other construction materials, wood has different rates of shrinkage and expansion in different directions.

The second characteristic of wood is that it comes from many species of trees with various properties. Attaining a measure of standardization for this material is not easy. There are two basic classes of trees: hardwood and softwood. Hardwood trees are typically deciduous, such as oak and maple. Softwoods are usually evergreens: fir, pine, and spruce.

The third unique feature of wood is that it behaves differently under short- and long-term loading. Its strength diminishes significantly under sustained loading.

Wood has demonstrated great dependability around the world: Many structures, from residential timber buildings in Greece to pagodas and temples in China and Japan, have survived hundreds of years of use. In the environmentally demanding and seismically active regions of China there are more than a dozen timber buildings at least 1000 years old.[1]

With wood buildings being so prevalent, their rehabilitation is a common challenge, whether strengthening an existing structure or repairing decayed wood members. Prior to embarking on any rehabilitation or strengthening activity, an assessment of the load-carrying capacity of the structure and the extent of any deterioration must be made. This chapter first discusses the historical methods of wood design and construction, proceeds to the methods of detecting wood deterioration and preventing it from progressing further, and finally examines the most effective techniques of restoration and strength enhancement.

8.1.2 Wood framing: Some perspective

The first structures built by the U.S. colonists were framed with wood, a natural solution in a land abundant in timber. Today, wood is still the material of choice for single-family houses in most areas of the country, because it is readily available, inexpensive, and easy to work with.

Even buildings framed with concrete or steel usually have some wood elements, ranging from flooring to window trim. Among the main weaknesses of timber is its susceptibility to damage caused by insects, fungi, and fire. If protected against these factors, wood-framed buildings can last a long time and withstand most natural disasters. Wood houses and small buildings are typically quite rigid, boxlike structures that keep their shapes during earthquakes, landslides, and hurricanes (although blown-off roofs somewhat mar this record).

A few colonial houses have survived to this day. The Fairbanks House (Fig. 8.1) in Dedham, Mass., is one of the oldest wood-framed buildings in the United States. Built in 1636, it had most of the clapboard siding of white pine replaced in 1903 and in 1999. One of its owners was Charles W. Fairbanks, Teddy Roosevelt's vice president.

The flip side of the ease of working with wood is a lack of barriers to entry into this trade and, consequently, variable quality of construction. Some old buildings exhibit meticulous workmanship, while others seem to have been put together by amateurs. The problem is compounded by the fact that few old timber buildings were rationally designed or had detailed drawings prepared for them. Timber connections were made based on the prevalent practices of the time and the

Figure 8.1 Fairbanks House in Dedham, Mass., (c. 1636) is one of the oldest wood-framed houses in the United States. Most of the clapboard siding of white pine was replaced in 1903 and in 1999.

worker's experience. For these reasons, field investigation of wood-framed buildings should not be taken lightly, as is unfortunately common (who hasn't heard such sentiments as "all these buildings are the same," "haven't you seen enough two-by-fours," etc.?). A typical horror tale of one house renovation is worth reciting.

A homeowner, who was also a structural engineer, wanted to change the wallpaper in the kitchen. The old wallpaper wouldn't tear off easily, and large chunks of plaster were coming off with it. Before long, the wall surface was full of craters, and it was decided to remove the plaster and replace it with plasterboard. However, when the plaster was removed, the existing wood studs were exposed—and found to be in a rather sorry condition. It appeared that the house walls had been built with scrap lumber without any rhyme or reason, with multiple vertical splices made within a story's height, and a mishmash of sizes used. To the homeowner's chagrin, the walls had to be rebuilt—when the original plan was simply to change the wallpaper.

8.1.3 Early American construction practices

The very first wood buildings in North America were constructed of rough lumber; they were exemplified by the log cabin. An abundance of forests was countered by a lack of the industrial tools necessary to convert trees into usable lumber. Another problem was transportation: It was just too difficult to move large poles by means other than sea or river. The timber used in the vicinity of seaports was as likely to be delivered by ship from hundreds of miles away as to be hauled from the local forest a few miles inland.[2]

Eventually, the technology of making hewn and square framing from tree trunks became available, and buildings were framed like those in the colonists' parent countries—with *post-and-beam frames*. In this popular European construction method, the full-height exterior posts were spaced 3 to 5 ft on center, with beams spanning the distance in between. The space between the wood framing was filled first with solid masonry—a common European design that did not fare well in harsh American winters—and then with wood siding and brick veneer. Lateral resistance was provided by diagonal bracing. Both the frames and the bracing were often left exposed. (Some architectural styles, such as Tudor, attempt to replicate that look.)

There were many regional variations of post-and-beam construction, reflecting the origins of the settlers; there were many chronological changes in the style as well. Fischetti and Lewandoski[3] differentiate between preindustrial and industrial construction. Preindustrial framing was more hewn and riven than sawn, and often had a shaped (tapered or flared) configuration. The framing layout was often irregular. A frame was typically laid out by the scribe rule, in which wood pieces were placed on top of, or near to, the intersecting pieces, scribed to fit, and cut. Each member and each joint was unique and had to be marked for identification. Therefore, restoration of preindustrial buildings must be done on a piece-by-piece basis. Industrial framing, on the other hand, was generally sawn, straight, and of a repetitive nature. The members were fit together by a square rule, their lengths determined by calculations rather than by physical layout and marking.

Post-and-beam frames reigned supreme until *balloon framing* was introduced around 1830. The new framing utilized much smaller members—2 × 4 and 2 × 6 studs spaced 16 or 24 in on center—that appeared extremely light in comparison with their predecessors (hence the name). Balloon framing became very popular for residential and low-rise commercial construction: By 1840, it had replaced post-and-beam construction in the Midwest; by 1870, it had spread to both coasts. As in post-and-beam buildings, the studs of balloon framing extended the full height of the structure. The floor joists were nailed to the sides of the studs and could be also supported by let-in ribbon boards. Lateral resistance was provided by let-in bracing, diagonal blocking, or rigid wall finishes—plaster or wood-board sheathing. Balloon framing generally provided poor lateral-load resistance because floor diaphragms, made of diagonal or straight sheathing boards, stopped at the interior faces of the exterior wall studs, and no reliable connection was made to them.[4] Current building codes frown on this type of construction.

Around 1910, *platform framing,* also known as western framing, was developed and quickly displaced balloon framing. In platform framing, floor and roof joists bear on top of studs. Each lower floor is completely framed and covered with floor sheathing before the next story is erected. Floor sheathing extends over the exterior studs, and direct connection between the two can be easily made for adequate transfer of lateral loading. Platform framing remains the dominant method of constructing light wood buildings.

We should mention in passing one more type of construction—the so-called single-side wall framing, which is sometimes used in auxiliary buildings and in rural areas. There are no wall studs here; all loads are supported by exterior vertical siding boards (1 × 10s or 1 × 12s). The gaps between the boards are covered by 1 × 2 vertical battens.[4]

8.1.4   The mill building

Another type of framing was developed for industrial applications. It took slightly different forms in various areas of the country and evolved over time. The framing followed the same principles as post-and-beam construction, but used larger spans, different details, and unreinforced masonry bearing walls at the exterior. The various versions of this style were called mill-building construction, heavy-timber framing, or slow-burning construction.

In all of these, heavy timber beams, typically of southern pine, were spaced 8 to 11 ft apart (a 10-ft spacing was most common) and spanned 16 to 25 ft between their supports. The interior supports were wood or cast-iron columns; the exterior, masonry walls. The wall thickness varied from floor to floor. The walls had large windows with brick pilasters in between, where heavy timber beams were supported. In standard mill-building construction, the secondary floor framing was 3- or 4-in solid wood decking spiked directly to the beams. In the style sometimes called semimill construction, 6-in-thick intermediate wood beams were used. In yet another design, timber columns supported iron girders that carried intermediate wood beams. A wearing floor of $^7/_8 \times 2^1/_4$-in maple boards was commonly used; it was toenailed to the planking, sometimes with waterproof paper in between.[5,6]

The term *slow-burning framing construction* highlighted the fire safety of buildings made of heavy timbers. Those buildings, though combustible and often containing hazardous industries, were designed to retard the spread of fires and were considered better insurance risks than an average building of the time.

The typical "standard mill" design was widely adopted in the 1880s. In these three-story buildings, brick pilasters were 32 or 36 in thick at the lowest story and decreased 4 in in each upper story. The design live load on the second and third floors was 50 psf; that on the roof, 40 psf. The beam sizes were 12×16 in at the floors and 10×16 in at the roof.[6] Where required, mill buildings were constructed to resist larger live loads and to be taller. For example, a 1919 seven-story mill building in Chicago was designed for 175-psf loading.[5]

An interior view of a mill building, one of the thousands constructed throughout the Northeast and elsewhere, is shown in Fig. 8.2. The exterior view can be seen in Chap. 9, Fig. 9.2. Multistory mill buildings are among the most frequently renovated wood structures, because most of them are so ingrained in the urban fabric that they carry historic significance. Accordingly, municipalities that take pride in their industrial past encourage the adaptive reuse of old mill buildings; they have been turned into offices, retail establishments, and laboratory space, to name a few uses.

Despite their apparent solidity, wood-framed mill buildings slated for reuse often require extensive renovation efforts. Some of the typical deficiencies are discussed in other chapters: a lack of connections between wood-plank floors and masonry walls, insufficient rigidity of

Figure 8.2 The interior view of a mill building.

wood-board diaphragms, poor ductility of unreinforced-masonry walls, and too few ties between the wall masonry layers. The area of main interest for this chapter is the design of typical beam-to-column and beam-to-wall connections.

One common beam-to-column connection is shown in Fig. 8.3. Note that the end of the column in the upper tier is tapered, as are the beam ends. This design allows the upper column to avoid bearing on the beams, since doing so would introduce undesirable cross-grain compression into them. Also, beam shrinkage across the grain would present a problem with the column sinking down. In Fig. 8.3, both the beams and the column bear on a cast-iron cap plate on top of a lower-tier column. Alternatively, steel angle or timber brackets were used. In yet another design, a funnel-shaped metal pintle was placed between the two column tiers. To keep the abutting beam ends in line, they were often connected with wrought-iron straps that were spiked or bolted into the beam sides or placed on top of the beams and connected with iron dowels ("dogs").

In a typical beam-to-wall connection, the end of a heavy timber beam was placed into a U-shaped metal bracket, sometimes with a matching metal cap. In many cases, this design was "simplified" by

Figure 8.3   A beam-to-column connection common in mill buildings.

having the beam bear on the masonry directly. So-called government anchors of bent rods were occasionally provided to make positive connections to walls. The beam ends were often fire-cut—cut at an angle that would facilitate beam rotation during a fire, so that individual wood members could burn and fall without pulling down the walls with them. Many of these direct-bearing connections have eventually rotted, or have been degraded by moisture and chemicals in the wall.

8.1.5   Changing technology

The development of building materials proceeded independent of changes in framing types. The floors and walls of most timber-framed buildings constructed prior to 1945 were sheathed with diagonal or straight boards. Plywood sheathing became popular after 1945. The early nails were wrought by hand, and cut nails with tapered rectangular shanks were in common use from around 1800 until the early twentieth century. By the 1880s, wire nails had become available.[4]

Common lumber dimensions did not stay constant through the centuries. The nominal and actual dimensions of wood members were approximately the same in buildings constructed prior to the late 1800s, when framing was generally hand-hewn. As machine-surfaced lumber became available (about 1940), the nominal and actual sizes started to diverge. The actual sizes of modern lumber are based on the standard dressed sizes of wood surfaced on four sides by planing, but even these dressed sizes have undergone changes. Prior to the 1970s, the dressed dimensions for the *width* of sawn lumber 2 to 4 in thick were determined by subtracting $^3/_8$ in from the nominal size; for thicker lumber, $^1/_2$ in was subtracted. The dressed dimensions for the *depth* of lumber 4 to 6 in deep were determined by subtracting $^3/_8$ in from the nominal size; for deeper members, $^1/_2$ in was subtracted. So, for example, the pre-1970 dressed size of a 2 × 4 was $1^5/_8$ × $3^5/_8$ in. Around 1970, the actual sizes were decreased, so that a 2 × 4 actually measures $1^1/_2$ × $3^1/_2$ in.

The standard dressed sizes for today's lumber are available in most books dealing with wood design. Making calculations using old framing sizes can sometimes be a nuisance, as the structural properties of those timbers are no longer published. The simplest solution is to compute the properties by hand, based on the actual field-measured lumber dimensions.

The sources of U.S. lumber have also changed. The native old-growth forests of the East Coast were largely exhausted by the end of the nineteenth century, and timber from the South, the Rocky Mountains, and the Pacific Coast became widely available.[2] Eventually, even those seemingly limitless sources were threatened, and temporary lumber shortages—and price spikes—periodically occurred in the last decades of the twentieth century. Those who drive around the magnificent Olympia National Park in the state of Washington are greeted by countless miles of cleared and reseeded woods. There are many more trees being planted today than being logged, but it will take time for the new growth to reach the prime sizes demanded by the construction industry. Meanwhile, the availability of premium lumber grades has steadily declined.

The void has been partly filled by technological advances, as glued laminated beams (glulams), laminated veneer lumber (LVL), and hybrid engineered lumber products were developed in the 1930s and later. Instead of using a single piece of premium lumber, these products optimize the now-scarce resource by placing the best pieces in the most demanding locations, by using premium glues, and by making timber I-beams with webs of oriented strand board or similar materials.

Curiously, since the early 1970s, heavy timber construction has enjoyed a revival of sorts. According to Brungraber,[7] a number of heavy wood structures have recently been built in New England, the birthplace of the industry. These structures utilize the traditional methods of connecting heavy timbers to each other, such as pegged mortises and tenons and through-spline joints. Of course, these nostalgic endeavors require rare expertise and expensive heavy lumber. They are largely educational in nature and cannot possibly supersede modern methods of framing wood structures.

8.1.6   Evolution of design methods and allowable stress values

Most early wood structures were not based on rational engineering designs. Their builders followed so-called conventional construction techniques—the traditional methods developed by carpenters during many decades of working with wood. This conventional practice sufficed for houses and small buildings (although even there roofs tended to be blown off in hurricanes), but it offered little help when long spans had to be covered or unusual layouts were involved.

At the end of the nineteenth century, rational design procedures began to come into use. Some general design principles for timber construction and allowable stresses for various wood species were included in the first building codes. (A brief history of these codes is given in Chap. 1.) Unfortunately, the permissible stress values were inconsistent from code to code, especially among the city-enacted codes at the turn of the twentieth century. For example, look at the allowable stresses for yellow pine, one of the most popular wood species: The maximum bending stress allowed by the cities of Chicago (1898) and Boston (1899) was 1250 psi, while Philadelphia (1899) permitted 1600 psi. Similarly, the maximum shear stress allowed in New York (1900) and Philadelphia was 70 psi, while in Chicago and Boston it was 100 psi.[8]

The U.S. government helped provide a measure of uniformity. An important publication, the *Guide to the Grading of Structural Timbers and the Determination of Working Stresses,* was jointly issued in 1933 by the Forest Products Laboratory, an agency of the U.S. Department of Agriculture Forest Service, and the National Lumber Manufacturers Association (NLMA). It was published in 1934 as U.S. Department of Agriculture Miscellaneous Publication 185. This guide, based on extensive test data, became the basis for many subsequent standards establishing working stresses in various wood species.[9]

Another seminal publication by the U.S. Department of Agriculture was *Wood Handbook* (1935). Among other design information, it included formulas establishing a relationship between the specific gravity and the strength of various wood species.[10] The handbook was subsequently revised,[11,12] and remains a definitive source of information on timber design. A related publication, *Wood Structural Design Data,* was issued by the NLMA in 1934.[13] In addition to a lot of reference material, including information on timber fasteners, it featured extensive span and load tables for beams and columns of various sizes. The publication was revised in 1939.[14]

These publications and the accumulated results of other research laid the foundation for the first comprehensive national standard of wood design. The *National Design Specification for Stress-Graded*

Lumber and Its Fastenings was published by the NLMA in 1944.[15] The specification was updated and revised numerous times, and to this day remains the most authoritative national standard for wood design and construction. In 1971, the specification was expanded to include wood products other than stress-graded lumber. To reflect its broadened scope, in 1977 its title was changed to *National Design Specification for Wood Construction,* and this name remained current at the turn of the twenty-first century. The load and resistance factor design (LRFD) version of the specification, AF&PA/ASCE 16-95, followed in 1995.

Meanwhile, the NLMA changed its name twice: In 1968, it became known as the National Forest Products Association, and in 1993, after merging with the American Paper Institute, it became the American Forest & Paper Association (AF&PA).

A supplement to the *National Design Specification* contains allowable stresses for various grades of most American wood species. Since the 1920s, the stress values had been based on tests conducted on small clear wood specimens. However, by the mid-1960s it became possible to obtain test results from full-size lumber pieces. A comprehensive study of the actual lumber produced in North America—in contrast to testing small pieces sawn in the lab, as was done earlier— was launched in 1977 and continued for over a decade. The results of this "in-grade" testing were incorporated in the supplement to the 1991 and later editions of the *National Design Specification.* The new values of allowable stresses for bending and tension are in some cases different from those used earlier. So, the allowable bending stresses for no. 1 southern pine were unchanged, but those for no. 2 grade were decreased 13 percent; the stresses for compression parallel to grain were substantially increased for both grades.[16]

In general, the allowable tension stresses used in the first half of the twentieth century are now considered to be overly optimistic. This means, for example, that the sizes of the bottom chords of some existing trusses might be insufficient by today's standards. Prior to 1965, the allowable tensile stresses in lumber were the same as the allowable bending stresses, but after 1965, the allowable tensile stresses were gradually reduced. These stresses were also found to be size-dependent: The bigger the size, the smaller the allowable tension stress it can carry. For example, for 2×6 and larger members of kiln-dried no. 1 southern yellow pine, the allowable tensile stresses stated in the various editions of the *National Design Specification* were 1750 psi in 1962, 1100 psi in 1971, and 900 psi in 1991 and 1997. The stresses in larger members have been reduced even more drastically.[17]

How can one determine the allowable stresses for timber framing in an existing building? Timber properties can vary widely. Even the first step of the process—identifying the species of wood—is difficult for most of today's engineers and contractors.

In many cases, the existing lumber will bear a stamp by a grading agency identifying the wood species and grade. If it does not, a laboratory specializing in lumber testing or grading can help. Another source of wood identification is a public-service laboratory called the Center for Wood Anatomy Research, located at 1 Gifford Pinchot Dr., Madison, Wis., 53705-2398. The center was created by the U.S. Forest Service as a branch of the U.S. Forest Products Laboratory in 1914. It will identify wood species if a sender ships lumber in 6-in pieces and encloses a letter explaining where the wood came from and what information about it is desired.[18]

The values of allowable stresses for old timber were generally higher than those found in today's codes, especially if the lumber came from an old-growth forest. As the old stands of trees disappeared and the age of the lumber used in buildings decreased, the allowable stresses also gradually decreased. So, if the original design drawings are not available, no grading stamp can be found on the wood, and laboratory testing is not possible for some reason, the last resort is to simply make a reasonable assumption about the species of lumber and take the allowable stresses for existing wood as those specified today.

One exception to this general rule is timber buildings constructed during World War II. At that time of grave national emergency, the need for massive amounts of structural-grade lumber overwhelmed the available supply, and the armed services had to cut a great many corners to produce the required building volumes. For example, there were cases in which lumber with a normally allowable bending stress of 1200 psi was arbitrarily assigned a stress grade of 1800 psi, or when ungraded or unseasoned lumber was purchased, or when the design snow loads were simply reduced to 15 psf from the much higher prewar levels.[19] Those involved in renovations of wartime wood buildings should beware.

8.2 Wood Deterioration

8.2.1 Introduction

Unprotected wood deteriorates—for without degradation the Earth would be littered with the remnants of all the trees that have ever grown. Fortunately for the new seedlings, but unfortunately for timber users, nature has numerous forces that can break down wood. These forces can be placed into two primary classes: biotic (living) agents and physical (nonliving) agents. In many cases, the first agents that attack wood also provide favorable conditions for other agents.

The effectiveness of any inspection of deteriorated timber depends greatly upon the inspector's knowledge of the agents of deterioration. A poorly trained inspector can sometimes miss even obvious signs of trouble. For this reason, we briefly describe these agents.

8.2.2 Wood deterioration caused by bacteria, insects, and marine borers

Biotic agents of wood deterioration include bacteria, insects, marine borers, and fungi (the last are discussed separately because of their variety and importance). Being living organisms, all of them require certain conditions for survival: moisture, oxygen, proper temperature—and food. When these basic living conditions are provided, biotic agents proliferate, but if any one of them is removed, the wood is safe from further biotic attack.

Bacteria. Bacteria can colonize untreated wood in very wet environments. Bacterial damage can include softening of the wood surface, increased permeability, and even degradation of chemical preservatives by some tolerant organisms.[20] The process of bacterial attack is usually very slow, but with extensive exposure, damage can be significant.

Insects. Many insect species use wood as food or shelter. Termites, beetles, bees, wasps, and ants are the most common insects causing the deterioration of wood by making cavities or tunnels in it. Termites can totally degrade wood from the inside, eating everything but a paper-thin exterior shell (Fig. 8.4). The longitudinal cross section of a termite-damaged timber shows feeding galleries extending parallel to the grain and filled with frass (insect feces) and soil. Carpenter ants, on the other hand, do not eat wood—they make tunnels in it for their nests. For this reason, they may attack even preservative-treated wood.[21] The nesting galleries of ants typically cut across the grain and are free of residue.[22] The presence of insects in the wood can lead to further damage by inviting woodpeckers. As many homeowners around the country—including the author—have learned, woodpeckers hunting for insects (or so they think) can cause more damage than the insects themselves. Woodpecker damage can be especially severe in wooden utility poles, for obvious reasons.[23] Insect infestation can be betrayed by the presence of small piles of wood powder or frass near the outside of the wood.

Marine borers. Marine borers can affect timber substructures located in salt or brackish water. In 1965 the U.S. Navy reported that these organisms cause over $250 million in damage annually.[24]

8.2.3 Wood deterioration caused by fungi

When conditions are favorable, most types of wood become attractive food sources for a variety of decay-producing fungi. These biotic organisms become active when moderate temperatures and oxygen

Figure 8.4 Termites can totally degrade wood from the inside, leaving only a paper-thin exterior shell.

are present, and the moisture content is 19 percent or greater. Decay progresses most rapidly at temperatures between 10 and 35°C (50 and 95°F). Outside this range, decay slows down; it ceases completely when the temperature drops below 2°C (35°F) or rises above 38°C (100°F). Wood can be too wet for decay to take place: In water-soaked wood, the supply of oxygen is inadequate to support the development of typical decay fungi.[12] Thus, if the appropriate temperature, oxygen, and moisture content are absent, decay will not start, and any ongoing decay will not progress further.

Examples of wood preservation by environmental conditions are abundant. The 1000-year-old timber pagodas in China mentioned in the beginning of this chapter have survived this long because their wood was kept dry. Conversely, some entrepreneurs in the United States are recovering old-growth wood from sunken transport ships and selling it, because this water-saturated timber has been well preserved.

Decay fungi can be placed in two main categories: brown rot and white rot. They can be told apart by the appearance of the damaged wood surface. *Brown rot* appears darker because its fungi attack the cellulose and hemicellulose in wood fiber, but not the lignin, which remains and gives the wood its brown color. Wood with advanced brown-rot decay tends to crack along and across the grain, forming small cube-shaped sections. Brown rot accounts for most damage by all fungi because it includes a particularly pernicious variant—dry rot.

Dry rot is the most common type of decay. The attacked wood becomes brown and crumbly in an apparently dry condition, but the name "dry rot" is a misnomer: As just stated, decay requires some moisture. Dry rot is difficult to eradicate for several reasons. First, it can stay dormant when the timber is dry and come back to life when moisture returns.[25] Second, some fungi have water-conducting strands (hyphae) that are capable of carrying water, usually from the soil, into buildings or wood piles, and thus introduce enough moisture to initiate rotting.[12] Dry rot fungi can even travel through masonry and plaster to attack the adjacent dry wood.[21] And third, the fungus produces water as a by-product; in some cases this makes it self-sustaining.[26]

White rot appears light in color and does not cause cracking across the grain until a very advanced stage. In contrast to brown rot, white rot fungi consume both lignin and cellulose and leave the surface appearing intact but looking bleached or white. The appearance is deceiving, as damaged wood possesses little strength. White rot prefers hardwoods—and therefore is more common in flooring and trim than in structural elements—while dry rot attacks both hardwoods and softwoods.

The last fungus of interest is the so-called wet rot. It requires a continual supply of moisture and thus grows in wet areas, such as cellars, vaults, and leaky roofs. It dies when moisture disappears.[26]

Interior decay can take place even when the wood surface is covered with a protective coating. The reason? Any cracks formed in surface-treated wood will extend into the untreated core material and allow water to enter. Water can also get into the core of the "protected" wood through fungi hyphae. In either case, if enough water penetrates into the core, decay fungi will flourish.

8.2.4 Effect of decay on mechanical properties of wood

The primary effects of fungal attack on wood can be characterized by the following changes[27]:

- Change of color
- Change of odor
- Decreased weight
- Decreased strength
- Decreased stiffness
- Increased hygroscopicity (easier absorption of water)
- Increased combustibility
- Increased susceptibility to insect attack.

The incipient stage of fungal attack might bring only a change of color (and perhaps of odor). This first stage can be very difficult to detect visually, and, since changes in the strength and hardness of the wood have not yet occurred, the pick test and other surface tests will not help.

Unfortunately, significant changes in strength and stiffness can take place before any rotting is detected. Visual clues appear mostly during the advanced stages of fungal attack, when the wood becomes soft, punky, or crumbly. When decay is discovered by visual inspection or inferred from a change in weight, the damage has already been done. (Reduction in unit weight leads to a decrease in strength and nearly every other mechanical property, especially compression perpendicular to grain.) Brown rot can reduce mechanical properties by 10 percent before any significant weight reduction is noticed. When weight loss is only between 5 and 10 percent, the reduction in mechanical properties might be 20 to 80 percent.[27] Specifically, this change in weight could mean a 50 to 70 percent reduction in bending strength and a 60 to 70 percent reduction in moduli of rupture and elasticity.[28]

Wood members with high stresses in compression perpendicular to the grain, such as beams at supports, are often affected by a common manifestation of decay—the so-called mushrooming, or bulging. Untreated beams that protrude past the exterior walls, either because they cantilever from the inside or because they are supported by exterior columns, are particularly susceptible. Mushrooming decay starts when water is trapped at the bottom surface of the wood beam. One such case is shown in Fig. 8.5, where water collected on top of the steel support invited decay in the bearing area. (The areas of decay are indicated by discoloration at the sides and bottom of the beam.)

Figure 8.5 Water trapped between the steel support and the bottom surface of the glulam wood beam allowed decay to occur in the bearing area and started the process of "mushrooming."

A similar problem can occur in heavy-timber connections made with steel U-brackets, common in mill buildings. If water gets inside the brackets, it tends to stay there and to enter the wood through the end grain or from the sides, following the fastener holes. Even worse is the case where wood is built into masonry—the whole section is then exposed to moisture. As decay progresses, the wood weakens at supports and other places where perpendicular-to-grain compressive stresses are high. The result is significant vertical deformation or total disfigurement of the members (Fig. 8.6).

Figure 8.6 Undergoing advanced stages of decay (and perhaps chemical attack from alkalies in mortar), this wood beam in historic Ft. Knox, Me., is weakened to the point of total disfigurement.

8.2.5 Deterioration caused by physical agents

Physical agents of wood deterioration include abrasion, mechanical impact, by-products of metallic corrosion, highly acidic or alkaline substances, and ultraviolet light. They are not as common as biotic agents, but sometimes their impact can become quite serious. Physical agents can degrade both wood and preservative treatments in it, resulting in an increased susceptibility to attack by biotic agents. Here is a brief description of the most common physical agents.

Water. Wood can buckle if water-related swelling cannot be accommodated by the adjacent construction. Depending on the severity of the buckling, wood can become permanently distorted, especially if the water damage invites rot. The wood flooring of Fig. 8.7 was swelled by persistent roof leaks and, since its expansion was prevented, buckled. Because of rapidly progressing white rot (note the stains in the picture), the flooring has lost much of its strength and remains in the buckled position. Another problem: Moisture can weaken or destroy interior-grade glues in the finger joints of glulam beams. The adhesives in glulams produced prior to the late 1940s were not waterproof, so old glulam buildings may suffer more than their share of problems. There have been cases in which even contemporary glulams have deteriorated because the roof was not installed in time and they were left exposed to several months of driving rain.[19] Extended exposure to water can swell plywood and oriented-strand board sheathing, resulting in the connecting nails' heads popping out when the sheathing dries. Water condensing at the underside of wood planks and on wood framing can also lead to rotting, although high humidity alone usually does not make wood wet enough to decay.[22]

Mechanical damage. Mechanical damage is caused by a number of factors and varies considerably in its structural significance. Abrasion, vibration, overload, and foundation settlement are the most common causes of mechanical damage. Figure 8.8 shows chipping and abrasion at the bottom of a timber column in a mill building, a result of many decades of abuse.

Metallic corrosion. When metal fasteners in wood corrode, they release ferric ions that attack the walls of wood cells and cause their degradation. The attack can severely reduce wood strength in the affected area and make the lumber dark in color and soft to the touch. (The extent of wood-metal corrosion can be limited by using galvanized or nonferrous fasteners.)

Chemical degradation. Strong acids and alkalies can cause chemical degradation of wood. Acids degrade cellulose and hemicellulose, resulting in weight and strength loss; strong alkalies degrade hemicellulose and lignin.[20] Acid-damaged wood looks dark, almost as if it had been charred by fire. By contrast, wood exposed to alkalies looks bleached.

Figure 8.7 Water damage can lead both to swelling of wood and to its rotting. This wood flooring was swelled by persistent roof leaks and, since its expansion was prevented, buckled. Note the water stains and white rot.

Figure 8.8 Mechanical damage to the bottom of a timber column in a mill building.

Strong chemical exposure is rare, except in cases of accidental spills, but moderate chemical attack can occur when water penetrates into bearing pockets of old masonry walls. When this happens, the water reacts with lime in the mortar, and calcium hydroxide forms—a caustic solution with a pH as high as 12.4. Timber beams bearing in these masonry pockets will typically suffer from discoloration and loss of mass, and will appear shrunken. Eventually, the bottom surfaces of the timbers—the areas most exposed to moisture—will lose strength and flatten out under load, allowing the beams to move down in the pocket or become loose.[29]

Degradation by ultraviolet light. Ultraviolet light reacts with lignin near the surface, and the resulting deterioration is highly visible. Ultraviolet degradation changes the color of the wood: Light-colored woods become darker, and dark woods lighter. However, the damage penetrates only a short distance below the surface[30] and results in little strength loss.

8.3 Detecting Deterioration

8.3.1 Overview of detection methods

Finding areas of deterioration in existing structures is both an art and a science. Not surprisingly, some available methods of detection rely on the subjective feelings of the person using them, while others produce objective documentation.

Detection methods can be classified as either exterior or interior, with different methods used for different types of damage and in different structures. There is no single method that can determine the condition of a given structure with absolute certainty, but a combination of methods and tools used by well-trained personnel can provide a reasonably accurate assessment.

Exterior detection methods investigate conditions at the surface of the wood. These methods, which include visual inspection, probing, and picking, are best at checking for obvious signs of decay and giving indications of probable rotting inside. The presence of hidden decay can then be revealed by interior detection methods—core sampling, moisture meters, and several modern sonic evaluation techniques.

This section outlines the detection methods germane to our later discussion; most of them are described in greater detail in Chap. 2. For those seeking additional guidance, Ref. 19 is a great source of information about the evaluation of various types of wood structures.

8.3.2 Exterior detection methods

The most common exterior detection methods include visual inspection, probing, and the pick test. All of them are easy to use, and they provide a basis for further interior investigation that will define the extent of damage. Surface decay can be identified by both visual and probing techniques.

Visual inspection is the simplest method of locating decay on the exterior surfaces of wood. It is most suitable for detecting advanced stages of decay; the incipient stages, as already noted, are almost impossible to identify visually. Decayed wood tends to have a very rough texture, with closely spaced cracks and grooves. Some other visual indicators of deterioration are[31]

1. *Fruiting bodies.* Some fungi produce fruiting bodies that appear on the surfaces of decaying wood. Fruiting bodies may not stay on the wood for a long time, being easily removed by weathering and other natural factors, but their appearance is a good indicator of extensive decay.

2. *Sunken faces.* Localized depressions in wood that appears intact or nearly intact are often signs of internal decay near the surface.

3. *Staining or discoloration.* Surface stains can indicate that the wood member has come in contact with water (see Fig. 8.5).

4. *Bulging at beam supports.* A decrease in specific gravity caused by an attack by fungi greatly diminishes the perpendicular-to-grain bearing capacity of wood, as discussed previously and as shown in Fig. 8.6.

5. *Insect infestation.* This can be identified by small exit holes, piles of wood powder, or frass. Subterranean termite activity can be detected by the termites' earthen access tubes on the exterior or interior of walls or posts (Fig. 8.9). Occasionally, these access tubes can be seen hanging under wood framing (Fig. 8.10).

6. *Plant or moss growth* indicates a relatively high moisture level—a condition conducive to decay.

The process of visual inspection is further described in Ref. 22.

Probing to locate soft areas of the wood surface can be done with a pointed tool. Timber that can be easily penetrated and partly removed with a pocket knife or a flat-headed screwdriver may be decayed or simply water-softened, so experience is necessary in order to interpret the results. Most often, however, these are the symptoms of decay. In

Figure 8.9 These earthen tubes on the exterior walls indicate subterranean termite activity.

Figure 8.10 What appears to be a thin rope hanging under the crawl-space framing is actually a termite access tube.

Fig. 8.11, a screwdriver can be easily pushed into the wood uncovered when aluminum siding was partly removed from a building. Water that gets behind the nearly impervious siding can stay in the wood for a long time and cause its total decay.

The *pick test* is a simple method of finding decay. It involves driving a metal pick or a screwdriver a short distance into the wood surface and bending the tool back to pry off a small area of wood. If the wood splinters in long slivers, it is probably sound; if it abruptly breaks or crumbles, decay is most likely present.

Figure 8.11 A screwdriver can be easily driven into the wood uncovered when aluminum siding was partly removed from this commercial building. If water can get behind the nearly impervious siding, it can stay in the wood for a long time and cause its total decay.

In lieu of these simple pointed tools, a special probing instrument, the *Pilodyn,* has been developed and is extensively used in Europe. The Pilodyn is a spring-loaded device that drives a pin into the wood surface. The depth of penetration, when compared to the moisture content and the wood species, provides a measure of surface decay.[32]

8.3.3 Interior detection methods

Deterioration deep inside the wood cannot be detected by exterior detection methods; to find it, other methods are required. Hammer sounding, moisture meters, drilling and coring, sonic evaluation, and x-ray devices investigate the inside areas of wood and are collectively called interior detection methods. These evaluation technologies can help provide information about the internal condition and residual load-carrying capacity of a damaged wood member. Again, these methods are further described in Chap. 2.

Hammer sounding is done by striking the wood surface with a hammer and evaluating the tonal quality. Dull or hollow sounds may indicate internal decay, but not with absolute certainty, because many other factors can influence the sound of wood struck with a hammer. Still, sounding is quick, easy, and inexpensive, and the suspected areas it identifies can be verified by other methods such as core sampling.

Moisture meters can help identify wood areas with high moisture content. Untreated wood with a moisture content higher than 20 to 25 percent indicates conditions conducive to decay.[33]

Drilling and coring are some of the most common methods of interior decay detection.[34] The inspector drills into the structure at different locations, observing the torque resistance and examining the drill shavings for evidence of decay. A hand drill allows for a better feel than a power drill. Both drilling and coring require sharp tools for proper use; crushed wood caused by a dull bit could be mistaken for decay. The drilled-out holes should be plugged with wood dowels or fillers to prevent future decay in these areas.[19]

Core samples are taken with an increment borer. An extracted solid-wood core can be examined for evidence of decay or voids. The cores can show the extent of deterioration and provide lab samples that can be studied to determine the presence of decay fungi, to analyze the specific gravity of the wood, and to provide an assessment of future risk.[35] Drilling and coring can verify whether deterioration exists in the areas brought under suspicion by moisture meters, visual inspection, or other methods.

In the past, *x-ray scanners* were used to locate internal voids in wood,[36] but the high cost of the equipment, the required safety precautions, and the labor necessary to operate them have curtailed their use. Today, a new generation of x-ray scanners is being developed to provide internal images of uncut logs in an attempt to locate defects in them, so that yield and cutting patterns can be optimized.[37] In Europe, x-ray scanners have been used to provide internal images of wooden poles.[31]

8.3.4   Sonic evaluation

Several different sonic-wave propagation methods have recently been developed, among them sonic-wave velocity, acoustic emission, and stress-wave analysis. Basically, these methods use different techniques to measure the velocity change of a sound wave moving through wood. The underlying principle is that a sonic wave is altered by wood defects and by various deterioration agents.[37] These nondestructive evaluation methods have attracted much research[38] and are gaining wide acceptance for assessing the performance of wood in structures. For example, researchers have shown that acoustic-emission techniques can detect the presence of termites in wood members.[39] One of the most celebrated American wooden ships, USS Constitution, built under orders from George Washington, was renovated for its 200th anniversary using ultrasonic and stress-wave testing to locate areas of deterioration.[40,41]

Are sonic evaluation methods practical enough to be used for routine inspection purposes? Yes, and some inspection professionals are already using commercial adaptations of these technologies in their work.[42] Coupled with a thorough visual examination, sonic evaluation methods can help determine the internal condition and residual load-carrying capacities of wood members.[37] The use of one of these methods, the stress-wave timer, is illustrated in Example 8.1 and in Example 8.3 in Sec. 8.6.

Example 8.1: Using Stress-Wave Velocity Timer.   Stress-wave measurement has recently become popular for locating internal decay in wood members. Stress-wave analysis involves sending a sound wave through a medium (wood in this instance) and measuring its velocity. The sound wave is introduced into the material by striking it with a hammer or blunt object. When the vibrations reach a nearby accelerometer, an accurate timer is started; when the sound reaches a second accelerometer at the opposite side of the member, the timer is stopped. The distance between the two accelerometers is then measured. With the distance and the time of travel known, the average velocity of the stress (sound) wave can be found.

The modulus of elasticity of the material is theoretically related to the velocity of the stress wave and the density, according to the following equation:

$$E = c^2/\rho$$

where E = modulus of elasticity
$\quad c$ = velocity of the stress wave
$\quad \rho$ = density of the material

The measured modulus of elasticity E indicates whether decay is present. Typically, the modulus of elasticity for sound Douglas fir ranges from 1.5×10^6 to 2.2×10^6 psi. (The range can be tightened if the exact grade of wood is known.) When decay is present, the modulus of elasticity measured with the help of a stress-wave timer will be significantly lower than the expected range. Why? The specific gravity of wood suffering from fungal decay decreases, and a decrease in specific gravity causes a decrease in the modulus of elasticity. This in turn decreases the velocity of the stress wave (i.e., increases the time of wave travel). Therefore, the longer the stress-wave time, the larger the probability of internal decay at that location. Mapping the measured stress-wave times on the side of the member can help identify potential areas of decay.

Proper calibration of the stress-wave timer is critical. The calibration is established by taking a sample of sound wood from the structural member and measuring the velocity of wave propagation through it. Several consistent results should be obtained before moving on to the decayed areas. The velocity must be measured in the same manner to minimize variation and false readings.

A calibration curve for the stress-wave timer, as used on Douglas fir beams and columns, is shown in Fig. 8.12. The curve indicates the relationship between the stress-wave time and the specific gravity. Generally, stress-wave times greater than 300 µs/ft (for Douglas fir in this example) indicate that fungal decay may have significantly degraded the strength and stiffness properties of the wood. The calibration curve was created by measuring stress-wave times on an existing glulam and then taking several samples with a core drill. The specific gravities of the samples were measured in a laboratory to develop the curve.

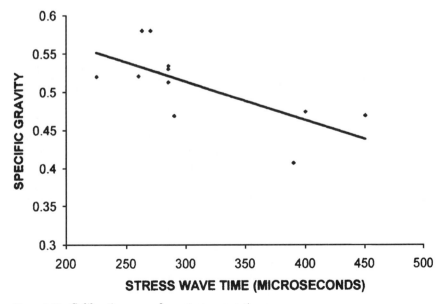

Figure 8.12   Calibration curve for a stress-wave timer.

8.4   Preventing Deterioration

8.4.1   Overview

Why go to the trouble and expense of replacing decayed timbers if the new wood will be as vulnerable to deterioration as the old was? Certainly, some steps must be taken to protect the investment in the replacement structure. The existing wood elements that are being repaired or strengthened deserve the same level of protection if they are to survive for as long as a replacement structure would. Some methods of protection are discussed in this section.

Measures to prevent wood deterioration can take many forms. Two of the most important ones are using preservative treatments and controlling moisture. Because of their toxicity to humans and the environment, preservative treatments should be used only when absolutely necessary. Good design and construction detailing is much more effective in controlling decay than the use of harsh, toxic chemicals. Perhaps the best solution lies in effective protection of wood from moisture—keeping it dry, with less than 19 percent moisture content. Keeping wood members dry goes beyond simply erecting a roof over them, as wind-blown rain will soak the exposed framing. Water can also find its way into wood through the adjacent construction materials, such as concrete in foundations and slabs or metals that are "sweaty" with moisture from condensation.

But what keeps unprotected living trees from decay? Trees use multiple defense systems to prevent or slow fungal growth. First, the bark provides an effective barrier against fungal attack. Second, sapwood can respond to fungal attack by terminating cellular metabolism in the affected areas, creating an adverse environment for fungi. Sapwood can also transport resins to seal off the infected areas, isolating them and reducing the extent of decay. Third, heartwood contains chemicals that are hostile to fungi. In fact, heartwood of some redwood and cedar species can be used in exterior decks without any chemical treatment because of the excellent fungus-resistant chemicals that it naturally contains. Also, heartwood can become plugged with growths called tyloses that restrict the movement of both water and fungi.

There are many types of human-made chemical preservatives available. The best known is creosote, often used to preserve wood utility structures. Pentachlorophenol is used to treat bridge girders and other wood members where human exposure is minimal, but because it can leach out and is toxic, its applications in buildings are quite limited. Chromate copper arsenate (CCA) is an effective wood preservative that is relatively safe for humans. Unfortunately, the treatment process for CCA uses water as a transport mechanism; this can cause splits and checks when the treated wood members, especially larger ones, dry out.

In addition, the effectiveness of CCA in heartwood is questionable because of its generally poor penetration (often caused by tyloses). Most chemical treatments require special pressure tanks to obtain the necessary penetration depth for effective decay resistance. Surface treatment is not nearly as effective as pressure treatment because a thin protective coating is easily broken by localized splits, checks, and cracks, creating avenues for fungal attack. For this reason, in-place surface treatment of existing wood structures or components with preservatives—the only possible method of applying them without taking the building apart—rarely solves the problem for long.

Alternatively, chemical treatment of existing structures can be provided by fumigants, as described in Sec. 8.4.3. Fumigants were first used for protection of wood utility poles, and their application was later expanded to beams and columns.

A simple means of decay prevention is oxygen deprivation with moisture, exemplified by storage and transport of logs in water. Historically, the ideal place to store logs prior to mill shipment was a body of water—a river or a lake. In water, fungi are deprived of oxygen, which is essential for their survival. Today, logs are often stored on land and continuously soaked by sprinklers to ensure saturation. Storing logs in a wet state also reduces their checking and the potential for fires.

8.4.2 Moisture control

Perhaps the most effective and least expensive method of preventing the decay of wood is lowering the wood's moisture content. When the moisture content drops below a certain value (19 percent is often mentioned), the fungi become dormant and decay stops. To lower the moisture content of wood, all entry of water from outside the structure must be completely eliminated, either by means of waterproofing and sealers or by appropriate construction details. A classic example of a design intended to prevent water from coming into contact with wood structures is that of covered bridges. This solution prevented rainwater from soaking the bridge timbers and often kept them at moisture levels below 19 percent.

In particular, effective moisture control measures must prevent water from being absorbed into the end grain of timbers. Because of the cellular makeup of wood, water is most easily transported along the grain. In most wood species, some water transport perpendicular to the grain can also take place, although at a tiny fraction (1 percent) of the parallel-to-grain transport rate. Water absorbed by end-grain fibers can travel great distances; it can raise the moisture content of wood located far away from the water source and provide the fungi there with moisture. So, an exposed cantilevered roof beam might experience bulging at the wall support—many feet away from the beam end—because of decay caused by water traveling from the end grain.

What about paints and sealers? These coatings can be excellent moisture barriers, although they introduce other problems. The most effective paints and sealers, generally those with high solids contents (above 30 percent), tend not only to keep out water, but also to seal in moisture. Also, painted surfaces tend to develop small cracks and pores, which provide entry routes for moisture, while its exit is constrained. The longer moisture is locked within the wood, the longer the fungus can continue its work. For this reason, heavy timber and glulam beams (those with nominal widths of 4 in and greater) are usually stained, rather than painted. Stains are generally free from the problems associated with paints. Protective coatings are best used in combination with construction details designed to prevent water from reaching the wood in the first place.

8.4.3 In-place preservative treatment

In-place preservative treatment methods can arrest existing decay or allow the moisture content of undamaged wood to exceed 19 percent without a risk of fungal attack. In-place treatment can significantly extend the life of wood structures. Several case studies[20] have shown that in-place preservative treatment of timber bridges can extend their life by 20 years or more. There are two basic methods of applying preservatives: surface treatment, typically used for decay prevention, and fumigants, used to treat existing internal decay.

Surface treatment. Surface treatment is used for decay prevention rather than cure, because its shallow penetration makes it ineffective against internal decay. The treatment materials are available in liquid, gel, and paste consistencies. Liquid preservatives can be applied by brushing, squirting, or spray-flooding the surface. Semisolid greases or pastes are helpful for treating vertical surfaces. Information about commonly used chemical preservatives is widely available,[12,31] and their possible health risks must be fully understood before use.

Surface treatment is most effective for relatively dry wood, but some tests suggest that even wet timber can be treated if the recommended preservative concentration is doubled.[43] Field tests have shown that surface treatment can prevent decay for as long as 20 years,[44] although manufacturers generally recommend reapplication every 3 to 5 years. (Also, recall the caveat about cracks and splits reducing the effectiveness of surface-applied preservatives.)

One proven surface-treatment product is a proprietary sodium fluoride paste. The paste has been shown to work well in penetrating the wood and halting decay in existing utility poles and other outdoor structures. It works best at low moisture contents. Sodium fluoride is supplied in gel, rod, or paste form to suit the particular repair need. Unfortunately, like most effective preservatives, sodium fluoride is toxic not only to the wood-destroying organisms, but also to humans.

Fumigants. Fumigants are preservative chemicals supplied in liquid or solid form and placed in predrilled holes. They are most effective for stopping internal decay when they are applied to relatively sound wood. Over time, fumigants vaporize into gas and move through the wood, eliminating decay and insects in the process. Fumigants can diffuse as far away as 8 ft from the point of application in vertical members and 2 to 4 ft in horizontal members.[31] Different fumigants diffuse at different rates, but all of them will eventually evaporate and require reapplication. Research at the Forest Products Laboratory[45] has demonstrated the effectiveness of such fumigants as sodium N-methyl dithiocarbamate (Vapam), trichloronitromethane (Chloropicrin), and methylisothiocyanate (MIT) in stopping internal decay when applied through drilled holes.

Decay often attacks heavy timbers through deep seasoning splits and checks, and treatment holes should be placed on each side of a crack. Maximum hole spacing should not exceed 4 ft. If a liquid fumigant is placed, the holes should be plugged with dowels of preservative-treated wood or with rubber plugs.[46]

Two other effective fumigant materials are boron and the already mentioned sodium fluoride. Boron is slightly less toxic to humans than other chemical preservatives. Boron rods are typically inserted into predrilled holes in structural wood members and dissolve slowly over time, relying on natural wood moisture to transport the preservative through the pores. The moisture content of the wood needs to be greater than 40 percent for adequate boron transport through Douglas fir heartwood.[47] Therefore, treatment with boron rods may not be appropriate in areas where the wood is dry. They should not be used in structures that were previously affected by rot, but have since become dry owing to improved construction detailing, flashing, or previously made roof repairs. Boron rods are ideal for exposed beams or structural members in contact with the ground.

Sodium fluoride can be used not only as a surface-treatment material, but also as a diffusing fumigant. It functions similarly to boron: Embedded sodium fluoride rods slowly dissolve, and the material travels through the wood's cellular structure with natural moisture flows.

Which material should be specified for toxic-sensitive environments—homes and schools? Some chemical preservatives of low toxicity can be applied in situ to slow or stop fungal decay in these areas. Again, boron is at the top of the list. Boron-based preservatives are essentially low-toxicity pesticides designed to penetrate wood and wood composites and protect them from termites, boring insects, ants, and fungi. They work on a slightly different principle from strong fumigants. Rather than using high toxicity to kill the insects directly, boron kills the microbes in the insects' digestive systems, leading to their death by starvation. Boron-based preservatives are available as solid rods, powders, gels, and sprays.

Example 8.2 shows how to calculate the application requirements for boron rods.

Example 8.2: Finding the Required Boron Dosage. The number of boron rods required for a specific application depends on the volume of the exposed wood. For effective long-term protection, 6 oz of boric-acid-equivalent (BAE) material are required for each cubic foot of wood to be treated. Since boron rods are supplied in many different sizes and potencies, the required number of rods depends on the rod characteristics. One way to determine the required dosage of a particular type of boron rod is as follows:

Type of rod used:	$3/_4$-in diameter $\times$ 3 in long (for this example)
BAE:	2.03 oz/rod (as stated in the rod manufacturer's specifications)
Required concentration:	6.00 oz BAE /ft^3 (as stated above)
Rods required:	(6.00 oz BAE /ft^3)/(2.03 oz BAE/rod) = 2.96 rods/ft^3
The volume of one rod:	$^\pi/_4$(0.75 in^2)(3 in) = 1.33 in^3 = 7.67 $\times$ 10^{-4} ft^3
The required dosage:	(7.67 $\times$ 10^{-4} ft^3)(2.96 rods/ft^3) = 0.00227 = 0.23 percent by volume

The amount of boron by volume required to treat the wood is 0.23 percent. So, if 100 in^3 of exposed wood need preservative treatment, 0.23 in^3 of boron rods is required. For best effectiveness, the rods should be evenly distributed throughout the treated wood.

8.5   Shrinkage and Defects

8.5.1   Shrinkage

Since lumber is made of fibers running longitudinally along the tree, wood is anisotropic—its structural properties depend on the fiber orientation. When wood dries out, it shrinks, but not uniformly in all directions. Typically, the length of the fibers changes little or not at all, but the loss of moisture brings the fibers closer together, making lumber shrink across the grain. The rates of shrinkage differ among various wood species.

Figure 8.13 illustrates the dimensional sensitivity, in three principal directions, to changes in moisture content of a Douglas fir member. Assuming that the moisture content stays between 0 and 30 percent,

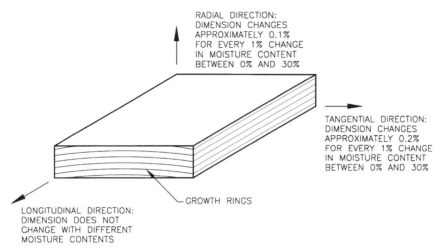

RADIAL DIRECTION:
DIMENSION CHANGES
APPROXIMATELY 0.1%
FOR EVERY 1% CHANGE
IN MOISTURE CONTENT
BETWEEN 0% AND 30%

TANGENTIAL DIRECTION:
DIMENSION CHANGES
APPROXIMATELY 0.2%
FOR EVERY 1% CHANGE
IN MOISTURE CONTENT
BETWEEN 0% AND 30%

GROWTH RINGS

LONGITUDINAL DIRECTION:
DIMENSION DOES NOT
CHANGE WITH DIFFERENT
MOISTURE CONTENTS

Figure 8.13 Dimensional sensitivity of wood to changes in moisture content in three principal directions.

for every 1 percent of its change, Douglas fir typically expands or contracts 0.1 percent in the radial direction and 0.2 percent in the tangential direction. For example, the depth of a 2 × 12 joist will lessen by almost $^3/_8$ in when its moisture content changes from 30 to 15 percent. There are no significant dimensional changes in the longitudinal direction in mature wood. Wood shrinkage in the tangential direction is responsible for the familiar longitudinal splits in heavy timbers.

Shrinkage causes several problems for wood structures:

1. Most obviously, it brings dimensional distortion, resulting in the wood defects described in the next section. Drying wood shrinks, warps, cracks, and splits at the ends.

2. Shrinkage requires careful consideration of the moisture content of structural members and accommodation of the expansion and contraction caused by it. For example, wood boards or planks installed at a high level of ambient moisture, not likely to be reached again in service, could be placed touching each other at the sides. Conversely, dry wood boards installed during a cold winter, when humidity is low, will absorb moisture in service and require separation from each other. Failure to consider these issues will result in either large permanent gaps or buckled floors.

3. Shrinkage of structural members supporting other elements tends to magnify the dimensional changes. Sometimes, the drooping floors that appear to result from foundation settlement are really a consequence of unequal shrinkage. Washington[48] describes several situations in which this occurred.

In one case, a 1-year-old house developed cracks in partition plaster, binding doors and windows, and floors sloping toward the center. The two-story house with a basement was framed with two spans of wood joists. At the exterior, the joists were bearing on a concrete foundation wall on one side and on 14-in wood beams on the opposite side. The center line of supports at the first and second floors was provided by 14-in wood beams. A survey revealed that floor elevations were lower along the center line of supports, raising the possibility of foundation settlement. However, the survey also found that the second floor was depressed more than the first, a fact that argued against the settlement theory. There could be only two other possible explanations: the influence of some construction defects and shrinkage of the 14-in-deep beams. Since no serious construction anomalies were found, shrinkage was concluded to be the cause of the distress. The problem was corrected by jacking up the floors to a level position and refinishing the partitions.

Another case described by Washington involved sagging cantilevered floor joists in a shopping center. The joists were supported by a partition at their interior ends and by a 20-in-deep solid timber beam near their cantilevered ends. The exterior walls were supported by the cantilevers some distance away from the ends; the joists extended further to form a canopy over an entrance. The total length of the cantilevers was 8 ft, with a 14-ft back span. When the 20-in-thick beam shrank, the cantilevered ends moved downward. Coupled with deflection from the wall weight and other service loading, the excessive sag of the cantilevered joist ends caused much alarm. The situation was remedied in a straightforward manner—by some jacking, shimming, and refinishing.

4. Shrinkage can cause problems with roof leakage. For example, when roof rafters shrink, any flashing attached to them will also move downward. If the movement is serious enough, flashing around vent pipes can pull out from under the counterflashing attached to the pipes that are supported by the sanitary equipment below and do not move. The result is a nasty leak around the pipes. When the same problem occurs at parapet walls, an even more serious leakage can happen.[48]

5. Shrinkage can damage old timber trusses framed with vertical members made of steel rods. As the wood chords shrink in the tangential direction, the rods slacken and the nuts connecting them to the chords loosen, making these members ineffective. Similarly, shrinkage can undermine the effectiveness of reinforcing made with bolted side pieces or clamps, as discussed later.

8.5.2 Wood defects

It is a rare piece of lumber that does not have any natural defects. Some of these defects occur in a living tree, and some are the result of weathering; some influence the strength of wood, and some don't. The most common defects in structural lumber are shown in Fig. 8.14. These defects are taken into consideration when lumber is graded, and their presence should not necessarily be alarming. Some cause for concern arises if a defect appears late in the service life of the structure. A recent defect will be of a different color; for example, the inside of a new crack will be lighter in color than the rest of the member.

A *split* is a relatively wide separation of wood fibers approximately parallel to the grain, cutting across the annual rings. The worst of these defects are supposed to be trimmed off when the lumber is cut. Splits often occur at member ends (Fig. 8.15) and at notches, where stress concentrations are present. A *shake* is a thin separation between the annual rings. A *check* is a crack or separation across (or occasionally along) the annual rings. Splits and checks appear when wood dries; shakes can occur either in living trees or during the drying process.

A *pitch pocket* is a void containing pitch, either solidified or still liquid. A *wane* is missing wood at the edges of the timber, where bark used to be. *Crossgrain* occurs when the fibers do not run parallel to the length of the member; this defect can seriously weaken wood. *Compression wood* is internal damage to fibers caused by a natural disaster or induced during logging; the damaged fibers can snap when placed under tension or flexure. Another lumber irregularity is sloped grain, which typically reduces strength in both flexure and compression. Grading rules spell out what the maximum allowable angle of slope deviation should be.

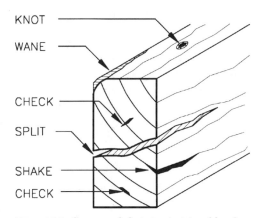

KNOT
WANE
CHECK
SPLIT
SHAKE
CHECK

Figure 8.14 Common defects in structural lumber.

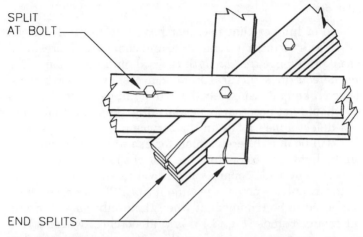

SPLIT
AT BOLT

END SPLITS

Figure 8.15 Splits in timber trusses often occur at the ends of members and at the fasteners.

A *knot* is a remnant of a branch enveloped by the later growth. It would seem that small, tight knots are harmless and that it is the loose large knots that can fall out that should be of concern. However, the situation is more complicated than that. Some tight intergrown knots introduce crossgrain in lumber, while some encased knots might indeed fall out, but the lumber around them is in fact stronger. In any case, the effects of knots on strength depend not only on their sizes, but also on their location and the type of the member.[49]

Lumber defects are becoming more and more prevalent as old-growth lumber becomes but a memory and younger trees of smaller diameters are harvested. Lumber from very young trees should not be used in engineered construction. This *juvenile lumber* can be distinguished by the presence of pith within its section. Juvenile lumber generally has lower strength than mature wood and is more susceptible to dimensional distortion after drying. It can even shrink appreciably lengthwise—up to 10 times more than mature lumber.[49]

These and other wood defects are further discussed and illustrated in Ref. 22. How serious are they? Some are obviously more critical than others. Compression wood and crossgrain are perhaps the most dangerous. A few knots, a small wane (its maximum size is limited by the grading rules), or a pitch pocket rarely affects strength. Lumber with sloped grain, compression wood, and juvenile lumber shrink and swell parallel to grain more than regular wood.[50] These excessive movements can put severe strain on the connections and lead to their distress—breakage of fasteners or splitting the wood.

The significance of splits, checks, and shakes is less clear; they can affect some members more than others. These defects certainly affect shear capacity, and wood design specifications often permit using greater allowable horizontal shear values for members with the ends free of them. The decades-old article by Ketchum et al.[51] offers guidance on the kinds of defects that are generally acceptable and thus require no repairs. Here is a sampling of the recommendations:

- Splits, checks, and shakes in columns and other compression members may be disregarded unless slippage at connections is evident.

- Members loaded in tension parallel to the grain are generally not weakened by splits, unless visible deformations at connections exist. But if a split shows the slope of grain exceeding the allowable for the grade, or if a part of the member is in danger of separating from the main piece, replacement or reinforcing is needed. Checks and shakes are generally much less critical than splits, but if they are very large and deep, they can be evaluated following the criteria for splits.

- End splits at connections may look menacing (Fig. 8.15), but they are allowed unless their maximum width exceeds $^3/_{16}$ in for a connector with a diameter of $2\,^5/_8$ in or less, or $^1/_4$ in for larger connectors. The combined width of multiple splits in the same member end should not exceed 1.5 times these values.

- Splits at bolts away from the member ends (Fig. 8.15) require further investigation. (They might be good candidates for stitching and clamping repairs, discussed in a later section.)

But what about flexural members? Large longitudinal splits running along heavy timber beams are a common sight in mill buildings. Traditionally, these splits have not been repaired unless they threatened to physically separate the beam in two. Numerous buildings with these longitudinal splits remain in service with no apparent ill effects, even though a rigorous investigation would most likely reveal that their horizontal-shear capacities were seriously impaired. When loading conditions change, when structural modifications are planned, and in other appropriate circumstances, longitudinal splits can be repaired by mechanical methods or with epoxy adhesives.

Before attempting repairs of timber with defects, it is prudent to exhaust all the available analytical methods of assessing severity of the problem. Some guidance can be found in AF&PA's *National Design Specification for Wood Construction*,[52] Appendix E, which includes a discussion on the horizontal shear resistance of checked beams. As noted there, distribution of shear in checked beams occurs in a more uniform manner than in unchecked beams. This fact helps explain why beams with serious defects carry loads that they theoretically shouldn't be able to. Incidentally, the typical shear design values shown in the supplement to the specification already allow for *some* splits, checks, and shakes. Indeed, the supplement contains tables of shear stress factor (C_H) that can be used to increase the tabulated shear values if the size of the defect falls within a certain limit.

8.5.3   Human-made defects: Notches and cuts

Natural wood defects pale in comparison with what careless people can do to lumber. Even a cursory inspection of many houses with exposed framing will find joists and studs that have been notched, drilled, or even completely severed to allow the passage of plumbing pipes and other utilities. Which of these "defects" should be fixed? Obviously, a completely severed joist or a rafter with a notch claiming one-half of the member deserves replacement. But is a 1-in notch harmful enough?

Various building codes stipulate the maximum allowable sizes of notches and holes and the limits on their placement. AF&PA's *National Design Specification for Wood Construction,* Paragraph 3.2.3, discusses notches. It notes that notches, especially in tension fibers, should be avoided. If absolutely necessary, notches should be of gradual tapered configuration, rather than square-cornered. Among other provisions, it stipulates that notches are not allowed in the middle third of the span of all members. For members of $3\frac{1}{2}$-in or larger actual thickness and for glulam beams, no notches in the tension face are allowed at all, except at supports. The maximum depth of a notch at a support is $d/4$, where d is the actual depth of the member. Beyond the support areas, the maximum depth of a notch, where permitted, is $d/6$, and the maximum length is $d/3$.

Wickersheimer[53] adds that the maximum size of a hole that can be drilled through a joist is $d/3$, and it should be located at least 2 in from the closest top or bottom edge. Holes at the ends of the member should be avoided. He also suggests that notches be avoided in the midsection of loadbearing studs; elsewhere, their diameter should not exceed $d/3$. The 1996 edition of the BOCA *National Building Code*[54] states that notches in solid lumber studs shall not exceed $d/4$; bored holes should not be bigger than $0.4d$ and should not be made closer than $\frac{5}{8}$ in from the edges.

Where these provisions are violated, reinforcement may be needed. But, as in our discussion of natural wood defects, we suggest that analytical methods be tried first. For example, Paragraph 3.4.4 of the *National Design Specification* includes equations for shear design of notched bending members. It might turn out that the weakened member is still able to carry the imposed loads. If repair is required, typical methods include "sistering" additional pieces of lumber, adding steel straps across the notches, and other techniques discussed in the next section.

8.6   Repairing Wood Members

8.6.1   Introduction

As in the other chapters, methods for repairing deteriorated members and strengthening basically sound structures are discussed separately. In this section, we outline the available means of repairing decayed or otherwise damaged wood.

If deterioration is the result of fungal attack—dry rot, for example—the first step involves stopping the ongoing decay. Structural members should be restored only after the decay fungi are eradicated as described in Sec. 8.4. Still, time and again, attempts are made to jump right to restoration, leaving the fungi alive and ready to pounce on both the new and the old timber. Beckmann[25] observes that a building with a leaking roof that stands vacant for many years and is eventually renovated can be in for trouble: The wood can be subjected to a ferocious attack by fungi that have been lying dormant for years of cold temperatures and have come to life once the building is heated again. (We would add that this problem is more likely in cold climates.)

There are three basic methods of repairing deteriorated wood: partial replacement in kind, mechanical reinforcement, and reconstitution with epoxy or wood fillers. These are outlined in the next section. A more complete discussion can be found in *Evaluation, Maintenance, and Upgrading of Wood Structures,*[19] an excellent publication by the American Society of Civil Engineers that inspired many illustrations in this chapter.

8.6.2   Replacement or partial replacement in kind

The most direct method of dealing with a damaged structural element is simply to replace it with a similar piece of lumber. So, a truss diagonal eaten up by termites can be removed (after shoring and unloading the truss) and exchanged for a new member of equivalent section. A wood column or a stud wrecked by moving equipment can also be removed and replaced in kind. Similarly, partly damaged wood flooring or siding can be cut out and replaced.

Replacement need not be confined to single members—it can be done on large floor areas or even complete floors. Figure 8.16 shows a section through a partly damaged wood-frame building in which this course was taken. Here, the old joists, with a nominal size of 2×10 (and an actual size of $1\frac{7}{8} \times 9\frac{1}{4}$ in) were spaced 16 in on center. The joists carried traditional double-board flooring. In some areas, the building floors had to be reinforced for a much higher live load than the existing joists could possibly support, even in perfect condition. Also, some of the supporting partitions had to be removed, and their place taken by steel beams and columns.

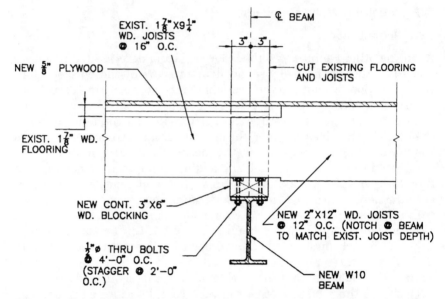

Figure 8.16 Partial replacement of the floor joists in a wood-frame building. The replacement joists at right are notched at top and bottom to fit with the existing joists at left. (*Maguire Group Inc.*)

The replacement structure included new 2 × 12 joists spaced at 12 in on center. Their ends and the ends of the existing joists were supported by new steel beams supplied with bolted-on 3 × 6 continuous nailers. The beams were framed into tubular steel columns bearing on new foundations. Note that the replacement joists at right were notched at the top and bottom to fit with the existing joists at left.

A complete replacement in kind may not be economical when damage affects only small areas of beams and columns of very large sizes, continuous members, or moment-resisting frames. In all these cases, partial replacement can be made. It is insufficient to remove only the visibly damaged areas of wood. To get rid of any unseen fungus, the apparently sound wood on each side of the visibly rotted areas must be removed as well. How much should be removed? Traditionally, about 3 ft,[25] although a lot depends on the actual circumstances. Freund and Olsen[26] recommend removing at least 2 ft of the apparently sound wood. They also advise carefully cleaning all surrounding brick and stone with a fungicide and vacuuming all dust, which could contain fungi spores.

A good example of partial member replacement is repair of laminated wood arches damaged by rot at the bearing points. At these points, the wood is in close contact with moisture because it is either embedded in concrete pockets or placed into fabricated steel brackets bearing on concrete. Both support details allow multiple paths of moisture entry into the end grain: absorption from concrete, seepage from the outside, puddling from floor washing, and accumulation of condensation. Whatever its route, water collected at the bottom invites rotting. Lavon[28] describes a case in which the bottom of a wood arch was completely decayed, with the roof loads apparently carried only by nonloadbearing wall cladding!

Decay typically affects only a short length of the arch leg, and it seems wasteful to replace the whole arch—or perhaps all the building arches—simply to repair the rotted ends. But the damaged area cannot simply be cut out and replaced with a wood stub, because doing so would introduce an additional hinge into the three-hinged arch and make the structure unstable. Also, when columns are spliced end to end, the fibers of the harder lumber tend to penetrate those of the softer lumber, as if two brushes were pushed against each other.[25] The correct way of addressing this difficult situation is to provide a moment-resisting splice between the replacement section and the rest of the member. One such design is shown in Fig. 8.17. There are other possible methods of repair, of course; some of them are examined in later sections.

8.6.3   Using mechanical reinforcement

Mechanical repair methods typically involve the use of fasteners and additional wood or steel components to restore the structural capacities of existing wood members. These repairs may be needed to fix serious defects in lumber or to restore the parts damaged by rot. The two main methods of repair are member augmentation and clamping and stitching.

Member augmentation involves locally reinforcing the damaged member with added pieces of wood, steel, or other materials. As with partial replacement of arches, connections between the added section and the rest of the member must be able to resist the design bending moments and transfer shear forces.

Perhaps the most common application of this method is reinforcing the deteriorated ends of wood members bearing on, or embedded into, concrete and masonry. At these bearing points, moisture can be trapped and absorbed by end-grain fibers. The typical repair procedure involves shoring the member to be repaired (usually a beam or a joist), removing all deteriorated wood, and bolting on two pieces of structural steel channels (Fig. 8.18). The bolted connections are designed to resist the combined effects of flexure and shear.

This kind of local reinforcement of damaged members need not be confined to member ends. Steel or wood side pieces are appropriate for repair of mechanical damage, including broken and notched members.

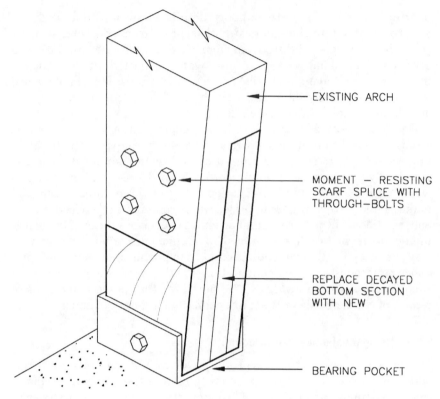

EXISTING ARCH

MOMENT – RESISTING
SCARF SPLICE WITH
THROUGH–BOLTS

REPLACE DECAYED
BOTTOM SECTION
WITH NEW

BEARING POCKET

Figure 8.17 The decayed portion of a laminated wood arch is replaced; there is a moment-resisting splice between the replacement section and the rest of the member.

They are also appropriate for fixing serious wood defects, such as large longitudinal splits caused by shrinkage, overloading, or poor design details. Figure 8.19 illustrates reinforcing weakened members by splicing the damaged areas with through-bolted steel and wood side pieces.

Ketchum et al.[51] recommend that the spliced member be fully cut through in order to equally distribute loads to the splice plates. So, if one piece of a double-member chord is cut and spliced, the other should be cut and spliced as well, to avoid differential deformations between the two. They also suggest that each splice piece of lumber be narrow enough to contain only one row of fasteners, to minimize the possibility of future splitting caused by shrinkage. Still, the sizes of the splice pieces should be sufficient to maintain the minimum edge distances required by code for the splice fasteners. (We should add that using only one row of fasteners per splice piece of lumber is not always required if the moisture content of both the splice pieces and the original wood is already low and is likely to stay that way.)

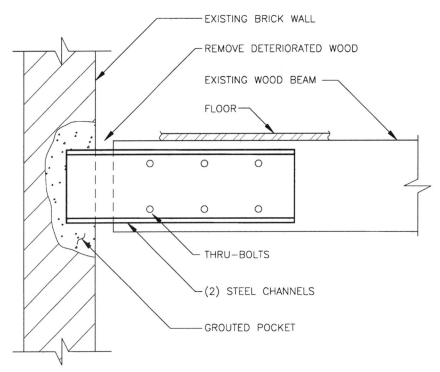

EXISTING BRICK WALL

REMOVE DETERIORATED WOOD

EXISTING WOOD BEAM

FLOOR

THRU—BOLTS

(2) STEEL CHANNELS

GROUTED POCKET

Figure 8.18 Member augmentation by bolted steel channels.

Other methods of repairing cracks, splits, and delaminations and preventing them from further development include clamping and stitching. *Clamping* involves the use of bolts and steel plate or angle assemblies to stabilize cracks and splits. Clamping is often specified for containing split and cracked truss members, such as those shown in Fig. 8.15, to maintain the effectiveness of the steel connectors. However, clamping cannot restore the horizontal shear capacity of flexural members. Clamps can be used in both single and double wood members (Fig. 8.20). Another version of clamping relies on bolted steel side plates, with bolts placed above and below the defect and beyond it to the side, for good anchorage.

Clamping can also be used to repair damaged columns. Ebeling[19] describes a case in which a glulam column fell down during erection, and several splits opened in it. The column was reinforced by clamping at 2-ft intervals with $^{3}/_{4}$-in bolts and $2^{1}/_{2} \times 2^{1}/_{2} \times ^{1}/_{4}$-in angle sections, similar to the detail shown in Fig. 8.20a.

Any time a timber is reinforced with bolted side members or by clamping, attention must be paid to the present and the probable future moisture content of the wood. Making these types of repairs in

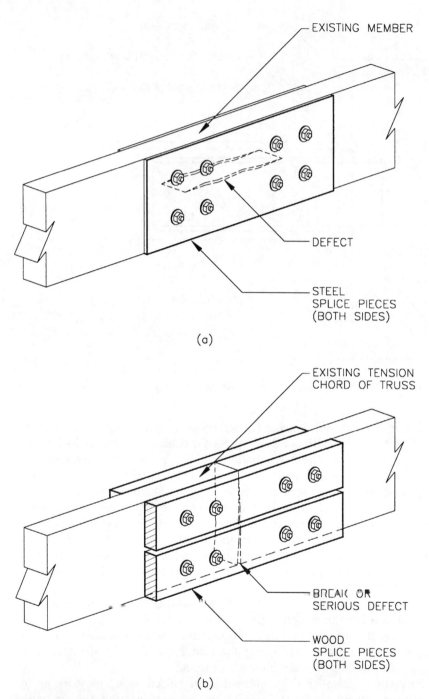

Figure 8.19 Reinforcing damaged members by adding side pieces: (*a*) Using a steel splice plate to reinforce a beam with a serious defect; (*b*) using wood side pieces to repair the broken bottom chord of a truss.

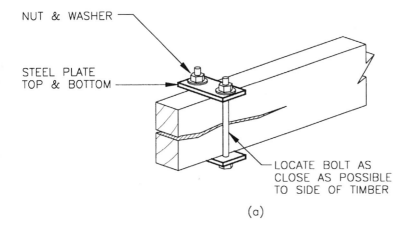

(a)

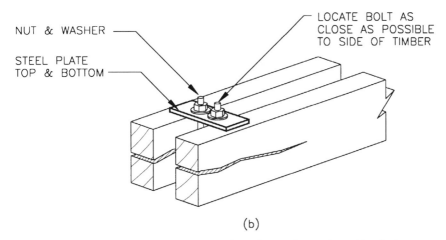

(b)

Figure 8.20 Clamping details: (*a*) single member; (*b*) double member. (*After Ref. 19.*)

relatively moist wood, which is likely to shrink in service, may lead to cracks at the bolt locations, especially if two or more rows of bolts are involved. Shrinkage can also lead to loosening of the clamps, undermining the effectiveness of this reinforcing method.

Stitching is placing through-bolts or lag bolts into the member to connect the partly separated areas (Fig. 8.21). According to Ketchum et al.[51] stitch bolts ought to be located between 2 and 3 in from the end for maximum effectiveness. To minimize the reduction in section area, it is best to use small bolts of $3/_8$- or $1/_2$-in diameter. The bolts should be tightened only to the point where they are placed in tension (i.e., where the crack closes only slightly). No attempt should be made to

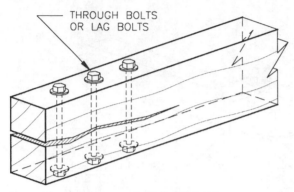

Figure 8.21   Stitching repair. (*After Ref. 19.*)

forcibly close the split, because doing so might lead to the crack's propagating to the other end of the member, as if a lever was used.

8.6.4   Epoxy repair of damaged wood

An alternative to partial replacement in kind and to mechanical reinforcement is to repair the damaged areas with epoxy. An epoxy compound can act as both a filler of hollowed-out areas and a glue stitching together the separated fibers or surfaces. Epoxy can be used to fix end splits in trusses and beams, longitudinal splits in sawn or glulam beams, broken members (perhaps in combination with mechanical splices), and partly decayed wood.

Epoxy repairs are most effective when the resin is injected between the lapped surfaces and placed in shear. The issues of calculating shear stresses in epoxy-repaired joints and testing the repairs can be found in *Wood Engineering and Construction Handbook*[50] and other sources and need not be repeated here. A typical procedure for fixing splits at the ends and along the length of a member involves several steps:[19]

Preparation.   When decay is involved, the totally deteriorated areas are removed, any required reinforcement is placed, and the surfaces are cleaned of all dust and debris. Temporary shoring is installed if circumstances warrant. All roof leaks and other probable avenues of water entry are repaired, and the moisture content of the wood is brought below 20 percent, so that decay does not continue in the adjacent areas. If a member is sagging, it can be carefully jacked up to a level condition, but no higher, so as not to damage the flooring or finishes above. If needed, splice plates are attached to broken wood pieces, unless the joint already contains lapped members.

Sealing the area.   After injection ports are installed, the area of repair is sealed to contain fluid epoxy. Sealant is applied around the ports and at the joint edges, deep cracks, and bolt holes. The sealant material is typically high-viscosity epoxy with the consistency of a putty. Wood surfaces are often coated with a thick epoxy paint to fill minor cracks and voids and to make the joint leakproof.

Epoxy injection and finishing. As in the concrete epoxy injection process described in Chap. 5, injection starts at one of the ports and proceeds until the resin comes out of all the other ports, which are then sealed. After the last port has been closed, injection continues for some time, to make certain that epoxy penetrates into the fibers. Any leaks are repaired with a quick-setting patching cement. The art of the process is not to overfill and swell the joint, so that the surfaces separate and break the seals. To that effect, it helps to keep the nozzle pressure below 40 psi. Epoxy sets in about 1 h and fully cures in a few days. After that, the shoring and injection ports are removed, and the surface is sanded smooth and painted if desired.

Another common application of epoxy repair is fixing the ends of members bearing on concrete or masonry. Because these areas often contain moisture, the ends tend to decay, while the rest of the member remains intact. As an alternative to the partial-replacement approach of Fig. 8.18, the damaged ends can also be repaired by removing the heavily decayed areas and filling them with epoxy compounds.

Figure 8.22 illustrates the procedure for repairing the end of a rotted beam bearing on a masonry wall. After the member is shored, the completely decayed wood is removed, holes are drilled at a shallow angle into the partly damaged part, and formwork is erected to confine the area of repair. Reinforcing rods are then placed into the holes, and epoxy mastic is poured around them. If the volume of the repair is large, the epoxy can be extended by adding sand aggregate. When the epoxy hardens, the rods become permanently bonded to the wood. Whether used neat or extended, the epoxy is much stronger than the wood it replaces. The epoxy resin used for this repair should be formulated for structural work, rather than for adhesive applications.[55]

Epoxy injection can also be used to repair deteriorated glulam members. As already mentioned, moisture can weaken or destroy interior-grade glues in finger joints of glulam beams. Ebeling[19] describes a case in which glulam beams deteriorated simply because the roof over them was not installed in time, so that they were left exposed for several months of driving rain. The repair included shoring the beams, jacking them to a level position, and injecting all broken joints with epoxy (it was found that the joints had to be opened about $1/16$ in to achieve full penetration). In addition, a continuous steel plate was scabbed to the bottom of the beams.

Another case of glulam repair is given by Silva,[19] who describes how partly deteriorated laminated wood arches spanning 175 ft were fixed.

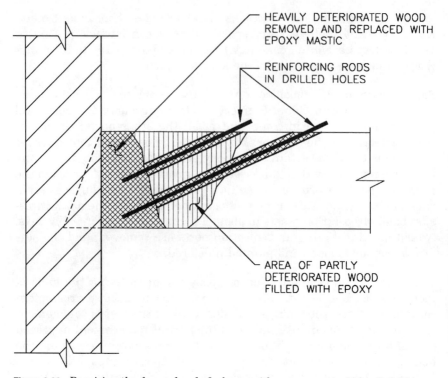

HEAVILY DETERIORATED WOOD
REMOVED AND REPLACED WITH
EPOXY MASTIC

REINFORCING RODS
IN DRILLED HOLES

AREA OF PARTLY
DETERIORATED WOOD
FILLED WITH EPOXY

Figure 8.22   Repairing the decayed end of a beam with epoxy mastic. (*After Ref. 55.*)

The arches were of World War II vintage, and decades of roof leaks had resulted in some wood decay at the top of the arches. (In addition, there was some insect and fire damage.) The repairs included cutting out damaged wood areas and replacing them with matching pieces of 1 × 8 wood that were glued together and to the arch with epoxy-based material. To reduce the amount of difficult field work in place, the replacement wood was carefully cut to fit the removed laminations, preassembled on the ground, and lifted into position. The injection process proceeded in a manner similar to that described previously.

The obvious advantages of epoxy repair is the absence of any "sistered" pieces, connecting bolts, or plates—the repair is essentially internal. Also, unlike metal, properly formulated epoxy is flexible and can move with the wood. Among the disadvantages of epoxy is its relatively high cost, the hazardous nature of the material, and the need for a specialized contractor's expertise. In addition, epoxy loses its strength in fire, and for this reason some engineers question the wisdom of its use in building renovation.

Epoxy repair has been popular for restoration of highly visible historic wood structures. In fact, one of the best sources of information on the subject is *Epoxies for Wood Repairs in Historic Buildings,* by the Office of Archeology and Historic Preservation of the U.S. Department of the Interior.[56] The publication discusses wood consolidation and patching with various epoxy compounds and includes suggested procedures for use, accompanied by case studies.

8.6.5 Slash-and-cut repair

A special case involves repair of the exposed cantilevered ends of wood beams. A design in which beam ends protrude through the exterior walls was frequently used in the past, either to achieve an architectural effect or to support roof canopies. Unfortunately, as discussed previously and illustrated in Fig. 8.5, the exposed ends of cantilevered wood beams often absorb water, inviting rotting.

Since a cantilevered beam end usually carries only a small level of stress, its repair need not develop the full flexural capacity of the beam. Quite often all that is needed is to remove the decayed areas and replace them with an appropriate wood filler. This procedure is called slash-and-cut repair. It requires that two key parameters be satisfied:

1. The remaining portion of the beam's end must be adequate to support the applied loads.
2. The exposed end grain must be of high enough quality to permit proper planing and sealing. (This may be determined using the stress-wave timer, as described in Example 8.3.)

An integral part of the slash-and-cut procedure is sealing the beam ends with a coating that can effectively prevent further water absorption through the end grain. As already mentioned, water is easily absorbed into the end grain and can be carried inside the member for a long distance, providing enough moisture for active fungal attack many feet away from the source. An especially harmful situation occurs when the moisture collected through an exposed beam end is transported to the support at the exterior wall, where stresses are high. Simply covering the ends with a regular paint or sealer is not the solution, because these would seal the moisture inside the wood without providing a very effective barrier to the outside water. Instead, the beam ends should be covered with a specially formulated sealant, which could include a high-solids coating, paraffin wax, and a preservative treatment.

Example 8.3 illustrates the use of stress-wave timing to determine the condition of partly deteriorated cantilevered beam ends.

Example 8.3: Use of Stress-Wave Timing for Slash-and-Cut Repair. Two timber structures in Dawson Creek, British Columbia, collapsed in 1994 and 1997. The failures were attributed to lateral instability and inadequate connections of the structural components, combined with a possible overload.[57] The two collapses initiated a structural review of all the schools within the local district. One of the Dawson Creek school buildings had cantilevered glulam beams with ends exposed to weather. Since the wood was untreated and open to moisture, decay was suspected.

A stress-wave timing evaluation involving the technique described in Sec. 8.3.4 was used to find the deteriorated areas. Figure 8.23 shows a contour map created with a stress-wave timer, drawn on the side of a beam. The very end of the beam had high stress-wave time values (greater than 300 µs/ft), meaning that it took a relatively long time for the stress waves to penetrate it. The abnormally high values correspond to the areas of low wood density, in this case most probably caused by fungal decay. The measured stress-wave time values tend to decrease toward the wall, indicating a higher density of wood there. The covered area next to the wall has the lowest stress-wave time values, because this region is partly protected and subjected to the least amount of moisture.

The heavily decayed areas identified by the testing were removed, and the beam ends, originally rectangular, became triangular in shape. Boron-rod fumigants were inserted into the beam sides at the areas of light deterioration to prevent further decay. Finally, the exposed beam ends were covered with an appropriate sealant (Fig. 8.24).

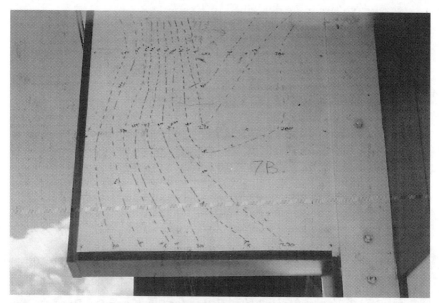

Figure 8.23 Stress-wave time contour map, drawn on the side of a beam, shows areas of decay.

Figure 8.24 Application of a protective coating to an exposed end of a cantilevered glulam beam.

8.7 Strengthening Wood Members

8.7.1 Overview of methods

Strengthening is used when damage to the wood member is extensive, or when its load-carrying capacity must be increased. Some methods of strengthening are similar to those of repair. There is no carved-in-wood border between the two terms, except that strengthening usually involves upgrading the whole member rather than its parts. The methods of strengthening wood structures are similar to those used for steel (Chap. 3) and concrete (Chap. 4).

Strengthening methods can be generally divided into active and passive. In *passive* methods, the combined section is designed to resist only the future loads on the member. Reinforcing does not help carry the existing loads; it becomes active only after some deformations take place in the section. So, when steel or wood side pieces are added to a timber beam, they do not in any way help to resist the loads already carried by the beam. In *active* methods, by contrast, the reinforced member is designed to resist both future and present loads. Active methods involve prestressing or jacking the original member to remove some or all the loads it carries.

Among the conventional methods of strengthening wood structures, *shortening the span* by adding knee braces, diagonal braces, and intermediate columns is commonly employed. For example, sagging wood-joist floors can be strengthened by adding a line of supports at the midspan of the joists. The supports could be built-up or solid wood beams, or perhaps steel wide-flange members, supported by new columns and foundations. The size of the new beams and the spacing of their supports is determined by analysis, taking into consideration the actual condition of the wood joists and the required deflection criteria for the new floor system. Depending on whether the existing sag is removed or not, the strengthening operation can be of active or passive type. If the sag is left in place, the new supports will receive only future superimposed load—a classic passive approach.

If leveling out the sagging floor is required, an active approach is used. It typically involves providing a line of temporary supports (jacks) of adjustable height parallel and next to the proposed permanent columns. Then, concrete foundations for new supports are placed and fully cured. Using the temporary supports, the floor is carefully jacked up to slightly above the level position and then slowly lowered onto the permanent beams when they are installed. Obviously, this method can be used only if the presence of the added supports in the space below is acceptable; it is ideal for strengthening wood floors over partly occupied basements.

Other conventional methods of strengthening wood structures include *adding members* and *enlarging member section (augmentation)*. The former involves placing additional wood pieces between the existing ones; the latter, increasing the size of the existing sections. Both these methods may require some trimming or shimming to match the depths of the new and existing wood members.

There are several methods of section enlargement, in which reinforcement is placed at the sides or at the bottom of the existing wood. These are described in the remainder of this section.

Occasionally, strengthening methods used mostly for concrete structures find their way in renovation of wood. For example, external post-tensioning, discussed in Chap. 4, has been used in wood structures as well. Figure 8.25 illustrates a design in which steel wires are placed under the steel deviator brackets and anchored at the ends of the timber beam. When tensioned, the wires exert upward forces, counteracting the downward loads on the beam. A similar approach was previously popular for strengthening undersized timbers, except that instead of steel post-tensioned wires, $^3/_4$-in round rods were used, tensioned by turnbuckles. The assembly was essentially a queen-post

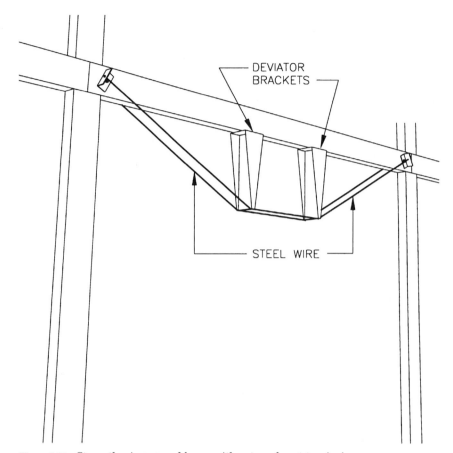

Figure 8.25 Strengthening a wood beam with external post-tensioning.

truss rather than a beam. Other augmentation methods used for utility poles and timber pilings have involved fiber-reinforced plastic (FRP) sleeves, or reinforced concrete placed within FRP jackets.[58]

8.7.2 Member augmentation

Augmentation is the addition of new material to an existing structural element with the purpose of increasing its strength. The added material may be wood, steel, or, more recently, FRP attached to the existing timber in the field. The attachment is typically made by steel through-bolts or lag screws, sometimes coupled with epoxy adhesives.

One common method of augmentation is "sistering," or placing additional wood members next to the existing ones. It is quite familiar to house builders, who routinely "sister" floor and roof joists to increase their load-carrying capacities, as can be required when an uninhabited attic is converted to bedroom space. While ideal for joists, using additional wood sections might not be practical or desirable for strengthening large timber beams; steel side members, usually plates or channels, may be more appropriate. Plates and channels—the two steel sections with flat sides—can be readily through-bolted to the sides of the augmented member (Fig. 8.26).

Despite the simplicity of this solution, there are many cases in which side members cannot be used. For example, they are impractical when floor joists are flush-framed into the reinforced beam or when utilities running at its sides are difficult to relocate. In these cases, reinforcement can be placed *inside* the member. Figure 8.27 illustrates three possible ways of doing so: (*a*) placing a steel plate in a sawn slot and bonding the plate to the reinforced member with epoxy mastic; (*b*) using a steel "flitch" beam (a through-bolted plate placed inside the member or between two adjacent wood members); and (*c*) placing stainless-steel or fiberglass reinforcing bars in a sawn slot and bonding them in place with epoxy.

These methods are especially appropriate for repairing termite damage, where wood members are easy to excavate. But, while appealing in principle, all these methods aim to rigidly join building materials with different coefficients of expansion. It is not clear how well the bond between wood, epoxy, and steel can survive thermal and moisture movements over the long term. The problem may be most pronounced in buildings without year-round air-conditioning located in four-season climates. It appears that, of the three methods, the through-bolted flitch beam offers the least amount of restraint because of slightly oversized bolt holes, the absence of rigid epoxy, and only a relatively small bolt area resisting wood movements.[25]

A conceptually simple method of augmenting wood beams is *scabbing*— adding reinforcement to their bottom surfaces. The reinforcement can

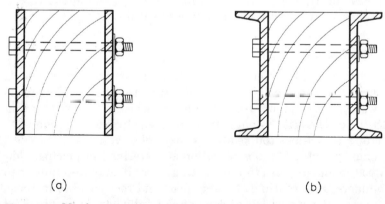

(a)   (b)

Figure 8.26 Member augmentation with steel side members.

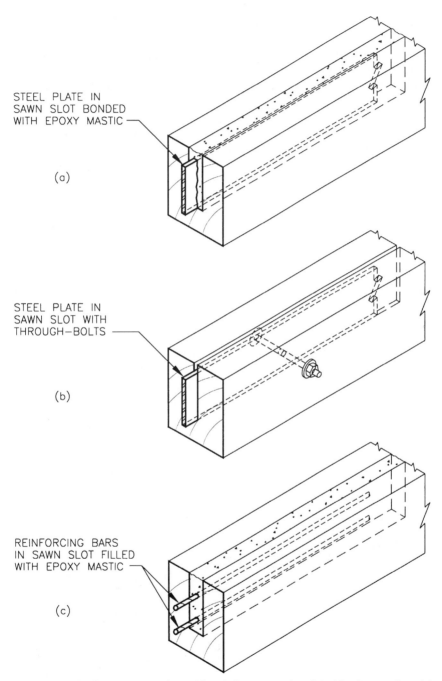

STEEL PLATE IN
SAWN SLOT BONDED
WITH EPOXY MASTIC

(a)

STEEL PLATE IN
SAWN SLOT WITH
THROUGH–BOLTS

(b)

REINFORCING BARS
IN SAWN SLOT FILLED
WITH EPOXY MASTIC

(c)

Figure 8.27 Member augmentation with reinforcement placed inside the member: (*a*) Steel plate placed in sawn slot and bonded with epoxy mastic; (*b*) steel flitch beam; (*c*) reinforcing bars placed in sawn slot and bonded with epoxy. (*After Refs. 25 and 55.*)

be made of wood, steel, or FRP, as seen in Fig. 8.28. Since the purpose of member augmentation is to increase flexural stiffness and strength, the reinforcement usually need not be extended over the ends of the member and under beam supports.

Designing with added wood pieces scabbed to the bottom is rather straightforward. It involves computing the moment of inertia of a combined section using transformed section methods and the theory of elasticity. The next step is finding the number of fasteners needed to carry the horizontal shear forces between the new and the existing wood sections. In cases of reinforcement with both wood and steel, the section properties of the added pieces must be transformed to account for the differences in the moduli of elasticity of the new and existing materials. Designing with FRP requires specialized expertise; it is discussed in a separate section.

8.7.3 Flitch beam

The flitch beam deserves a separate discussion. Many engineers are not certain how to evaluate existing flitch beams, when they need strengthening, and what reinforcement should look like. This venerable design has been used for well over a century, and the obvious pitfalls of combining the two vastly different materials have been known for at least that long. For example, Birkmire, in his 1891 book,[59] describes design methods for flitch beams. He cites the rule that the thickness of the wrought-iron plate—the material of choice

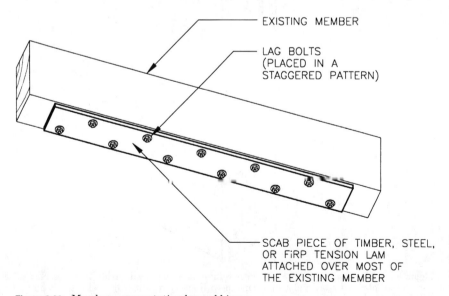

EXISTING MEMBER

LAG BOLTS
(PLACED IN A
STAGGERED PATTERN)

SCAB PIECE OF TIMBER, STEEL,
OR FiRP TENSION LAM
ATTACHED OVER MOST OF
THE EXISTING MEMBER

Figure 8.28   Member augmentation by scabbing.

at the time—should be equal to one-twelfth that of the entire girder. The explanation:

> The elasticity of iron being so much greater than that of wood, the beams should be proportioned so that they will bend at the same time as the iron; otherwise the whole strain might be thrown on the plate. The modulus of elasticity of wrought iron is about thirteen times that of hard pine; or a beam of hard pine one inch wide would bend thirteen times as much as a plate of iron of the same size under the same load.

It appears that little progress in flitch-beam design has been made since Birkmire's time. Weisenfeld[60] observes that flitch beams are not covered in either steel or wood design specifications. He proposes that loads be shared among the steel and wood members in proportion to their rigidities. The steel and wood components should be designed as separate structural members, as opposed to treating the flitch beam as a composite member. The design is most easily done on a trial-section basis.

The most difficult question relating to flitch-beam design is how often to space the bolts and of what size they should be. Weisenfeld suggests using $1/2$-in-diameter bolts for plates less than $1/2$-in thick, $5/8$-in bolts for plates up to $3/4$ in thick, and $3/4$-in bolts for plates up to 1 in thick. If thicker plates are required, it is better to use multiple plates or rolled sections (see Fig. 8.26).

Bolt spacing is a matter of debate. According to Weisenfeld, some engineers space them 2 ft apart regardless of plate thickness and other variables; others design the bolts for load transfer between the beam layers. The answer may depend on how the floor joists frame into the beam. If the joists are flush-framed (a common scenario), the bolts should be sized to transfer a portion of the load to the other beam layers, in proportion to their rigidities. Another function of the bolts is to provide bracing for the steel plate's compression flange. In analyzing one example, Weisenfeld concludes that, to be effective as bracing, the bolts should be spaced a maximum of 15 to 19 in apart. Consideration should also be given to the transfer of bearing stresses at the beam ends: If only some of the layers are supported, the load from the others needs to be transferred, requiring additional bolts. To put this discussion in perspective, there have been no reported failures of flitch beams, and elaborate design procedures for them may not be justified.

Despite some design uncertainty surrounding flitch beams, they often come in handy when flush framing precludes the use of deep beams. Borden[61] describes a case in which an addition to a house required removal of a 20-ft-long loadbearing wall, and no dropped beams were allowed. The solution? He used a flitch beam composed of the existing 2 × 10 rim joists, three new $3/4$ × 9 in steel plates, and a new 2 × 10. The "sandwich" beam carried flush-framed floor joists on each side, supported by joist hangers.

8.7.4 Member augmentation using high-strength plastics

High-strength fiber-reinforced plastics can be used to increase the strength and stiffness of existing glulam and solid sawn beams. FRP is typically applied in place by scabbing to the tension surface of the beam. In this case, instead of wrapping the member with FRP and leaving the material exposed, as is often done with plastic-reinforced concrete structures, FRP prebonded to pieces of lumber is often applied. The scabbed assembly (tension lamination) consists of single or multiple layers of high-strength FRP, typically carbon, aramid, or glass-fiber sheets, bonded under controlled conditions to the wide face of 2-in-thick or thicker high-quality sawn or laminated-veneer lumber. The thickness of each FRP layer is typically 0.07 in (1.8 mm).

A photograph of an FRP-reinforced tension lamination bonded to an existing glulam beam is shown in Fig. 8.29. The lamination is attached to the bottom of the beam with structural epoxy and lag screws. The composite FRP and lumber lamination product is marketed under the trade name FiRP* retrofit tension lamination.[42]

Why is lumber added to FRP sheets to make FiRP? The wood laminations add depth to the repaired wood member and thus increase the strength and rigidity of the combined section; they also facilitate installation of the thin layer(s) of FRP. The ends of the lumber laminations are finger- or scarf-jointed every 8 to 10 ft, so that theoretically they can be produced in any length. In practice, shipping limitations impose constraints, and for very long reinforced members the laminations require field splicing with steel plates and lag bolts (Fig. 8.30).

The design of FRP-reinforced tension laminations is an adaptation of high-strength fiber-reinforced glulam beam design (FiRP Glulams), developed after long-term research and many application studies. After nearly 1000 full-scale bending tests and tests of components, FRP-reinforced glulam was accepted by the ICBO (Evaluation Report No. 5100).[62]

FRP has several advantages over steel. Most importantly, it deforms in a manner more compatible with wood. The yield strain—the ratio of the elongation to the original length at the assumed yield point—for Douglas fir lumber is approximately 0.4 percent (it varies depending on grade and size). For mild steel conforming to ASTM A36, the yield strain is only 0.12 percent. This means that at the same level of deformation,

---

*FiRP is a registered trademark of Dr. Dan Tingley, Ph.D., P.E.

Figure 8.29 FRP-reinforced tension lamination attached to an existing glulam beam.

Figure 8.30 Field-splicing FRP laminations with steel plates and lag bolts.

as happens when wood and steel are bonded together in a composite beam, the steel reinforcement starts yielding well before the wood reaches its strength limit. Most energy absorption will take place in the steel, not the wood, and the steel plate will reach its maximum stress level when the stress in the adjacent wood is only at approximately 40 percent of its maximum.

Conversely, the aramid-reinforced plastic (ARP) often used in FRP applications has a yield strain of 2 percent—much greater than that of wood.[63] The high yield strain of ARP allows more energy absorption to take place in the wood. Therefore, the yield-strain properties allow a composite beam of FRP and wood to be stronger and more efficient than one made of steel and wood.

Bonding to wood is another area in which FRP reinforcement excels over steel. Douglas fir may expand 0.1 percent in the radial and 0.2 percent in the tangential direction for every 1 percent of moisture content change, without any appreciable dimensional changes in the longitudinal direction. (See the discussion in Sec. 8.5.1 and Fig. 8.13.) This seasonal expansion and contraction of wood must be accommodated no matter what type of scabbed reinforcement is used. Steel is about 15 times more rigid than wood; the stiffness properties of FRP reinforcement are much more compatible with those of timber. Therefore, as wood moves in response to changing environmental conditions, no major added stresses in the adhesive between the wood and the FRP are likely to develop. The practical result: Fiber-reinforced plastics designed to be compatible with wood can be attached with conventional adhesives.

Example 8.4 illustrates a FiRP retrofit of existing glulam beams.

Example 8.4: Glulam Retrofit with FiRP Tension Laminations. A gymnasium with a roof that was about 10 ft higher than the roof of an adjacent building was added to a grade school in Canada. The difference in the roof elevations required strengthening of the lower roof to allow it to carry the added load of drifted snow. An analysis showed that under the new loading, the design tension stresses in the glulam beams supporting the older roof would be exceeded by about 20 percent. The glulams were 5.25 in wide by 28.5 in deep and spanned 36 ft.

The existing roof was covering a busy workshop, so shortening the span by adding columns or significantly increasing the depth of the beams was ruled out. The design team decided to strengthen the existing glulams with FRP-reinforced tension laminations (FiRP).

The task was relatively easy in design and execution. The required amount of FiRP was found using methods of transformed sections and elastic analysis. The final design thickness was 0.28 in, or four layers of 0.07-in aramid-reinforced plastic attached to wood members. The FRP-reinforced tension laminations were attached to the bottom of the existing glulam beams over the entire span length using epoxy and $\frac{1}{4}$-in lag screws. They were left without added finishes, as shown in Fig. 8.29. The strengthening effort resulted in a 24 percent reduction of the tension stress in the original glulam beams, with only a 5 percent increase in depth.

Another application of fiber-reinforced plastic is FRP-reinforced plywood, which can be used as tensile reinforcement of roof decking and trusses and as shear reinforcement of beams. Figure 8.31 shows a typical candidate for strengthening: a partly deteriorated wood truss, one of many uncovered during a roof rehabilitation project. Instead of replacing the trusses, the designers opted to strengthen them with FRP-reinforced plywood applied to both sides of the trusses and attached to them by screws. The plywood reinforcement was coped at supports to maintain the depth of the seated ends of the original trusses (Fig. 8.32).

8.8   Renovating Wood Trusses

Wood trusses are one of the most cost-effective ways to span long distances with timber. The trusses can be divided into two general types: those made of heavy or multiple wood members connected with mechanical fasteners, and lightweight trusses connected by pressed metal plates. The two types of trusses tend to be affected by different problems and are examined separately.

Figure 8.31   These partly deteriorated wood trusses were uncovered during roof rehabilitation.

Figure 8.32   FRP-reinforced plywood used to strengthen partly deteriorated wood trusses.

8.8.1   Heavy and multiple-member trusses

Heavy timber trusses have been around at least since Roman times, although their configurations and connections have changed. In traditional heavy trusses, the individual timbers were typically connected by cutting them into each other and by steel rods and straps. These trusses typically supported wood purlins, and truss spacing was controlled by the spanning capacities of the purlins. Trusses carrying purlins were spaced 12 to 20 ft apart; they were often called girder trusses.

The twentieth-century trusses had their members connected by through bolts, often with split rings or shear plates. Wood or steel side plates were common at joints and splices. Trusses could be built of single members located in the same plane, or of multiple members. The elements of single-member trusses were typically connected by steel side plates. Multiple-member trusses usually had chords of two parallel timbers, with diagonals framed in between, as was shown in Fig. 8.15.

Trusses suffer from two main kinds of problems. The first is common to all wood structures: decay, damage by insects, fires, and natural and human-made defects. The second is endemic to trusses: loosening of connection fasteners, instability resulting from lack of bracing, and excessive stresses caused by eccentricities at joints.

Some truss problems, such as splitting, checking, and cracking, have already been described and illustrated. Most of these problems are caused by shrinkage of top and bottom chords, the heavy lumber in which was rarely kiln-dried. As the moisture content in the chords drops, they shrink in the transverse direction. In trusses with diagonal and vertical members made of wood, chord shrinkage typically causes end splits and distress around fasteners. Shrinkage causes particularly serious effects in trusses with vertical members made of steel rods. Here, as the wood chords shrink, the steel rods slacken and the connecting nuts loosen. As a result, the verticals lose their effectiveness, undermining the truss performance.

The splitting and checking often observed in the bottom chords of heavy trusses can have another explanation: the overly high allowable stresses for tension parallel to grain used prior to the late 1960s, to be further discussed later. In some cases, the previously permitted stresses for heavy sawn timbers were twice as high as those allowed today. The worst cases of splitting occur when overstress is combined with shrinkage.

A problem that affects all trusses is arching. Arching typically occurs when the moisture content in the top chord of the truss is higher than that in its bottom chord, as can happen when the top chord is subjected to leaks or condensation, or is located above the roof insulation. In arching, the top chord expands more than the bottom chord, resulting in the truss deflecting upward. The upward arching can crack finishes attached to the truss, although its structural effects are usually insignificant.[50] Trusses made of high-quality lumber may not suffer from arching, but those containing one or more of the defects that affect wood shrinkage parallel to grain may. Those defects include lumber with sloped grain, compression wood, and juvenile lumber, as described in Sec. 8.5.2.

Another common problem in heavy trusses is decay at their bearing ends, in the bolt holes and recesses around split rings, where moisture tends to collect. Any water that enters these areas gets absorbed into the end grain of the wood and can travel long distances from there, inviting rot in sometimes remote locations. This deterioration is nearly impossible to detect, either by visual means (for obvious reasons) or by stress-wave testing (because the members are physically separated). The solutions to this problem include repairing the joints with epoxy, as described in Sec. 8.6.4, and disassembling the joints and replacing the damaged elements, as illustrated in Case Study 1.

8.8.2 Bowstring trusses

Bowstring trusses were so popular from the 1920s through the early 1960s that they deserve a separate discussion. As the name suggests, these trusses have a curved top of parabolic shape and flat bottom. With spans from 50 to over 200 ft, bowstring trusses were extensively used for factories, warehouses, garages, and commercial buildings.

The top chord of a typical bowstring truss is made of two continuous layers of vertically stacked individual laminations. Each layer could include from 4 to 10 individual laminations made of 2×3 or 2×4 pieces of lumber, spiked together. The laminations were usually assembled on site. The top chords of bowstring trusses built in the 1960s were made of glulams. The bottom chords were typically built of two layers of sawn lumber, spliced in the field with side pieces. The ends of the bottom chords were enclosed by through-bolted steel U-shaped brackets (shoes), so the ends of the top-chord members were bearing against a steel surface. The single-piece diagonals and verticals were placed in between the double chord members.[17]

Bowstring trusses have had more than their share of problems and failures. There are two main reasons for that. First, many of the troublesome trusses were constructed during World War II and were intended for rather short service lives. As mentioned in Sec. 8.1, many corners had to be cut during those trying times, including artificially inflating the allowable bending stresses in lumber, using green lumber, and designing for smaller levels of snow loads than would normally be used. The price paid for these necessary measures was decreased reliability. Second, bowstring trusses were long and complex structures that suffered from the cumulative effects of shrinkage, expansion and contraction, sometimes sloppy workmanship, prolonged exposure to high temperatures, leaks, and the condensation common in the roof structures.[50]

Kristie and Johnson[17] investigated many failures of bowstring trusses in the Chicago area. They observed two most common design deficiencies:

1. *Overstress of bottom chord members.* As was noted previously, the allowable tension stresses used during the first half of the twentieth century were overly optimistic. Thus, despite the fact that the bowstring trusses were typically made of good lumber (such as no. 1 or dense no. 1 southern pine), the sizes of their bottom chords were insufficient by today's standards. Similarly, the previously allowable capacities of bolted connections were much larger than those permitted by today's codes. A typical 12-bolt bottom-chord splice-plate connection was allowed to carry about 30 percent less load in 1991 than in 1949. In addition to this "as-designed" lack of capacity, bottom-chord tension members were further weakened by natural and human-made wood defects.

2. *Overstress of web-to-chord connections.* Web-to-chord connections of bowstring trusses typically consist of only one or two through-bolts. This might be sufficient for carrying uniformly distributed loads, when the truss functions essentially as a tied arch, almost without reliance on the diagonals. But unbalanced loading, such as that caused by drifting snow at parapets and valleys, can induce forces several times larger—sometimes large enough to break the chords by splitting them at connections. A related issue is separation (referred to as "sagging" or "galloping") of the individual top-chord laminations caused by eccentricity of the load applied at web-to-chord connections.

Apart from the obvious remedy of removal and replacement, over-stressed bottom chords can be strengthened by adding post-tensioned rods or cables designed to carry the excess tension forces. The area of steel required can be determined from the compatibility of wood and steel elongations:[50]

$$\frac{P_{wood}}{A_{wood}E_{wood}} = \frac{P_{steel}}{A_{steel}E_{steel}}$$

Alternatively, if the bottom chord is seriously damaged, the new steel section can be designed to resist the whole tension force. In either case, the critical detail is the anchorage of the added steel to the truss timber. This can be done by attaching new end plates or brackets to the steel heel plates, if these are present at the ends of the truss, or by bolting the brackets to the sides of the bottom chord.

Web-to-chord connections can be strengthened or repaired simply by adding more through-bolts if there is space for them at the intersection of the chord and the diagonal. If sufficient space does not exist, as is often the case, another possibility would be to use a custom-made light-gage metal U-strap bolted to the diagonal and draped over (and also bolted to) the chord. This reinforcing strap can effectively transfer tension, and compression can be transmitted by timber blocks bolted on each side of the diagonal and cut to shape for bearing on the chord lumber.

As in all heavy trusses, decay resulting from roofs leaking at parapets, skylights, valleys, and penetrations is a common problem affecting top-chord lumber. Bowstring trusses add a complication: They have numerous horizontal gaps between the individual laminations, in which water running down the side can collect, inviting rotting. Another peculiarity of bowstring trusses is that they often have steel shoes at their ends, which moisture can easily get into, but cannot easily escape from. Not surprisingly, the most common areas of decay in bowstring trusses are at these steel-encased ends.[17]

Local decay, separation, or other distress in top-chord laminations can be repaired by carefully cutting out and replacing the affected laminations, similar to the case of glulam repair in partly deteriorated wood arches described in Sec. 8.6.4. (It might also be possible to use a whole new chord section, spliced with the existing section with bolts and side plates.) As already mentioned, any replacement wood should be kiln-dried and be at a moisture content similar to that of the existing lumber. Kristie and Johnson recommend a maximum moisture content of 12 percent, because the moisture content in the existing trusses, except those suffering from leakage, is typically 7 to 8 percent. Top chords are typically repaired with glulam lumber, because it can be produced to match the existing curvature. Separation of top-chord laminations can also be fixed by clamping and stitching. However, clamps require maintenance, and stitch bolts may be difficult to install if the existing spikes interfere.[17]

With any type of truss renovation, temporary supports and scaffolding are required, adding a major cost component to the repairs. The most stable method of shoring bowstring trusses is to support the top chord.

8.8.3 Lightweight plate-connected trusses

The introduction of repetitive lightweight plate-connected trusses was a long-awaited solution to the problem of the high labor intensiveness of traditional trusses. The new trusses were made of single members located in the same plane and connected by thin metal gusset plates at both sides, in contrast with the multiple members and bolted connections of traditional heavy trusses. Bolted chord splices were replaced by continuous chords produced by making structural glued connections.

Because of the limited load-carrying capacities of plate-connected trusses, they are spaced much closer that heavy trusses, from 1 to 4 ft on center—most commonly, 2 ft. The sheathing is attached directly to the top chord, dispensing with purlins. The plates are made of 20-gage galvanized steel with punched-out teeth that are simply pressed into the lumber. The plate's load-transfer capacity depends on its size. The larger the size, the more teeth are available, and the larger the load that can be transmitted by the plate. Design and construction procedures for metal-plate-connected trusses follow the recommendations of the Truss Plate Institute (TPI) and the Wood Truss Council of America (WTCA).

The simplicity and cost-effectiveness of mass-produced lightweight trusses has made them popular in competitive markets, such as commercial and residential applications. They do have a downside, and many problems have been reported. With leaky roofs, the punched-out teeth have been known to back out of the lumber after repeated cycles of wetting and drying, totally severing the connection. The plates themselves can eventually rust through, with similar results. Rust typically starts at the teeth locations, where punching exposes unprotected edges in the galvanized sheet. The punching also causes stress concentrations in the thin sheets.[64]

Perhaps most importantly, these trusses are very unstable during erection and require elaborate bracing for stability and full effectiveness. There have been numerous cases of trusses collapsing during poorly planned construction. Like any wood structure, these trusses are also susceptible to damage by decay and insects. Distressed trusses can show excessive deflections—more than, say, $L/180$—as a result of damaged connections, decay, or broken truss members.[22]

Methods of repair of deteriorated lightweight trusses include the same options discussed for other trusses and for wood members in general. Replacement of damaged members is risky, since it involves removing the existing connecting plates. The plates can be ruined by the removal operation and should not be simply pressed back into place. If replacement of separate members is undertaken, the assistance of a truss manufacturer should be sought to determine the plate requirements. Or, the damaged member can be augmented by side wood pieces, if these can be reliably connected to the rest of the truss structure.

Some inexperienced people are so discouraged by the connection difficulties that they opt for questionable tactics. Meeks[65] tells a remarkable tale of a truss damaged by termites which someone "repaired" by replacing the existing eaten-through members, except that some of the replacement pieces were simply omitted and others were not connected to the rest of the truss.

The most reliable course of action is to replace the whole truss, if that is possible. An overlay with reinforced plywood like that shown in Fig. 8.32 is another option.

8.9   Case Study 1: Repairing Termite Damage in Trusses*

Project background

Termite damage was detected at a single-story maintenance building in Hawaii. The owner hired a consultant to assess the extent of the damage and prepare a plan for repairs, should those be necessary. The owner also wanted to know if present conditions allowed for safe usage

---

*This case study is based on a project performed by Maguire Group Inc., which supplied the accompanying illustrations (Figs. 8.33 to 8.38).

of an interior bridge crane. The shop operations could not be interrupted, and the normal use, including that of the crane, was to continue unless an adverse decision was reached. Access to the roof structure for repairs or close inspection had to be scheduled in advance.

The shop, built at the end of World War II, measured 45 × 120 ft and was 22 ft high. Its framing included nine timber trusses spanning 45 ft between 8 × 10 timber columns (Fig. 8.33). The trusses carried 2 × 8 wood purlins supporting 1 × 8 sheathing boards. The walls were built of 2 × 6 wood studs and wood board sheathing. The trusses were braced near panel points of their bottom chords. A 2-ton underhung bridge crane was hung from transverse 8 × 10 beams spanning between the truss panel points. Fortunately, a set of the original design drawings was available.

Investigating existing conditions

As a first step, the consultant visited the site to conduct a visual inspection and to document any visible areas of distress. To determine the extent and severity of termite infestation, a reputable local extermination company was hired; it examined every building element and catalogued damage to structural members. Finding several active mud tunnels at the perimeter of the building, the exterminators concluded that the agents of attack were ground termites. The exterminators' report included estimated percentages of section loss for each member.

Figure 8.33 Interior view of the building in Case Study 1.

Damage to wood members, sometimes very severe, existed in the columns, trusses, bracing, wall studs, and sheathing. Figure 8.34 shows framing in one severely degraded corner, where a column, a diagonal wall brace, and a truss bottom chord were all damaged. (Another deteriorated vertical brace is shown in Fig. 8.4 at the beginning of the chapter.) Most damaged were building columns, some of them having lost 90 percent of section, because of their proximity to the ground.

Figure 8.34 Severely damaged framing in this corner includes a column, a diagonal wall brace, and a truss bottom chord. (The members damaged by termites are identified by spray paint.)

Meanwhile, a separate structural survey uncovered many instances of shrinkage-related and human-made defects in trusses and columns. Some split diagonals had been previously repaired by stitch bolts; truss bottom chords, by bolted-on splice pieces. In most places, however, the defects were left uncorrected (or perhaps occurred after the attempted repairs). Figure 8.35 illustrates one truss joint in which every member has one or more longitudinal splits. As noted in Sec. 8.5.2, end and side splits in truss members may or may not be critical—the verdict depends on their sizes.

One limitation of visual inspection was inability to check the condition of interior members at the joints. Unfortunately, moisture is most likely to linger at the recesses around split rings and other hidden locations—and this is where rot often begins. The most definitive method of evaluation would be to physically disassemble the joints and examine them (after shoring the truss, of course). That was clearly not practical for the whole building; instead, the designers made some common-sense assumptions concerning repair of areas of hidden decay.

Structural analysis of damaged framing

A computer analysis of the trusses and columns was made first for a typical truss in undamaged condition and then for all the damaged trusses. The trusses were analyzed for several load combinations involving dead, roof live, and bridge crane loads. Separate computer runs evaluated the

Figure 8.35 Every member in this truss joint has at least one horizontal split.

effects of loading applied at the ends and at various points of the crane. Since the trusses were statically determinate, the analysis was not greatly affected by the member sizes, and their undamaged properties could be used for computer modeling. However, if a member was 100 percent damaged, it was removed from that truss model.

As discussed in Secs. 8.1 and 8.8, higher allowable bending stresses in lumber were often used in buildings designed during World War II, and this building was no exception. The original design drawings indicated the allowable bending stress for 8 × 10 beams made of Douglas fir no. 1 as 1600 psi, whereas the value allowed at the time of the renovations was only 1300 psi.

Once the maximum forces in each damaged member were known, the actual reduced member sections could be checked. The percentage of the section area lost to termite damage, taken from the exterminators' report, was applied by proportionally reducing the member depth. For example, an 8 × 10 column damaged 40 percent was analyzed as an 8 × 6 member. This approximation was conservative for compression members, because it increased their slenderness ratios. (Another possible approach might be to use the whole section but reduce the allowable stresses by 40 percent.) Stress increases for short-term loading were used as allowed by the code.

Members with severe longitudinal splits appearing at both faces were conservatively checked as two separate members; those with splits only on one side were assumed to have lost 50 percent of section for horizontal-shear check. These simplified assumptions were made to speed up the analysis process, given the time constraints the consultant faced. Analysis of the weakened trusses identified the members that were significantly overstressed by the worst-case loading combination and required attention.

The analysis results indicated that the trusses were theoretically capable of safely resisting the original design loads, but that some of the heavy timber beams supporting the crane were not even in perfect condition. The analysis results indicated that the underhung crane could not be safely used until repairs were made. The crane was immediately taken out of operation, adding to the urgency of the situation.

Design of repairs

The members shown by analysis to be overstressed were slated for replacement. Realizing that the degree of combined termite and shrinkage damage was not fully known, the designers concluded that repairing the framing under these circumstances was not cost-effective. Another reason was that the governing building code did not require reanalyzing and upgrading the whole building for lateral loads for repair projects where structural members were replaced in kind. In some trusses, most members required replacement, in others, only a few (Fig. 8.36).

The truss diagonals could be replaced by shoring the truss, disassembling the affected joints, removing the damaged members, and using new sections in their stead, following the connection details of the original construction. Where a splice had to be made in the bottom chord, an appropriate detail was provided (Fig. 8.37).

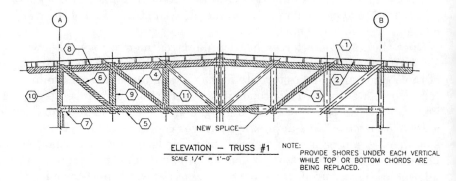

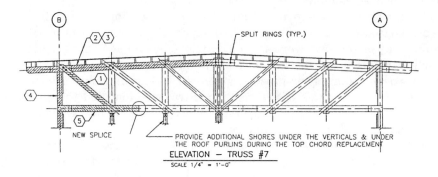

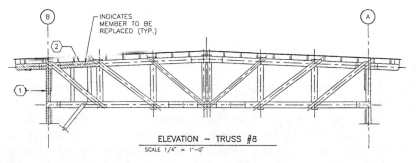

Figure 8.36 Schematic truss elevations showing members requiring replacement.

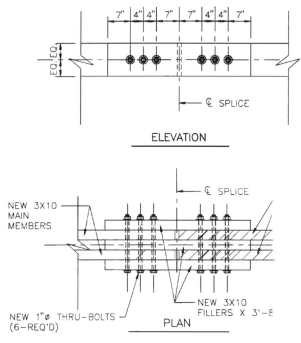

Figure 8.37 Typical detail of bottom-chord splice.

At every damaged truss, there was a column that had to be replaced—it was through the columns that the termites made their way to the trusses. Column replacement was not easy, because there were numerous pipes, conduits, equipment, and bracing attached to the exterior walls that had to be removed and rerouted to keep the operations uninterrupted. Figure 8.38 shows some of the work involved in replacing columns and provides some additional details of truss construction.

The most critical trusses were shored until the completion of repairs. Prior to the start of the work, ground treatment and fumigation took place to eradicate termite infestation and prevent it from recurring in the near future. (Interior fumigation required the temporary removal of some building equipment.) Future periodic treatments against termites—every five years and as needed—were strongly recommended.

The repair work was awarded to a local contractor, who performed it without incident. As was expected, many more members were found to be damaged when construction actually started and the trusses were taken apart. Any additional work was handled on a unit-price basis. The total repair cost was about $300,000.

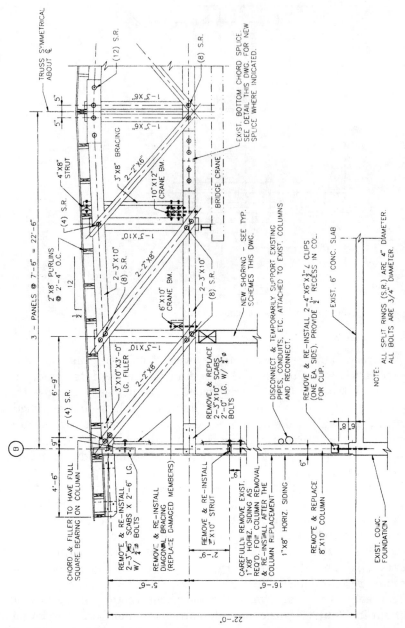

Figure 8.38 Details of replacing columns and repairing trusses.

TRUSS SYMMETRICAL ABOUT ℄

(12) S.R.

(8) S.R.

EXIST. BOTTOM CHORD SPLICE. SEE DETAIL THIS DWG. FOR NEW SPLICE WHERE INDICATED.

1-3"X6"

1-3"X6"

4"X8" STRUT

3"X8" BRACING

2-2"X6"

10"X12" CRANE BM.

BRIDGE CRANE

(4) S.R.

2"X8" PURLINS @ 2'-4" O.C.

1-3"X10"

3 - PANELS @ 7'-6" = 22'-6"

12

½

2-3"X10"
(8) S.R.

2-2"X6"

6"X10" CRANE BM.

2-3"X10"
(8) S.R.

NEW SHORING - SEE TYP. SCHEMES THIS DWG.

DISCONNECT & TEMPORARILY SUPPORT EXISTING PIPES, CONDUITS, ETC. ATTACHED TO EXIST. COLUMNS AND RECONNECT.

6'-9"

3"X10"X3'-0" LG. FILLER

1-3"X10"

2-2"X8"

REMOVE & REPLACE 2-3"X10" SCABS 2'-0" LG. W/ ¾" Ø BOLTS

REMOVE & RE-INSTALL 2-4"X6"X1½" ∠ CLIPS (ONE EA. SIDE). PROVIDE ½" RECESS IN COL. FOR CLIP.

EXIST. 6" CONC. SLAB

(4) S.R.

B

9"

CHORD & FILLER TO HAVE FULL SQUARE BEARING ON COLUMN

4'-6"

REMOVE & RE-INSTALL 2-3"X5" SCABS X 2'-6" LG. W/ ¾" Ø BOLTS

REMOVE & RE-INSTALL DIAGONAL BRACING (REPLACE DAMAGED MEMBERS)

REMOVE & RE-INSTALL 4"X10" STRUT

2'-9"

6"

CAREFULLY REMOVE EXIST. 1"X8" HORIZ. SIDING AS REQ'D. FOR COLUMN REMOVAL & RE-INSTALL AFTER THE COLUMN REPLACEMENT

1"X8" HORIZ. SIDING

REMOVE & REPLACE 8"X10 COLUMN

6"

8"

6"

EXIST. CONC. FOUNDATION

NOTE: ALL SPLIT RINGS (S.R.) ARE 4" DIAMETER. ALL BOLTS ARE ¾" DIAMETER.

5'-6"

16'-6"

22'-0"

492

8.10   Case Study 2: Restoring Fire Damage to the Exeter Street Theater*

Project background

In August of 1995, a severe fire engulfed the Exeter Street Theater, a 110-year-old architectural treasure in Boston, Massachusetts. This impressive historic masonry and timber structure with a slate roof and copper cupola is located at the corner of Exeter and Newbury Streets (Fig. 8.39). The devastating fire destroyed portions of the roof and cupola and severely damaged one of the heavy wood trusses that spanned nearly 90 ft across the building. The damage caused by fire compounded another problem: Years of deterioration had compromised the ability of the roof trusses to carry the weight of the original structure plus the loads added when a new floor was suspended from the trusses during a major renovation in the 1980s. In the aftermath of the fire, an extensive restoration effort repaired an array of structural problems that threatened the viability of this local landmark dating back to 1885.

Description of the building

The building's roof was framed with four wood trusses spanning over what in its original occupancy had been the sanctuary of the First Spiritualist Temple of Boston. The trusses supported the roof and also, by metal hanger rods, an occupied floor located below the trusses but above the sanctuary. There was a two-tiered attic, with one level within the depth of the trusses and another above their top chords, but below the original hip roof and a 35-ft cupola.

In the early 1900s, the building was converted into the Exeter Street Theater, and the sanctuary became an auditorium. Still later, the building was converted to office, retail, and restaurant occupancy, and two new floors were installed within the high-ceilinged auditorium space. During that conversion, the occupancy of the floor suspended from the attic trusses changed from office to storage and handling. The lightly used floor of the lower tier of the attic, at the level of the bottom chords of the trusses, was reframed and strengthened to create a new fifth-floor office.

As a result of the renovation, the building went from three to six occupied floors, and the trusses now supported the suspended fourth

*This case study is based on a project performed by Simpson Gumpertz & Heger Inc., Consulting Engineers. The author is grateful to Messrs. Donald O. Dusenberry, P.E., and Conrad P. Roberge, P.E., for supplying this material, including illustrations for Figs. 8.39 to 8.46. The case study was previously reported in *Structure,* Fall 1999 (Ref. 66), and Figs. 8.39 and 8.42 to 8.44 are reproduced with permission of both *Structure* and the case writers.

Figure 8.39   A view of the Exeter Street Theater (with trusses superimposed on the roof).

and fifth floors, the upper tier of the attic, and the roof. To make a bold visual statement and define the unique character of this first-class space, the architect integrated the original wood trusses into the décor of the fifth-floor office.

The four wood queen-rod trusses were 18 ft deep and 89 ft long. The top chords were constructed with four parallel 6 × 14 wood timbers separated by 2-in spaces for blocking at connections. The four timbers and the wood blocking were interconnected with through-bolts. The bottom chords of the trusses were constructed with four parallel 6 × 12 timbers (also with 2-in gaps for blocking in between) spliced with shear-key iron dowels. The dowels were installed vertically in drilled holes that were partly within the sides of the 6 × 12 timbers and partly within the blocking. The 2-in wood blocking members were the splice plates.

The compression diagonals were two 8 × 10 timbers. The verticals were three $1^3/_4$-in-diameter rods that extended between the top and bottom chords at each truss panel point. The two outside rods terminated at the bottom chord. The center rod continued down approximately 14 ft below the bottom chord, to support the occupied floor below.

Repairing fire damage in trusses

The fire was contained primarily within the roof, the attic, and the offices at the level of the truss bottom chords. The fire severely damaged the top chord of one wood truss (Fig. 8.40). Early structural assessments showed that the building could be restored by using methods of repair that would enhance the strength of the framing while remaining inconspicuous, thus preserving the architecturally significant appearance of the interior and exterior of the building.

The selected repair system actually took advantage of the damage caused by the fire. In one truss, the flames charred and consumed approximately 1 in on each surface of the four timbers that formed the top chord. In effect, the 6 × 14 timbers were reduced to 4 × 12s. At connections, the timbers were not seriously damaged because of the

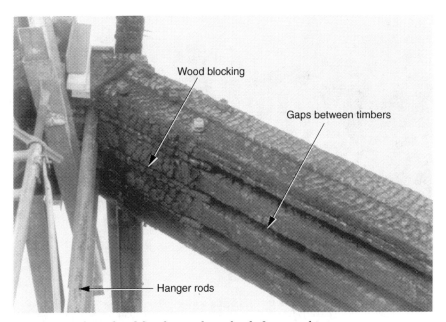

Figure 8.40 A charred and fire-damaged top chord of one wood truss.

protection provided by the solid wood blocking. The gaps between the timbers, when cleared of charred material, thus had increased in width from the original 2 in to approximately 4 in. These newly enlarged gaps and the remaining sound wood at the connections provided the opportunity to strengthen the damaged wood members by concealing new structural steel plates between them. Two 2 × 12 steel plates were placed in the gaps and bonded to the wood with shear keys and epoxy (Fig. 8.41).

This design recognized that the present dead load was being resisted by the remnants of the original wood members alone. The loads to occur later—the weights of the roof and interior finishes plus the live and snow loads—were to be resisted by the composite assembly of wood and steel plates. To keep the original wood structure from being overstressed, the steel plates needed to be stiff enough to support approximately 90 percent of the forces that would be added after repairs.

The installation sequence demanded temporary removal of the vertical steel rods that supported the bottom chord and of the wood diagonals at the first panel point. To avoid extensive and intrusive shoring to the ground below, a temporary beam was placed on the suspended fourth floor between the exterior masonry bearing wall and the rods at the second truss panel point. The temporary beam supported the shoring posts and transferred the forces internally within the existing system. In essence, the truss was "held up by its bootstraps" during

Figure 8.41 The top chord of the truss is being strengthened by insertion of 2 × 12 steel plates between the weakened wood members. (The plates are held in position by welded-on cross bars.)

this operation (Fig. 8.42), and the work could be done without interfering with the occupants below.

The new 2 × 12-in steel plates needed to reinforce the damaged chord were installed in five pieces. This avoided the need to unbolt the compression top chord, which risked local buckling of the damaged timbers, still carrying load, while repairs proceeded. The plates were fitted with intermittent steel shear keys, placed into precisely cut slots in the sound wood, with enlarged ends ("hammer heads") to create the most effective and least scarring of the various possible connections. The new plates were inconspicuously hidden in gaps between the top-chord timbers to preserve the appearance of the wood trusses.

Repairing truss deterioration

Preliminary investigations, made to evaluate the truss capacity during emergency stabilization and design of fire repairs, revealed that some truss members that had not been damaged by fire were overstressed as well. A subsequent complete review of all the trusses revealed a number of serious conditions that added to the complexity of the truss repair:

1. The trusses were visibly distorted, a condition that was attributed to shrinkage. As the truss chords shrank across their depths, the trusses distorted to maintain contact between the chords and the compression diagonals, which had become too short. This distortion loosened the counterbrace diagonals.

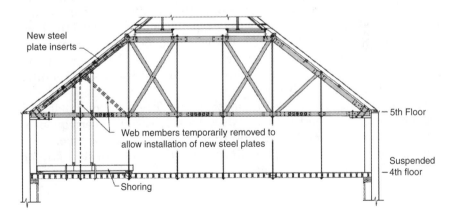

Figure 8.42 Repair of fire damaged truss top chord. A temporary beam placed between the exterior bearing wall and the second truss panel point supports the shoring posts, transferring the forces internally within the existing system.

2. The dowel splice connections in the bottom chord, which transferred tension between timbers and the splice plates, had slipped. The reason was clear: As the wood had dried over the years, gaps had opened up between the timbers and the splice plates. As a result, the wood did not fully bear on the dowels, making the connection capacity inadequate.

3. As might be expected from Sec. 8.8, serious wood decay was present at the bearing pockets in the masonry walls, where the bottom chord and the sloped top chord met and were supported. The decay was most extensive at the interface of the main timbers and the blocking, where it was hidden from view. At two of the eight such locations, the deterioration was so severe that at least one of the four members that made up the connection had failed in horizontal shear.

Therefore, major repairs were needed to correct not only the fire damage, but structural deterioration as well. Again, shoring through the occupied spaces below was out of the question.

The solution for correcting the deterioration of the truss bottom chords included installing two large steel beams (W24 × 176, Gr. 50) below and slightly to each side of the bottom chord (Fig. 8.43). The beams spanned the full width of the building and supported the trusses while the decayed truss bearing regions were removed (Fig. 8.44). To keep the weights to be handled during construction to the practical minimum, each steel beam was fabricated in seven sections. Each section was hoisted by a crane, rigged through a window, and temporarily hung from the bottom chord of a truss. The seven steel sections were then welded in place to form a continuous beam with a 6-in camber.

New steel bearing assemblies (seats) were installed to replace the decayed wood bearing ends (Fig. 8.45). The ends of the steel beams were attached to the seats with wide wing plates that received the horizontal thrust from the top chord. With the two steel beams connected as new truss bottom chords in the repaired trusses, the original wood bottom chords, with their deteriorated wood and weak splice joints, could be abandoned in place.

The new beams also helped in another way: The dead load that had previously been supported by the wood trusses was now permanently carried by these beams. This meant that the remaining wood members were free of loads during repairs, increasing their reserve strength for loads applied after construction.

Transferring the weight

Once the steel beams were installed, the challenge was to transfer the existing dead load from the trusses to the beams. This was accom-

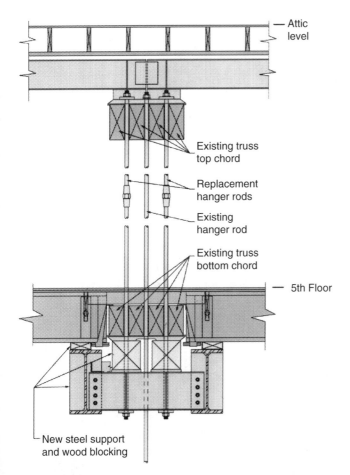

Figure 8.43 Section through a truss, showing original and new elements.

plished by jacking the trusses onto the beams. To do so, the top chord of each truss was shored with an 8 × 12 header and two 8 × 8 timber posts placed between the six truss panel points and the steel beams. Screw jacks below each post transferred the load from the truss to the beams.

During jacking, the transfer of load caused the beams to deflect 6 in (the amount of camber) and the truss to rebound 1 in upward in response to being unloaded. This meant that the "ground" under the shores—the steel beams—was constantly sinking, and the forces in all shores were changing as any one shore was being jacked. To control the process and avoid overstressing wood members during the jacking operation, an

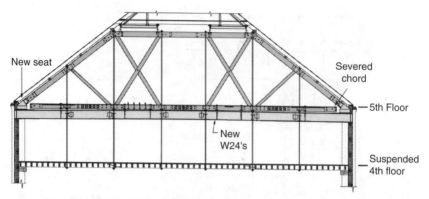

Figure 8.44 Steel beams span the full width of the building and serve as platforms for truss support while the decayed truss bearing regions are removed.

Figure 8.45 New steel bearing seats replace the decayed wood bearing ends.

electronic monitoring system was installed. The system not only monitored the shoring forces and beam deformations, but also helped determine when to stop jacking. The monitoring system consisted of custom-made load cells with a 25,000-lb capacity, placed between the screw jacks and the wood posts and wired to a field computer (Fig. 8.46). Using the preprogrammed results of stress analysis for imposed distortions, the computer permitted a real-time determination of theoretical stresses in truss members during jacking operations.

Figure 8.46 Truss shoring assembly. Note the load cells placed between the screw jacks and the wood shoring posts.

The jacking began by simultaneous turning pairs of screw jacks in accordance with the computed deflection of the steel beams. After a pair of jacks was turned, the other screw jacks were adjusted until the load in the shores equalized. The jacking operation continued sequentially, working from the middle of the beam toward both supports; it was stopped when the top and bottom chords of the existing truss started to separate.

Once the beams fully supported the truss loads, workers removed the decayed truss seats, installed new steel seats, tightened the compression diagonals and counterbraces, and repaired other observed problems. After these repairs were completed, longer hanger rods were installed to engage the new beams, and new steel shear keys transferred forces between the truss diagonals and the steel beams. Finally, the jacks and shoring were unloaded to make the new assembly work as a truss again—except that the original wood bottom chord had been replaced with steel.

References

1. S. Hu, "The Earthquake-Resistant Properties of Chinese Traditional Architecture," *Earthquake Spectra*, vol. 7, no. 3, 1991, pp. 355–389.
2. R. Bruce Hoadley, "The Timber Resource and the Evolutionary Change in Structural Products, *Proceedings of Structures Congress XIII*, Boston, 1995, pp. 1061–1064.
3. D. Fischetti and J. Lewandoski, "Historic Timber Structures," Lecture Series on Rehabilitation/Restoration at MIT, BSCES/ASCE Structural Group, Boston, 1993.
4. *NEHRP Commentary on the Guidelines for the Seismic Rehabilitation of Buildings*, FEMA-274, FEMA, Washington, D.C., 1997.
5. Almon H. Fuller and Frank Kerekes, *Analysis and Design of Steel Structures*, Van Nostrand, New York, 1936.
6. Sara Wermiel, "Slow-Burning Construction," *Building Renovation*, September-October 1992, pp. 66–69.
7. Robert L. Brungraber, "Recent Heavy Timber Construction in New England," *Proceedings of Structures Congress XIII*, Boston, 1995, pp. 1069–1080.
8. "Structural Renovation and Rehabilitation of Buildings," Lecture Series at MIT, BSCES/ASCE Structural Group, Boston, 1979.
9. T. C. R. Wilson, *Guide to the Grading of Structural Timbers and the Determination of Working Stresses*, Miscellaneous Publication 185, U.S. Department of Agriculture, Washington, D.C., 1934.
10. Glenn Murphy, *Properties of Engineering Materials*, International Textbook Co., Scranton, Pa., 1957, pp. 316–318.
11. *Wood Handbook: Wood as an Engineering Material*, Agricultural Handbook 72, U.S. Department of Agriculture Forest Products Laboratory, Washington, D.C., 1987, pp. 4–43.
12. *Wood Handbook: Wood as an Engineering Material*, Agricultural Handbook 72, U.S. Department of Agriculture Forest Products Laboratory, Washington, D.C., 1999.
13. *Wood Structural Design Data*, National Lumber Manufacturers Association, Washington, D.C., 1934.
14. *Commentary on the National Design Specification for Wood Construction*, American Forest & Paper Association, Washington, D.C., 1993, pp. 1–3.
15. *National Design Specification for Stress-Graded Lumber and Its Fastenings*, National Lumber Manufacturers Association, Washington, D.C., 1944.
16. *Southern Pine: Empirical Design Values*, Southern Pine Marketing Council, Southern Forest Products Association, Kenner, La., 1991.
17. Richard J. Kristie and Arne P. Johnson, "Investigating and Repairing Wood Bowstring Trusses," *Practice Periodical on Structural Design and Construction*, vol. 1, no. 1, February 1996, pp. 23–30.
18. Christopher Knight, "Just What Kind of Wood Would a Woodchuck Chuck? Ask Here," *The Wall Street Journal*, October 22, 1997, p. B1.
19. Alan Freas, ed., *Evaluation, Maintenance, and Upgrading of Wood Structures*, American Society of Civil Engineers, New York, 1982.
20. E. L. Ellwood and B. A. Eklund, "Bacterial Attack of Pine Logs in Storage," *Forest Products Journal*, vol. 9, 1959, pp. 283–292.
21. Ronna T. Eisenberg, "Identifying Wood Rot," *Building Renovation*, November-December 1993, pp. 35–39.
22. *Inspection of Wood Beams & Trusses*, NAVFAC MO-111.1, 1985.
23. Osmose Wood Preserving, Inc., 980 Ellicott Street, Buffalo, New York, 14209.

24. *Marine Biological Operational Handbook: Inspection, Repair, and Preservation of Waterfront Structures*, NAVDOCKS MO-311, U.S. Navy, Washington, D.C., 1965.
25. Poul Beckmann, *Structural Aspects of Building Conservation*, McGraw-Hill International, Maidenhead, Berkshire, England, 1995, Chap. 5.
26. Eric C. Freund and Gary L. Olsen, "Renovating Commercial Structures: A Primer," *The Construction Specifier*, July 1985, pp. 36–47.
27. J. Bodig and B. A. Jayne, *Mechanics of Wood and Wood Composites*, Krieger Publishing Co., Fla., 1993, pp. 586–589.
28. Benjamin Lavon, "In Rehab We Trust," *Civil Engineering*, September 1994, pp. 73, 74.
29. Randall Noon, "Historic Downtown Dilemma," *Civil Engineering*, December 1997, pp. 2A–5A.
30. W. Feist, "Weathering and Protection for Wood," *Proceedings, American Wood Preservers Association*, no. 79, 1983, pp. 195–205.
31. Michael A. Ritter, *Timber Bridges Design, Construction, Inspection, and Maintenance*, EM 7700-8, U.S. Department of Agriculture Forest Service, Washington, D.C., 1992.
32. S. M. Smith and J. J. Morrell, "Correcting Pilodyn Measurements of Douglas-Fir for Different Moisture Levels," *Forest Products Journal*, vol. 36, no. 1, 1986, pp. 45–46.
33. W. L. James, *Electric Moisture Meters for Wood*, General Technical Report FPL 6, U.S. Department of Agriculture Forest Service, Forest Products Laboratory, Madison, Wis. 1975.
34. R. R. Maeglin, "Increment Cores—How to Collect, Handle and Use Them," General Technical Report. FPL 25, U.S. Department of Agriculture Forest Service, Forest Products Laboratory, Madison, Wis., 1979.
35. J. J. Morrell et al., *Marine Wood Maintenance Manual: A Guide for Proper Use of Douglass Fir in Marine Exposure*, Research Bulletin 48, Oregon State University, Forest Research Laboratory, Corvallis, Ore., 1984.
36. J. S. Mothershead and S. S. Stacey, "Applicability of Radiography to Inspection of Wood Products," *Proceedings, 2nd Symposium on Non-Destructive Testing of Wood*, Spokane, Wash., 1965.
37. R. J. Ross et al., "Nondestructive Evaluation of Wood," *Forest Products Journal*, vol. 48, no. 1, 1998, pp. 14–19.
38. R. J. Ross and R. F. Pellerin, *Nondestructive Testing for Assessing Wood Members in Structures: A Review*, General Technical Report 70, U.S. Department of Agriculture Forest Service, Forest Products Laboratory, Madison, Wis., 1994.
39. R. L. Lemaster et al., "Detection of Termites with Acoustic Emission," *Forest Products Journal*, vol. 47, no. 2, 1997, pp. 75–79.
40. P. W. Witherell et al., "Using Today's Technology to Help Preserve USS Constitution," *Naval Engineers Journal*, No. 104, no. 3, 1992, pp.124–134.
41. L. Mardin, "Restoring Old Ironsides," *National Geographic*, vol. 191, no. 6, 1997. pp. 38–53.
42. Publications by Wood Science and Technology Institute, Ltd., 1600 SW Western Blvd., Suite 190, Corvallis, OR, 97333, telephone: (541) 753-4548. On the web at: http:/www.wsti-wce.com
43. J. W. Clark and W. E. Eslyn, *Decay in Wood Bridges: Inspection and Preventative & Remedial Maintenance*, U.S. Department of Agriculture Forest Service, Forest Products Laboratory, Madison, Wis., 1977.
44. T. C. Scheffer and W. E. Eslyn, "Twenty-Year Test of On-site Preservative Treatments to Control Decay in Exterior Wood of Buildings," *Material und Organismen*, vol. 17, no. 3, 1982, pp. 181–198.
45. T. L. Highley, "Longevity of Chloropicrin and Vapam in Controlling Internal Decay in Creosoted Douglas-Fir Timber Exposed Above Ground," *Material und Organismen*, vol. 22, no. 3, 1987, pp. 225–233.
46. "Eradicating Decay in Exterior Timbers," *Techline*, United States Department of Agriculture Forest Service, 1998.
47. J. J. Morrell et al., "Effect of Moisture Content of Douglas-Fir Heartwood on Longitudinal Diffusion of Boron from Fused Borate Rods," *Forest Products Journal*, vol. 40, no. 4, 1990, pp. 37–40.

48. George Washington, "Wood Shrinkage Problems," *The Construction Specifier,* August 1985, pp. 26–27.
49. Robert Stroh, "Is Lumber Quality Slipping?" *The Journal of Light Construction,* April 1989, pp. 62–64.
50. Keith F. Faherty and Thomas G. Williamson, eds., *Wood Engineering and Construction Handbook,* McGraw-Hill, New York, 1989.
51. V. Ketchum et al., "Are Timber Checks and Cracks Serious?" *Engineering News Record,* July 27, 1944, pp. 90–93.
52. *National Design Specification for Wood Construction,* American Forest & Paper Association, Washington, D.C., 1997.
53. David Wickersheimer, "Straight Talk about Wood Structures," *The Journal of Light Construction,* April 1989, pp. 22—27.
54. *BOCA National Building Code,* Building Officials and Code Administrators International, Inc., Country Club Hills, Ill., 1996.
55. John Leeke, "Structural Epoxy Repairs," *The Journal of Light Construction,* April 1989, pp. 70–72.
56. *Epoxies for Wood Repairs in Historic Buildings,* U.S. Department of the Interior, Office of Archeology and Historic Preservation, Washington, D.C., 1978.
57. Brad Shipton, G.H. Cook and Associates, Inc., Dawson Creek, B.C., Canada.
58. "Bridge Pilings Can Be Protected; FRP Jackets Stop Deterioration," *Better Roads,* May 1980, pp. 20–25.
59. William H. Birkmire, *Architectural Iron and Steel, and Its Application in the Construction of Buildings,* John Wiley & Sons, New York, 1891.
60. J. D. Weisenfeld, "Glitches in Flitch Beam Design," *Civil Engineering,* September 1989, pp. 65, 66.
61. Eric Borden, "Flitch Beam Retrofit," *The Journal of Light Construction,* September 1999, pp. 39–43.
62. D. Tingley, "Over a Decade of Research Results in New, Improved Glulam," *Canadian Consulting Engineer,* March/April 1996.
63. D. Tingley, "High-Strength-Fiber-Reinforced Plastic Compatibility with Structural Wood Composites," Paper #24, Wood Science and Technology Institute, Ltd., Corvallis, Ore., 1995.
64. H. Dagher et al., "Using Lightweight MPC Wood Trusses in Bridges," *Proceedings of Structures Congress XIII,* Boston, 1995, pp. 9–12.
65. John E. Meeks, "Re: Termite Damage??" *Practice Periodical on Structural Design and Construction,* vol. 2, no. 4, November 1997, pp. 164–167.
66. Conrad P. Roberge et al., "Saving the Exeter Street Theater," *Structure,* Fall 1999.

Renovating Masonry*

9.1 Masonry as a Construction Material

9.1.1 Introduction

Masonry is one of the oldest construction materials, and in some sense masonry represents civilization. It is the Roman masonry ruins that tourists crave; it is the medieval stone cathedrals and castles that attract many of us to Europe; it is masonry that architects turn to when permanence, solidity, and beauty are sought. From the first rubble walls of the appropriately named Stone Age to contemporary brick veneer systems, the evolution of masonry materials and designs closely parallels the levels of human scientific sophistication.

The term *masonry* is usually defined by dictionaries simply as the work of a mason. It includes several loosely related materials—brick, stone, concrete, and tile—that are manufactured as relatively small units and bonded together in the field, usually by mortar. All these various kinds of masonry can provide strength, protection from weather, security, and fire and sound resistance, although their structural behavior is slightly different.

Masonry can carry structural loads or be used for enclosure walls. In this chapter, we limit our discussion to structural renovation of load-bearing masonry of all kinds. Renovation of exterior masonry walls is addressed in Chap. 13, although a few issues necessarily overlap those discussed here.

---

*The author is grateful to David B. Woodham, P.E., of Atkinson-Noland & Associates, Inc., for providing much of the information for this chapter and the photographs for Figs. 9.9 through 9.14 and 9.17.

The types of masonry units, the mortars used, and the methods of placement have all evolved over the centuries. For example, clay masonry construction, which started as unreinforced massive walls, has changed to modern brick veneer and to reinforced single-layer oversize brick units. Knowledge of the approximate age of the structure is useful for its assessment and analysis, and for making decisions regarding renovation options.

In the United States, masonry buildings have been used in the southeastern and southwestern parts of the country since the 1500s, in the central and eastern parts since the 1770s, and in the western half since the 1850s. Most masonry structures encountered in the United States today were built in the last 150 years.[1]

9.1.2 Stone masonry

The first masonry structures were carved into the sides of stone cliffs or made of dry-laid boulders. Later, stones were cemented together with mud. Later still, stone walls were built of specially cut interlocking units, some of them carved with great precision. A few of these old solid-stone structures survive to this day, despite the numerous natural disasters that have befallen them (the Pyramids of Egypt come to mind).

A technological challenge facing stone builders was the need to make wall openings, a difficult task prior to the invention of steel and concrete. The Romans solved this problem by inventing a self-supporting semicircular arch, the symbol of their construction style. Arches have since become inseparable from monumental stone masonry. In relatively recent times, they were especially popular when the Romanesque Revival architectural style, often associated with H. H. Richardson, came into fashion. One structure representative of this style is the 1892 Flour and Grain Exchange Building in Boston (Fig. 9.1). This building was originally used for meetings of the Boston Chamber of Commerce and was restored in 1988. This majestic structure, with tiered arches and rock-faced masonry, unabashedly celebrates the beauty of stone.

The issues of stone deterioration and repair are examined later in this chapter and in Chap. 13.

9.1.3 Clay masonry: Brick

Clay masonry includes brick and tile. These materials are made by burning clay in the oven to give it hardness and durability. Brick was known to many ancient civilizations, including the Romans, and it certainly was around in Colonial America. Records indicate that there was an established brickmaker in Jamestown, Virginia, as early as 1610.[2] If suitable clay deposits existed, brick was often produced near

Figure 9.1   The Flour and Grain Exchange Building in Boston is a tribute to the versatility and beauty of stone masonry.

the building site to reduce transportation costs. Later, brickmaking was industrialized, and most towns had local brick plants to supply the surrounding communities. By the late 1800s, mechanized brick production and downdraft kilns became widely available. These technologies were largely responsible for the steady improvement in the quality and uniformity of the brick produced at that and later times.

In the last decades of the nineteenth century, extrusion or stiff-mud brickmaking machines were invented, and some brick manufacturers started producing bricks with holes to decrease weight and speed up drying time. Various states attempted to standardize the brick sizes used within their jurisdictions, although no uniformity existed nationwide. Thus, in 1883, New Jersey established a standard brick size of

$9^{1}/_{2} \times 4^{1}/_{2} \times 2^{3}/_{4}$ in, while in the District of Columbia a standard brick measured $9^{1}/_{4} \times 4^{5}/_{8} \times 2^{1}/_{4}$ in.[3]

Brick terminology developed even earlier, and these terms are still with us today. The bricks oriented at 90° to the wall plane are called *header* courses. The similarly positioned bricks placed with their narrowest sides horizontal are *rowlock* courses; these have been popular for capping the tops of parapets and for framing arches. The vertical joint between the *wythes* (layers) is called the *collar joint*; the horizontal surface where brick is laid is the *bed joint,* and the vertical interface between the adjacent bricks in a row is the *head joint.*

During the late nineteenth century and the first half of the twentieth century, brick was usually used in loadbearing wall applications, where it supported wood, steel, or concrete framing. Separate wythes of bricks were typically interconnected with brick headers at every sixth or seventh course. Loadbearing brick was particularly popular in mill building construction, described in Chap. 8. Numerous mill buildings with brick exteriors, wood floors, and wood or cast-iron columns were constructed throughout the Northeast and elsewhere. These structures featured large windows with brick pilasters in between, where heavy timber beams were supported (Fig. 9.2).

The quality of brick was largely dependent on its firing temperature. Many old bricks were underburned, and because of their pink color were called salmon brick. The overburned "clinker" bricks were used largely for heat-resisting structures. Bricks produced before the twentieth century tended to be quite soft, typically less than 1500 psi in strength. By today's standards, they were extremely absorbent: Some underburned bricks could absorb water up to 35 percent of their weight.[4] Bricks produced at that time typically had very hard exterior surfaces imparted during firing, but were soft and porous inside.

Brick strength markedly increased in the early twentieth century, even for underburned units. A 1929 study reported that the flat-wise compressive strength of U.S. bricks averaged 7246 psi for both hard and salmon grades, and 40 percent of the bricks had compressive strengths in excess of 8000 psi. The gradation was as follows: About 6 percent had compressive strength between 1250 and 2500 psi, 20 percent between 2225 and 4500 psi, and 74 percent over 4500 psi.[3]

Brick construction is extremely labor intensive, one of its few but serious disadvantages. When labor cost began to matter more than the cost of materials, the age of solid brick walls was over. More and more, brick was used as a mere facing on the less-expensive concrete block walls, and brick veneer eventually displaced solid brick. Beginning in the mid-1960s, brick veneer has been often used in combination with steel-stud backup, a system described in Chap. 13.

Figure 9.2 Exterior of a typical mill building.

9.1.4 Clay masonry: Hollow tile, terra cotta, and adobe

Hollow clay tile is similar to brick in composition but not in construction. Tile, which is made by extrusion, is much lighter than brick, owing to internal voids contained within relatively thin ribs. Hollow clay tile has been used in three main applications: rectangular tile blocks, primarily found in partitions and nonbearing exterior walls; hollow tile floor units spanning between steel beams; and column fireproofing. The first U.S. production of structural clay tile took place in New Jersey in 1875.[3]

At one time, *hollow clay tile* was different from *terra cotta*. Terra cotta was originally made by mixing clay and wood particles (sawdust, for example) and firing the molded units in a kiln. The intense firing burned out all the wood particles, leaving the terra cotta blocks light and porous. By contrast, hollow clay tile was made only from burnt clay; it contained no combustible additives. Later, this distinction became somewhat blurred, and "terra cotta" has become nearly synonymous with "clay tile masonry," as combustible particles are no longer added in either material.

Structural rectangular hollow tile became quite popular for a time after the famous 1871 Chicago fire and other conflagrations highlighted the need for noncombustible building products and fireproofing. The tile was especially in demand during World War I, when lumber was in short supply. Hollow tile was prized for its light weight and relatively good insulation properties. The units could be produced with glazed surfaces, making them virtually maintenance-free and therefore desirable for partitions in industrial applications. Alternatively, they could have rough surfaces that provided good adhesion for mortar or plaster, making the tiles suitable for inclusion in multiwythe walls.

Rectangular tiles often replaced heavier and more expensive bricks and concrete masonry units, as illustrated in Case 2 of Chap. 12. The main drawback of rectangular hollow tiles was their brittleness and lack of solidity, making attachment to them difficult and cutting openings treacherous, as can be seen in Fig. 9.3. Some representative allowable loads on tile walls, based on the earlier design codes, are given in Sec. 9.2.

At the end of the nineteenth century and in the early twentieth century, hollow tile was used in floor construction, to span between iron or steel I-beams. (Later, this role was taken over by concrete slabs.) There were two general designs: the *flat arch* and the *segmented* (curved) *arch*. Both types of arches usually carried timber sleepers, to which wood flooring was nailed. The space between the sleepers was filled with plain concrete, cement cinders, or similar noncombustible material. The tiles were cemented together.

The tile blocks in flat arches had skewed ribs and totally filled the space between the beams (Fig. 9.4a). Special tile units with cutouts for beam flanges were sometimes used, a design that allowed for a complete encasement of the beams for increased fire resistance (Fig. 9.4b). The stability of the flat arch hinged on the center wedge shaped tile, which acted as a keystone.[5] Or, horizontal metal tie rods were provided at the bottom for a tied-arch design (Fig. 9.4c). The personnel performing field investigation of tile-arch floors should be alerted not to cut these tie rods during exploratory demolition, as is very easy to do.[6]

The bottom surface of the tiles was usually roughened in a dovetail profile to support a plaster finish. According to the 1891 book by Birkmire,[5] the smallest tile arch listed for a 4-ft span was 6 in deep and had a weight of 29 psf; it could support a safe load of 1000 psf. The largest tile arch available for a 7-ft span was 12 in deep, had a weight

Figure 9.3 Rectangular hollow tile with glazed surface used to be popular for partitions in industrial applications. (*a*) Cross section of tile partition; (*b*) the brittle nature of hollow tile makes attachment difficult and cutting of openings treacherous.

of 48 psf, and could carry a safe load of 1800 psf. These extremely high listed load-carrying capacities might be the result of small applied factors of safety.

Forty years later, the allowable loads listed in the literature were a bit smaller. The 1936 book by Fuller and Kerekes[7] reproduces the National Fireproofing Corporation's load tables for flat and segmental arch floors. These tables, based on a factor of safety of 7, list the safe load for the same flat 6-in-thick arch spanning 4 ft as 236 psf, and that for a 12-in-thick arch spanning 7 ft as 206 psf. Most other combinations of thickness and span could support even higher allowable loads, so properly built tile arches in good condition theoretically should be strong enough for all but the heaviest building loads.

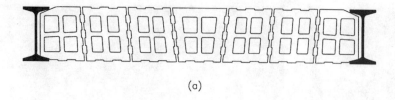

(a)

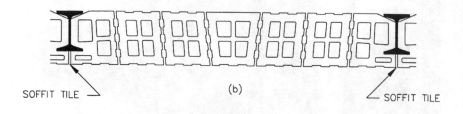

SOFFIT TILE

(b)

SOFFIT TILE

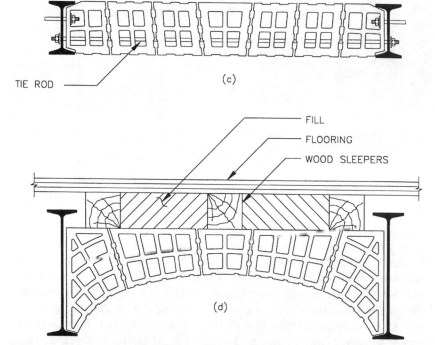

TIE ROD

(c)

FILL

FLOORING

WOOD SLEEPERS

(d)

Figure 9.4 Floor arches of hollow clay tile. (*a*) Flat arch; (*b*) detail of special tile units with cutouts ("shoulders") for encasing the I-beam flange; (*c*) tied arch; (*d*) segmented arch. (*After Refs. 5 and 6.*)

Segmented (or segmental) arches were made of hollow tile blocks with varying depths (Fig. 9.4*d*). This type of construction is more reliable and less susceptible to accidents stemming from making floor openings than flat arches. Before hollow tile was used for this application, segmented arches were built with ordinary bricks spanning between steel I-beams—a much heavier construction that made fireproofing of the bottom beam flanges difficult. The evolution of hollow-tile floor arches, from the early segmental and flat designs to the late versions, is described by Wermiel.[8] According to Fuller and Kerekes, the allowable load for a 6-in-thick segmental arch spanning 12 ft with a rise of $^3/_4$ in—the smallest listed—was 285 psf (again, with a factor of safety of 7). If the rise was 2 in, the same arch could safely support a much higher load—727 psf.

Another application of hollow tile blocks was to serve as formwork for reinforced-concrete beam-and-slab construction. The tiles were temporarily carried on wood forms, and concrete was placed on top of the units and between them, where spaces for concrete beams or joists were formed. The bottom surface of the tiles was eventually coated with plaster, which needed no lath because of its excellent adhesion to tile.[9] The contribution of the tile sections was neglected in computing the load-carrying capacities of these floors. However, since the sides of the tile blocks were scored and concrete was extended into them, the assumed effective shear area of a concrete joist included one-half of the thickness of the vertical tile webs, which were typically $^3/_4$ in thick.

Architectural terra cotta has been widely used for exterior ornaments in cornices and column capitals. The material has been known since the Middle Ages, but its biggest—and last—spurt of popularity occurred in the mid-nineteenth century, as European designers realized that architectural terra cotta could replicate the appearance of cut stone at a lower cost and with lower weight. According to Plummer,[10] the material has been produced in the United States since 1879. It was manufactured in both glazed and unglazed varieties.

Unfortunately, design standards for architectural terra cotta were lacking, and the product quality was largely dependent on the manufacturer's expertise. As the years passed and the units aged, many terra cotta façade elements have become unstable and required rebuilding or resecuring, as discussed in Chap. 13. By the 1920s, when the National Terra Cotta Society started its comprehensive research program, it was too late—the popularity of the product was already waning. By the end of World War II, architectural terra cotta had essentially vanished, having been displaced by other, more austere types of products.[11]

One more clay-based material is *adobe,* unfired blocks of clay. Adobe walls covered with a wood roof structure constitute the so-called adobe construction used (mostly in the past) in the Southwest. The rather obvious problem with unfired units is their lack of cohesion and their deterioration after prolonged exposure to water. The current masonry codes differentiate between stabilized adobe, to which emulsified asphalt has been added to reduce water absorption and increase durability, and unstabilized adobe without such admixtures.

9.1.5 Concrete masonry units

Concrete masonry units (CMUs), also known as concrete blocks, are extremely common today. It is difficult to realize that they became readily available only in the late nineteenth century, when the first commercial process for their manufacture was developed. Among the first recorded uses, concrete blocks were used in construction of a 1897 residence, where they were produced on site to reduce transportation costs and breakage. Concrete blocks became especially attractive at the turn of the twentieth century, when material prices for brick and timber skyrocketed. According to Bell et al.,[12] in the short time from 1898 to 1906, the prices for brick increased 59 percent, and those for timber 64 percent. CMUs were marketed as a fireproof and durable construction material.

In 1900, Harmon S. Palmer patented a machine to cast hollow concrete blocks. The invention proved wildly popular, because the units could now be produced inexpensively and, being larger than bricks, installed faster. Within six years of Palmer's invention, there were over a thousand places where blocks were made. During this so-called backyard phase, similar machines were marketed to homeowners (by Sears, Roebuck and Company, among others).[13] In 1905, the blocks were used by the U.S. government for buildings in the Panama Canal Zone and in the Philippines. At first, the units were cured by sprinkling them with water for 7 days; later (around 1912), high-pressure steam curing became common.[3]

Early concrete blocks did not resemble their contemporary cousins. First, the block-making machines allowed the users to make the units with a rock face—the appearance of cut stone—and with other elaborate designs. The rock-face finish was especially popular from 1900 to about 1930; this fact can help identify the age of many surviving block buildings. Second, the blocks were typically much larger. The first Palmer machines made blocks with dimensions of 10 × 8 × 30 in, and other suppliers were producing 12 × 9 × 32-in blocks weighing about 180 lb. Obviously, those blocks were difficult to handle.

Inventors were working on making smaller and lighter blocks, and the 8 × 8 × 16-in CMU size that we know today became common by 1930. In 1917, F. J. Straub patented his "cinderblock," which was promoted as a light, fireproof, and nailable product.[13] The trade name has outlived the company, which did not survive the fierce competition with many other makers of lightweight CMUs.

The National Association of Cement Users (NACU) was formed in 1904. In 1913, it became the American Concrete Institute. In 1908, the NACU proposed the earliest specification for hollow concrete block units; this specification was adopted in 1910. The NACU specification called for the units used in bearing walls to have an average strength of 1000 psi, with a minimum strength of 700 psi. The amount of air space in the block was limited to 33 percent, and the average absorption to 15 percent. In 1922, the first specifications for nonloadbearing concrete masonry units were proposed; they called for a minimum required block strength of 300 psi. In 1925, the ACI adopted three levels of the minimum required block strengths: 250 psi for nonloadbearing walls, 700 psi for "medium-loadbearing" walls, and 1200 psi for "heavy-loadbearing" walls.[3]

Automated vibrating machines ended the backyard phase of block production in the 1930s. The semicustom rock-face and other patterned blocks were displaced by the mass-produced utilitarian plain gray units that are ubiquitous today. Several decades later, block variety returned at another level of sophistication, with a palette of architectural choices—split-face, scored, and glazed units—in an array of colors.

9.1.6   Mortar

Mortar and grout are sometimes confused, even though these two materials play different roles in masonry construction. Mortar is the glue that binds masonry units together; grout fills the voids in them. The differences between mortar and grout involve not only the function, but also the composition and consistency of the materials. Grout is not "soupy concrete" or "thin mortar"—adding excessive water to mortar will not turn it into grout, as some masons believe.[14] (Unfortunately, the practice of adding water to mortar to make grout has been so common that it has even been suggested in some respected trade literature.)

The oldest mortar of sorts is thought to have been used in Egypt in 4000 B.C., where stacked adobe bricks were coated with mud to hold them together.[15] A millennium later, the Babylonians and Assyrians used bitumen to join units of burnt brick and alabaster. The Romans used *pozzolana*—finely ground volcanic ash containing siliceous materials and, often, alumina—to make bedding sand and cement. Mortar made of pozzolana, lime, sand, and water was produced in Rome and Greece in 500 B.C. This basic formula was used for more than two millennia, although pozzolana was not available outside the volcanic regions and was eventually dropped from the mix, and lime assumed a leading role in mortar making.

What exactly is lime? Speweik[16] explains that lime is made by heating limestone (essentially, calcium carbonate) to more than 1650°F, a process that removes carbon dioxide and water and produces calcium magnesium oxide, the so-called quicklime (unslaked lime). In a slaking process, quicklime is mixed with water, and this lime putty is allowed to mature for months or even years. The final product contains hydroxides of calcium and magnesium and is known as slaked lime. The lime putty, after excess water is removed, is mixed with sand to make mortar, using one part slaked lime putty to 3 parts sand.

Some types of lime absorb more water than others during slaking. So-called fat lime absorbs nearly one-half of its volume in water and produces a rich mortar, oily in consistency and prized by masons. In contrast, *hydraulic lime* is made from limestone containing some clay; it absorbs relatively little water and produces less lime paste. Early American buildings most likely used mortars based on weak hydraulic lime.[4]

Lime mortar remains plastic for several hours after mixing. As lime crystallizes and stiffens (sets), it gains strength and becomes able to carry the weight of the masonry above. However, it takes a long time for the chemical reaction between carbon dioxide in the air and hydrated lime to form hard calcium carbonate and for true hardening to occur. (The hardening reaction is the opposite of making quicklime from limestone.) Because the stiffening process depends on an airborne chemical, the surface layers of the mortar harden first, while the interior hardens very slowly. Depending on the porosity of the mortar, carbon dioxide may *never* reach the interior of a thick brick wall, meaning that the mortar there never really hardens. That dried but not properly hardened interior mortar may be little more than a mixture of sand and lime, possessing almost no strength and being a prime source of efflorescence.[2,4]

Lime mortars have very low 28-day compressive strengths—between 50 and 300 psi—but they have good workability, tensile bond, flexibility, and self-healing properties. Being the weaker component of the wall assembly, lime mortar is able to absorb some movement of the bricks and thus prevent their cracking. In the United States, mortar of sand and lime was used until around 1871, when portland cement first became domestically available from a Pennsylvania factory.[15] Portland cement was of much better quality than the so-called natural cement made by burning the natural cement rocks and grinding the resulting lumps into powder. Natural cement was occasionally used to make masonry mortar for monumental-quality buildings, but it was largely replaced by portland cement after the latter's arrival.

In the 1870s, masons began combining cement, lime, sand, and water to make mortar. Portland cement was initially introduced into mortar to speed up setting; its contribution to mortar strength was discovered later. In the last decades of the nineteenth century, most buildings were already being constructed using portland cement and lime mortars. Lime was still necessary because bricks joined with mortars made only of portland cement and sand tended to crack. In 1951, the first edition of ASTM C270, *Standard Specification for Mortar for Unit Masonry,* was published. ASTM C270, slightly revised over the years, is still the definitive standard on masonry mortar.

A variation of masonry mortar is *hydraulic mortar,* which is intended to set and harden under water. Hydraulic mortars may contain hydraulic lime, sand, and water or be cement-based.

How can one tell which type of mortar is present in the building? In addition to the age, another clue is color. All mortars are gray, of course, but cement-based mortars are darker than lime mortars. The definitive answer can be obtained by testing.

9.1.7 Masonry grout

Grout consists primarily of portland cement, aggregate, and water. Masonry grout conforming to ASTM C476, *Standard Specification for Grout for Masonry,* is allowed to contain hydrated lime or lime putty up to one-tenth of the volume of cement. While mortar is weaker than masonry, grout is stronger. As far as consistency is concerned, mortar normally contains only enough water to make it workable and "buttery." Adding more water to mortar weakens it and increases the rate of shrinkage. By contrast, masonry grout is typically fluid, so that it can flow into cavities and fill all the voids. A grout slump of 10.5 to 11 in is typical.

Masonry grout should not be confused with a related material, cementitious or epoxy-based nonshrink grout. The main function of nonshrink grout is to fill the spaces between steel and concrete—under column base plates, for example—or within concrete members. Nonshrink grout is expected to have minimal shrinkage and maximum strength. The function of masonry grout is different, and there are reasons that it contains so much water: Without plenty of water, masonry grout cannot be fluid; a lot of the water is immediately absorbed by the porous masonry; and enough water must be left to properly hydrate the cement in the grout. If the grout is too stiff because it has too little water, or if regular concrete is used instead of masonry grout, the cement in the grout may not fully hydrate. The result is a weak assembly with some unfilled voids—hardly a desirable outcome.

Depending on the aggregate used, grout can be coarse or fine. Coarse grout typically contains pea-gravel aggregate with a maximum size of $^3/_8$ in, while fine grout cannot include sand particles larger than $^3/_{16}$ in. Coarse grout is used to fill voids wider than 2 in; fine grout is specified for smaller spaces. A typical fine grout conforming to ASTM C476 is made of 1 part cement and 2 $^1/_4$ to 3 parts sand, plus up to $^1/_{10}$ part hydrated lime or lime putty. A typical ASTM C 476 coarse grout contains coarse aggregate equal to 1 to 2 times the volume of cement.

9.2   Evolution of Masonry Design Methods

9.2.1   Tried-and-true rules

For millennia, masonry buildings around the world were constructed using established guidelines taught by master masons to their apprentices. These tried-and-true rules of the past constitute the empirical approach to masonry construction, in contrast to rational design, which relies on engineering calculations. The empirical rules specify the minimum required bearing wall thickness for a given height, assuming that the loads are applied near the centerline of the wall and neglecting any role of steel reinforcement.

The masonry construction of the nineteenth century had its share of innovation—such as Marc Isambard Brunel's use of reinforced brick in the building of the Thames Tunnel (England, 1825)[17]—but without a scientific explanation. Brunel's invention was not properly explained until a century later, when reinforced masonry beams were built and tested by Hugo Filippi in the United States in 1913 and by L. J. Mensch in 1919, and technical research took place in India (published in 1923). Pioneering research in California, starting in 1935, developed and tested several ideas that have since become widely accepted. Among them was using sand-and-cement grout to bond the layers of brick in lieu of header courses, and placing steel reinforcement into the grout space for flexural resistance.[3]

Rule-of-thumb building design was prevalent in the United States until building codes started to emphasize the rational design approach.

9.2.2   First building codes

Prior to 1900, building codes in the United States were developed by each individual city or jurisdiction. As discussed in Chap. 1, the first attempt at establishing a model code was made in 1905, when the National Board of Fire Underwriters published the National Building Code. The main focus of that code, as could be surmised from the name of the publishing organization, was prevention of fire losses.

Contemporary building codes trace their origins to Herbert Hoover's Department of Commerce efforts of the 1920s. The department established a Building Code Committee that developed suggested code formats and published several technical bulletins. In 1924 the committee prepared a report entitled *Minimum Live Loads Allowable for Use in Design of Buildings,* considered by many to be the first model code. This document was later renamed ANSI Standard 58 and then ASCE 7.[18] Masonry construction was specifically addressed in the same year, when the National Bureau of Standards produced its Publication No. BH6, *Recommended Minimum Requirements for Masonry Wall Construction.*[19]

The Pacific Coast Building Officials Conference (now the ICBO) produced the first true comprehensive building code in 1927. This code is now known as the *Uniform Building Code* (UBC). The *Southern* (later *Standard*) *Building Code* first appeared in 1945, and in 1950 the *Basic* (later *National*) *Building Code* was first published. The UBC has been adopted primarily by municipalities in the west and central states, the *Standard Building Code* mainly by those in the southeast, and the *Basic/National Building Code* primarily by those in the northeastern United States.

The first model codes embraced the empirical approach to masonry design. They specified the minimum unsupported height-to-thickness ratios (h/t) for various wall materials (or length-to-thickness ratios for walls spanning horizontally between pilasters or intersecting walls) and established minimum absolute wall thicknesses. For example, the 1983 edition of the UBC[20] required a minimum h/t of 25 for reinforced grouted or hollow-unit masonry, of 20 for solid and grouted masonry, and of 14 for stone masonry.

9.2.3 Specialized masonry codes

The NACU and ACI specifications mentioned in Sec. 9.1.5 could be considered trade documents, but the first nationwide U.S. standard developed for masonry was ASA A41.1, later renamed ANSI A41.1, *American Standard Building Code Requirements for Masonry.*[21] This standard appeared in 1944 and dealt only with empirical design of unreinforced masonry. The code writers were wise enough to preserve the old empirical rules, and some of those rules have survived to this day and are included even in the latest masonry codes. In 1960, ANSI A41.1 was renamed *Building Code Requirements for Reinforced Masonry* and was expanded by adding design procedures for reinforced masonry construction.[22]

According to ANSI A41.1, the unbraced height for nonloadbearing partitions was generally allowed to be twice that of loadbearing walls. For example, the maximum unbraced height for a nominal 6-in-thick tile wall (with an actual thickness of $5^3/_4$ in) was 18 ft if the wall was a nonloadbearing partition, but only 9 ft if it carried loads. The allowable loads depended on a net area of the wall and on a type of mortar. For type M mortar, the maximum working stress was 85 psi; for type S mortar, 75 psi, and for type N, 70 psi. Using these numbers, a 6-in-thick tile wall could carry 5870 lb/lin. ft with type M mortar, 5180 lb/lin. ft with type S mortar, and 4830 lb/lin. ft with type N mortar. An 8-in-thick tile wall (with an actual thickness of $7^3/_4$ in) could carry 7900, 6980, and 6510 lb/lin. ft, respectively.[23]

In 1966, the Brick Institute of America (now Brick Industry Association) published its *Building Code Requirements for Engineered Brick Masonry,* a well-regarded document that was referenced in many building codes. The "little red book," as it was affectionately known, contained both empirical rules and the rational design method for clay brick masonry. BIA's counterpart in the concrete-block industry was the National Concrete Masonry Association (NCMA), which in 1970 published *NCMA Specification for the Design and Construction of Load-bearing Concrete Masonry.* Like the BIA code, this specification contained both empirical rules and formulas for rational design. The BIA and NCMA codes remained authoritative design standards on brick and concrete masonry for two decades.

Meanwhile, the American Concrete Institute (ACI) was developing its own standards for concrete masonry. In 1976, the ACI published *Specification for Concrete Masonry Structures* (ACI 531.1) and in 1979, *Building Code Requirements for Concrete Masonry Structures* (ACI 531). These documents contained procedures for rational design of concrete masonry that were slightly different from NCMA's and did not apply to clay brick.

The simultaneous existence of all these competing ANSI, BIA, NCMA, and ACI masonry standards caused a good deal of confusion. It was clear that a consensus document was needed.

9.2.4 MSJC Code and Specification

A joint committee of ACI, ASCE, and TMS (The Masonry Society) was formed to prepare a consensus code. In 1988, after 10 years of work, the joint committee unveiled ACI 530/ASCE 5/TMS 402, *Building Code Requirements for Masonry Structures,* and ACI 530.1/ASCE 6/TMS 602, *Specification for Masonry Structures.* These standards became known as the Masonry Standards Joint Committee (MSJC) Code and Specification, although many simply call them ACI 530 and ACI 530.1. The former established requirements for rational and empirical design of both brick and concrete block; the latter dealt with construction of masonry. Indeed, one of the distinguishing features of these documents is a clear demarcation between design- and construction-related subjects.

The MSJC code and specification assumed that all masonry work would be inspected, in contrast with some earlier masonry codes (including the UBC), which allowed masonry construction with or without proper inspection. If the work was not inspected, those codes stipulated that the allowable stresses in masonry had to be halved. Another change (questioned by some engineers) was the elimination of a requirement for the minimum area of steel reinforcement, long a widely accepted provision of the NCMA specification.

Among other new provisions were replacement of wall height-to-thickness (h/t) ratios with slenderness ratio (h/r); a formula for combined axial and bending stresses; and equations for shear stresses in collar joints (useful for the design of composite walls). The MSJC code even changed the way the net area of a CMU block was measured. Instead of simply using the block's net cross-sectional area, which was the previous practice, the new method was to use the mortar-bedded area. The bedded area typically excluded the webs and therefore made the calculated compressive stresses higher.[22]

The empirical design provisions of the MSJC code are similar to those found in ANSI A41.1. Why was this approach preserved in the new document? The code writers felt that many masonry structures were small, had plenty of rigidity, and were in no need of sophisticated engineering analysis. But the applicability of the empirical design provisions was quite circumscribed: They could not be used in seismic zones 3 and 4, in areas with basic wind loads exceeding 25 psf, and in shear-wall buildings higher than 35 ft or without a certain minimum number of shear walls. The masonry had to be laid in running bond only.

Rational design of engineered masonry allows the construction of taller and thinner walls that are far outside the limits allowed by empirical design. The slenderness ratios of the so-called tall slender walls exceed those permitted previously by a wide margin. These walls make full use of higher-strength masonry units, increased amounts of vertical reinforcing steel, positioning of rebars in two rows rather than in the middle of the wall, and provision of base fixity at the foundations.[24]

The MSJC code and specification were revised in 1992, 1995, and 1999. The revisions included a new chapter on prestressed masonry, new requirements for adhered masonry veneer, and an expanded section on quality assurance. It is noteworthy that the 1995 and later editions again allowed the height-to-thickness ratio of a cavity wall to be based on the sum of the nominal thicknesses of the two wythes, a provision dating back at least as far as ASA A41.1. (The 1992 edition has attempted to be more restrictive.)

The MSJC documents, which are expected to be revised every three years by the joint committee, have finally become the authoritative sources on all aspects of masonry design and construction. The *International Building Code* refers to their 1999 edition for the design and construction of masonry structures.

9.3 Evaluation of Masonry Structures

9.3.1 Introduction

It should be obvious by now that in renovation work, "existing masonry" is too vague a term to accurately describe the existing construction. At the very least, it is necessary to know what kind of masonry it is (clay-based or concrete), and its composition, condition, and strength. Some of this information can be readily obtained, some requires extensive investigation and testing. A recently built single-wythe CMU wall without finishes, for which the original design and inspection documents are available, might need little investigation other than a visual confirmation. On the other hand, a plastered multiwythe wall of unknown origin located in a building undergoing seismic strengthening might require quite a bit of studying.

The actual condition of masonry will help determine what renovation steps to take—to leave it as is, to reinforce it, or to replace it. Therefore, before we get into the methods of renovation, the available means of masonry evaluation ought to be discussed. Our goal is not to repeat the information on assessment techniques contained in Chap. 2, but to explain where and how to use those techniques and to offer some amplification.

As with any renovation project, the first steps should include the review of available documentation. More often than not, adequate plans and specifications for old masonry buildings are not available. Still, it is well worth spending some time attempting to locate the original construction documents—it can save considerable time in the end. If found, these documents should be reviewed prior to the initial site visit to determine details, load paths, and critical connections that need site investigation or confirmation. It is also wise to determine the original use of the building and any changes in its occupancy that have taken place; speaking with the current and past owners or tenants might be helpful.

9.3.2 Field investigation of masonry: General considerations

As discussed in Sec. 2.2.2 of Chap. 2, field investigation includes a walk-through (or more than one, if necessary), visual evaluation of the structure, and subsequent testing, if that is needed. Visual examination is the simplest and most important way of finding problems, as no other method of assessment can simultaneously scan for multiple types of distress and cover large areas quickly. It need not rely on the naked eye alone: Optical aids, such as binoculars and spotting scopes, can greatly increase the range of the observer.

Before these efforts take place, consideration should be given to some of their practical aspects. Will there be noise, vibration, or dust during the assessment? Must the area of the investigation be cleared out and some occupants temporarily relocated?

One of the objectives of a field investigation is to verify that the masonry was constructed as shown on the existing drawings, if they are available. Another is to visually assess its condition. Are there problems and defects dating from the original construction or caused by aging, water damage, fire, hurricanes, or earthquakes? Are there problems created by improper modifications or additions to the original building? Most masonry troubles are caused by

- Volumetric changes in the masonry that either were never taken into consideration in the original design and construction or were compromised by subsequent repairs

- Differential movement between adjacent masonry materials, or between the masonry and the building structure, that was not properly accommodated

- Water entry, either because of poor design details or because of improper maintenance

The extent and nature of damage can be documented with notes, sketches, and photographs. A walk-through with a video camera is also very useful, especially for mapping the relative orientation of the building elements to one another. This is the time to decide whether subsequent investigations are necessary, to consider which unknown conditions ought to be studied, and to decide which tests are appropriate. This is also the time to determine the methods of access for subsequent work and to verify, for example, that proper anchorage exists for swing staging. In newer construction, it is of interest to know whether the building owner plans a legal action against the original building designers, suppliers, or contractors for the defects observed.

If a legal action is looming, a more detailed and thorough investigation and documentation efforts are generally required.

Issues of building safety should be among the primary goals of the survey. A key to successful investigation is finding whether the observed damage is in critical masonry members (CMU bearing walls, for one) or in noncritical elements (partitions). If the defects are severe enough to threaten the safety of the building occupants, a written note should be promptly sent to the owners, advising them of the observed conditions and the suggested steps to be taken.

9.3.3 Field investigation of masonry: Some specifics

Here are a few specific items related to field investigation of masonry:

Equipment. General equipment for masonry evaluation can include a notepad, camera with flash, video camera, binoculars, flashlight, tape measure, plumb bob, level or laser level, chalk, tool belt, and crack gage. Optional equipment could also include a mechanic's mirror, portable borescope (an instrument to be discussed later), cordless drill, and steel locator.

Access. Access requirements should be determined in advance. Is access from below (ladder, boom lift, scissors lift, or man lift) or from above (boatswain's chair, swing stage) more convenient? How can damage to landscaping or sidewalks be minimized when some of the heavier lifts are used at ground level?

Taking measurements. The building may be measured in order to develop accurate drawings or to verify the dimensions shown in the existing design documents. Measurements may also be needed to locate some important building elements or to map the areas of deterioration. While the measurements are being taken, a check should be made for out-of-plumb or bulging walls, sagging lintels, and similar areas of distortion.

Detecting corrosion activity. Corrosion of steel elements in masonry cannot be determined or quantified by the same methods used for steel in concrete (e.g., half-cell testing), because steel elements in masonry are generally discontinuous (loose lintels, for example) or not positively connected to other steel members (rebars). Therefore, detection of corrosion in masonry is generally done by visual means: observing rust stains, examining embedded materials through a borescope, or making limited wall openings. Deterioration of metal ties and anchors in brick-veneer walls can result in quite noticeable bowing of walls or separation of wythes. In some instances, pullout or other destructive tests may help determine the ultimate capacity of corroded elements.

Looking for signs of water penetration. How well does the building resist water entry? Dampness, previous water damage, water stains, and efflorescence indicate ongoing problems, unless repairs have just been made. Moisture damage in masonry walls can often be traced to such mundane deficiencies as blocked or failed downspouts or torn flashing (see Fig. 13.23 in Chap. 13). Water entering into the wall tends to migrate to the exterior; it can spall a thin outer layer of masonry if the wall is coated with an impermeable paint (Fig. 9.5). The issues of moisture penetration in exterior walls are discussed briefly in a later section and in more detail in Chap. 13.

9.3.4 Evaluating masonry cracks

The appearance of cracks can often reveal the type of structural problem causing them. Most *vertical cracks* occur when moisture and/or thermal movements of the wall are restrained or when incompatible materials are rigidly joined. Incompatible materials, by the way, include brick and CMU. As discussed in Chap. 13, these two masonry materials move differently with changes in moisture and temperature. For this reason, vertical cracks are common in the accent courses of CMU placed in the middle of brick-veneer walls.

Figure 9.5 This crumbling brick is a result of water vapor buildup behind the impervious paint.

Some vertical cracks in brick-veneer walls start at the upper corners of large openings; these may be caused by expansion and contraction of steel lintels. Vertical cracks in brick can occur when long masonry panels above and below the openings, unbroken by expansion joints, are restrained by short masonry piers or wall corners (Fig. 9.6).

Vertical cracks in walls of concrete masonry units are typically caused by expansion and contraction coupled with inadequate horizontal reinforcement and lack of control joints (see Fig. 13.4 in Chap. 13). Vertical cracks at the corners of walls retaining soil are usually the result of structural failure (Fig. 9.7). Similarly, large vertical cracks in a masonry lintel, wider at the bottom and closing at the top, generally indicate insufficient flexural capacity; hairline cracks at this location are normal.

Diagonal cracks in masonry near the bottom of the walls generally indicate differential vertical movement caused by settlement or other loss of support (see Fig. 2.21 in Chap. 2). Diagonal masonry cracks at the top can be caused by restricted thermal movement or can have other origins, such as poorly installed lintels or elastic distortion of spandrel beams and exterior columns.

Horizontal cracks are typically the result of differential movement between the wall and the frame, in-plane or out-of-plane expansion or shrinkage of the building frame, or flexural overload of the wall caused by lateral loading (Fig. 9.8). Some examples include shrinkage of concrete slabs or timber beams—both of which tend to pull the attached

Figure 9.6 Vertical cracks in brick can occur when long masonry expanses above and below the openings are restrained by wall corners.

Figure 9.7 Vertical crack in a stone retaining wall caused by structural failure.

Figure 9.8 Horizontal and diagonal cracks in this brick-veneer wall were attributed to differential movements between the wall and the building structure.

walls inward—and expansion of steel beams or bar joists, which tends to push the walls out. Another source of such cracks is a horizontal thrust exerted by trusses or arches supported by the masonry. Also common are horizontal cracks at the bottom of brick parapets; these result from temperature expansion and contraction of the parapets relative to the walls below (see the discussion in Chap. 13). Still another possibility is rusting of continuous steel lintels or other embedded steel members.

9.3.5   Nondestructive tests of masonry

When visual inspection identifies some areas of distress, further information can be obtained by nondestructive tests in the field or by destructive testing in a laboratory. Nondestructive tests of masonry are similar to those used for concrete (Chap. 2), with some peculiarities that are briefly described here.

Surface hardness.   Surface hardness is an indicator of material uniformity. Observed variations in hardness can be used to find the boundaries between deteriorated and sound masonry. The Schmidt hammer, typically used for concrete, is a relatively fast and inexpensive method for evaluating the hardness of masonry as well.

Stress waves.   The travel time of a stress wave is a good predictor of masonry density and quality of construction. Stress waves can be introduced by several methods.

Ultrasonic testers are used to accurately measure the transit time of an ultrasonic pulse through a masonry wall or a pier. The path length divided by the transit time yields the velocity of the wave. Ultrasonic transducers for these units are generally available in the 40- to 60-kHz range, and the masonry thicknesses (the path lengths) that can be measured are limited to 1 to 3 ft. Typical velocities in older masonry made with soft fired brick and lime mortar may be about 3000 ft/s or less, while more modern masonry construction may have ultrasonic velocities of 10,000 ft/s.

Mechanical pulse velocity equipment includes a hammer fitted with instrumentation and a high-speed data acquisition system, both containing accelerometers. The impact of the hammer activates the instruments and prompts the data acquisition system to begin recording the data provided by its accelerometer. The arrival of the stress wave can be determined from the trace recorded by the data acquisition device. The advantage of the mechanical pulse method is that hammers of various sizes and with interchangeable impactors can be used; some of those can induce stress waves that travel through 6 to 12 ft of masonry—a much longer distance than that possible with ultrasonic testers.

Figure 9.9 shows mechanical pulse velocity measurements on a thick masonry pier. One person is repeatedly striking the pier with a hammer connected to the portable computer, and another holds a receiver recording the resulting stress waves. Because the equipment is portable and easy to use, the area can be thoroughly checked and a variety of possible wave paths investigated. The results are used to create a velocity map of the tested member, in which internal anomalies can be detected.

The borescope. The borescope is a fiber-optic device that can be inserted into small holes drilled in mortar joints to investigate internal wall details, verify the presence of voids, observe wall ties, etc. It is also valuable for inspecting embedded flashing or finding the path of moisture entry into the wall. The borescope typically contains a source of light for illuminating the interior of the wall. The image can be viewed in the borescope eyepiece or on a small monitor; photographs and video images can also be made through the device. In Fig. 9.10, the investigator examines the condition of a collar joint in stone masonry using this equipment.

Thermal imaging. Thermal imaging using commercial infrared cameras can reveal information about the interior condition of masonry walls. It has had considerable success in detecting the presence or

Figure 9.9 Taking mechanical pulse velocity measurements on a thick masonry pier.

Figure 9.10 Collar-joint investigation using a fiber-optic borescope.

absence of grout in reinforced-concrete masonry walls and other masonry elements. Thermal imaging finds changes in composition or irregularities of construction by detecting the temperature differentials they create. For example, a temperature gradient across the wall exists when the heated structure is observed on a cold winter night: The temperature of the grouted cells on the exterior wall surface is slightly higher than that of the ungrouted cells, because grouted cells conduct heat faster. Some of the better infrared cameras have a resolution as small as 0.1°C, quite sufficient for precise observation of minor thermal variations. The main advantage of thermal imaging is that it allows whole buildings to be investigated, whereas other techniques check masonry only at a few isolated locations.

Magnetic methods of locating rebars. The magnetic methods of investigation utilize pachometers, cover meters, and rebar locators. A pachometer includes a horseshoe-shaped pad that is pressed against the masonry surface and generates a magnetic field. Since the field is affected by the presence of reinforcing steel, bar locations can be identified by the resulting anomalies. Some inexpensive pachometers have recently been selling for less than $100. Cover meters and rebar locators operate on the same principle; they are further described in Chap.

2, Sec. 2.4.1. Figure 9.11 illustrates a pachometer being used to scan a granite-veneer wall for the presence of iron ties. We should note that the magnetic methods are inexpensive to perform, but are not nessarily very reliable because they will show any embedded steel, including nonstructural bars.

Petrographic examination. The petrographer examines masonry under a microscope to establish its chemical composition and obtain other information—evidence of alkali-silica reactions or microcracking, for example. Petrographic examination is conducted in accordance with the standard procedures contained in ASTM C856. This in-depth masonry investigation tends to provide definitive information but is costly and requires experienced personnel.

Water penetration tests. In addition to the tests discussed in Chap. 13, one more test is worth mentioning here, because it deals specifically with measuring water leakage through masonry. ASTM E514, *Standard Test Method for Water Penetration and Leakage through Masonry,* is intended to quantify the water permeability of masonry walls. The test was originally developed as a laboratory method, but it has been successfully used in the field to evaluate the effectiveness of various masonry materials, designs, and construction techniques in resisting water penetration.

The ASTM E514 test simulates the effects of wind-driven rain. The test setup includes a chamber attached to the wall surface,

Figure 9.11 Using a pachometer to scan a granite-veneer wall for the presence of iron ties.

through which water is circulated from a separate reservoir at a rate of 3.4 gal/ft^2 of the wall area. The chamber is pressurized to 10 psf. The amount of moisture lost into or through the wall is measured by periodic weighing of the reservoir or by collection of the water that penetrates the wall by a flashing-and-trough system. Figure 9.12 gives an overall view of the testing procedure conducted on a granite-clad building and a close-up of the test fixture. The test is usually conducted for 4 h, a long enough period of time to stabilize the rate of flow.

The permeability of masonry depends on all the elements of the assembly, but especially on the condition of the joints. The rate of water penetration is also affected by the bond between masonry units

(a)

Figure 9.12 Water-penetration testing of a granite-clad building before and after low-pressure grouting. (*a*) The overall view.

(b)

Figure 9.12   (*b*) A close-up of the test fixture.

and mortar, joint tooling, fullness of mortar joints, and other aspects of workmanship.

9.3.6   In situ material tests of masonry

In situ (in-place) tests of masonry walls include those for shear, flexural bond, stress, and material deformability. These tests can establish the critical material properties while the wall remains in place and virtually undisturbed—a major advantage for renovation of occupied buildings.

The *in situ shear test* is valuable for finding the shear capacities of existing unreinforced masonry walls. The test is conducted by removing the head joints surrounding the tested masonry unit and applying horizontal force to the unit until it slides. The test can use hydraulic rams, in which case the unit adjacent to the test unit is removed to provide space for the ram. (Small flatjacks can also be used for this test; these are described later.) The in-place shear test procedure can be found in the UBC under "Uniform Building Code Standard 21-6, In-Place Masonry Shear Tests." Its use is illustrated in Case 2 of Chap. 12.

The *in situ flexural bond test* measures the flexural bond strength at the interface of masonry units and mortar. The test requires removing several units above the tested one and drilling out the head joints on either side of it. The testing device consists of a channel clamped over the tested unit, to which a torque wrench is attached. A torque (moment) is then applied to the unit in the direction normal to the wall until the bond in the bed joint is broken (Fig. 9.13). The value of the ultimate flexural bond stress is found by dividing the applied bending moment at failure by the section modulus of the tested section. If the

Figure 9.13 In situ flexural bond test using a bond wrench.

break occurs in the masonry rather than in the mortar, the test will have measured the flexural tensile strength of the masonry.

The *in situ stress test* helps determine the existing compressive stresses in the wall, an important piece of information for those considering adding load to it. The stress test is conducted in accordance with ASTM C1196, *Standard Test Method for In Situ Compressive Stress within Solid Unit Masonry Estimated Using Flatjack Measurements.* The test is based on the principle of stress relief. First, gage points are attached above and below the tested bed joint and accurate measurements of the gage-point locations are taken. Then the bed joint is removed. Because of the existing compressive stresses in the wall, the distance between the gage points will shorten, and this movement is measured in the test.

The existing compressive stress is measured by inserting a *flatjack* into the bed joint and pressurizing the jack until the original gage-point locations are restored. The flatjack is a small, round, flat envelope made of ductile metal, with a dumbbell cross section. When pumped with a hydraulic fluid, a flatjack can exert great forces, while expanding only up to 1 in in thickness.[25] The pressure in the flatjack, modified by two correction factors—one based on the ratio of the slot size to the jack area and the other based on the losses in the jack—should approximately equal the existing compressive stress in the wall.

The in situ stress test can also be used to indicate the presence of flexural stresses in masonry walls and piers by conducting the test on both sides of the member. If flexural stresses exist, the stress measurements taken on opposite sides will be unequal. The accuracy of the test has been repeatedly checked on piers placed in a compressive-test machine, where compressive stresses were controlled. The results indicate that in situ stress tests yield compressive stress values within 15 to 20 percent of the actual.

The in situ stress test can be used in combination with other helpful tests, such as *in situ determination of masonry compressive strength.* The test setup is similar to that for the stress test, but with two flatjacks inserted in horizontal slots located above each other, four to six courses apart. The jacks are gradually stressed, placing the masonry between them in compression, until the material reaches its peak compressive stress. How can one identify that point? For relatively weak masonry, by observing the stress-strain curve plotted as the test progresses—the point of maximum stress is where the slope flattens out. For stronger masonry, there is a potential problem: Typical flatjacks may not be strong enough to reach the ultimate strength of the material.

The *in situ masonry deformability test,* also known as *in situ determination of the elastic modulus of masonry,* measures the stress-strain behavior of masonry. ASTM C1197, *Standard Test Method for In Situ*

Measurement of Masonry Deformability Properties Using the Flatjack Method, describes the test procedure. This test can be conducted after a stress test or after a masonry compressive strength test, using the previously made slots. Alternatively, two new slots located one over the other and separated by at least five courses of masonry can be made for the test (Fig. 9.14). The masonry between the two slots is compressed by flatjacks inserted there, with the surrounding masonry acting as a reaction frame. The jacks are loaded to less than half the masonry strength, to preserve the elastic behavior of the material. The amount of masonry compression between the slots is measured with gage points or electrical distance-measuring devices, and the strain is found by dividing this amount by the spacing of the gages. The applied vertical compressive stress is found from the measured hydraulic pressure, modified by correction factors. Dividing this stress by strain yields the elastic modulus of the masonry.

9.3.7 Destructive tests of masonry

If nondestructive and in situ tests do not provide the desired information, or if the equipment for making them is not available, a few relatively simple and definitive destructive tests can answer questions about the structural condition of masonry. The tests involve taking

Figure 9.14 In situ masonry deformability test using flatjacks.

rather large masonry samples and testing them in a laboratory. Generally, the samples are taken from both intact and deteriorated areas to arrive at the upper and lower bounds for the material properties. If the building has been remodeled or expanded over time, it is helpful to sample the materials from each construction.

Destructive tests can determine the structural properties of masonry, such as compressive strength, modulus of elasticity, and modulus of rupture, as well as nonstructural properties, such as freeze-and-thaw resistance and rate of moisture absorption. Brief descriptions of these tests are provided in Chap. 2, Sec. 2.8.4, and are not repeated here.

In addition to testing samples, full-scale load tests can be performed in place to measure the response of masonry walls or other elements to external loads. Load testing is usually done to determine the allowable compressive or lateral load on the masonry wall, and it can also help in establishing the allowable diagonal tension and flexural bond stresses in masonry. The procedures for making load tests can be found in local building codes and industry standards. For example, the *Uniform Building Code* requires that the tested member be subjected to twice the design live load plus one-half the design dead load for a period of 24 h. Load tests are expensive to perform and run the risk of permanent damage to the structure, but in some cases there is no other way to obtain the desired information.

9.3.8 Using reference data in lieu of testing

What if only a low-budget feasibility study is needed, and there is neither time nor money to conduct the tests just discussed? If a relatively small building is involved, some idea of its structural characteristics can be gleaned from assuming the "worst-case scenario," i.e., masonry strength at the lower end of the probable spectrum. FEMA-274[3] suggests that

> [d]efault values of compressive strength are set at very low stresses to reflect an absolute lower bound. Masonry in poor condition is given a strength equal to one-third that for masonry in good condition, to reflect the influence of mortar deterioration and unit cracking on compressive strength.

If this approach yields overly conservative or unacceptable results (the structure is found to possess little strength), testing is unavoidable. Indeed, much higher compressive-strength values could probably be justified if at least *some* testing is done.

The easiest form of testing would be to remove some bricks or blocks and test them in compression, but the information thus provided relates to the compressive strength of the *units,* not that of the whole assembly. The MSJC code contains tables correlating the compressive strength of the units with the compressive strength of masonry, f'_m, but it is written for new construction. Is it possible to determine the compressive strength of existing masonry by using these tables? Perhaps not; FEMA-274 cautions that the MSJC code tables are based on data from masonry constructed after the 1950s, so that the tables are not applicable to buildings constructed prior to 1960. As already noted, the early mortars were lime-based and weaker than the cement-based mortars assumed in the tables. Even if design designation of the mortar was available, direct comparison with today's codes would be problematic, because earlier mortar classifications differed from today's.

What, then, should one do? FEMA-274 suggests using an approximate method for using MSJC tables:

> Expected masonry strength should be determined by multiplying Table 1 values [compressive strength of clay masonry] by a factor of 2.0 or Table 2 values [compressive strength of CMU] by a factor of 1.5. These approximate factors are based on estimated ratios between expected and lower bound compressive strengths, as well as on correction factors for clay brick and concrete block prisms.

Further, the default value of the elastic modulus should be based on the expected masonry compressive strength multiplied by a factor of 550. What are the default values for tensile strength? "Low" even for masonry in good condition, and zero for masonry in poor condition. For shear strength, the default factors range from 27 psi for masonry in good condition to 13 psi for masonry in poor condition.

9.3.9 Arriving at a diagnosis

The goal of field investigation is to find the root cause of an observed problem and verify the proposed diagnosis with static analysis, instrumentation (crack monitors, strain gages, tilt meters, etc.), destructive or nondestructive testing, or load tests. A related structural analysis should use the actual material sample data, if possible. Because of limited available information and material data, the analysis will generally be restricted to simple methods.

In many cases, the probable cause of visible masonry distress will be obvious; in others, it can be determined only by systematic investigation. The experienced investigator does not make a snap diagnosis and then force the observations to comply. On the contrary, he or she takes care to verify all assumptions and to prove that the final diagnosis is the only one that is consistent with all the observations and measurements.

9.4 Masonry Repair

9.4.1 Introduction

Once the problem has been diagnosed, repairs or strengthening can be undertaken. In this section, we examine the most common problems of masonry structures and methods for their repair. The successful repair solutions will be consistent with the history of the building and its future use. For example, designers converting nineteenth-century warehouses to urban condominiums might favor strengthening masonry with exposed steel frames and treating the frames as interior design elements. In historic buildings, less visible structural repairs are more appropriate.

Solving one problem should not create another. One way of ensuring that strengthening one building component does not cause harm to others is to monitor the building for signs of distress during repair work. The actual conditions encountered during construction will probably be different from those assumed in the design, and the frequent presence of the engineer on site to resolve these construction issues will go a long way toward assuring a successful outcome. Also, some quality control procedures for the repair work should be established. These procedures might include conducting pullout tests for the installed anchors, material testing of grout, or ultrasonic pulse velocity testing for verification of grout placement. All atypical conditions should be documented at the completion of rehabilitation work in as-built drawings.

The following masonry problems are common:

1. Cracking

2. Deterioration

3. Loss of connection between the wythes

4. Insufficient strength of structural elements

Each of these is examined in turn. (Evaluation of various masonry cracks is covered in Sec. 9.3.4.) In general, the repair should attempt to fix the cause of the distress before fixing the symptoms and may require modifying components other than masonry.

9.4.2 Repairing vertical cracks in walls

Vertical cracks are often active, and fixing them begins with elimination of the underlying problems. If the cracks are caused by moisture or thermal movement coupled with a lack of expansion or control joints (see the discussion in Chap. 13 about the difference between the two), the repair should include cutting these joints into the walls or introducing other means of reducing wall movements.

Repairing a wall with vertical cracks involves removing cracked bricks, blocks, or stones, inserting new units of matching color and texture, and repointing or replacing cracked mortar. Very narrow cracks can be filled with crack-repair materials (epoxy, for example). The repair procedure itself is rather straightforward—the main challenge is obtaining the matching bricks and mortar.

When investigation finds that vertical cracks at the corners of large openings in brick-veneer walls are caused by expansion and contraction of steel lintels, these cracks can be repaired in a similar way—by accommodating the lintel movement and fixing the wall. However, when these cracks are the result of rusting steel lintels, the lintels must be isolated, cleaned, and given a corrosion-protective coating (or replaced if corrosion has gone too far) before the brick is repaired. This difficult operation involves removing some of the brick on top of the lintels and perhaps shoring the brick above, if the openings are large.

Vertical (and some diagonal) cracks can be stabilized by inserting small-diameter reinforcing bars in the bed joints intersecting the cracks. In this procedure, the mortar joints on either side of the crack are first raked out for a distance sufficient to adequately develop the reinforcing bars. The bars are then inserted either in every course or farther apart, as determined by engineering analysis. Finally, the bed joints containing the steel bars are refilled, to waterproof the wall and to confine the steel. The joints are generally filled with the same mortar as all other joints, but if a higher strength is needed, the joints can be repointed with mortar only at the surface, with masonry grout injected behind the mortar to envelop the steel.

9.4.3 Repairing vertical cracks in retaining walls and piers

Vertical cracks at the corners of retaining walls (Fig. 9.7) are usually caused by insufficient structural capacity of masonry. Such damage to retaining walls cannot simply be patched, because the forces that caused it are still at work and are likely to break the repair. A retaining wall cracks when the buildup of lateral pressure behind it exceeds its structural capacity, usually because the original wall design was inadequate. It is common to see retaining stone walls that are poorly bonded and have no footings; a lack of weep holes can lead to a buildup of water pressure behind the wall, adding to the lateral soil forces.

To be effective, the repair should include excavation of at least some backfill behind the wall and replacement of this backfill with a free-draining material. The wall itself should be removed and replaced, or at least extensively modified by drilling weep holes and filling all internal voids. To prevent the new weep holes from being clogged by silt and soil fines, they require protection at the back of the wall; this was traditionally done with a sack of crushed stone, and now is often done with filter fabric. Only after these steps are taken can the cracks be repaired by filling them with fluid nonshrink grout, epoxy, or other suitable repair material.

A large vertical crack in a masonry pier (Fig. 9.15) can occur when the supported structural member bears on only a small part of the pier and induces excessive local stresses in the masonry. Again, the underlying condition must be remedied before the crack is repaired, either by extending the supported member so that it bears on the middle part of the pier, or by extending the pier—and its footing—to achieve the

Figure 9.15 Vertical crack in a loadbearing masonry pier.

same result. Vertical pier cracking and splitting can also result from the lateral pull exerted by the supported members. Thus, an unreinforced masonry pier can be literally torn apart if a beam connected to one side of it pulls it horizontally, while the rest of the masonry is restrained by the beams framing from other directions.

Once the underlying cause of splitting has been eliminated, the pier can be fixed, as shown in Fig. 9.16 for a basement pier. Here, the topmost part of the pier, containing loose and deteriorated brick, is rebuilt, while the vertical crack is "stitched" by U-brackets and filled with fluid nonshrink grout. Instead of U-brackets, exterior perimeter hoops or FRP wrapping (see Chap. 4) could be used.

Masonry piers and buttresses can also be stitched by drilling horizontal holes through them and inserting steel rods into the grouted holes.

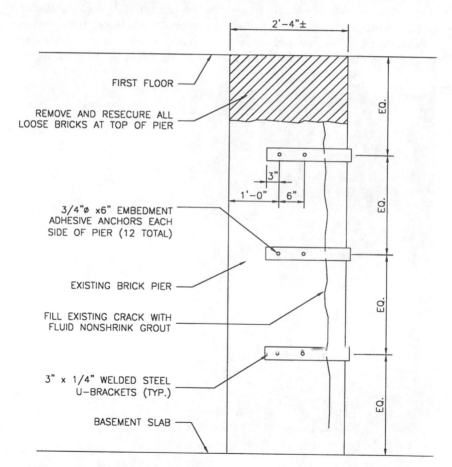

Figure 9.16 Repair of vertical crack in a masonry pier. (*Maguire Group Inc.*)

The rods may be attached to anchor plates at the ends and post-tensioned, if necessary. In exterior applications, it is wise to use stainless-steel rods and anchors for protection against corrosion. A variation of this technique is to remove mortar at every third or fourth horizontal joint and insert stainless-steel strands in the repointing mortar.[25]

9.4.4 Repairing diagonal and horizontal wall cracks

As already mentioned, diagonal masonry cracks near the bottom of the wall are typically the result of settlement. At the top, such cracks could be caused by restricted thermal movement or by other factors. When settlement is suspected, a simple test described in the next subsection can help determine whether settlement is ongoing or has already stopped. If settlement has stabilized, as is often the case, diagonal masonry cracks can be fixed in the same way as vertical cracks.

Horizontal cracks are likely to be the result of a flexural overload of the wall or shrinkage of the structural frame. Their repair is difficult; it might involve cutting continuous horizontal soft joints into the walls at regular intervals (see a case study in Chap. 13) or even a complete stripping and rebuilding of the façade.

9.4.5 Which masonry cracks should be repaired?

As discussed in Chap. 5, there is no definitive rule that states which cracks must be repaired, and hairline cracks in concrete are often left alone. The situation in masonry is somewhat different: Very narrow cracks, which might be autogenously repaired in concrete, will not heal by themselves if they occur in masonry units. However, hairline cracks in lime or cement-lime mortar will usually heal. The reason? Rainwater dissolves some hydrated lime from the mortar, and the calcium hydroxide in it reacts with carbon dioxide in the air, producing calcium carbonate—the main ingredient of limestone. The absence of cracks in many historic masonry buildings is attributed to the large amounts of lime in their mortars.[26]

Cracks as small as 0.005 in in masonry units can allow significant amounts of moisture into the wall; thus, they are harmful and are often repaired. It is important to determine whether the crack is dormant or active (again, see Chap. 5 for a discussion of this). A simple method of monitoring cracks is to plaster over the crack and then write the date on the patch with a marker. If the crack is dormant, the plaster will remain uncracked for a long period. Commercial gages are available for visual monitoring of cracks, and more sophisticated vibrating-wire strain gages can be used for continuous monitoring.

9.4.6 Repairing deteriorated masonry units

Like all materials, masonry deteriorates with age and requires maintenance or repair. The most common types of masonry deterioration are

1. Crumbling, flaking, or cracked masonry units

2. Crumbling and eroded lime-based soft mortar in exterior brick walls

Masonry units can deteriorate as a result of water penetration and freeze-thaw cycles. Despite the excellent overall durability of masonry, this does occasionally occur, as illustrations in Chap. 13 demonstrate. Like any other material, masonry can sustain physical damage caused by impact. Masonry can also be damaged by fire: Marble and limestone become calcined, and pieces of granite break off.[4] There are a number of options for restoring the function and appearance of damaged masonry, among them replacement, partial replacement, removal and reorienting, and patching.

The most direct method of dealing with masonry deterioration is to replace all the damaged units. The feasibility of replacing brick, block, or stone obviously depends on the current availability of the original materials. Most large cities have building component recyclers, who stock materials from demolished buildings and often have units that can be used for replacement. The replacement units should match the original not only in appearance, but also in material properties. For brick, these properties include compressive strength, elastic modulus, and initial rate of absorption (IRA). Still, even if units from the original source are located—the most the restorers can hope for—they will not be weathered to the same degree as the units in the wall and will never fully match them. Attempts to artificially make the replacement units look weathered may backfire in the long run.[4] The new masonry should be "toothed" into the existing as seamlessly as possible.

In stonework, partial replacement is common. Here, the deteriorated part is removed, and the replacement stone (the "dutchman") is placed to fill the resulting void. Another approach is to remove the stone with surface deterioration, retool the surface, and reset the stone. Or, in a removal and reorienting technique, the stone is removed and simply turned 180°, so that another face of the stone (hopefully, one that is undamaged) is exposed.

For unique historic buildings, replacement bricks or blocks can be extremely difficult to obtain. When neither partial replacement nor removal and reorienting is practical, patching of the damaged units can be considered. The procedure for using patching mortars to repair badly weathered brownstone façades is described in Chap. 13.

One procedure that *should not* be used in masonry restoration, especially for brick, is sandblasting. Sandblasting removes the hard outer layer of the brick, exposing the softer inner material and accelerating its degradation. Sandblasting can also erode softer stones and speed up their weathering.

9.4.7 Repairing deteriorated mortar

Deteriorated mortar is usually repaired by repointing. Repointing requires removal of the mortar joints to a depth below the face of the wall of at least two times the joint thickness or a minimum of $3/4$ in. Removal to a larger depth may be needed if the mortar condition is extremely poor. The correct method of mortar excavation is to expose the bedding surfaces of the units, so that the area to be repointed is rectangular in cross section. After the damaged mortar is removed, the joint surfaces are cleaned by compressed air, brush, or stream of water, and the new mortar is placed. The new mortar should match the original with respect to hardness, color, and texture, and should be as dry as is practical.

Joints are usually filled in three passes of equal depth. So, for the common joint thickness of $3/8$ in, the mortar is removed to a depth of $3/4$ in and repointed in three separate $1/4$-in-deep passes. Each mortar layer is installed only after the previously placed mortar has set.

The new mortar joints are densified and sealed by tooling—compressing their surfaces to a configuration matching the original joint profile. Head joints are tooled first. The joints are then cured, preferably for 48 h, because proper curing is important for reducing water loss in the mortar. Curing methods include covering the joints with plastic sheeting or with periodically moistened burlap.

The new mortar must match the existing mortar in strength and appearance. The strength of the existing mortar should be determined by testing, not guessing. In historically significant structures, the existing mortar is often analyzed by petrographers. These trained specialists can develop a mortar mix that closely matches the original and provides the required durability. Using a strong repointing material in combination with the soft and crumbly mortar in the rest of the wall is counterproductive. The new mortar will become a hard point, a wedge on which the whole weight of the wall may end up being concentrated. If that happens, the brick underneath the repointed area may become locally overstressed in bearing—and spall. A repointing mortar for historic buildings with soft brick may contain some amount of cement for durability, but not more than one part cement to two or three parts lime; even the softest standard cement mortar (type N) is too hard to be used here. However, cement-based mortars may be used for repointing relatively new buildings.

Repointing can be used not only for walls with cracked and deteriorated mortar and for those with significant settlement cracks, but also for walls that were previously repointed with excessively hard mortar. In addition to strengthening the masonry wall, repointing restores its water-shedding properties and improves its appearance. Repointing is best done by hand, especially in historic buildings. The operation is further described in Chap. 13.

How should repointing of wide joints in ashlar stone masonry be handled? Krogstad[27] recommends following essentially the same procedure just outlined and limiting the thickness of the mortar layers to $1/2$ in. He suggests using sand that is more coarse than usual, but still well graded and within the limits of ASTM C144, *Standard Specification for Aggregate for Masonry Mortar*. To reduce shrinkage, he recommends prehydrating the mortar and using the softest mortar allowed for the application.

9.4.8 Providing connection between the wythes

Until brick veneer was invented, masonry walls contained several layers (wythes) interconnected by header courses or, less frequently, by metal ties. For brick walls, it was quite common to use the best bricks in the exterior layers and to fill the space between them with brick of inferior quality—or sometimes even with loose rubble barely spiked with mortar. Occasionally, terra cotta tiles were used on the interior surfaces and in the middle, with brick covering only the exterior. Some of this construction is illustrated in older books dealing with brick design (Plummer's *Brick and Tile Engineering,*[10] for example).

The separate masonry layers in these composite walls were often interconnected by header courses—bricks turned with their long dimension perpendicular to the wall. The header courses could be installed at various intervals; in the *common bond,* they were placed at every sixth course. This sensible lay-up pattern was temporarily abandoned in the middle of the nineteenth century, when the *running bond* became fashionable. The header courses were out, and the exterior bricks were usually connected to the rest of the wall by such pathetic means as cutting off their interior corners and placing diagonal bonding courses within the notches. Even this so-called bonding was done infrequently, in some cases only in the middle of the wall. The results were predictable. Figure 9.17 shows a failure in a Central City, Colorado, building. Here, the outer wythe has partly collapsed,

Figure 9.17 Partial collapse of the outer wythe of this brick wall reveals the only connection it had to the interior—a course of diagonally placed bricks in the middle. Note the scarcity of mortar in the wall.

revealing the only connection it had to the interior—a row of diagonally placed bricks in the middle of the wall.

Sometimes, the masonry wythes were connected by bent metal wire ties placed in the joints. Unfortunately, these ties were usually spaced too far apart to meet the requirements of contemporary codes. Also, the ties typically had no special coatings for corrosion protection, and many have since rusted to a degree that makes them ineffective.

But even worse was construction without *any* brick bonding at all, where the separate wythes were held in place only by mortar placed in the collar joints at their interface. As aging and water damage took their toll on the mortar, the wall layers could start to separate.

Similarly, multilayer walls connected by metal ties could be in danger of separation if the original ties have corroded, especially if too few of them were installed to begin with.

The likely result of poor interwythe bonding is that the façade moves away from the structural backup and becomes unstable. This interlayer separation is not easy to detect until it progresses to such an advanced stage that facade bowing and cracking at the sides of the windows are easily noticeable. Tragically, the problem may become obvious only after an earthquake or a strong hurricane literally peels the exterior wythe away from the rest of the building, as happened to the building in Fig. 9.17.

Another problem stemming from the loss of connection between the wythes occurs when the floor and roof framing is bearing only on the interior wythe, which becomes overstressed and eventually separates from the other masonry layers.

Masonry layers that are separated or are in danger of separating can be interconnected by retrofit ties designed to anchor them to the rest of the wall. These ties, produced by several manufacturers, rely on epoxy, friction, grout, or expansion to develop an adequate anchorage. Some of the ties are shown in Chap. 13. The challenges of reattaching wythes of different materials are further discussed and illustrated in Case Studies 1 and 2 of Chap. 12.

9.5 Strengthening Masonry Structural Elements

9.5.1 Introduction

The monumental masonry buildings of old were generally well built and rarely require any strengthening. Some other masonry structures are not so fortunate. Numerous commercial and industrial buildings were constructed of unreinforced CMU walls 18 to 20 ft high, with or without brick veneer. Any rational analysis of these walls would show them to be incapable of resisting the wind and seismic loading stipulated in today's codes. (The contentious topic of whether these buildings must be upgraded at all is discussed in Chaps. 1 and 11.) Other masonry members may require strengthening because the loading on them has changed or because of modifications to them that were previously made or are being proposed. Several methods of strengthening masonry are outlined here.

9.5.2 Adding steel reinforcement to masonry walls

The flexural and shear capacities of the just-mentioned 20-ft-tall CMU walls would probably be adequate if the walls contained enough reinforcing bars. Is it possible to add bars into existing walls? Yes, it is possible, and it has been done. There are two methods of placing the bars inside the wall: by inserting them from the side and by center coring. There are also methods involving placing the bonded reinforcement outside the wall.

Inserting rebars from the side. One way of gaining access to the inside of a CMU wall is by cutting vertical slots in the shell at the desired intervals (4 ft on centers, for example) and placing and grouting reinforcing bars into these slots. The operation is usually done from the interior to avoid interference with exterior finishes, a particularly important consideration for CMU walls with brick veneer. The obvious problem with this approach is that the interior finishes must be removed. It is also possible to cut the slots from the outside. Doing so reduces disruption to the building operations and to interior finishes, which may be more important to the owner than preservation of the exterior veneer.

Removing parts of the wall weakens the masonry, because the added grout in the cells does not participate in resisting the loads that are already present. Therefore, the wall should be relieved of the existing loading to the greatest extent possible before this strengthening takes place. The process of cutting the slots from the inside is illustrated in Case 1 of Chap. 12.

Center coring. In this repair method, vertical cores are drilled from the top of the wall and reinforcing bars are grouted into them. As in the previous method, the spacing of the cores is determined by analysis, and is typically from 4 to 5 ft on center. The 4-in-diameter cores could be grouted with regular nonshrink grout. A polyester-sand mixture can also be used, as suggested by FEMA-172,[28] because of its excellent workability and low shrinkage. The method is shown in Fig. 9.18. The center coring technique, although new, has already been successfully used on several California seismic-strengthening projects.

Post-tensioned masonry. Post-tensioned masonry can be used in combination with the center-coring technique. Instead of regular reinforcing bars, greased and coated unbonded monostrands, enclosed in plastic or steel ducts, are used. The tendons are placed in drilled vertical holes and are anchored at the bottom with special self-activating dead-end anchors. At the top, the anchorage is made by the stressing anchors bearing on grouted masonry or on prefabricated concrete pads.

Some case studies of post-tensioned masonry, including seismic strengthening of a 104-year-old church in Santa Cruz, California, are described by Ganz and Shaw.[29] Abrams[30] reports the results of load tests performed on post-tensioned masonry assemblies. He observes that,

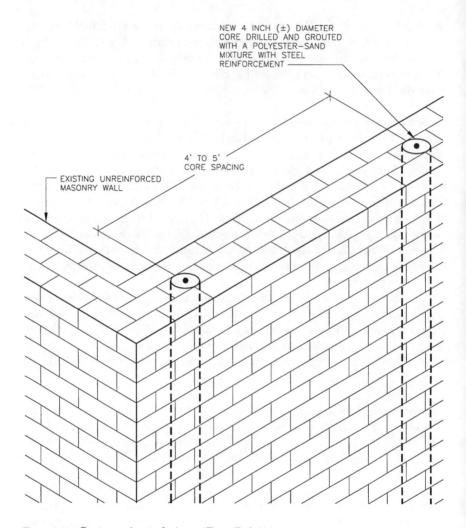

NEW 4 INCH (±) DIAMETER
CORE DRILLED AND GROUTED
WITH A POLYESTER–SAND
MIXTURE WITH STEEL
REINFORCEMENT

4' TO 5'
CORE SPACING

EXISTING UNREINFORCED
MASONRY WALL

Figure 9.18 Center coring technique. (*From Ref. 28.*)

because prestressing forces counteracted all tension from lateral loading, the system remained elastic, and its stiffness and lateral deflections
could be estimated using conventional computation methods

With post-tensioned masonry, there is no need to grout the cored
openings, and they can be as small as 2 in in diameter, although larger openings are still needed for the anchors at the top and bottom of
the wall. The number of drilled holes may be less than that needed
with conventional center-coring technique. An incidental benefit of not
grouting the vertical cores is that they can be filled with insulation.
Post-tensioned masonry systems originated in Europe and are now

offered in the United States by a leading manufacturer of masonry reinforcement and accessories.[31]

Post-tensioned masonry is especially attractive for renovation or addition of free-standing interior partitions. Other promising applications include strengthening of retaining walls. A design table for post-tensioned walls that are unbraced at the top, along with explanations of the methodology, can be found in the booklet *Engineered Masonry*, published by the Masonry Advisory Council.[32] Additional design information about post-tensioned masonry is included in the 1999 edition of MSJC and in NCMA publication TEK 14-20, *Post-Tensioned Concrete Masonry*.[33]

Other reinforcing methods. In some cases, adding rebars inside the wall is insufficient to improve its out-of-plane flexural and shear capacities. For this situation, there is another method: Applying a layer of shotcrete to one or both faces of an unreinforced masonry wall. If the shotcrete contains reinforcing bars, the final result approximates a regular reinforced-concrete wall with bars at both faces. The main challenge in this operation is to assure that the shotcrete bonds to the existing wall. To that effect, the shotcrete can be keyed into the wall at regular intervals, typically from 6 to 8 ft.

For badly deteriorated masonry walls, these vertical ribs of shotcrete can be reinforced as concrete columns, with four vertical bars and closely spaced lateral ties. The shotcrete layers span horizontally between the ribs, without any reliance on the existing wall. If the wall is in a decent condition, a composite structure can be assumed, and the shotcrete can be considered an outer shell of the masonry. The technique for applying layers of shotcrete to masonry is illustrated in Chap. 11.

Some other emerging methods of strengthening masonry walls are also given in Chap. 11, including increasing the shear strength of masonry walls with carbon and glass-fiber wraps attached to existing walls with epoxy. Another method for out-of-plane strengthening of reinforced masonry walls is described by Bhende and Ovadia.[34] The researchers tested a method involving narrow steel plates bolted to each face of a CMU wall. A pair of plates $^5/_8$-in thick and 6 in wide was provided for a 4-ft-wide masonry wall. The attachments were made only at the top and bottom of the wall by means of several 1-in-diameter through-bolts. The result was that the reinforced wall segments had a tenfold increase in their out-of-plane flexural capacities. This method has not yet been widely used, but it holds some promise for certain situations, such as those in which extensive reconstruction methods are not practical.

9.5.3 Consolidation and strengthening of masonry with injection grouting

Low-pressure grout injection has been used in historic and modern masonry buildings to solve a number of structural and serviceability problems. Structures in need of stabilization or strengthening and those experiencing excessive water infiltration can benefit from this procedure. Grout injection can bond the inner and outer wythes together and establish a composite action between the wythes for improved out-of-plane moment capacity. Grout injection into wall collar joints can also remedy water penetration problems. Water penetration tests conducted before and after grouting, using a modified version of ASTM E514, *Standard Test Method for Water Penetration and Leakage Through Masonry* (see Sec. 9.3.5), have shown 80 to 99 percent reduction of water penetration rates.

Grout injection can be used when a system of internal voids exists within the wall, allowing grout to enter and propagate. Different grout formulations are appropriate for different types of masonry structures. The type of grout used depends on the desired strength and bonding properties and on the sizes of the crack network or void system. Fine grouts can fill cracks as small as 0.005 in, while coarse grouts are useful for filling wide collar joints between the wythes.

Kariotis and Roselund[35] report that a grout consisting of 1 part portland cement, $1/_2$ part type S hydrated lime, and $1/_2$ part type F fly ash was used for repairing earthquake-damaged unreinforced masonry buildings in southern California. This grout was used to fill collar joints $1/_2$ to 1 in wide, as well as cracks in multiwythe brick walls.

Grout injection is a powerful rehabilitation technique, but it can cause harm if it is not handled properly. The most obvious problem arises from the lateral pressure exerted by the grout: If the wall wythes are poorly interconnected, the wall being grouted can split apart and collapse. Therefore, before a collar joint or a cavity is grouted, the presence of adequate ties should be verified. Low-lift grouting is preferred, because it reduces the hydrostatic pressure and can prevent the outward thrust of the fluid grout from displacing one of the wythes.

9.5.4 Using additional structure

A rational analysis of older masonry walls would often find them inadequate for combined out-of-plane and axial loads. In addition to the reinforcing and strengthening techniques just discussed, the structural capacity of walls can be increased by using additional structure, typically to reduce wall slenderness, or height-to-thickness ratio (h/t). This can be done either by reducing the effective height of the wall (decreasing the h) or by making the wall thicker (increasing the t).

How can the effective height of a wall be reduced? One method is to use diagonal steel bracing extending to the floor or roof levels, in much the same way as masonry-veneer hangers are braced by steel angle kickers. Another alternative is to use external steel frames spanning vertically between the floors. The masonry then spans horizontally a much shorter distance between the frames, rather than vertically from floor to floor. If the existing floors cannot provide adequate bracing, the frames can be designed to span horizontally between new concrete or masonry piers or counterforts; the masonry will then span vertically between these girts.

The steel members can be tied to the existing masonry by drilled-in anchors. If the exposed steel frames are visually undesirable, they can be placed within the wall and be bonded to it by grout. Or, cast-in-place concrete frames can be used instead, also placed within the wall.

One important caveat: To be effective, the added bracing—whether internal or external, vertical or horizontal—must be rigid enough to be able to brace the brittle masonry. Otherwise, the "braced" masonry may crack well before the steel structure becomes effective. So, 6-in steel channels can hardly prevent cracking of a 12-in brick wall spanning 20 ft. How sturdy should the bracing be? There are no guidelines cast in stone, so to speak, but it seems that the same criteria used to select the stiffness of steel-stud backup behind a brick veneer—a maximum lateral deflection of $H/600$, for one—would be appropriate. (A similar situation often comes up with metal building systems, where some designers attempt to laterally brace tall CMU walls with light-gage steel girts. This topic is discussed by the author elsewhere.[36])

Instead of reducing the wall height, the wall can be made thicker. A wall can be thickened by adding new layers of masonry compatible with the existing material and rigidly bonded to it by metal ties and collar-joint grouting. A more economical alternative is to use shotcrete wall overlays, discussed in Sec. 9.5.2 and in Chap. 11.

The method of using additional material can be applied to the strengthening of all kinds of masonry structures, from simple walls to complex ancient domes. Some case studies of reinforcing monumental masonry dome structures are described by Gregorian.[37]

9.5.5 Some removal and replacement may be unavoidable

Despite the best intentions of designers aiming to preserve existing masonry, some selective removal and replacement may be unavoidable and should be planned for. A common situation involves removing some nonloadbearing veneer to allow strengthening of the structural backup.

A case in point is a school in Wyoming, where construction was nearing completion when cracks began to appear in the concrete masonry walls. Initial investigations of the walls discovered that the reinforcing steel was missing in many places and that a number of the reinforced cells were not properly grouted. Additional investigations uncovered other workmanship problems. It was ultimately agreed that every reinforced cell was to be checked for its full height by pachometer for the presence of steel bars and by sounding and borescope for presence of grout.

However, one 29-ft-high wall constructed with 12-in CMU backup, insulation, and 4-in CMU veneer could not be tested with these conventional methods either from outside or from inside the building. In this wall, the steel studs and finish materials on the interior made checking the wall from the inside impossible. Checking from the outside was even less practical, because pachometers could not read through the layers of CMU veneer and insulation.

After some lengthy discussions of other testing possibilities, such as using infrared techniques, radar imaging, and even structural load tests, it was concluded that removal of the 4-in veneer was the most cost-effective and direct method of conducting the investigation. It was also concluded that the CMU cells that were missing reinforcing steel and/or grout could best be repaired from the exterior. The replacement masonry units were ordered immediately, so that they would be at least 28 days old by the time of installation. Finally, the 4-in veneer was removed, testing was done, the uncovered defects were repaired from the exterior, and the veneer was rebuilt.

As already noted, when partial rebuilding is required, it is desirable to obtain replacement materials with matching color and texture. In the case of exterior walls, the large exposed areas tend to magnify minor color differences. In addition to visual compatibility, consideration should be given to obtaining replacement masonry with matching compressive strength, elastic modulus, absorption properties, thermal expansion coefficient, and other relevant characteristics.

One last point: The more masonry is "selectively" removed, the easier it is to justify a complete tearing down and rebuilding the walls with new materials. Where is the fine line beyond which partial replacement should become complete rebuilding? Perhaps when more than 50 percent of the wall structure needs replacement, it is time for the whole masonry system to be retired.

9.6   Repairing Masonry Arches

9.6.1   Typical problems

The first masonry arch is thought to have been built in Babylonia around 1400 B.C.[38] Since Roman times, the arch has been a staple of masonry buildings, and well-proportioned stone arches have survived for millennia. The Romans developed the theory and practice of building semicircular arches—the only kind they knew. Later, other types of arches were tried, and the successful ones were adopted as part of the arsenal of masonry construction.

The arches that worked well shared two common attributes. First, they had a deep rise, a large keystone, or a keystone-like layout of the *voussoirs* (masonry units of the arch) to make them stable. Second, they had a sufficient amount of masonry at the abutments to resist the horizontal thrust generated by the arches. Failures have occurred when one or both these rules have been flouted.

Boothby et al.[39] studied the modes of failure of unreinforced masonry arches and concluded that, most commonly, abutment deficiencies were involved. The troubled abutments were those that moved laterally because of moisture and thermal changes, those that lacked sufficient stiffness, and those bearing on settled foundations.

Structural engineers have long known that the horizontal thrust exerted by segmental arches can vastly exceed the magnitude of their vertical reactions. As the rise-to-span ratio of segmental arches decreases, the thrust increases dramatically. Abutments that are too weak, either too slim or built with inferior masonry, lack sufficient strength to resist the horizontal thrust—and crack. The amount of horizontal thrust an abutment can resist in shear is roughly equal to its cross-sectional area multiplied by the number of shear-resisting planes (usually two, radiating at an angle from the base of the arch), multiplied by the allowable value of the shear stress. Abutments can also fail from overturning caused by the thrust, from excessive building sway during an earthquake, and from compressive overloading.

Boothby et al. observe that in arches in masonry buildings constructed prior to 1890s, the most common failure mechanism was slippage of voussoirs or of whole sections of the arch ring. Why? Moisture and thermal expansion of masonry are usually responsible. As the wall above the arch expands with increased temperature or moisture—the phenomenon of "brick drift," discussed in Chap. 13—the arch abutments tend to move with it, away from each other. This movement is in addition to the lateral spread of the abutments as a result of lateral thrust. When the spreading movements reach a critical level, the arch cracks, typically at the bottom surface near the midspan, and the voussoirs slip relative to each other. In some cases the lateral spread is so pronounced that the arches end up being supported by the window frames!

Figure 9.19 shows a failed segmental arch with slipped voussoirs. Note the evidence of prior attempts to repair the cracks with mortar. In this particular case, the slippage appears to have been caused both by moisture and thermal movements and by the overly shallow configuration of the arch. True, segmental arches with even shallower configurations—and the totally flat jack arches with spans less than 6 ft —have been used numerous times without apparent ill effects. This includes terra cotta arches spanning between steel beams, discussed at the beginning of this chapter. But all those shallow and flat arches were wedged in place by well-defined keystones or, in the case of some terra cotta arches, contained tie rods. The failed arch of Fig. 9.19 had neither.

The shallower the arch, the more vulnerable it is to spreading of its abutments. (Think what would happen to a simply supported beam if one of its supports moved away.) According to empirical rules, the minimum rise-to-span ratio of segmental arches is 0.15. Another rule of thumb holds that the skewback angle—the angle between the inclined surface where the arch joins the abutment and the horizontal line—should be less than about 65 degrees from the horizontal. Unit slippage is less likely to occur when the units or mortar joints are tapered and when the shear bond strength between them is good.[26]

9.6.2 Methods of repair

How can these failures in masonry arches be repaired? Usually, the best solution is to remove the damaged brick and rebuild the arch. In the case of the shallow arch of Fig. 9.19, rebuilding in kind may not work unless the existing arch profile is changed. A curved steel angle lintel, shown in Fig. 9.20, has a better chance of success. (In fact, similar steel lintels are generally used to support arches in contemporary

Figure 9.19 A failed shallow segmental arch. Note the evidence of attempted prior repairs.

buildings.) For major spans, the lintel is likely to be a tubular steel member. If a wide soffit is required, two veneer arches can be built at the sides, forming a U shape with a curved brick bottom attached to the sides with closely spaced metal ties. An alternative is to rebuild the arch with reinforced masonry or with reinforced concrete placed between the outer brick wythes.

The replacement arch structure should be designed not only for gravity, but also for lateral (out-of-plane) loading. The load path for wind loading acting perpendicular to the wall surface should include the abutments spanning vertically between the floors. A long arch should never simply terminate in a masonry wall without any provision for resisting these out-of-plane loads. In cases where the abutments are not strong enough to span vertically between the floors and where the arch is slender, diagonal bracing (kickers) from the bottom of the arch to the underside of the floor might be required.

When brick-veneer arches are used to replace original arches of solid brick, another problem arises: While the old arches were of the barrier-wall type and resisted water penetration by their sheer bulk, brick-veneer arches require flashing. Figures 9.20 and 9.21 illustrate some difficult but necessary details of providing flashing for various types of brick-veneer arches. Those interested in further information on this subject should review *Brick Masonry Arches: Introduction, BIA's Technical Note No. 31.*[38]

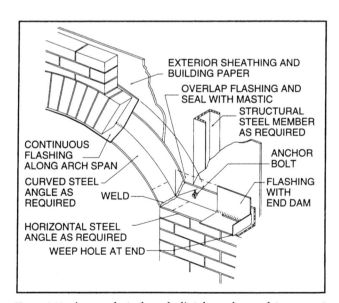

Figure 9.20 A curved steel angle lintel can be used to support replacement for a shallow arch. (*Brick Industry Association.*)

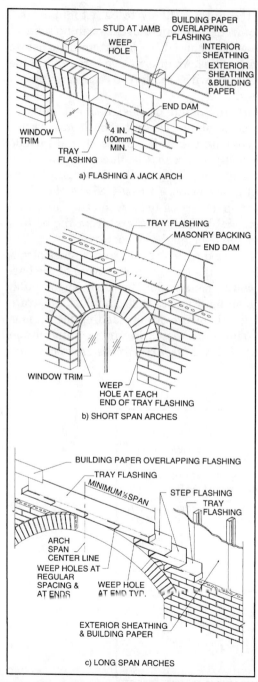

a) FLASHING A JACK ARCH

b) SHORT SPAN ARCHES

c) LONG SPAN ARCHES

Figure 9.21   Proper methods of providing flashing in various types of brick-veneer arches. (*Brick Industry Association.*)

Distressed masonry arches can also be repaired or partially replaced with matching units— brick by brick, if necessary. Soft cement-lime mortars should be used, to give a better bond and to permit absorption of minor cracks and movements, as discussed previously.

A displaced or cracked abutment can be removed and rebuilt with reinforced solid masonry or with brick enclosing a reinforced-concrete core. If space permits, a failed abutment can also be strengthened by adding more masonry or by enveloping it in concrete. Naturally, the arch must be temporarily shored during the pier replacement or, if it is also distressed, rebuilt together with its abutments. When the abutment troubles result from settling foundations, the adequacy of those foundations and the soil below them must be investigated. If it is determined that settlement is continuing, repair of the superstructure will be fruitless until the foundations are strengthened or the soil improved.

9.7   Other Masonry Renovation Tasks

A few masonry renovation tasks tend to occur regularly. Some of these are described here.

9.7.1   Making openings in existing walls

When an existing masonry building is being renovated, its interior layout usually changes. Some interior partitions may be removed, some added, some given new openings. How can one safely make these openings in existing masonry, loadbearing or not? Obviously, relatively small holes can be simply cut through without incident (Fig. 9.3), but it is equally obvious that making a 20-ft-wide opening for an overhead door in this manner is rather unwise.

It has been long accepted that a masonry wall can span over an opening by behaving as an arch, provided that there is enough masonry above the opening and beyond its sides. Therefore, brick and CMU lintels conforming to these limitations are typically designed for the weight of masonry of a triangular configuration, the base of the triangle being equal to the design span of the lintel and the height to one-half that. (There are special rules for dealing with superimposed uniform and concentrated loads.) This arching behavior is applicable to making wall openings.

The masonry above the imaginary triangle capping the opening is assumed to be able to support itself, but if no lintel is provided, the material within the triangle is held in place only by mortar tension, suspended as it were from the "arch" above. The allowable tensile stress of masonry and mortar is quite modest, but it is not zero (although this assumption is made in the design of reinforced masonry), and it may be sufficient to resist the weight of the masonry triangle for small openings. If the actual tensile strength of a given masonry material were known, one could theoretically determine the maximum opening size that could be made without any lintels. However, real-life material and workmanship variations argue against this approach. Instead, the common practice is to provide lintels for all masonry openings above some minimum size, such as 2 ft.

In CMU walls, the edges of cut openings are generally formed with concrete to provide smooth surfaces. Or, if metal window or door frames are used, all the irregular spaces between the frames and the masonry edges can simply be grouted. The edges of openings made in brick walls can be given a border of new bricks carefully "toothed" into the existing ones.

But what if an opening must be made in an existing reinforced CMU wall, requiring cutting a few vertical bars? In this case, some reinforcing bars should be added at the edges. One simple rule is to add at each side of the opening an amount of steel equal to one-half of the interrupted existing bars. For the general case of walls spanning vertically, these full-height vertical bars are placed at the jambs, while the horizontal edges of the opening are reinforced with bars spanning between the jamb reinforcement (Fig. 9.22). The jamb bars can be inserted by cutting the wall from a side, as discussed previously. The horizontal bars can be hooked at the ends or can extend past the opening's edges for a distance adequate to develop them.

In Fig. 9.22, the lintel above the opening is made by placing reinforcing bars into a bond beam, but the same objective can also be met by placing two or more angle lintels ("loose" lintels) over the opening and grouting the space between them and the masonry edge. The vertical legs of the angles can be placed inside the wall, which involves some masonry cutting but also hides the angles from view, or outside it, if cost is the primary factor and appearance is less important. From a structural standpoint, two angles placed back-to-back inside the wall are preferable, because they brace each other against the torsional forces that arise because the load is not applied at the shear centers of the angles.

How are the lintels installed? The simplest, and most dangerous, method is to simply make the opening in the wall before the lintel is placed. Quite often no ill effects follow, and lintel installation is easy, but this procedure can result in some cracked masonry above the opening that could fall down and injure the workers. A safer approach is to use one of three methods to provide some protection.

The first is to cut the opening, let the masonry crack, remove all loose and cracked units, and then rebuild that area after the lintel is placed. There is still a danger that, if the bond between the units is poor, crack-

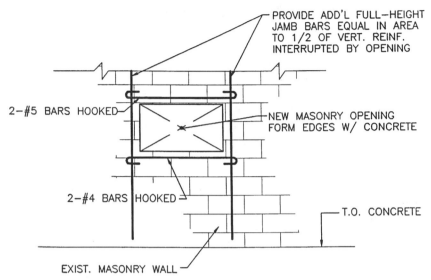

PROVIDE ADD'L FULL—HEIGHT
JAMB BARS EQUAL IN AREA
TO 1/2 OF VERT. REINF.
INTERRUPTED BY OPENING

2—#5 BARS HOOKED

NEW MASONRY OPENING
FORM EDGES W/ CONCRETE

2—#4 BARS HOOKED

T.O. CONCRETE

EXIST. MASONRY WALL

Figure 9.22 Added reinforcing bars at the edges of an opening cut into an existing CMU wall.

ing could be widespread, and collapse of the masonry above cannot be ruled out. The second is to shore the wall by placing so-called needle beams above the proposed opening. These are beams placed perpendicular to the wall and inserted through the holes cut in it. The beams pierce the wall like needles. The needle beams are supported at each end by posts placed inside and outside the building. The third method works best with loose lintels: The opening is made only to one-half the wall thickness and an angle lintel is placed above the cut area. The uncut half of the wall still provides support for the masonry above, so there is little danger of collapse, assuming that the wall is not carrying extraordinarily heavy loading. Once the first angle is placed, the second half of the wall thickness is removed, and the second lintel angle is installed. By that time, the first angle already carries the wall weight.

9.7.2 Fixing previous poorly made repairs

Poorly made prior renovations provide a steady source of work for masonry consultants. Unfortunately, masonry repairs made without proper understanding of the issues tend to do more harm than good. A case in point: As discussed in Chap. 1, the 200-year-old Octagon Building in Washington, D.C., was rather carelessly renovated in 1954, and the effects of that renovation had to be undone in the early 1990s. One relevant aspect of that renovation was fixing the previously repaired brick flat arches, where the replacement brick was cracking.

An investigation found that the brick did not match the original material and, more importantly from a structural standpoint, was placed in a hard portland cement mortar instead of the lime mortar used originally. As discussed previously, lime mortars tend to cushion brick, allowing for gradual movements and healing minor cracks, something cement-based mortars cannot do. The repairs to the cracked jack arches involved shoring the brick, raking out the hard renovation mortar, raising the arches to a level position with screw jacks, repairing cracked bricks with epoxy or replacing them with original matching units found on site, and filling the joints with lime mortar.[40]

One serious mistake that is often made in thoughtless brick renovations is to sandblast the old brick to remove accumulated paint or grime. True, sandblasting may be the easiest method of coating removal, but it also removes the hard exterior layer of brick that is produced during its firing, exposing the relatively soft and porous interior. The unintended result is to make the brick much more absorbent than before. After this operation, the brick may look good for a while, but it tends to quickly deteriorate under freeze-thaw cycles and weathering. Similarly, chemical washing to remove efflorescence and stains can do more harm than good if the acid is not thoroughly rinsed off.

9.7.3 Seismic strengthening

Older masonry buildings often lack adequate connections between the floor and roof diaphragms and the exterior walls. These connections are crucial to lateral-load resistance of buildings; if they are deficient, it is important to establish them. In most existing buildings, the floor and roofing joists have only gravity connections with the walls—typically, direct bearing with sparse anchorage. These gravity connections are nearly useless in resisting out-of-plane wind suction and seismic forces.

Making proper connections between floor or roof diaphragms and shear walls involves installing special drilled-in or bolted anchors, through-bolts, or clip angles. Local strengthening of wood-framed diaphragms is often necessary to help distribute the load to other members. Some of these connections are illustrated in Chap. 11. Other seismic-strengthening tasks include strengthening or bracing wall parapets, bracing chimneys, and adding new or enlarging existing masonry shear walls. All of these tasks are also discussed in Chap. 11.

9.7.4 Stabilizing bulging walls

One of the most frustrating tasks in masonry renovation is dealing with buckled or bulging walls. Walls can bulge for a variety of reasons, including eccentrically applied floor or roof loads, lateral soil pressure, shrinkage or expansion of concrete floors tied into the walls, and expansive pressure of water collecting in the cavity behind the face brick. As usual, the first step is finding out what caused the distress. This is not always an easy task, and sometimes brick walls seem to bulge for no apparent reason.

Assuming that the reasons for bulging or buckling are found and eliminated, the remedial methods are rather straightforward. The options are:

1. Leaving the wall as is, if the designer is certain that the situation is not likely to get any worse and the wall can safely carry the imposed loads

2. Removing the wall and replacing it in kind or with another type of structure, such as modern brick veneer

3. Tying the wall to intermediate floors, if it is not already connected to them

4. Tying the wall to existing intersecting masonry walls or to new walls or piers built specifically for this purpose

The first two options are self-explanatory. The third, tying the wall to intermediate floors, can help because it reduces the wall slenderness and allows it to carry more loading. The connections can be made with bolts or threaded rods anchored to (or through) the masonry and to the floor structure. Another possibility is to use steel brackets bolted to the wall and to the floor members.

The last option, tying the buckled wall to intersecting masonry walls, is practical if the spacing of these walls is closer than the distance between the floors or if the bulge occurs at only one spot, which happens to be near a perpendicular wall. Today's good practice dictates that intersecting masonry walls and partitions be tied together by hooked bond-beam reinforcement or by T-shaped horizontal joint reinforcement, but such connections were not always part of previous practice. If it is desired to stabilize the exterior wall by tying it to a perpendicular wall, there are a number of ways of doing it.

Perhaps the simplest is to connect the two walls by bolted structural-steel clip angles spaced at regular intervals (Fig. 9.23). If a more aesthetic solution is desired, the connections can be made *inside* the walls by removing some masonry units and replacing them with L-shaped cast-in-place concrete bands containing reinforcing bars. Beckmann[25] recommends extending these bands for a distance of 750 mm (30 in) from the intersection point and spacing them vertically at 450 mm (18 in) on centers, alternating the sides. He also suggests that precast concrete elbows can be used for this purpose. Still another approach would be to drill and grout U-shaped bars through the face wall and into the intersecting wall.

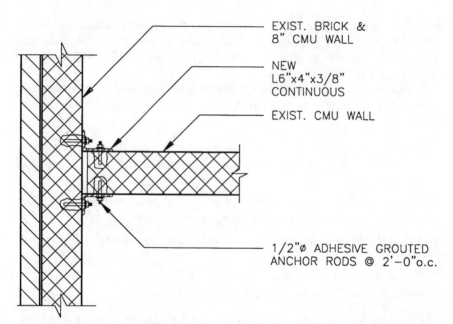

Figure 9.23 Stabilizing a bulging wall by tying it to an intersecting wall. (*Maguire Group Inc.*)

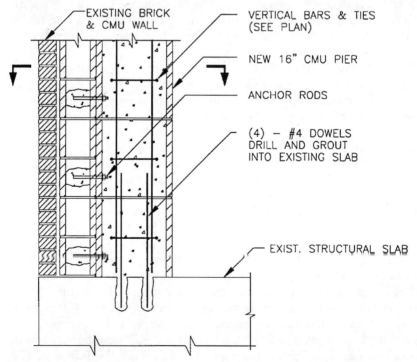

Figure 9.24 A reinforced masonry pier is added to brace a bulging wall. (*Maguire Group Inc.*)

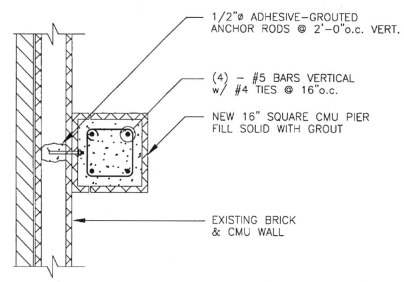

1/2"⌀ ADHESIVE-GROUTED
ANCHOR RODS @ 2'-0"o.c. VERT.

(4) - #5 BARS VERTICAL
w/ #4 TIES @ 16"o.c.

NEW 16" SQUARE CMU PIER
FILL SOLID WITH GROUT

EXISTING BRICK
& CMU WALL

Figure 9.24 (*Continued*) A reinforced masonry pier is added to brace a bulging wall. (*Maguire Group Inc.*)

The last remedy is to build some new intersecting walls, piers, or counterforts to laterally support the bulging wall. The anchorage can be done in the manner just described. Figure 9.24 shows a reinforced masonry pier added to brace a wall that was bulging for no apparent reason. Obviously, this is an expensive solution, from both the construction-cost and the space-utilization standpoints. It is important to realize that when these strengthening members are added, they will not immediately become effective in bracing the wall, and it will take some further bulging to stress the connections to the desired level.

References

1. FEMA-273, *NEHRP Guidelines for the Seismic Rehabilitation of Buildings*, Federal Emergency Management Agency, Washington, D.C., 1997.
2. Harley J. McKee, *Introduction to Early American Masonry: Stone, Brick, Mortar and Plaster*, National Trust for Historic Preservation, Washington, D.C., 1973.
3. FEMA-274, *NEHRP Commentary on the Guidelines for the Seismic Rehabilitation of Buildings*, Federal Emergency Management Agency, Washington, D.C., 1997.
4. Melvyn Green, "Masonry Rehabilitation," Lecture Series on Rehabilitation/Restoration at MIT, BSCES/ASCE Structural Group, Boston, 1993.
5. William H. Birkmire, *Architectural Iron and Steel, and Its Application in the Construction of Buildings*, Wiley, New York, 1891.
6. Donald Friedman, "Taking Stock," *Building Renovation*, Winter 1995 , pp. 15–17.
7. Almon H. Fuller and Frank Kerekes, *Analysis and Design of Steel Structures*, Van Nostrand, New York, 1936.
8. Sara Wermiel, "Structural Hollow Tile," *Building Renovation*, Spring 1994, pp. 41–44.

9. Theodore Crane, *Architectural Construction,* 2d ed., Wiley, New York, 1956.
10. Harry C. Plummer, *Brick and Tile Engineering,* Brick Institute of America, McLean, Va., 1962.
11. Theo H. M. Prudon, "Architectural Terra Cotta," *Building Renovation,* May–June 1993, pp. 57–60.
12. J. Bell et al., *From the Carriage Age...To the Space Age...The Birth and Growth of the Concrete Masonry Industry,* National Concrete Masonry Association, Reston, Va., 1969.
13. Pamela H. Simpson, "Blocks Like Rocks," *Building Renovation,* Spring 1995, pp. 49–53.
14. Carolyn Schierhorn, "Getting a Good Grout," *Masonry Construction,* January 1998, pp. 19–23.
15. John P. Speweik, *The History of Masonry Mortar in America 1720–1995,* National Lime Association, Arlington, Va., 1995.
16. John P. Speweik, "Lime's Role in Mortar," *Masonry Construction,* August 1996, pp. 364–368.
17. Harry C. Plummer and J. A. Blume, *Reinforced Brick Masonry and Lateral Force Design,* Structural Clay Products Institute, Reston, Va., 1953.
18. ASCE Standard 7-95, *Minimum Design Loads for Buildings and Other Structures,* American Society of Civil Engineers, New York, 1995.
19. *Recommended Minimum Requirements for Masonry Wall Construction,* Publication No. BH6, National Bureau of Standards, Washington, D.C., 1924.
20. *Uniform Building Code,* International Conference of Building Officials, Whittier, Calif., 1983.
21. ASA A41.1, *American Standard Building Code Requirements for Masonry,* American Standards Association, New York, 1944.
22. "Brick and Block Get Together in New Code," *The Magazine of Masonry Construction,* April, 1988.
23. John Hancock Callender (ed.), *Time-Saver Standards for Architectural Design Data,* 6th ed., McGraw-Hill, New York, 1982, pp. 2-212–2-213.
24. James E. Amrhein, "Constructing Tall Slender Walls," *Masonry Construction,* December 1998.
25. Poul Beckmann, *Structural Aspects of Building Conservation,* McGraw-Hill Europe, Maidenhead, Berkshire, England, 1995.
26. Elizabeth Keating, "Building Arches," *Masonry Construction,* April 1997, pp. 193–200.
27. Norbert V. Krogstad, "Repointing Ashlar Joints," an answer in Troubleshooting column of *Masonry Construction,* October 1997, p. 516.
28. FEMA-172, *NEHRP Handbook for Seismic Rehabilitation of Existing Buildings,* FEMA, Washington, D.C., June 1992.
29. Hanz Rudolf Ganz and Gerry Shaw, "Stressing Masonry's Future," *Civil Engineering,* January 1997, pp. 42–45.
30. Daniel P. Abrams, "Masonry Carries the Load," *Civil Engineering,* February 1993, pp. 66–67.
31. "SURE-STRESS™: A Post-Tensioning System for Masonry," Technical Bulletin 99-1, Dur-O-Wal Co., Aurora, Ill., 1999.
32. Mark Meyers, *Engineered Masonry,* Masonry Advisory Council, Park Ridge, Ill., undated.
33. TEK 14-20, *Post-Tensioned Concrete Masonry,* National Concrete Masonry Association, Herndon, Va., 1997.
34. Datta Bhende and David Ovadia, "Out-of-Plane Strengthening Scheme for Reinforced Masonry Walls," *Concrete International,* April 1994, pp. 30–34.
35. J. C. Kariotis and N. A. Roselund, "Repair of Earthquake Damage to Unreinforced Masonry Buildings," *Proceedings, the Second Joint USA-Italy Workshop on Evaluation and Retrofit of Masonry Buildings,* Los Angeles, August 19–29, 1987.
36. Alexander Newman, *Metal Building Systems: Design and Specifications,* McGraw-Hill, New York, 1997.

37. Zareh B. Gregorian, "Evaluation, Rehabilitation, and Innovative Design Procedures for Masonry Structures–Case Studies," SP-85-4, in G. Sabnis, ed., *Rehabilitation, Renovation, and Preservation of Concrete and Masonry Structures,* ACI, Detroit, Mich., 1985.

38. *Brick Masonry Arches: Introduction,* Technical Notes on Brick Construction No. 31, Brick Industry Association, Reston, Va., 1995.

39. Thomas E. Boothby et al., "A Visual Classification System for Masonry Arch Failures," *Proceedings, 10th International Brick Masonry Conference,* Calgary, Alberta, Canada, July 1994, as quoted in Ref. 26.

40. Raul A. Barreneche, "Octagon's Progress," *Architecture,* November 1993, pp. 107–113.

Renovating Metal Building Systems

10.1 Introduction

Metal building systems, also known as pre-engineered buildings, are structures designed and fabricated by their manufacturers. These buildings, which are often unassuming and simple in appearance, dominate the low-rise nonresidential market. Metal building systems are found in over two-thirds of all new one- and two-story buildings with areas of up to 150,000 ft². In 1997, members of the Metal Building Manufacturers Association (MBMA) covered 388 million ft² of space, 46 percent of it in manufacturing applications. Commercial uses, such as warehouses, small office buildings, garages, supermarkets, and retail stores, represented 31 percent of the 1997 sales. Community buildings, among them schools, town halls, and even churches, took 14 percent of the sales, and agricultural and miscellaneous buildings accounted for the rest. Large industrial facilities with areas of over 150,000 square feet added another 41.3 million square feet of new space.[1] Figure 10.1 illustrates a typical industrial building with metal building system.

Metal building systems enjoy this popularity because of such advantages as faster occupancy, lower construction cost, flexibility of expansion, easy maintenance, and single-source responsibility. The structure can be custom designed by the manufacturer to fit the project's requirements and minimize material waste. The flip side of this is a certain lack of reserve strength and therefore some difficulty in modifying the framing for new loading conditions. Another problem is the proprietary

Figure 10.1   A typical industrial building utilizing metal building systems.

design of these structures and the variability of design practices among the manufacturers, both of which make design and analysis of metal building systems something of a mystery to many design professionals. To be able to deal with the proprietary nature of these systems, we must first become familiar with some structural basics of their design and with typical practices of pre-engineered construction.

A typical metal building structure includes primary frames, secondary members (roof purlins and wall girts), roof and wall bracing, metal roofing, and siding (Fig. 10.2). A metal building system is indeed a system—an interdependent group of items forming a unified whole, with all the components designed to work together.

Metal buildings are dimensioned in a peculiar manner. *Frame width,* also called *sidewall structural line,* is the distance between the exterior surfaces of girts, rather than between the column centerlines, as is common in conventional construction. *Clear span* is the distance between the interior faces of columns. Some manufacturers measure clear span between the widest parts of the columns (for rigid frames it is usually the *knee,* a joint between the beam and the column), while others use the distance between the interior faces of the column bases. *Eave height* is measured from the bottom of the column base plate to the top of the eave strut, discussed in Sec. 10.6.1. *Clear height* is the distance between the floor and the lowest point of the *rafter.*

Renovations and modifications of existing pre-engineered buildings involve significant structural engineering challenges of types rarely found in rehabilitation of conventional structures. Typically, pre-engineered buildings need renovation because their metal roofs start to leak. Less commonly, the old wall siding becomes an eyesore and needs

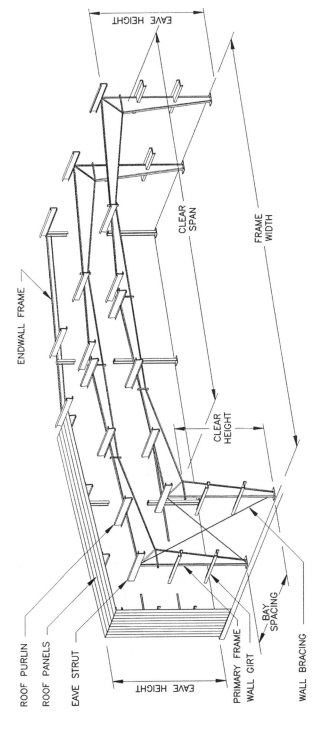

EAVE HEIGHT

CLEAR
SPAN

FRAME
WIDTH

ENDWALL FRAME

CLEAR
HEIGHT

ROOF PURLIN

ROOF PANELS

EAVE STRUT

EAVE HEIGHT

PRIMARY FRAME

WALL GIRT

BAY
SPACING

WALL BRACING

Figure 10.2 Typical components of metal building systems.

refinishing or replacement. The framing may also need structural modifications to accommodate a change in use or in equipment layout, or because a building extension or addition is required. A metal building system may also require strengthening simply because of extensive nonstructural renovations: The local building code might require that all buildings where the monetary value of renovations exceeds a certain limit be upgraded in accordance with the code for new construction. Most often, a structural engineer's involvement may be needed when the owner wants answers to one of the following common questions:

- Changes in the process equipment layout require that a new overhead door be located at the bay containing cross-bracing. Can the bracing be moved to another bay?

- A new piece of equipment can fit into the existing facility only if some primary-frame rafters are reduced in depth by a few inches or the building height is increased. Can the existing framing be modified in this fashion?

- A tall addition is proposed next to an existing metal building system. As one consequence, drifted snow will probably accumulate on the existing roof and theoretically overload it. How can the roof structure be strengthened for this additional loading?

- A heavy piece of mechanical equipment is proposed to be placed on the existing roof in the course of the HVAC system renovations. Can the roof framing safely support it?

- An overhead crane or monorail suspended from the frame rafters is moved or its capacity increased. Can the frame accommodate this unanticipated loading?

- A new floor pit needs to be installed—right where the existing foundation tie rod is located below the slab. Can the foundation be changed?

The last question in this list deals with foundation design for metal building systems, a subject not covered in this chapter. Those readers seeking information about foundations for pre-engineered buildings, as well as those who wish to learn more about this fascinating type of construction, are referred to another book by the author, *Metal Building Systems: Design and Specifications.*[2]

10.2   Evolution of Metal Building Systems

The first buildings with prefabricated metal components were used for garages. According to Butler Manufacturing Company, its first prefabricated building was developed in 1909 to house the famous Ford

Model T. The building was actually framed with wood, but covered with corrugated metal. To improve fire resistance, the company later switched to a metal-framed structure sheathed with corrugated curved steel sheets.[3]

The next step was taken by the Austin Company of Cleveland, Ohio, which in 1917 started offering ten standard designs for factory buildings. The Austin buildings, which could be ordered from a catalog, were framed with predesigned steel columns and roof trusses. Thus, a tradition of pre-engineered construction had begun. Soon, prefabricated factory buildings were also offered by Liberty Steel Products Company of Chicago and by other firms.

The fledgling industry received a further boost during the oil boom of the 1920s and 1930s, when several major metal-building companies were formed to satisfy the need for inexpensive equipment storage.

The framing for those early metal buildings commonly included steel trusses spanning between trussed columns. The wall covering consisted of 8- by 12-ft corrugated galvanized sheet sections spanning vertically and bounded by riveted steel angles.

Metal buildings of similar design were constructed for aircraft hangars during World War II. The roofs of those buildings were framed with bowstring trusses bearing on columns made of laced $6 \times 4 \times \frac{3}{8}$-in angles. The tall walls of the hangars required intermediate girts for siding support, an innovation that has become a staple of metal buildings. Also during World War II, another type of prefabricated building became a household word: Quonset huts were produced by the hundreds of thousands to provide inexpensive and standardized shelter for American soldiers. Some of those structures survive to this day (Fig. 10.3).

After the war, a new crop of metal buildings helped sustain the postwar economic boom by filling the need for affordable and quickly constructed factory space. These inexpensive structures used gabled steel trusses with a typical 4:12 roof pitch and were clad in corrugated, galvanized sheet metal siding and roofing. They generally lacked windows, insulation, and elaborate mechanical systems.

The first metal buildings with single-span rigid frames were introduced in the late 1940s. At first, they could span only 40 feet. As the technology improved over the next decade, 50-, 60-, and 70-ft-wide buildings became available, and by the late 1950s rigid frames with 100-ft spans were being produced. This breakthrough and several other major innovations of the late 1950s and early 1960s brought the industry to a new level of sophistication. Some of these innovations included ribbed (rather than corrugated) metal siding, colored panels, factory-insulated walls, continuous-span cold-formed Z purlins, the first UL-approved metal roof, and the first computer-designed metal buildings.[4]

Figure 10.3   Quonset hut. (*Photo: David Nacci.*)

To differentiate its new products from the unsophisticated "prefabricated buildings" produced earlier, the metal-building industry started calling them "pre-engineered buildings" in the 1960s. This scientific-sounding name implied that a limited number of off-the-shelf structural configurations had already been engineered prior to sale and therefore could save the purchaser the time that would normally be allocated to design. However, by the 1980s this term had become obsolete as well. With the wide availability of computers, it was no longer necessary to restrict the available building sizes to a few standard dimensions, and each building could be custom designed. To reflect this freedom of design, and perhaps to hint at the role of its computer capability, the industry now prefers to call its products "metal building systems." As already stated, this definition is quite proper, because a well-designed and well-constructed metal building is indeed a system, an interdependent group of items forming a unified whole.

The major developments of the 1950s took place against the background of a disorganized metal-building industry handicapped by problems with building codes, insurance, and labor unions. Each building manufacturer was using its own engineering assumptions and methods of analysis, a situation that resulted in variable dependability of the early metal buildings (and that makes it difficult to analyze them today). When a trade organization, the Metal Building Manufacturers Association (MBMA), was founded in October 1956, development of common engineering standards was among its first priorities.

The first edition of the MBMA manual appeared in 1959, and since then this book has been the main reference source for metal building manufacturers, their engineers, and builders. The manual has been periodically reissued and updated to reflect the advances in research and design practices. The 1996 edition[5] was issued in a convenient loose-leaf binder format that allows for easy updates.

Another major design standard that is useful in the analysis of cold-formed components of metal building systems is the *Specification for the Design of Cold-Formed Steel Structural Members* by the American Iron & Steel Institute (AISI).[6] The first edition of the specification, largely based on AISI-funded research in the late 1930s and early 1940s, was published in 1946. Since then, this document has undergone frequent and drastic revisions corresponding to the rapid developments in the field of cold-formed steel design. The design of hot-rolled structural members in metal building systems follows the provisions of the AISC specification and manual.[7]

10.3 Primary Frames

10.3.1 Types and construction

Primary frames are the main load-resisting elements of metal building systems. Contemporary primary frames are usually made from high-strength steel conforming to ASTM A572 (eventually superseded by ASTM A992) with a minimum yield strength of 50,000 psi, or from ASTM A36 steel. The grade of steel in existing buildings constructed prior to 1960 can be determined by examining the information provided in Chap. 3; the framing is most likely to have a yield strength of 33 or 36 ksi. The most definitive way of discovering the actual strength of steel is, of course, by testing. Some popular types of primary frames are illustrated in Fig. 10.4.

The tapered beam system, shown in Fig. 10.4*a,* is also called wedge beam or slant beam. This framing type looks like a metal building system's version of conventional post-and-beam construction. The main difference between the tapered beam system and a built-up plate girder supported by two wide-flange columns is the tapered profile, which follows the moment diagram of a simple-span beam. Usually, the slope is at the top to allow for easy water runoff; the bottom is kept horizontal for installation of the ceiling. Sometimes, architectural requirements call for the opposite. In roofs with steep pitches and low-slope cathedral ceilings, both flanges can be sloped, as in a scissors truss. If required, splices are made in the middle of the beams. Primary frames of metal building systems are typically spliced with bolted end-plate connections using high-strength bolts conforming to ASTM A325.

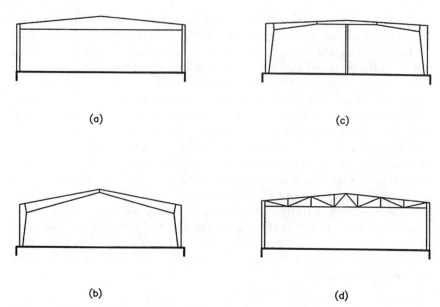

(a)

(c)

(b)

(d)

Figure 10.4 Common types of primary frames: (a) Tapered beam system, (b) single-span rigid frame, (c) multispan rigid frame, (d) single-span truss.

There is another difference between the tapered beam system and a built-up plate girder on columns. The tapered beam is often assumed to be attached to the columns with "wind connections," intended to be rigid enough to resist lateral (wind and seismic) loads, but flexible enough to allow for single-span behavior under gravity (snow and live) loads. Unfortunately, this manufacturer's assumption is rarely accompanied by the appropriate detailing. Instead of the typical semirigid connection used in structural steel design—a pair of flexible clip angles at the top and bottom of the beam—the tapered beam "wind" connections are made with bolted end plates. Does the end-plate connection provide sufficient inelastic rotation capacity to prevent failure under combined gravity and wind loads? It is doubtful. Another question: Are the columns designed under this scenario overstressed by the end moments resulting from the near fixity of the beam ends? If the tapered beam framing is being renovated for new loading, it is wise to investigate these issues.

Another type of primary framing commonly found in existing metal building systems is the single-span rigid frame (Fig. 10.4b). This system is designed to take full advantage of the connection rigidity and to approximate the shape of the moment diagram for two- or three-hinge frames. The member splices occur at the knee and in the rafter.

The multispan rigid frame, sometimes called a post-and-beam, modular, or continuous-beam frame, is similar to the single-span rigid frame but has two or more spans (Fig. 10.4c). Multispan rigid frames typically have tapered exterior and straight interior columns and tapered rafters. This framing type is common in large industrial buildings and warehouses.

The single-span truss system (Fig. 10.4d) in some respect resembles the tapered beam system. This type of framing is common in pre-engineered buildings of older vintage. Continuous trusses, which are somewhat less common, are more akin to multispan rigid frames. Buildings with trusses not only are framed in a cost-efficient manner, but also allow pipes and ducts to be placed in the web space and decrease the eave height. A lower building is less expensive to build and maintain. A design example of renovating one of these buildings is included in Sec. 10.11.

There are a few other primary framing systems that are not shown in Fig. 10.4. Among them is lean-to framing, essentially a tapered beam attached at one end to another building. Since the top surface of the beam has a single slope, taper is usually provided at the bottom flange. Lean-to framing is useful for small wings, additions, and other minor attached structures. Another framing system that is not specifically shown is the single-slope rigid frame, of single- or continuous-span type; it is similar in behavior and details to the gable rigid frame. Sometimes, existing buildings contain truly unique proprietary systems that defy classification and must be carefully measured and analyzed if it is necessary to alter them.

Primary framing relies on the bottom-flange bracing for maximum efficiency under wind uplift, when the bottom flange is in compression (Fig. 10.5). Such bracing usually consists of diagonal angle sections bolted to the bottom flange and to purlins (see Fig. 10.6a for details). Flange bracing is also required for rigid-frame columns, where the interior flanges are normally in compression under wind loading acting away from the building (Fig. 10.6b). The column flange bracing is made to the webs of wall girts. Flange bracing is not always required, but it is so common that its absence in an existing building should be a cause for investigation.

The spacing of primary frames (the bay size) depends on the load-carrying capacities of the purlins and girts they support. A 25-ft spacing is most common, because it represents the maximum practical distance for both girts and purlins acting in a single-span fashion. Bay sizes of 25 to 30 ft normally require continuous-span secondary members, and 30 ft is presently considered the limit of cost-effective cold-formed construction.

Unlike hot-rolled structural steel beams, the primary frames of metal building systems are usually made with single-side (and sometimes even intermittent) web-to-flange welds. Some engineers distrust one-sided

Figure 10.5 The bottom flange of this multispan rigid frame is braced to the purlins above.

welds and consider them structurally deficient.[8] There is no specific prohibition of this kind of welding in either the AISC or the AWS specification. Also, an extensive testing program undertaken by Prof. Thomas M. Murray has found that single-sided welds did not reduce the ultimate strength of the frames, with one notable exception: End-plate connections welded in this manner did not fare well under simulated seismic loading, because repeated local buckling induced fracture of the single-sided welds in the rafters near the end plates. One conclusion that could be drawn from the tests is that single-sided welds may be considered acceptable for static loading, but double-sided welds are better suited for framing that must resist lateral loading or concentrated loads, or that is subjected to fatigue.[9]

10.3.2 Strengthening

Strengthening of primary framing may be required for two main reasons: changes in structural loading and member deterioration. Changes in structural loading often follow a change in ownership or changes in occupancy, manufacturing process, or mechanical systems. Some common circumstances when a structural engineer's help might be needed are listed in Sec. 10.1. Many of them require changes in, or at least an analysis of, existing pre-engineered framing. Similarly, any program of strengthening required because of member deterioration must begin with an assessment of the design strength of the old framing. But unlike structural steel buildings, where beams and columns can be identified with some degree of certainty and checked for new loads, the proprietary nature of metal-building framing makes the analysis difficult. It is not impossible, however, with a well-planned effort on the part of the engineer, the cooperation of the owner, and some luck.

The simplest way to determine framing sizes is to examine the original drawings and calculations prepared by the metal-building manufacturer. This information may be available in the owner's files or other sources discussed in Chap. 2. The framing dimensions can then be spot-checked in the field. But such luck is rare. More likely than not, if anything is found at all, it will consist only of basic erection plans lacking any detailed information. Even so, the drawings may contain at least the manufacturer's name, the date of construction, and the order number (this could also be marked on unpainted framing). The engineer can then call the original building manufacturer, if the manufacturer is still in business, to see whether any original design files are available. This too is not very likely, but the manufacturer can at least provide some information about the design assumptions and grades of steel used by the company at the time of construction. Then again, the manufacturer may be unreceptive because of liability concerns, as has happened in at least one renovation project.[10] In the author's experience, manufacturers are rarely interested in investigation of other manufacturers' existing pre-engineered buildings, reasoning that their main product is manufacturing, not engineering.

If this search for information fails, the engineers can field-measure the framing sizes, take test coupons to determine the grade of steel, and analyze the framing themselves. This analysis may seem intimidating to many structural engineers who are unfamiliar with pre-engineered building design, and they might wish to call an expert for assistance. (Ironically, such extensive—and expensive—analysis efforts are rarely welcomed by the owners of old pre-engineered buildings, who tend to operate on tight budgets. Still, as with any existing building, the engineers should be familiar with the framing they intend to change—or refuse the assignment.)

The analysis of tapered sections can be done with a specialized computer program that can handle these sections. Or, the engineer can simply measure the actual frame dimensions at various locations and analyze a typical frame. A simplified approach is to use a general frame analysis program to make a model with, say, 5-ft-long segments of constant section rigidly connected to each other. Obviously, a realistic computer model depends on accurate field measurements, which should include the locations of all flange and web splices, because the splices in pre-engineered frames do not necessarily develop the full member capacities. As Johnson[11] points out, the flange segments used by some manufacturers can be as short as 5 ft, and welded web splices are easy to overlook.

If the analysis indicates a deficiency, or if the existing metal structure has deteriorated, strengthening or repairs are needed. In the worst case, the existing framing may be removed and replaced. The general methods of strengthening pre-engineered framing are similar to those used for structural steel buildings (see Chap. 3).

The most direct method of strengthening is to install additional primary frames between the existing ones, with new foundations. This method not only reduces the loading on the existing frames by one-half, but also cuts down the span of the existing purlins, quadrupling their flexural capacities. It requires precise execution to be effective, as the new frames must either be placed at exactly the same elevation as the existing ones for direct attachment of purlins, or be erected slightly lower for attachment by bearing clips (see Fig. 10.6). The latter approach will allow for field erection tolerances and is more practical than aiming for a precise fit. Also, using the clips will help protect the purlins against a web-crippling mode of failure, a weak point of the purlins in this type of installation because the purlins have only a single-web section at the new bearing points, as opposed to a doubled section at the original supports.

The span can be shortened by adding intermediate columns under the existing frames, if these columns do not hamper the operations and are acceptable to the owner. The added columns will result in redistribution of bending moments in the frame, and rafter reinforcement, or at least some new flange bracing and web stiffeners, could be needed.

The section modulus of a primary frame rafter can be increased by welding additional pieces. As shown in Fig. 3.7 in Chap. 3, a wide-

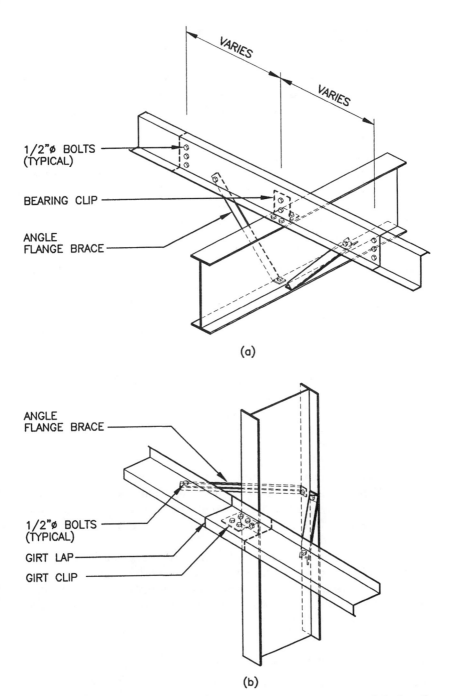

Figure 10.6 Flange bracing usually consists of diagonal angle sections bolted to the frame and to secondary members. (*a*) Bottom-flange bracing of frame rafters, (*b*) bracing of interior flanges of rigid-frame columns.

flange section can be reinforced for flexure by adding a bottom plate, by adding both a bottom plate and two narrow bars welded to the underside of the top flange, or by welding a WT section to the underside of the rafter. There are other possible solutions, and also combinations of several methods. For example, instead of welding two narrow bars in the second scenario, two continuous angles could be used, with one leg welded to the underside of the top flange and another to the web.

What if the shear capacity of the frame is insufficient? The webs of primary frames in metal building systems are normally quite thin and deep. They rarely have intermediate web stiffeners and infrequently rely on tension field action. Reinforcing these members for shear can be done by adding stiffeners at the locations of new concentrated loads and wherever shear stresses are excessively high. As Johnson states, the stiffeners can relatively inexpensively increase the shear capacity of the web by a factor of two or three, although in some cases there is no alternative to using web doubler plates, a more expensive solution.

The case of moving a suspended crane or monorail to a new location deserves a special mention. The frame analysis may well indicate that the frame section is adequate for the new position of concentrated loads representing the crane reactions. However, the rafter welds are another matter, and the rafter section in the proposed location should be checked to see if the welds attaching its bottom flange to the web require reinforcing. A built-up rafter that is welded only on one side should be locally reinforced with double-sided welds, and perhaps with web stiffeners, to resist the suspended loads, as was probably done in the original hanger location.

Built-up framing that resists any other types of concentrated loads or is subjected to fatigue should probably be reinforced in a similar manner. In a broader context, any one-sided weld that causes the joint to open on the opposite side (because of weld shrinkage, for example), may be considered to be contributing to notch development in the weld, as illustrated in Fig. 8-33 of the AISC LRFD manual.[12] In some circumstances, this is unacceptable and requires reinforcement. In seismic regions, the end-plate connections with single-sided welds should also be reinforced with additional welding on the other side of the web. The welds should extend from the plate for a minimum distance equal to the depth of the rafter.

10.4   Expansion of Metal Building Systems

Ease of expansion is among the theoretical advantages of metal building systems. (Keeping to structural issues, we omit a discussion of code-related requirements imposed when additions are made. These are examined in Chap. 1.) Certainly, it should not be difficult to extend the existing pre-engineered building by adding several more bays of like materials and splicing them at the endwall. It isn't difficult, that is, if the building was designed with expandable endwalls—those provided with regular interior frames and foundations at the endwall locations. The endwall framing in this case does not support vertical loads and is intended to be removable.

Unless the expansion was planned in advance, however, the existing building was most probably supplied with nonexpandable post-and-beam endwalls (Fig. 10.7). In this typical case, the elimination of an old endwall is more difficult: It requires temporarily shoring the purlins and girts at the endwall location, taking down the endwall framing, including the corner columns; placing new foundations, and erecting a new frame. The old endwall framing can be reused at the end of the addition.

Figure 10.7 Nonexpandable post-and-beam endwall framing.

Sometimes the owners attempt to save money by allowing the existing nonexpandable post-and-beam endwall framing to stay in place and support the addition's purlins on the existing building. This rarely helps, because the existing endwall framing and its foundations are likely to become overstressed by the loading from the addition. One possibility in this case is to separate the new and existing framing by several feet—enough for installation of the new foundations—and to cantilever the roof purlins over the gap.

Is it easier to place the addition alongside the existing building? Not really. In northern climates, where snow loads control roof design, two gable buildings placed side by side create a valley that collects drifted snow (Fig. 10.8). The weight of this snow may vastly exceed the design snow loading on the roof and result in the existing framing's being overstressed. (Framing for the addition can, of course, be designed for any load.) So, a side-by-side building expansion requires careful evaluation and a possible structural upgrade of the existing roof. According to Tobiasson,[13] the losses from failures attributable to neglecting this issue are measured in hundreds of millions of dollars.

Even in those locations where snow is rare, a side-by-side expansion is difficult to make. If the adjacent columns are placed in line with each other, there will be insufficient space for new column footings. Placing the new columns on the existing footings is hardly an alternative: It doubles the load on the footing, and strengthening foundations is not an easy matter. Staggering the new and existing frames eliminates the foundation challenges, but at the cost of limiting access from one building to another. There is also a problem with column loadbearing capacities, as discussed in the next paragraph for lean-to additions.

Whenever a lean-to addition is made to an existing metal building system, the existing sidewall is typically removed to "open up" the combined space. Again, the capacities of the existing column foundations will probably be insufficient, and these foundations may require strengthening. In addition, the loadbearing structural capacities of the existing frame columns must be evaluated. These columns have a double problem: On the one hand, the lean-to imposes additional loads on them; on the other hand, the girts that used to brace the columns in their weak direction have been removed, drastically decreasing their strength. In all likelihood, the columns will require reinforcement—for example, by added flange plates or by other methods described in Chap. 3. It might seem that the easiest thing to do is to leave the sidewall in place—or to remove every other sidewall bay, which will preserve the flange bracing—but even then the existing columns and footings will have to be checked for increased loads and will probably require reinforcement.

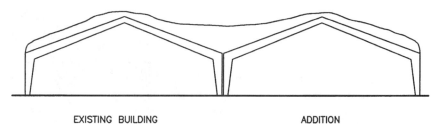

EXISTING BUILDING   ADDITION

Figure 10.8 A valley formed by two gable buildings placed side by side collects drifted snow.

One possible solution for this situation has been just mentioned for the case of endwall expansion—separate the new and existing framing by several feet, install the new foundations, and cantilever the roof purlins between the two adjacent rows of columns.

How can the clear height of an existing metal building system be increased? This problem arises when a new piece of equipment cannot fit into the existing facility. In this case, the first obvious thing to check is whether the equipment can be positioned at the point of the maximum clear height and between, rather than under, the primary frames. If the height is still insufficient by just a couple of inches, it might be possible to replace some roof purlins with wide-flange beams positioned on each side of the equipment's tallest point, with steel deck spanning between the beams. Or, the existing floor slab could be locally depressed to provide the necessary headroom. If none of this works, or if the height must be increased throughout the whole building, a more radical solution is necessary.

The most drastic approach is to replace the whole building—a practical thing to do if the building is in poor shape and requires a lot of renovation work anyway. If this is out of the question, it might be possible to erect a new building outside the existing one, completely enveloping the old structure in the process, and remove the old structure piece by piece when the enclosure is finished. This way, building operations can continue uninterrupted during construction (Fig. 10.9).

If the existing building is in good shape, it could be cost-effective to move the roof to a higher elevation. There are two methods of performing this major surgery, neither of them inexpensive. One is to simply dismantle the whole roof, remove the frame rafters from the columns at the knee splices, extend the columns or erect new, higher ones, and reassemble the roof on top of the extended or new columns. The second method also involves severing the frame rafters at the knee splices, but then lifting the whole roof assembly to the desired elevation with the help of a specialized contractor. While the roof is shored at the proper elevation, the existing columns can be extended

Figure 10.9   A new, higher metal building system constructed above the existing one.

or replaced with new, higher columns. A column extension can be made by adding a stub column with the same top and bottom plates as the existing splice plate at the knee. The stub-column method will work only if the bolted knee splice is made in a horizontal plane, as shown in Fig. 10.2. Some manufacturers make the splice in a vertical plane, within the rafter, in which case new columns are required. Still others make diagonal knee splices, the most difficult to handle.

Increasing the frame height will alter the frame's moment diagram. Will the extended frame be still structurally adequate? According to Johnson,[11] the changed bending moment at the knee of a rigid frame may be tolerable if the height extension is minor, but the moment at the ridge might change enough to justify reinforcing that area. The changed moment diagram can affect the splices and necessitate their strengthening. An analysis of a typical bolted end-plate splice can follow various methods, some of them described in Refs. 7 and 14.

If the analysis indicates that the forces on splice bolts are excessive, the existing ASTM A325 bolts can be replaced with stronger ASTM A490 bolts of the same diameter, or with A325 bolts of a larger size, in which case the holes must be enlarged by reaming. A load path from tension bolts to the web must be checked: If the tensile stresses in the web area near the bolts are excessive, a welded horizontal web stiffener may be needed. The stiffener should be welded to the end plate and to the web and should be long enough to develop the tension force.

The weld between the end plate and the web should also be checked and reinforced if needed. And finally, the end-plate thicknesses should be checked as well.

10.5   Lateral Stability

Typically, the lateral stability of metal building systems against wind and earthquake loads is provided by moment-resisting frames in one direction of the building and by braced frames in the other direction. Single- and multispan rigid frames can be designed to resist both gravity and lateral loads, but other frame systems must rely on braced frames along the perimeter of the building and on roof diaphragms capable of spanning the distances between them.

A typical roof diaphragm of a metal building system includes diagonal steel rods or cables designed as tension members and roof purlins designed as compression struts. Less commonly, roof diaphragms consist of corrugated roof decking. Certain kinds of through-fastened metal roofing can also provide diaphragm action, although of a lesser rigidity than a typical steel deck. For the lateral loading acting parallel to the primary rigid frames, the diaphragm span is short—the distance between the frames. In buildings without rigid frames, the diaphragm span is from endwall to endwall. For the loading acting perpendicular to the frames, the diaphragm span is from sidewall to sidewall.

A typical sidewall bracing bay consists of steel rod or cable diagonals, eave struts, and frame columns on each side (Fig. 10.2). The diagonals are typically connected to the column webs via so-called hillside washers (Fig. 10.10). Some problems with these washers are explored by the author elsewhere.[2] When renovation of wall bracing is involved, it is wise to use a better detail for attaching the rods to the columns. For example, the bracing rods or cables could be attached to a reinforcing plate welded to the column flanges. Figure 10.11 illustrates another approach: a special wall bracing clip anchored directly to the foundation. These clips can be used only when the existing foundations are heavy enough to resist wind uplift forces from the brace.

Where wall bracing must be avoided for architectural or functional reasons, the usual substitution is *portal frames*—small rectangular rigid frames that fit between building columns. Lateral stability can also be provided by concrete or masonry shear walls, if these materials are already used in the building exterior.

Occasionally, lateral stability of metal building systems is provided by flagpole-type framing fixed into the foundations. These cantilevered columns do not need wall bracing or diaphragms for stability. However, foundations and columns designed in this fashion are quite

Figure 10.10 Hillside washer attaching horizontal bracing to the column web. Note that the rod coming toward the column is bent, an undesirable condition.

expensive, and lateral drift of the building may be difficult to control. Thus, the flagpole-type framing is rarely used as a primary framing scheme, but it is often encountered as *wind posts* to replace exterior wall bracing or portal frames. Any time wind posts are found in an existing building that requires modifications, it is wise to investigate the adequacy of their foundations, since foundation design is beyond the scope (and quality control) of metal building manufacturers.

Various types of lateral-load-resisting elements, such as shear walls and cable bracing, should not be combined along the same wall. The lateral rigidity of shear walls vastly exceeds that of wall bracing. The bracing becomes useless in buildings with concrete or block masonry walls (except, perhaps, for assuring stability during construction). Unfortunately, it is common to see the two used together (Fig. 10.12).

To answer one of the questions posed in the beginning of the chapter—whether existing cross-bracing can be moved to another bay—one must know whether the columns and eave struts in the existing and proposed brace locations are of the same section or not. If they are, one brace location is as good as the other provided that the lateral forces can be transmitted by the eave struts to the new location. When the detailed drawings are unavailable, as is typical, this determination must be made by field measurement. If the sections are slightly different, the column in the proposed location can be analyzed and, if need be, reinforced using the methods outlined in Sec. 10.3.2. The existing eave struts can be reinforced by additional members.

Figure 10.11 Wall bracing clip attached directly to the foundation, instead of to the column web.

10.6 Secondary Framing

10.6.1 Basic structure

Secondary framing for metal building systems has traditionally utilized cold-formed C or Z sections. The C and Z configurations reflect the only two practical ways to bend a sheet of steel into a section with a web and two flanges. The secondary members are used as roof girts, wall purlins, and eave struts. The eave strut, also called an eave purlin and eave girt, is on the border between the purlins and the girts, supporting roofing with its top flange and wall siding with its web.

Cold-formed roof purlins run above the primary framing, being either directly bolted to rafters or connected to them by bearing clips (Fig. 10.6a). The method of attachment depends on the magnitude of the web-crippling stresses in the purlins. Obviously, direct bolting is cheaper, but bearing clips provide a desirable measure of lateral stability.

A 5-ft purlin spacing is common. The purlins can be designed as simple-span or continuous members. The continuity can be achieved by

Figure 10.12   Loose wall bracing in front of a CMU wall. (Also note serious water damage on the interior surface of the wall. This is the same wall shown in Fig. 10.14.)

overlapping the purlins and bolting them together, as shown in Fig. 10.6a. The longer the lap, the larger the degree of continuity, up to a point. The C sections are lapped back to back, and Z sections are nested one inside another, although it is difficult to do this with Z sections of thicker gages. The stiffening lips at the edges of the flanges are shaped with nesting in mind: They are formed with 90° bends for C shapes and with 45° bends for Z shapes.

Light-gage wall girts are normally designed as simple-span members. Unlike purlins, girts can be placed relative to columns in three different ways, called insets. In the bypass inset, the girts are located outside the columns, much like purlins (Fig. 10.6b). In the semiflush inset, some, but not all, of the girt section continues past the column. In the flush inset, the girts are framed into the column web and typically extend about 1 in past the exterior column face.

A minimum yield strength of modern light-gage sections made of 16-gage and heavier steel is usually 55,000 psi. The strength of existing cold-formed secondary members made after 1960 is likely to be 50,000 or 55,000 psi. The strength of cold-formed members produced earlier could be lower, in the range of 33,000 to 37,000 ksi. The actual yield strength can be found by testing.

Both C and Z shapes are not symmetrical about their vertical axes, and their shear centers are located eccentrically to the points of load application in the middle of the top flange. Therefore, even in a perfectly horizontal roof, any vertical load tends to twist the purlins and make C and Z sections unstable. A sloped roof can actually improve the situation by moving the center of rigidity closer to the line of load, if the purlins are oriented the right way. The optimum roof slope, where the line of load passes through the center of rigidity, equals the ratio of the purlin's flange width to the purlin's depth. This optimum slope is almost never present, and the secondary members are typically subjected to torsional loading that they are ill equipped to resist.

To stabilize purlins and girts made of C and Z sections, member bracing is normally required. For purlins, bracing is needed at the top flanges for downward loads and at the bottom flanges for upward loads, such as wind uplift. The top-flange bracing can be provided by light-gage angles or strapping. Most kinds of through-fastened roofing and steel deck also qualify as purlin bracing. There is an ongoing controversy about whether standing-seam roofing can serve as lateral bracing, given the fact that the roofing by definition is not directly attached to the purlins and can move relative to them. The most advanced "articulating" clips (Fig. 10.13) can even accommodate the slight slope changes resulting from movement of the roofing. Logically, this system also provides the least degree of purlin bracing.

Obviously, only angles and strapping can be used for bottom-flange bracing. The reader is referred to Newman[2] for the recommended details and spacing of braces. For most circumstances, the appropriate bracing located at 5 or 6 ft on centers and at supports will suffice. A line of bracing should also be provided at each concentrated load applied to the purlin, such as a pipe hanger. Incidentally, such hangers should never be attached to the purlin's stiffening lips—the proper place to attach them is the web.

The bracing for girts is similar to that for purlins. The exterior flanges of girts are normally braced by wall siding, and the interior flanges can be braced by liner panels or interior gypsum board. If the interior flanges are exposed, they should be braced in the same manner as the purlins.

Figure 10.13   Articulating clip.

10.6.2   Upgrade

Purlins and girts most commonly require strengthening because of member deterioration or changes in design loading. Our guide to this topic is again Johnson.[11]

When structural upgrade of secondary members is needed in only a few places, they can be reinforced with additional flange plates or angles. Adding two angles to a Z section can approximate a very efficient wide-flange section. The attachment can be made by welds, bolts, or self-drilling screws. It is also possible to reinforce a C or Z section with a channel that fits within the purlin and is attached to its web, but the channels that can practically be added are relatively shallow. Reinforcing with two angles would improve the section modulus to a greater degree. In continuous-purlin systems, reinforcement may be required only in the critical bays. The negative-moment capacity typically controls the design, and reinforcement is most needed at the supports, where complications arise because of splices. As discussed previously, the longer the lap, the larger the degree of continuity. To find out whether the existing lap length is sufficient, one may use specialized computer software devoted to the design of cold-formed framing.

When the whole roof must be strengthened, reinforcing each piece of cold-formed framing is usually not cost-effective. It may be more practical to place additional members between the existing purlins or girts. Special care should be given to bracing the new purlins, because it may be impossible to attach them to the roof, which in some cases could provide lateral bracing for the existing purlins. Because the new purlins are difficult to maneuver into place, they are limited to single-span lengths.

When strengthening is required not only for secondary framing, but also for the primary framing, adding intermediate frames (see Sec. 10.4) might prove to be the most economical method. But when new roofing or siding is also needed, the obvious approach is to replace the existing secondary framing, or to add the new members, along with the metal panels. In this case, continuous framing can be used, since roofing or siding will not be in the way.

Sometimes, strengthening is required simply because a concentrated load, such as a heavy pipe hanger, must be carried. Instead of reinforcing the purlin intended to carry the load, it might be easier to spread the load to the adjacent purlins by hanging a rigid spreader beam under them.

Strengthening of C and Z sections includes not only an increase in moment-resisting capacity, but also the means of load transfer at supports. As already mentioned, purlins can be attached to frames either by direct bolting or by bearing clips. Unless a time-consuming analysis of web-crippling, shear, and combined stresses is performed, using a bearing clip at each support might be a good idea. If manufactured clips are not available, pieces of large enough hot-rolled angles will do. The clips will not only transfer the load from the C and Z sections to the primary frames, but will also provide lateral stability for these secondary members at supports. If sheet-metal clips are used, so-called antiroll clips may be helpful in bracing the purlins at supports. These are essentially short diagonal braces attached to the purlin near its top flange and to the rafter.

10.7 Wall Materials

10.7.1 Basic types

The first pre-engineered buildings had walls of plain corrugated galvanized steel panels. By the late 1950s and early 1960s, panels with sprayed and baked-on paint had appeared, and in a few years the first factory-insulated wall panels were being made. Also in the early 1960s, ribbed metal siding started to supersede the tired corrugated design, and in the 1970s a variety of siding profiles became available. The evolving technology made it possible to form metal panels from factory-coated coils, and eventually a variety of durable finishes with a life expectancy of 20 years or more were produced.

The basic wall structure has remained the same throughout the years. Then, as now, metal wall panels were typically made of 24-, 26-, or 28-gage galvanized steel. They could be shop- or field-assembled and could have either exposed or concealed fasteners. The siding panels were laterally supported by cold-formed C or Z girts; the typical span has stayed at about 7 ft.

A field-assembled metal panel of the type most likely to be encountered in existing metal building systems includes steel siding and, perhaps, fiberglass insulation. Occasionally, it also has a liner sheet on the interior surface. A properly assembled panel has a bottom closure strip made of rubber, metal, or foam to seal off the wall from intrusion of insects, dust, and vermin. Window openings are framed with wall girts or framing channels.

A shop-assembled panel may include the same three components, preassembled and delivered as a unit. Shop assembly saves field labor, provides better fit, and allows the use of rigid insulation with better R values than fiberglass.

Both field-assembled and shop-assembled panels can be attached to structural girts with exposed or concealed fasteners. Exposed fasteners of various designs are typical, the best being self-drilling screws with neoprene washers, which are standard today. The concealed-fastener design, where screws are covered by adjacent panels, offers better corrosion protection and better appearance than siding with exposed fasteners. Metal walls with concealed fasteners have gained popularity in recent years but are relatively rare in existing buildings.

Some pre-engineered buildings have nonmetal ("hard") walls, most commonly of concrete masonry units (CMU). Walls made with brick veneer or with concrete, whether cast in place, precast, or tilt-up, are occasionally encountered. Hard walls are used for toughness, fire resistance, or appearance. Incorporating them in metal building systems involves special challenges, because flexible metal building systems are intended to work with forgiving metal finishes rather than rigid hard walls. To function properly, concrete and masonry walls require careful detailing and a higher level of rigidity of the metal framing than is usually provided

Unfortunately, hard walls in metal building systems historically received little design attention—with sad results. Figure 10.14 illustrates one such building with CMU walls that had no reinforcement in either direction and had control joints spaced as much as 50 ft apart. Each wall corner was cracked from restrained expansion and contraction, and cracks were also present midway between the existing control joints. The interior of the wall had serious water damage, shown in Fig.

10.12. This building also had loose wall bracing in front of the CMU— a poor design, as discussed in Sec. 10.5. Whenever hard walls are encountered in existing metal building systems, their performance and construction should be critically examined during field investigation.

10.7.2 Renovation of metal walls

Metal siding deteriorates in a familiar fashion, with gradual fading of the finish and eventual rusting. The fasteners corrode and loosen, leaving the siding vulnerable to blow-off during a hurricane (Fig. 10.15). The details of every accidental impact are imprinted and saved for future generations.

Figure 10.14 This CMU wall in a pre-engineered building had not received proper design and construction attention.

Figure 10.15   Rusted and missing wall fasteners.

The most basic kind of wall rehabilitation is repainting with a product specifically formulated for this application. As with any repainting project, proper surface preparation is critical. The siding should be prepared in accordance with the paint manufacturer's instructions, which generally require removal of the proverbial loose rust, dirt, oil, and other contaminants. Special attention should be given to preparing the panel edges, especially the bottom edge, even though this may be difficult, because rust originating in these areas tends to spread to the back of the panel.

Repainting will not help siding that is rusted through, such as that shown in Fig. 10.16. Here, a fascia panel above the industrial garage door has been perforated by rust at several spots along the bottom to such a degree that sunlight shines through, while the rest of the siding is barely affected. How could this happen? One explanation of this common occurrence is that the bottom of the fascia is covered with metal trim that collects moisture, and the moisture attacks the adjacent areas first. Also, rust typically starts at the cut panel ends, where the metal finish is interrupted, even though galvanizing theoretically has some protective properties that extend slightly beyond its area of application. Panels that are seriously deteriorated are best replaced.

The replacement siding must be compatible with the existing support system and must be able to span between the existing girts. It is wise to evaluate the structural condition of the girts before proceeding with siding replacement. The girts should be checked for proper size, spacing, extent of corrosion, presence of lateral bracing, and adequate connections. If the girts are deficient in any respect or are spaced too widely for typical wall siding (that is, their spacing exceeds about 7 ft), new matching girts can be added between the existing ones, or extra-

Figure 10.16   This metal siding is characteristically rusted through at the bottom.

deep panels can be specified. This approach applies to all kinds of metal siding, even to curved wall panels, which can be useful in dressing up tired metal buildings and in rehabilitation of old Quonset huts.

In some renovation projects, architects looking for visual impact may want to run wall siding horizontally, rather than vertically. While horizontal siding may indeed be more interesting, it requires that the girts run vertically, from the foundation to the eave strut, rather than from column to column. This introduces two complications. First, the light-gage eave strut now has to support unanticipated wind loading from the vertical girts, which usually means that it must be removed and replaced with a new heavy structural steel section. Second, the frame columns are now substantially weakened, because they have lost the bracing that the existing girts provided. Here, the best solution may be to run short pieces of verticle channels between the existing horizontal girts.

A typical exterior-wall renovation project includes replacement of windows, doors, flashing, and perhaps even framing around the openings for a completely coordinated and tight-fitting exterior envelope.

10.7.3   When hard walls are required

A change in the use or ownership of the building may require major changes in its appearance. Sometimes, there will be a desire to replace exterior metal siding with hard walls because of functional, security, fire-resistance, or durability concerns. The replacement should never be made without a structural analysis of the existing building to determine whether it is rigid enough to laterally support hard walls. Hard walls demand a much higher level of rigidity in the metal framing than that acceptable for flexible metal siding. Existing pre-engineered buildings are usually not rigid enough to provide proper lateral support for masonry or concrete exteriors.

An attempt to increase the lateral stiffness of the existing metal building can be made, but is not likely to be cost-effective. When hard walls are required, it might be more economical to replace the whole building, especially if it needs some other major work. Metal building systems can be replaced without interrupting the operations. One method is to build new exterior loadbearing hard walls and new roof purlins that enclose the existing building. Then, the old structure can be removed piece by piece. Naturally, this solution works only for buildings of moderate size. Another possibility is to build the hard walls inside the building perimeter, if space permits, and to completely isolate them from the metal structure. This design involves a significant loss of floor area, mainly because the walls must be cantilevered from the newly constructed foundations for them, which take space.

It is important to remember that brick veneer is also a hard wall that demands a rigid structure for lateral backup. Common practice dictates that brick be backed up by a system (steel studs, for example) that deflects no more than the height divided by 1/600 or 1/720 (the $H/600$ or $H/720$ deflection criterion). This level of rigidity is not likely to be afforded by typical wall girts. Therefore, to change the look of a metal building to brick, major strengthening of wall supports is required. Unfortunately, a common "solution" in some places is to simply attach brick ties to the existing metal siding and use it as a backup for brick. It goes without saying that brick that is laterally "braced" by the flexible siding, which can tolerate lateral deflections of $H/60$ and even larger, is headed for trouble.

10.8 Metal Roofing

10.8.1 Classification of metal roofing

Pre-engineered buildings have traditionally been covered with metal roofing. There are many kinds of metal roofing available, and designers involved in structural renovation of metal building systems should be familiar with at least the most common types. As we shall see, the type and slope of the roof should be appropriate for the building size and structure; otherwise, neither repairs nor replacement in kind will be successful.

Metal roofing can be classified in several ways. The first is by the degree of water resistance: Metal roofs can be divided into water-shedding and waterproof. Water-shedding roofing acts similarly to roof shingles: It needs a steep slope, such as 4:12 or 3:12, to rapidly remove rainwater. Water-shedding roofing requires underlayment (typically a 30-lb roofing felt) and sometimes a paper slip sheet to reduce friction during temperature movements. In contrast, waterproof, or water-barrier, roofs are intended to resist standing water, but not necessarily to be completely leak-free under a long-term immersion. The weak points of this roofing are the valleys, eaves, ridges, rakes, and penetrations.

To function properly, especially in cold regions where ice dams are common, waterproof roofing requires a minimum slope of about 1:12; this is substantially larger than the commonly provided slope of ¼:12. Why the disparity? A slope of ¼:12 is theoretically sufficient for drainage but leaves no room for many practical occurrences that may result in a roof that is flat in some places or even cause local depressions. Here are some of them: deflection of primary and secondary structural members under load; slightly uneven settlement of foundations; tolerances in the fabrication and installation of purlins and roofing panels; and local deflection of purlins under concentrated loads, such as mechanical equipment or suspended piping and light fixtures. As a result, a metal roof with this minimal pitch can become flat in a few spots, at which ponded water will collect. Then, any minute imperfection in roofing seams in these locations will invite leaks.

The second method of classification is by support requirements, with metal roofing divided into architectural and structural. Architectural roofing needs continuous structural support provided by decking or by closely spaced furring channels. Structural roofing is capable of spanning the distance between the roof purlins, typically 5 ft. As it happens, architectural roofing is frequently of water-shedding design (Fig. 10.17), and structural roofing often functions in a waterproof manner.

And finally, metal roofing can be classified on the basis of the method of attachment to supports. Through-fastened roofs are connected directly to purlins by screws or rivets. Standing-seam roofing is attached to purlins indirectly by concealed clips formed into the seams. The seams of standing-seam roofing, being the weak point of the system, are elevated 2 to 3 in above the flat part that carries water.

Most existing metal building systems are covered with galvanized through-fastened roofs with lapped seams (Fig. 10.18). This traditional design is inexpensive and easy to install, but it has two serious flaws.

First, as the name implies, this type of roofing is penetrated by fasteners, and each penetration invites water leakage. Self-tapping screws with neoprene washers help protect the fastener holes to some degree, but how effectively they do so depends on the skill of the installer. If the

Figure 10.17   Architectural roofing of water-shedding design.

screws are overtightened, the neoprene washers get squashed and become useless. If the screws are not tightened enough, the seals are not completed and the penetrations are not properly protected.

Second, the panels are firmly held by their fasteners and cannot easily expand and contract with temperature changes. On large roofs that undergo drastic temperature swings, through-fastened roofing can buckle, but far more typical damage is tearing of metal around the fasteners. The resulting irregular holes are unprotected and allow water intrusion. Still more dangerous is loosening of the screws' grip on the purlins: It could lead to the roofing being blown off during a hurricane.

On a relatively narrow roof, these disadvantages are not as harmful as on a wide one. A common rule of thumb is to limit the width of buildings with through-fastened roofs to about 60 ft, measured in the direction of the roofing span.

10.8.2   Evaluation of existing through-fastened roofs

Leaking roofs are among the most common complaints concerning vintage pre-engineered buildings. Indeed, roofs are the most frequently upgraded elements of metal building systems. Leaks generally start around the holes of through-fastened roof screws, around roof penetrations, at side laps, and at roof transitions such as ridges, valleys,

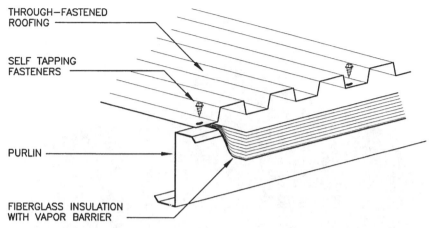

Figure 10.18   Through-fastened metal roofing.

and eaves. Metal roofing may require rehabilitation not only because of leaks, but also because of such characteristic problems as backed-out fasteners, rusted panel ends, opened-up seams, displaced metal trim (Fig. 10.19), and failed finishes.

Prior to attempting any remedial work on an old metal roof, a comprehensive roof evaluation should be made in accordance with the general procedures outlined in Chap. 2. The roof's condition can be assessed during a walk-through, supplemented by observation from within the building, by interviews with the building personnel, and by testing and analysis, if those are needed. The first item to check is whether the size and slope of the roof are appropriate for through-fastened roofing. The width of a building with through-fastened roofing normally should not exceed 60 feet; for a wider building, a standing-seam roof should be considered. As just discussed, a minimum roof slope of 1:12 is preferred. If the existing pitch is only ¼:12, and the roof is in such poor condition that replacement must be made, the slope can be increased by the methods described in Sec. 10.8.6. Even if the roof deterioration is not excessive, a slope buildup is beneficial, especially for buildings in snow regions and those subjected to massive amounts of rain.

The next item to check during a walk-through is the spacing and condition of the existing fasteners. The minimally acceptable size and spacing of the fasteners depends on the roof design loading and the diaphragm needs. For a preliminary check, the fastener spacing required by the U.S. Army Corps of Engineers specifications can be used: a maximum of 8 in at end laps and at all intermediate supports, and 12 in at side laps.

The condition of the fasteners can be evaluated visually, perhaps with some trial pullout tests. The goal is to determine whether the

Figure 10.19   Metal trim often fails before the rest of the roof.

screws are rusted and whether they are coated with a corrosion-resisting finish. The most durable fasteners are those made of stainless steel or aluminum, but at the very least the fasteners should be galvanized or cadmium-plated. Determination of the fasteners' condition is critical to making the decision about the roof's future.

Depending on the results of the roof evaluation, repairs can range from recoating to replacement. To be effective, repairs must be done properly, and the lowest bidder will not necessarily devote the time and money to study and solve the problem. Indeed, repairs of metal roofing often cost much more than the existing roof did. Stanford[15] mentions one project where the original roof construction cost was $700,000 and the cost of repairs was over $1.6 million; he also gives other similar examples. He attributes this disparity in cost to several factors. The first is the need to find and undo the effects of previous botched renovations, which in many cases have worsened the original problem. This task may take longer than to install a roof in a new building. Also, meticulous construction simply costs more than shoddy work. And finally, repairs in an occupied building require disassembly of piping, lights, and finishes, as well as a need for continuous weather protection of the building's interior—the constraints that are not present in new construction.

10.8.3   Renovation of through-fastened roofs by recoating

Recoating can make sense for adequately insulated roofs of proper size and slope that suffer from fading and discoloration of finishes, and perhaps from some surface rusting, but have fasteners that are only slightly affected by corrosion.

The available recoating products include many high-performance elastomeric coating systems that are specifically formulated for repair of aging metal roofs. These products, applied by brushing or spraying, require various degrees of surface preparation, ranging from a power wash to complete stripping of finishes. A roof of moderate size can usually be recoated in one seamless application. Some systems, but not all, rely on a primer coat for proper adhesion and fabric reinforcement for strength. The best membranes are made of almost pure rubber; others are "rubberized"—made of asphalt with just a bit of rubber. One system consists of a silicon membrane sprayed on top of polyurethane foam insulation, which is also sprayed in place. All these products are relatively new, and the specifiers should investigate the actual service performance of the proposed system before committing to it.

Like any refinishing project, a successful roof recoating requires proper surface preparation and repair of deteriorated areas. This includes a close examination and possibly replacement of the components that invite most leaks, among them the trim and flashing at roof slope changes and penetrations. Particular attention should be directed to the edges of holes around the existing screws—in the purlins as well as in the roofing. As discussed previously, temperature swings in through-fastened roofs may cause tearing of metal around the fasteners and allow water intrusion into those areas. The ensuing corrosion can affect not only roofing panels and roofing fasteners, but also the purlins. When the purlins are rusted, the holding capacity of the existing screws is compromised and new fasteners are required. If deterioration has gone that far, installation of new roofing, with or without the removal of the old, may be more economical than recoating.

In general, then, recoating is not the best answer in situations where a lot of surface rust exists in the roofing, where some panels are rusted through, and where fasteners are corroded. While it is possible to replace a few worst-case panels and to abandon the existing rusty screws to their inevitable fate and install new fasteners in between, this course of action is usually not as cost-effective in the long run as a brand new roof, which will come with a new-roof warranty. Still, even in such circumstances, recoating can be useful for temporary protection when absolutely no disruptions of the building operations can be tolerated.

10.8.4 Reroofing with through-fastened panels

Reroofing can be done in two ways: with a complete removal of the existing roofing, or with installation of new metal roofing over the old.

The removal option can make sense if the existing insulation needs replacement, as is typical. Removing the roofing facilitates insulation replacement but opens up the building to the elements.

Prior to specifying removal of the old roofing, the designers should investigate the condition of the roof purlins to check whether corrosion has spread to their upper flanges. To do this, selective roofing removal may be needed. Apart from rusting at the fastener hole locations, which was just discussed, purlin rust can be caused by repeated roof leaks, or even by serious condensation. For example, the roof in Fig. 10.20 was rusted through to such a degree that sunlight could be seen inside; major leakage had led to serious rusting of the adjacent purlins.

Corroded purlins with diminished load-carrying capacities can be replaced or augmented with new purlins. Replacement of both roofing and purlins not only is a major inconvenience in terms of building operations, but might also trigger a requirement of the local building code that the whole building comply with the code for new construction. In that event, the local building official could require a complete structural upgrade of the building. If a structural evaluation finds that the primary framing system is also deficient and needs strengthening, building replacement should be seriously considered. As stated already, it is possible to construct a new metal building enclosure around the existing facility and then remove the old building piece by piece—probably the least intrusive and the fastest course of action.

This scenario is of course extreme, and in most cases the purlins need not be replaced. All that is typically needed is a new layer of metal roofing, through-fastened or standing-seam, as determined during the previous analysis. (Working with standing-seam roofing is covered in the next subsection.)

If the existing metal panels are removed, the new through-fastened roofing must be able to span the distance between the existing purlins. The installation of new roofing is preceded by placement of fiberglass insulation, as in a new metal building system. Now is the chance to use high-quality roof fasteners with durable corrosion protection.

If the existing metal roof stays, new roofing can be installed on top of it, usually with added insulation in between, and with new fasteners placed between the existing ones. The critical detail is to make certain that the new roofing is attached to the existing purlins, not to the old panels, even though this may require careful field measurements and marking of purlin locations on top of the insulation.

Whenever insulation is added, roof slope is increased, or new roofing does not match the existing roofing profile, the new metal panels will not fit within the corrugations of the existing roofing. The new roofing must then be elevated some distance above the existing roofing with the help of new structural framing, typically light-gage hat channels

Figure 10.20 When sunlight is showing through the roof, the top flanges of the purlins could be seriously rusted.

or pressure-treated wood 2 × 4s laid on top of the existing roofing directly above the existing purlins.

To be sure, there are special roofing panels on the market that are designed to fit into the corrugations of some popular roofing profiles. These "retrofit" roofing panels are specifically made for reroofing applications, and some of them even include factory-applied self-adhering vapor retarders. The major selling points of these products are that they can be attached directly to the purlins without any additional structural framing and that they come with new-finish warranties, but they have the obvious limitations of not allowing for placement of added insulation or any increase in the roof pitch. Also, their structural capacity is limited by the original profiles, and some products may not be able to support the heavy wind and snow loads prescribed by modern building codes.

Placing structural framing on top of existing roofing also has major limitations. The most serious is the existing roofing itself, since a thin and rusted old roof makes for a doubtful support. Some field and analytic checking of the roofing—especially its web's compressive capacity—should be made before proceeding with this option. When the numbers are run for both downward and uplift loads, this method of attachment may end up being structurally unacceptable. Another problem: The self-drilling screws connecting the new support framing to the purlins cannot be driven tight because they would tend to crush the existing roofing, and their compressible washers may not become effective, inviting leaks.

A better connection bypasses the old roofing and is made directly to the purlins via blocking that fits within the roofing corrugations. The blocking can consist of short pieces of the same hat channels or pressure-treated wood 2 × 4s that will run on top of it. Another possible solution is to install structural channels with their webs horizontal and on top to span over the existing roofing corrugations. The channel flanges can be field-welded to the purlins through the existing roofing (which is usually made of 26-gage steel). Or, there is a new proprietary product designed for this purpose, Roof Hugger,* a short galvanized Z purlin notched at the bottom flange and at the web to match common roofing profiles.

Whichever way the new through-fastened roofing is installed, it should be installed correctly. The new screws must be driven to the proper depth perpendicular to the roof, so that gasketed washers are uniformly and properly compressed. All panel joints should include the sealants recommended by the roofing manufacturer; all sidelaps should be laid away from the prevailing winds. The exposed fasteners can be supplied with a color-coordinated head finish or be provided with colored plastic caps to reduce their visibility.

10.8.5 Replacing old roofing with standing-seam panels

Standing-seam roofing is superior to through-fastened in many respects. Most importantly, instead of being directly fastened to the purlins, standing-seam panels are attached to them with concealed clips that allow the roof to move during thermal expansion and contraction. Also, while sheets of through-fastened roofing are simply lapped together with a bead of sealant in between, the best standing-seam joints are elaborately formed by seaming machines. The most durable seams are of the so-called Pittsburgh double lock design, a 360° roll-formed seam that rivals in quality the seams in cans used for food. Other seam designs include roll-and-lock seams made with hand seam-making tools and seams that are simply snapped together.

Standing-seam roofing is old enough to be encountered in metal buildings requiring renovations, although the old designs are quite different from those used today. The first standing-seam roofing, intro-

---

*Roof Hugger is a registered trademark of Roof Hugger, Inc.

duced by Armco Buildings in 1934 had exposed fasteners; the concealed-clip design was introduced only in 1969 by Butler Manufacturing Company.

Standing-seam replacement roofing should be considered for buildings wider than 60 ft. The maximum size of an uninterrupted standing-seam membrane is about 200 ft, and expansion joints are needed beyond that. This maximum length is limited both by the expansion capacities of the standard roofing trim and by the amount of movement the concealed clips can provide. The expansion joints in newly constructed metal building systems with standing-seam roofs are typically of stepped design to allow for proper flashing details. This solution is obviously impractical for existing large pre-engineered buildings, and custom details must be used for reroofing. Some details for standing-seam roofing can be found in Ref. 2.

One nuance should not be overlooked when an old through-fastened roof is removed and replaced with new standing-seam roofing: The former provided lateral stability for the purlins, and perhaps even some diaphragm action for the building, while the latter is probably not capable of either. As we mentioned already, the most advanced "articulating" clips (Fig. 10.13) are especially unsuitable for purlin bracing. We recommend that whenever standing-seam roofing is specified, the existing purlins be provided with new lateral bracing. Also, a new roof diaphragm may be required if the building has no horizontal rod or cable bracing.

The process of evaluation and rehabilitation of existing through-fastened metal roofs can be rather laborious. One way of approaching it in a systematic manner is to follow the flow chart of Fig. 10.21, which summarizes the steps discussed in this section.

10.8.6 Methods of building up slope

For very small roofs, minor slope adjustments, such as changing the slope from $\frac{1}{4}$:12 to $\frac{1}{2}$:12, can be made by varying the sizes of new structural framing (light-gage framing or pressure-treated wood) laid on top of the existing roofing above the purlins. In most cases, however, the framing sizes required by this method quickly become impractical. The only other possible way of increasing the slope is to erect a light-gage framework of varying height.

The retrofit framework, as it is called, is typically built of light-gage purlins, columns, and bracing (Fig. 10.22). The new supports are located directly above the existing roof purlins. The new framework can provide any required slope and still add no more weight than 2 to 4 psf, including the new metal roofing. Usually, this small additional load can be safely accommodated by the existing roof structure; if not, a system of trusses can be designed to span directly between new stub

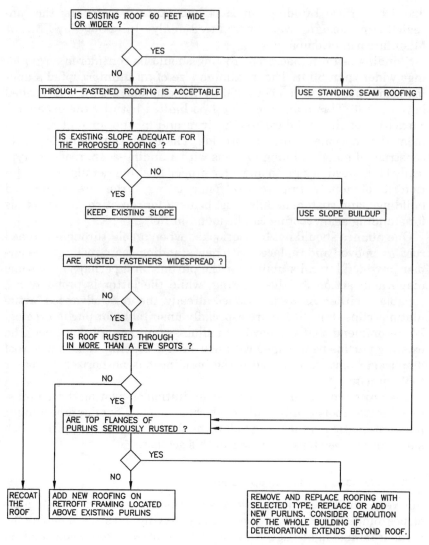

Figure 10.21   Flow chart for rehabilitation of metal roofing in metal building systems.

columns erected on top of the existing primary framing columns, without any reliance on the existing roof purlins. In any case, the structural capacity of the existing roof should be evaluated by the engineer of record before proceeding with reroofing design and system selection.

There are many manufacturers of metal reroofing systems, and most of them use their own proprietary framing and details. None of them will usually perform a structural evaluation of the roof or its capacity to

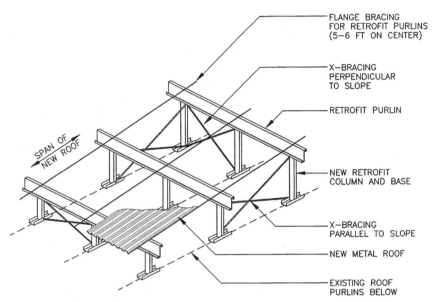

FLANGE BRACING
FOR RETROFIT PURLINS
(5-6 FT ON CENTER)

X-BRACING
PERPENDICULAR
TO SLOPE

RETROFIT PURLIN

SPAN OF
NEW ROOF

NEW RETROFIT
COLUMN AND BASE

X-BRACING
PARALLEL TO SLOPE

NEW METAL ROOF

EXISTING ROOF
PURLINS BELOW

Figure 10.22   The retrofit framework for slope changes.

support additional loads. Indeed, the manufacturers prefer to limit their responsibilities to the design of the metal components and place disclaimers in their literature to that effect. Still, the engineer of record must know the spacing of the retrofit columns and the loads they impose on the roof, or else base the calculations on the uniform-load capacity of the roof, an unconservative approximation. What can be done? One way of solving this dilemma is to perform the analysis based on a tentatively selected system and to involve that manufacturer in the design process. The analysis will then be based on the layout of supports proposed by the manufacturer, and that fact will be communicated to all the bidders, who would then be obligated to follow the same general scheme.

The retrofit framework should be as strong and rigid as a typical metal roof structure, with the same maximum purlin spacing—5 ft, and perhaps less in areas subjected to high winds, such as those near the eaves, rakes, and ridge. Like any cold-formed members, the retrofit purlins need lateral flange bracing for stability under both downward and upward loading. We recommend bracing the top and bottom flanges of the retrofit purlins at intervals not exceeding 5 to 6 ft, as discussed in Sec. 10.6.1.

The framework must be designed not only for the loads acting on the new roof itself, but also for the wind loads on the new gabled end walls. The minibuilding supported by the retrofit framework requires the same kind of vertical wall bracing as any metal building system. It also needs a horizontal diaphragm in the plane of the new roof, provided by steel rod or cable bracing or by a steel deck. To be effective against wind uplift, the bracing and retrofit columns must be properly anchored to the existing roof and wall structure. The anchorage details should be designed by the metal system manufacturer for the actual existing roof structure and carefully checked by the engineer of record. Any structural problems with retrofit framework are most likely to occur at the interface of the new and existing roofs.

A problem of classification often arises in dealing with the new "attic space" between the two roofs. It needs to be carefully assessed in terms of code requirements for fire safety, ventilation, and egress. A new fascia made of architectural wall panels is typically needed around the perimeter of the building to enclose the framework.

Another common question is what to do with existing rooftop heating, ventilation, and air-conditioning (HVAC) equipment. Small curb-supported units and exhaust fans can be moved to new curbs on the retrofit roof surface without incident. Alternatively, there are special metal extension curbs that can go on top of the existing curbs. Vent pipes can be extended. But the heavy equipment—chillers, air conditioners, and cooling towers—is very difficult to support on the lightweight retrofit framing. There are two possible solutions: to include structural steel beams in the retrofit framework (the beams will probably have to span between the stub columns placed on top of the existing building columns), or to leave the equipment where it is and totally or partially enclose it by the retrofit roofing. In the second case, the space between the existing and retrofit roofs may be considered as a mechanical penthouse of sorts and the new gable walls or fascia filled with louvers.

A final note of caution. When the slope buildup is drastic, the effects of the new "penthouse" on the existing building structure should be carefully assessed. Moreover, if a very extensive roof buildup is undertaken in combination with other major building renovations, the combined value of the upgrade could potentially exceed the limit beyond which the building must be upgraded to conform to the code for new construction.

10.9 Insulation and Vapor Retarders

Insulation and vapor retarders are admittedly not structural components. Still, they are part of metal building systems and should be at least acknowledged here. The first pre-engineered buildings and the factories of the 1940s were not insulated at all. Insulation continued to suffer from a lack of attention until the oil crisis of the 1970s woke up building owners to the perils of ignoring it. Since then, the prevalent method of insulating metal building systems has been to sandwich fiberglass blanket insulation between the metal roofing or siding and the supporting secondary members (Fig. 10.18). The squashed insulation does not perform very well, and heat transfer through the points of attachment is barely slowed. Later, thermal blocks made of rigid insulation were introduced to mitigate the problem of thermal bridging. In the 1980s and 1990s, several more refined insulation systems were developed. For those who are interested, this topic is further explored in Ref. 2.

One insulation-related factor that directly affects structure (by ruining it) is condensation. Condensation occurs when warm air, which carries more moisture than cold air, meets a cool surface. Condensation not only leads to corrosion of framing and metal covering, but also causes growth of mold and mildew, damages finishes, and degrades insulation. In humid climates and in buildings in which moisture is released, condensation can be so heavy that it can be mistaken for roof leaks!

Condensation on the underside of metal roofing can be reduced by a vapor retarder, a relatively impermeable material that slows down the movement of moisture through insulation. In metal building systems, vapor retarders are laminated to the insulation and installed on the warm side of the wall. (Condensation on the roof structural members cannot, of course, be prevented by a vapor retarder located above them—the only effective design would be to place the retarder *below* the members, a possible but very rare solution.)

The measure of the effectiveness of a vapor retarder in slowing down the transfer of moisture is reflected in its permeability. Permeability is measured in perms in the British system (a perm is one grain of water transmitted per hour per square foot for one inch of mercury vapor pressure differential) or in nanograms per pascal per second per square meter in SI. The lower the perm rating, the more effective the vapor retarder is. The perm ratings of commonly used materials range from 1.0 for vinyl to 0.02 for the newest composite materials, such as foil-scrim-kraft or metallized polypropylene scrim-kraft. When a new vapor retarder is added to an existing building, it is foolhardy to opt for the cheapest material (usually plain vinyl), as it will have a large perm rating. If this effort is made at all, it is wise to specify a retarder with a perm rating of 0.02.

To be effective, a vapor retarder should have all its edges and all penetrations properly sealed. A torn or degraded vapor retarder, such as that shown in Fig. 10.23, compromises not only the insulation properties of the building, but also its structural durability. Even the most minute imperfections in the vapor retarder can degrade its performance. Lotz[16] points out that a single staple increases the permeability of a foil-kraft laminate vapor retarder from 0.02 to 0.34 perms. He even states that it is better to use the cheapest vapor-retarder material that is properly sealed than the best product with unsealed joints or holes. For best results, sealing tapes specifically formulated for the particular laminate material should be used.

10.10   Renovation Checklist

The chances for successful renovation of a metal building system will be improved if the contractor is given adequate information about the existing structure with proprietary framing and about the design intent of the engineer of record. The contractor will typically be affiliated with a metal building manufacturer and will be tempted to apply the design practices of that manufacturer, which may or may not coincide with the expectations of the engineer. Therefore, we recommend that construction documents for renovations of pre-engineered buildings be as specific as possible, unless the project is of a true design-build type. At a minimum, the following information should be included (adapted from Newman[2,17]):

- The governing building code and edition. Do not list too many codes—they may contain conflicting design criteria, a conflict that could be exploited by some bidders who are hungry for work.

Figure 10.23 A deteriorated vapor retarder hastens the degradation of roof insulation and structure.

- Design live, snow, wind, and seismic loads and load combinations, if these are not included in the code or if additional combinations are to be considered. If these are not provided, the contractor will probably use the provisions of MBMA *Low-Rise Building System Manual*.[5]

- Information about the existing building, including the structural scheme and dimensions (use the terminology of measurement in Sec. 10.1). Include the original drawings, if available.

- Contract drawings for all work to be done. For reroofing, include a plan and sections of the proposed roof, with dimensions, profile, desired slope, and finish; specify structural requirements for retrofit framework, such as column spacing, bracing, and general diaphragm construction.

- For reroofing, indicate whether the existing roof is to be removed or retained and provide required UL and insurance ratings for the new roof.

- For reroofing with slope buildup, require the contractor to submit detailed shop drawings and structural design calculations for the retrofit framework. For critical applications, require a seal of a professional engineer on these documents.

- For exterior wall replacement, specify materials, allowable deflection or drift criteria, finishes, corrosion protection requirements, and insulation.

- If primary and secondary framing is to be strengthened, list the allowable vertical deflection criteria.

- If primary framing is to be strengthened specifically to reduce lateral drift, list the maximum allowable drift value.

10.11   Case Study*

10.1.1   Background

A government agency wished to renovate a maintenance shop located on a waterfront pier. The single-story pre-engineered building, constructed in 1953, measured approximately 217 ft by 43 ft. Widespread corrosion of the framing, buckled structural members, and general disrepair strongly suggested that the building was well past its useful life and should be replaced. Unfortunately, no replacement funds were

---

*This case study is based on a project designed by Maguire Group Inc. of Foxborough, Mass., which supplied the accompanying illustrations.

available, nor were any anticipated for several years. The agency asked for a short-term renovation program that would enable the building to be used for another 5 to 7 years.

10.11.2   Field investigation

The first step in rehabilitation consisted of visual inspection and condition assessment. During two field visits, the inspecting team discussed the history of the building and the loading it had seen. The team learned that the building had survived a few hurricanes and at one time was flooded with up to 2 ft of salt water. The existing drawings were located and taken for office analysis.

According to the drawings, the structure was manufactured by Stran-Steel Corp., a pioneer of metal building systems that invented many now-common metal building components. It was no longer in business at the time of the renovation. The drawings indicated a design yield strength of light-gage steel (F_y) of 40,000 psi. The roof structure was framed with pre-engineered light-gage metal trusses spaced at 4 ft on centers and supported at each end by built-up light-gage steel studs. The bottom chords of the trusses were braced by channels anchored into the endwall studs. Truss members consisted of unstiffened light-gage channels welded together. A typical truss is shown in Fig. 10.24.

The roof diaphragm consisted of horizontal cross-bracing at the top of the trusses. This diaphragm was designed to transfer wind loading to vertical bracing bents (called "wind panels" on the drawings) located in the endwalls and at four interior locations near the sidewalls.

The 12-ft-high sidewalls were framed with $3^5/_8$-in 16-gage light-gage steel studs. The building was clad in the original solid panels, which were made of a mixture of cement and asbestos. The panels were laterally supported by light-gage metal girts of the same size as the studs and spanning between the studs that supported the trusses. At the roof, the $1^9/_{16}$-in-thick solid cement-asbestos boards were supported by 2 × 4s spanning between the panel points of the trusses, a design that helped avoid local bending of the top chord. The building structure was largely covered with wall and ceiling finishes and was not accessible for a full visual evaluation. Therefore, the assumption had to be made that structural members in the areas that could not be readily accessed were in a condition similar to that of the ones that were accessible.

The main structural problems observed during field investigation included

1. Buckled and distorted vertical, diagonal, and bottom-chord truss members (Fig. 10.25). The unstiffened edges of many diagonal and vertical channels showed waviness. The bottom chords of two trusses were particularly distorted, probably because of damage by occupants

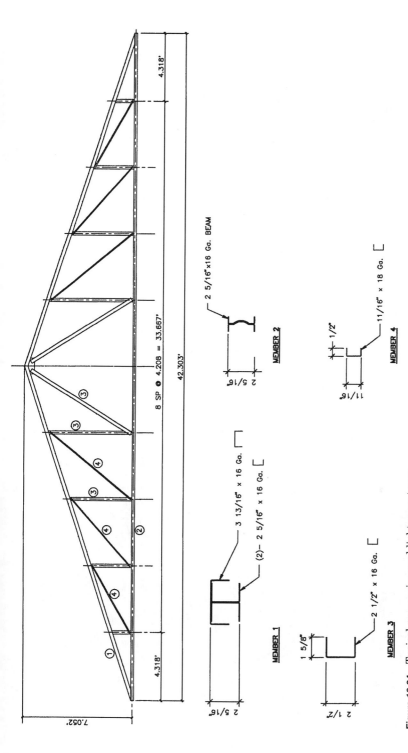

Figure 10.24 Typical pre-engineered light-gage truss.

615

or moving equipment. The bottom chords of the trusses were being used for storage of building materials, even though the trusses were not designed for any bottom-chord loading.

2. The bases of most wall studs supporting the trusses were rusted through to the point where the studs were structurally nonexistent (Fig. 10.26). The building appeared to be supported by the siding alone.

3. At three of the four interior locations, the wind-panel bracing had been removed, severely weakening the building's resistance to lateral loads in the short direction. In one case, only the top of the bracing remained; the bottom half had apparently been cut off during previous alterations (Fig. 10.27).

10.11.3 Analysis and evaluation

Field investigation confirmed that the maintenance shop was extremely deteriorated and needed replacement. It was obviously cost-prohibitive and impractical to reinforce or replace every structural member that did not comply with the current building codes. The owner well understood that any proposed repairs would not turn the building into a brand-new facility and that the purpose of the renovation was to increase the level of structural safety while funding for the replacement was being sought.

Figure 10.25 Buckled diagonal and bottom-chord truss members.

Figure 10.26 The rusted-through bases of loadbearing light-gage studs.

A typical roof truss was analyzed using the present-day code loads for wind and snow, as follows:

Design roof snow load: 25 psf

Design roof wind load: 12.6 psf suction plus 4.2 psf internal pressure (a total of 16.8 psf), assuming the truss to be a "main structural member"

Design dead load:

Asphalt shingles	2
$1^9/_{16}$-in roofing panels	10
Own weight of truss	2
Mechanical, miscellaneous	3
Total	17 psf

Figure 10.27 The bottom half of this wind panel had been cut off during previous alterations.

The existing drawings indicated a design dead load of only 8.6 psf. This number was difficult to accept, but to simply cut out a piece of asbestos-containing roof board and weigh it was equally difficult. Therefore, a dead load of 17 psf was used in combination with the design snow load, but, conservatively, the 8.6-psf loading was used in combination with the wind uplift loading. The truss analysis shown on the original drawings did not consider any wind loading at all, a common practice in the 1950s.

A computer analysis was performed for the loading combinations consisting of dead + snow and dead + wind uplift loads just listed. All the loads were applied at the panel points of the trusses. The results indicated that the bottom chord and diagonal members, which had been originally designed only for tension, became the compression members under wind uplift loading. The analysis results produced the following maximum forces:

- The bottom chord: 9.46 kips tension or 2.18 kips compression
- The diagonals: 2.05 kips tension or 0.46 kip compression
- The verticals: 0.39 kip tension or 1.76 kips compression
- The wall studs: 3.53 kips compression or 0.81 kip tension

10.11.4   Design of reinforcement

For practicality of construction, only two sizes of reinforcement members were selected, one for the bottom chord and one for the diagonals and verticals.

The reinforcement for the bottom chord was designed for a compressive force of 2.48 kips and an unbraced length of 10.5 ft—the distance between the bottom-chord bracing channels. It was also checked for a tensile force of 9.46 kips. A steel stud $3^5/_8$-in deep with 2-in-wide flanges 14-gage thick was good for 2.48 kips in compression and 17.85 kips in tension and therefore was acceptable (Fig. 10.28). It needed about 12 in of welding to the existing 16-gage material to develop the design forces.

The reinforcement for the diagonals was designed for a maximum compressive force of 0.46 kip and an unbraced length of 8.21 ft, or for 2.05 kips in tension. The reinforcement for the verticals was designed for a maximum compressive force of 1.76 kips and an unbraced length of 5.64 ft, or for 0.39 kip in tension. A steel stud $2^1/_2$-in deep with $1^5/_8$-in wide flanges 16-gage thick, good for 1.27 kips in compression and 10.74 kips in tension, was acceptable for both diagonals and verticals. It required 3 in of weld to develop the forces (Fig. 10.29).

(Information on the axial-load capacities of unbraced steel studs may be obtained from reputable steel-stud manufacturers. Be certain that it pertains to totally unbraced steel studs, not to regular wall studs, which are assumed to be braced in the weak direction by sheathing or wall bracing. In this case, the data for the allowable stud and weld capacities were taken from Ref. 18.)

The bases of approximately 100 wall studs supporting trusses were reinforced by adding steel channels on each side of the stud and anchoring the channels into concrete, as shown in Fig. 10.30. To transfer a 3.53-kip force into the foundation, 3 through-bolts of $1/_2$-in diameter were specified. The design was as follows:

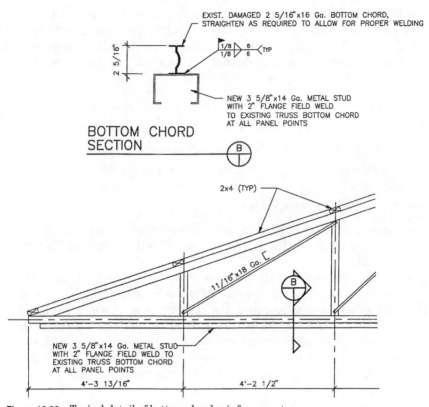

Figure 10.28 Typical detail of bottom-chord reinforcement.

Using AISI Specification, 1989 ed., Sec. E3.3, the maximum allowable bearing value of a $\frac{1}{2}$-in bolt in 16-gage steel is

$$P_{all} = F_p dt/2.22$$

where: $F_p = 3.00\ F_u$ (the ultimate strength of the stud material, assumed to be 50 ksi for this case)
$d = 0.5$ in (the bolt diameter)
$t = 0.0566$ in (the thickness of a single stud web)

$P_{all} = 3 \times 50 \times 0.5 \times 0.0566/2.22 = 1.91$ kips per stud web per bolt or a total of

$$1.91 \times 4\ \text{webs} \times 3\ \text{bolts} = 22.9\ \text{kips} > 3.53\ \text{kips} \qquad \text{OK}$$

The conservatism in provided bolt capacities was justified by the fact that the existing studs appeared to be somewhat corroded in the areas of the proposed bolt locations, even though the typical bracket length was set to be long enough to avoid the obviously rusted places.

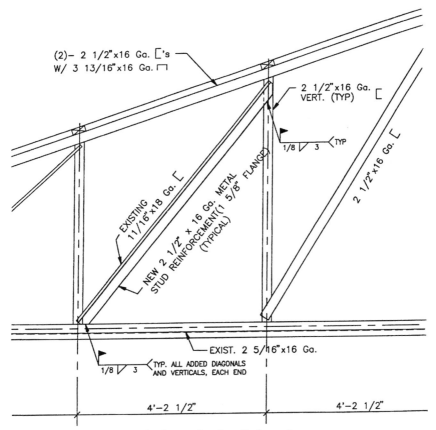

(2)– 2 1/2"x16 Ga. ⎣'s
W/ 3 13/16"x16 Ga. ⎤

2 1/2"x16 Ga. ⎣
VERT. (TYP)

1/8 ⟍ 3 ⟨TYP

2 1/2"x16 Ga. ⎣

EXISTING
11/16"x18 Ga. ⎣

NEW 2 1/2" x 16 Ga. METAL
STUD REINFORCEMENT(1 5/8" FLANGE)
(TYPICAL)

EXIST. 2 5/16"x16 Ga.

1/8 ⟍ 3 ⟨TYP. ALL ADDED DIAGONALS
AND VERTICALS, EACH END

4'–2 1/2"

4'–2 1/2"

Figure 10.29 Typical repair for diagonal and vertical members.

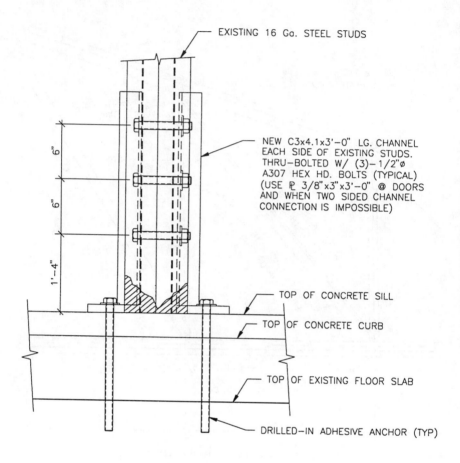

EXISTING 16 Ga. STEEL STUDS

NEW C3x4.1x3'-0" LG. CHANNEL
EACH SIDE OF EXISTING STUDS.
THRU-BOLTED W/ (3)-1/2"∅
A307 HEX HD. BOLTS (TYPICAL)
(USE ℙ 3/8"x3"x3'-0" @ DOORS
AND WHEN TWO SIDED CHANNEL
CONNECTION IS IMPOSSIBLE)

TOP OF CONCRETE SILL

TOP OF CONCRETE CURB

TOP OF EXISTING FLOOR SLAB

DRILLED-IN ADHESIVE ANCHOR (TYP)

6"

6"

1'-4"

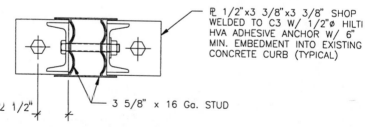

ℙ 1/2"x3 3/8"x3 3/8" SHOP
WELDED TO C3 W/ 1/2"∅ HILTI
HVA ADHESIVE ANCHOR W/ 6"
MIN. EMBEDMENT INTO EXISTING
CONCRETE CURB (TYPICAL)

3 5/8" x 16 Ga. STUD

2 1/2"

Figure 10.30 Typical repair of deteriorated stud base.

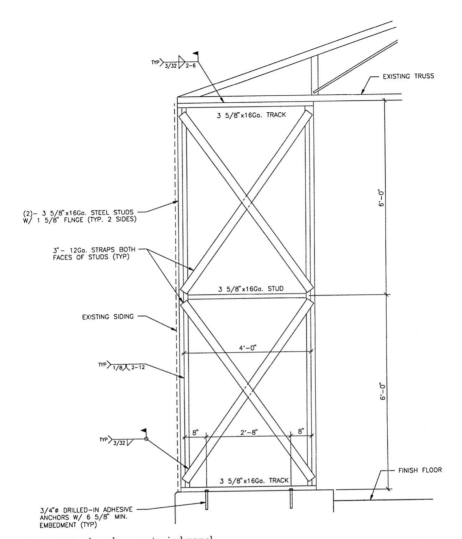

Figure 10.31   A replacement wind panel.

The missing interior cross-braced "wind panels" were restored using the original design (Fig. 10.31).

The final cost of the project was only about $70,000.

References

1. *1997 MBMA Business Review,* Metal Building Manufacturers Association, Inc., Cleveland, Ohio, 1998.
2. Alexander Newman, *Metal Building Systems: Design and Specifications,* McGraw-Hill, New York, 1997.
3. *Design/Specifiers Manual,* Butler Manufacturing Company, Roof Division, Kansas City, Mo., 1995.
4. "MBMA: 35 Years of Leading the Industry," a collection of articles, *Metal Construction News,* July 1991.
5. *Low-Rise Building Systems Manual,* Metal Building Manufacturers Association, Inc., Cleveland, Ohio, 1996.
6. *Specification for the Design of Cold-Formed Steel Structural Members,* American Iron & Steel Institute, Washington, D.C., 1995.
7. *Manual of Steel Construction, Allowable Stress Design,* American Institute of Steel Construction, Inc., Chicago, 1989.
8. Jeffrey S. Nawrocki, "How Fabricators Can Combat Metal Buildings," *Modern Steel Construction,* May 1997, pp. 78–81.
9. Letters to the editor (regarding Ref. 8), *Modern Steel Construction,* November 1997, pp. 29–31.
10. Thomas Fisher, "Giving a Plant Light and Air," *Building Renovation,* Fall 1995, p. 36.
11. Donald L. Johnson, "Reinforcement Design for Metal Building Systems," *Modern Steel Construction,* March 1998, pp. 71–78.
12. *Manual of Steel Construction, Load and Resistance Factor Design,* American Institute of Steel Construction, Inc., Chicago, 1993.
13. Wayne Tobiasson, "Some Thoughts on Snowloads," *Structure,* Winter 1995, pp. 14, 15.
14. Thomas M. Murray, *Extended End-Plate Moment Connections,* AISC Steel Design Guide No. 4, Chicago, 1990.
15. Robert Stanford, "The High Cost of Remedial Repairs on Metal Roofing...Why?" *Metal Construction News,* January 1998, pp. 46–50.
16. William A. Lotz, "Specifying Vapor Barriers," *Building Design and Construction,* November 1998, pp. 50–52.
17. Alexander Newman, "Procedures for Renovating Pre-engineered Metal Buildings," *Plant Engineering,* May 1998, pp. 49–54.
18. "Light Gauge Steel Framing," a catalog by DALE/INCOR, Dearborn, Mich., 1997.

Chapter

11

Strengthening
Lateral-Load-Resisting Systems

11.1 Lateral-Load Basics

11.1.1 Introduction

In today's engineering lexicon, the term *lateral loads* usually describes the effects of wind and earthquakes, even though in the recent past it included any horizontally applied forces. This terminology seeks to differentiate lateral loads from downward-acting gravity loads, even though in reality wind and seismic forces can act in both horizontal and vertical directions. Indeed, modern building codes require that wind be applied perpendicular to the roof surfaces—nearly upward for shallow roofs—and that a percentage of earthquake loading be applied vertically.

The ghastly pictures of damage caused by hurricanes and earthquakes testify that the builders of yesterday did not quite know how to make their structures resistant to these natural disasters. Even today, we are still learning how to make buildings withstand these extreme and largely unpredictable forces of nature. As we shall see, the building code provisions dealing with seismic design are in a state of *Sturm und Drang,* with major changes occurring every few years. New codified design approaches, usually even more confusing than the ones they replaced, and quasi-official advisory documents proliferate, each aiming to explain the complex reality better. Still, these design guidelines are but a gross oversimplification of what actually happens during an earthquake or hurricane.

In this chapter, we do not attempt to undertake an in-depth study of the specific requirements of any particular building code or design document; many of them are likely to have become obsolete by the time you read this book. Instead, we outline the salient concepts of design for wind and seismic forces and review the practices and the codes of yesterday. We also examine the palette of the most common practical approaches for strengthening the lateral-load-resisting systems of existing buildings. Specific design applications using this information can be found in the case studies of Chap. 12.

Despite being lumped together, earthquakes and hurricanes are fundamentally different in nature. The wind forces produced by hurricanes are external, and their effects depend on the shape and dimensions of the building. Earthquakes, on the other hand, damage structures because of the internal inertial forces, which depend on such factors as the building's mass, the type of construction, and the underlying soil. The magnitudes of both wind and earthquake forces are greatly influenced by the building's location.

The study of the effects of wind on structures is in the domain of wind engineers; the behavior of structures during earthquakes is researched by earthquake engineers. The latter should not be confused with seismologists, who deal with the magnitude and duration of earthquakes rather than with structural matters.

11.1.2 Wind loads

The fury of high winds has been known to humans since the beginning of our civilization. It is recorded in countless stories, legends, and fairy tales. Most people know about the enormous property damage from this common natural disaster, but many do not realize that fatalities and property losses caused by hurricanes rival or exceed those of the earthquakes, floods, and other calamities. According to one 1999 insurance industry estimate, the natural catastrophe resulting in the largest amount of insured losses up to that date was hurricane Andrew in 1992 ($16.5 billion). The runner-up, the 1994 Northridge earthquake, resulted in $12.5 billion of insured losses,[1] Curiously, the research budgets for mitigation of earthquake damage dwarf those for wind.[2]

The effects of wind on buildings are still not perfectly understood, and our knowledge in this area is constantly improving. Hurricane damage investigations and wind research lead to periodic revisions of the applicable building code provisions, although these revisions are not as frequent as building code revisions dealing with seismic design.

The model codes typically compute the magnitude of the wind force at a given location as a function of the design wind speed in miles per hour with an annual return probability of 0.02 (that is, a hurricane of this magnitude might be expected to return, on average, every 50 years). Design wind speed has historically been defined as the fastest-mile wind speed measured at 33 ft above the ground. Recently, ANSI/ASCE 7-95[3] and other codes changed the method of wind-speed measurement to monitoring the fastest 3-s gust speeds at 33 ft above the ground, because of changes in wind-speed measurement by the National Weather Service. The values of design wind speed can be found in wind maps or tables in the building codes. Once determined, the design wind speed is converted, with the aid of code-provided formulas, to the design velocity wind pressure on the building as a whole, taking into consideration the height and exposure category of the building.

High winds can cause four types of building damage[4]:

1. *Collapse,* when the building simply disintegrates, falling apart like a house of cards

2. *Partial damage,* when only the most vulnerable part of the building fails, as when the roof is blown off or windows are shattered

3. *Overturning,* when the building topples over but does not necessarily break up

4. *Sliding,* when the building moves horizontally away from its original position

Of these, partial damage occurs most frequently. It typically affects the areas of walls and roofs located near the building corners and roof eaves. Therefore, secondary structural members, roofing, and wall materials in those areas are designed for much higher wind loading than their counterparts elsewhere in the building. The exact definition of these "salient corner" areas varies among the codes, but, in general, they extend for a distance of 10 percent of the smallest dimension of the building but not more than 10 ft. Hurricanes frequently result in partial failures of roofing, wall cladding, and windows as a result of the combined attack of wind gusts and flying debris. Once the integrity of the building envelope is violated, rain blown inside can easily destroy the interior. The cost of the ensuing damage to finishes and of lost time often vastly exceeds the cost of fixing any structural problems.

One example: During the 1992 hurricane Andrew, the interior of the six-story headquarters of Burger King in Miami was totally soaked, and relocation of the personnel for over a year was required. The reinforced-concrete structural frame was undamaged, and even the windows withstood the 175-mph wind. The weak link was the window frames, which were bent by the wind and allowed the rain inside, causing property damage estimated at $25 to $30 million.[5]

Wind forces were historically assumed to act horizontally on the projected area of the building. Today's codes acknowledge a more complex picture of their distribution on gable buildings: Wind forces (acting either toward or away from the surfaces) are applied perpendicular to all roofs and walls, and both internal and external wind pressures are considered.

11.1.3 Earthquake loads

Earthquakes occur when two segments of the Earth's crust collide or move relative to each other. This movement generates seismic waves in the surrounding soil that diminish as the distance from the earthquake epicenter increases. Because seismic energy is transmitted in a wavelike pattern, earthquakes are cyclical and repetitive in nature. We perceive these waves as ground shaking. Seismologists differentiate between the P (pressure or primary) and S (shear or secondary) waves. Pressure waves generate surface movements of a push-and-pull kind; the shear waves act normal to pressure waves. In addition, there are the Rayleigh waves, often the most devastating to buildings; these travel at the surface of the ground.

Seismic forces are caused by the inertia of the structure, which tries to resist ground motions. As the shifting ground carries the building foundations along with it, inertia keeps the rest of the structure in place for a short while longer. The movement between the two parts of the building creates a force equal to the ground acceleration times the mass of the structure. The ground acceleration depends on the magnitude of the seismic event. The still-common method of measuring earthquake magnitude was invented in 1932 by Richter, who proposed a logarithmic scale of measurement. In a Richter scale, an earthquake with a magnitude of 8 is 10 times as large as one with a magnitude of 7 and 100 times as large as one with a magnitude of 6.

In addition to the earthquake magnitude and the mass of the building—the heavier it is, the larger the force—the value of the seismic force also depends on the type of soil under the building. Some soils tend to amplify seismic waves and can even turn to a liquidlike consistency during an earthquake (the liquefaction phenomenon). And there are other variables involved, including the type of lateral-load-resisting system of the building and its degree of rigidity. Building codes typically contain *design response spectra,* plots of maximum acceleration as a function of the building's natural period of vibration and its damping ratio (a measure of the building's ability to extinguish vibrations by internal friction among its elements). An example of a seismic response spectrum can be found in Chap. 12, Case 4.

In design practice, for most conventional rectangular buildings, the dynamic wavelike inertial forces are approximated by the equivalent static forces acting horizontally, plus a small vertical component. Here, the design seismic force is determined by multiplying the weight of the structure by several coefficients that account for the factors just mentioned. The coefficients differ among the building codes and among the various code editions.

Why is the method of equivalent static force more common than other methods? First, this approach is much more practical and suitable for routine office use than cumbersome dynamic analysis methods. Second, the forces so computed can be readily compared with the results of a wind-load analysis, and the larger of the two used in the design of the lateral-load-resisting system of the building. Still, for unique buildings of irregular sizes, those with plan asymmetries, or those with grossly unequal weight distribution among the floors, building codes typically require a dynamic analysis, because the seismic behavior of these structures might not be accurately predicted by the equivalent static force method.

Will a building meeting the seismic requirements of the code come out of an actual earthquake totally unscathed? Not necessarily. To summarize the philosophy of most contemporary building codes, buildings designed in accordance with the code's seismic provisions should be able resist minor earthquakes without damage, moderate earthquakes without structural damage, but with some nonstructural damage, and major seismic events without collapse.

The goal of preventing collapse under repeated overloads resulting from unpredictable earthquake forces favors structural systems that can deform without breaking—that is, that are ductile in nature. A ductile structure dissipates the earthquake-generated energy by stretching the material past its elastic limits into its inelastic region. The assumption that ductility will be present—that the building will be able to undergo inelastic deformations during strong earthquakes—lies at the base of the contemporary seismic design philosophy. Ductility is not just desirable, it is fundamental to the process of arriving at the magnitude of seismic forces. Without ductility, the prescribed design forces would be much higher than those presently used. The building codes specify smaller design seismic forces for ductile lateral-load-resisting systems, such as special moment-resisting frames, than for those with less ductility, such as shear walls and braced frames.

Present-day building codes are filled with many prescriptive requirements for design details that are intended to achieve ductility and establish the load path in lateral-load-resisting systems. The new trend, however, is to emphasize the final result, rather than the required ingredients. What works in the western United States may not be the best practice in the East, given the differences in the types of ground motions in these two areas. This notion of performance-based design will be revisited later.

11.2   Past Methods of Resisting Lateral Loads

11.2.1   Some common lateral-load-resisting systems

The destructive effects of hurricanes and earthquakes have been known since ancient times. So why have we not developed any 100 percent effective methods of resisting them? It is not for lack of trying, but rather for lack of research and understanding of the complex phenomena. Also, the builders of yesterday had to contend with the relatively primitive structural systems at their disposal. The most popular structural building materials of today—structural steel and reinforced concrete—are only 100 to 150 years old (see Chaps. 3 and 4), and the most cost-effective proprietary systems are barely 50 years old. Even so, many old buildings made of timber and masonry have survived, albeit with some damage and repairs.

The most obvious solution available to our predecessors was to make the unreinforced masonry *shear walls* (Fig. 11.1*a*) of their buildings thick and solid—in other words, "strong." This was a reasonable approach for improving hurricane resistance, as a 4-ft-thick wall could survive stronger winds than a 1-ft wall, but for earthquake resistance, increasing the mass is counterproductive. A variation on the shear wall idea was *counterforts,* essentially exterior shear walls designed to resist the lateral thrust of arches, a common design in medieval cathedrals (Fig. 11.1*b*).

Other traditional methods of resisting lateral loads included using *knee braces,* a typical fixture in open-wall wood buildings, *cross-bracing,* *K-bracing,* and the *rigid frame.* All these are illustrated in Fig. 11.1. The last of these, the rigid frame, was an outgrowth of the masonry arches developed by the Romans. Eventually, the rigid frame became a staple of pre-engineered metal buildings, as discussed in Chap. 10, and of structural steel framing, but it took a long time to find acceptance. Other systems were tried first, as discussed later.

Most wood buildings have traditionally, if implicitly, relied on exterior shear walls and interior partitions for lateral stability. These elements were not rationally designed, but their performance was acceptable. A contemporary rigorous study of wood-framed partitions has confirmed that they "resist a substantial amount of seismic forces depending on their stiffness and the aspect ratio of the horizontal diaphragm."[6] The design data for analysis of partitions framed with

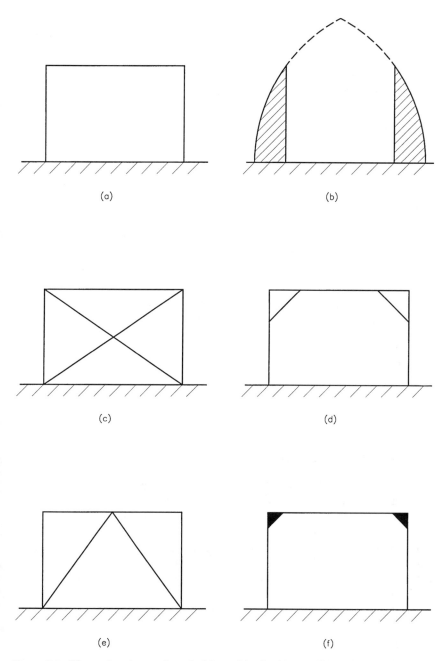

Figure 11.1 The various types of vertical lateral-load-resisting elements. (*a*) Shear wall; (*b*) counterforts; (*c*) braced frame; (*d*) knee brace; (*e*) K brace; (*f*) moment-resisting frame or frame with semirigid connections.

drywall or stucco can be found in many building codes and trade publications. The first concrete buildings also used shear walls for lateral stability.

11.2.2   Diaphragms

The term *diaphragm* reflects one of the most important concepts in lateral-load resistance. Indeed, the modern methodology of designing buildings for wind and seismic forces developed when designers discovered the concept of the lateral-load-resisting *system*—the floor framing and vertical wall elements acting together. A realization that floor and roof decks did not just resist gravity loads, but also acted as horizontal beams (diaphragms), was pivotal to the rational lateral-load analysis of structures.

Diaphragms distribute lateral loads among the vertical rigid elements, such as shear walls, braced frames, or moment frames. Diaphragms can consist of horizontal steel bracing, steel deck, plywood, wood sheathing, concrete deck, and even metal through-fastened roofing, as long as the components are properly attached to each other and to the vertical rigid elements. Depending on the material and attachments, diaphragms can have various degrees of flexibility. The most flexible diaphragms are those made with wood sheathing, plywood, and metal deck or roofing; the most rigid, those made with thick reinforced concrete. Unlike rigid diaphragms, flexible diaphragms are assumed not to be able to distribute torsional stresses among the vertical lateral-load-resisting elements.

11.2.3   Lateral-load-resisting systems for iron- and steel-framed buildings

When structural steel buildings came into being, their designers realized that simply stacking beams on top of columns would not be helpful in resisting lateral loads. The first steel-framed buildings depended on unreinforced masonry exteriors for lateral stability. The masonry shear walls were strong enough to provide the required shear resistance and heavy enough to resist overturning, but they were not tied into the floors and, as a result, could separate from the building during an earthquake.

Another solution for assuring stability in the older iron and structural steel buildings involved partially restrained (semirigid) connections. These "wind" connections, as they were called, were supposed to be rigid enough to resist lateral loads, like the moment connections of today, but flexible enough to allow the beams to behave as simply supported members under gravity loads. *Manual of Steel Construction*, vol. II, *Connections,*[7] by the American Institute of

Steel Construction, outlines the criteria that partially restrained connections must satisfy in order to be viable. Perhaps most importantly, "the connection material must have sufficient inelastic rotation capacity" to prevent it from failure under combined gravity and wind loads.

Figure 11.2 shows typical "wind" connections made with a pair of hot-rolled angles (Fig. 11.2a) and with tee sections (Fig. 11.2b) attached to the beam and the column by rivets. These connections have been experimentally demonstrated to possess the desired flexibility and have recently attracted renewed interest on the part of engineers looking for an alternative to expensive welded moment connections. Unfortunately, while the system has proven its mettle over decades of successful performance, it is difficult to analyze, and rigorous design methods have yet to be fully developed. Research on semirigid connections has been conducted by Prof. Roberto T. Leon, one of whose studies is summarized in Ref. 8, and by others. A good source of information is a publication by FEMA, *NEHRP Commentary on the Guidelines for the Seismic Rehabilitation of Buildings,* FEMA-274.[9]

The simplicity and low cost of partially restrained connections is counterbalanced by their relatively poor energy dissipation, because the joint is weaker than the connected members and plastic hinges cannot be formed in the beam, as is desirable. Also, buildings that use these connections tend to undergo large story drifts. The researchers point out that the behavior of semirigid connections is not fully predictable, as the connections seem to function differently under various types of ground motions.

Roeder,[10] in a study of the seismic performance of older steel frames with semirigid connections, concluded that, theoretically, these frames could not provide the level of seismic resistance required by the modern building codes. How is this conclusion reconciled with the relatively good real-life experience with partially restrained connections? Roeder's explanation is simple: None of the buildings he studied had actually had to withstand an extreme earthquake in its lifetime. Also, in all those buildings, the framing was encased by concrete fireproofing and surrounded by unreinforced masonry walls—elements that were much more rigid than the steel frame. These rigid elements could probably resist moderate earthquakes without any reliance on the frame's stiffness.

As steel-frame building systems were developed further, fully rigid (moment) connections between beams and columns appeared. Moment connections were made first with rivets (see Fig. 3.12 in Chap. 3) and then with bolts and welds. Welded moment connections similar to the one shown in Fig. 11.3 became commonplace in steel-framed buildings in the last third of the twentieth century. Among their presumed

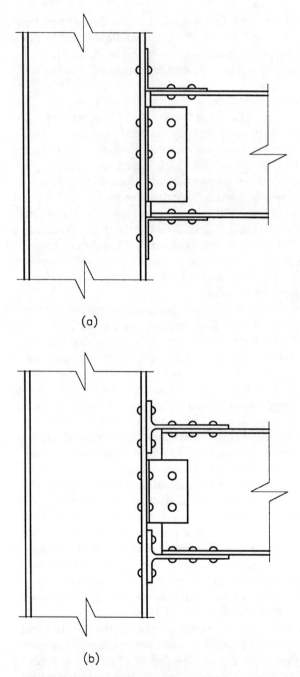

(a)

(b)

Figure 11.2 Typical "wind" connections made with (a) a pair of hot-rolled angles and (b) tee sections.

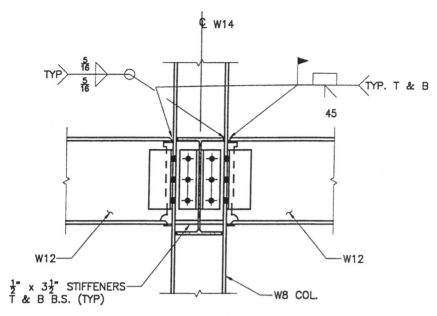

Figure 11.3   A typical welded moment connection common until the late 1990s.

advantages were excellent energy dissipation and the ability to form plastic hinges, because the connection was supposed to be as strong as the beam.

But this popular design was called into question after the 1994 Northridge earthquake. Northridge, considered to be the first earthquake with an intensity comparable to that assumed in the contemporary seismic codes that struck an urban area of the United States, was also the first large-scale load test of the seismic capabilities of modern steel-framed buildings. Damage investigations found that some of the welds in these moment connections cracked, even though none of the buildings collapsed or caused bodily injuries. The issue of cracked welds between the flanges of columns and girders was hotly debated in many technical publications; in one of these, Malley[11] reported that over 100 moment-resisting frames suffered structural damage. The Kobe earthquake in Japan the next year caused similar connection damage. Welded steel moment connections in at least two structures were found to be distressed as a result of the 1989 Loma Prieta earthquake.[12] The debate has culminated in adoption of the recommendations for welded moment connection details with extra reinforcement of the welds. Figure 11.4 shows one post-Northridge welded moment connection detail.

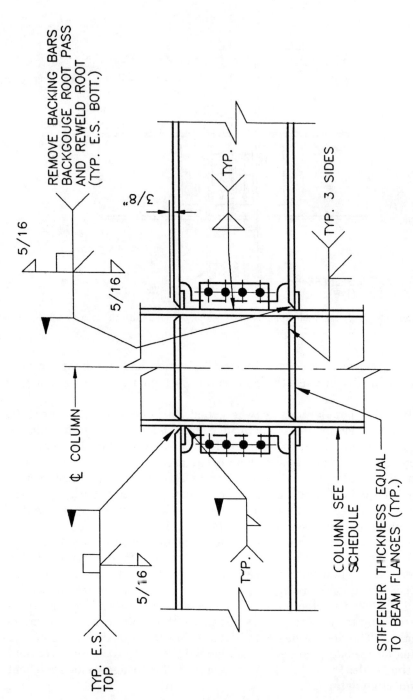

REMOVE BACKING BARS BACKGOUGE ROOT PASS AND REWELD ROOT (TYP. E.S. BOTT.)

5/16

5/16

3/8"

TYP.

TYP. 3 SIDES

℄ COLUMN

TYP. E.S. TOP

5/16

TYP.

COLUMN SEE SCHEDULE

STIFFENER THICKNESS EQUAL TO BEAM FLANGES (TYP.)

Figure 11.4 A typical post-Northridge welded moment connection.

11.3   History of Wind and Seismic Codes*

Many times, the only design information about a building slated for renovations is its location and its approximate age. This section gives a very brief historical overview of wind and seismic codes. Our goal is to allow the renovation designer to get a rough idea of whether the building is likely to have been designed for lateral loads, and if so, what procedure might have been followed.

11.3.1   Wind codes

Rational calculations for wind-load resistance were rare until the beginning of the twentieth century, when the first building codes started to mention lateral loads on buildings. One of the first to do so was the San Francisco building code, enacted after the destructive earthquake of 1906. That code required all buildings to be designed for a horizontal wind force of 30 psf. The objective, presumably, was to mandate at least some level of lateral stability for Bay Area buildings, even though the understanding of seismic issues was still in its infancy and seismic forces could not yet be reliably quantified. The wind forces were easier to comprehend — and to write standards for.

Throughout the first half of the twentieth century, the curricula of engineering colleges included little on the subject of wind design, and structural engineers only vaguely understood how wind affected buildings. "Wind engineering was not a separate discipline, but bits and pieces showed up from time to time," recalls William J. LeMessurier, one of the most respected engineers of our time, of his college experiences.[15] (He received his graduate degree in 1949.) According to LeMessurier, the *City of Boston Building Code,* 1938 edition, required a 10-psf design wind load on the first 40 ft of the building height, 15 psf for the next 40 ft, and 20 psf above that, while the City of New York at that time specified zero wind for heights below 100 ft. In the late 1950s, LeMessurier participated in a major revision of the wind provisions of the Boston building code that substantially increased the prescribed loads from the previous extremely low levels.

Over the next several decades, building code provisions dealing with wind loads were further refined, and this process is still continuing. The evolution can be best illustrated by changes in ASCE 7 (formerly known as ANSI 58.1), *Minimum Design Loads for Buildings and Other Structures,*[3] an authoritative standard that is cited as a reference by many contemporary model and local building codes.

---

*The main sources of information on the previous code editions cited in this section were Green[13] and Harris.[14] Other sources are acknowledged separately.

The design provisions and wind maps of ASCE 7 have undergone a few revisions during the last decades of the twentieth century. The 1993 and earlier editions contained a hurricane-coast importance factor of 1.05 to account for the increased wind speeds in the hurricanes that typically strike coastal regions. Some modifications were included in the wind-speed maps of the 1993 edition. The 1998 edition incorporated a fundamental change in the way the wind-speed maps are composed: For the first time, the importance factor was incorporated into the wind speeds. The map was designed to yield, after multiplying the wind loads derived from it by the ultimate load factors, factored loads with approximately the same return period for both hurricane and nonhurricane winds. The ultimate hurricane winds were assumed to have a return interval of 500 years and nonhurricane winds, one of 50 years.[16]

The main areas of wind research today concern the actual local distribution of wind loads across structures and experimental validation of code provisions for buildings of various heights.

11.3.2 The first seismic codes

Seismic requirements were not included in building codes as early as those for wind, although some experimentation had taken place in Europe and even more in Japan, which suffered from frequent seismic activity. Some of the early approaches yielded little result, but that did not stop curious minds from experimenting. The first application of Newton's first law to building codes dealing with seismic design was reportedly made in Italy following the 1911 Messina earthquake. Mindful that the force is equal to the mass times the acceleration, the regulations there started to require that all buildings be designed for a static horizontal force equal to 10 percent of their weight.

Striving to meet the demand for seismic features in buildings, Frank Lloyd Wright announced in 1923 that his Imperial Hotel in Tokyo was earthquake-resistant. The architect's measures to ensure seismic safety included making the exterior walls structurally independent of each other and supporting the building on cast-in-place concrete piers penetrating through the soft clay into the firm stratum. The Imperial Hotel indeed survived the strong 1923 Kwanto earthquake, which occurred shortly after the building's completion and reduced most of Tokyo to ruins. (But the architect then characteristically proclaimed that his building was *the only one* that survived in that city.)

In the United States, the brunt of seismic activity was borne by California, and the West Coast was the logical place to develop the first seismic codes. The first edition of the *Uniform Building Code* (UBC) in 1927 included an appendix with provisions for seismic design of buildings. (Chapter 1 includes a brief history of the model building codes.) The UBC called for buildings constructed on "firm ground"—soil with an allowable bearing capacity of at least 2 tons/ft² — to be designed for a horizontal force equal to 7.5 percent of both dead and live loads. A larger force (10 percent of dead and live loads) had to be used for buildings constructed on weaker soils and on piles. The allowable stresses for members resisting seismic forces could be increased by an amount ranging from one-third to one-half, depending on the material.

The first mandatory seismic regulations appeared after the 1933 Long Beach earthquake, which prompted the development of the Los Angeles building code and codes in other municipalities. The Los Angeles code contained a simple expression for the earthquake force as equal to the seismic coefficient C times the weight of the structure W:

$$F = CW$$

The seismic coefficient C was, again, taken as a fixed percentage of the gravity force acting on the building. The gravity force was to include the dead load plus one-half of the live load. The coefficient C was set as 0.08 for all types of buildings except schools, for which it was increased to 0.10. The UBC soon included this formula for buildings on good soils, but increased C to 0.16 for buildings on poor soils.

Afterward, California passed the Field and Riley Acts, which required a similar design approach, with various seismic coefficients C used for various buildings. The Riley Act required that most buildings, other than schools and hospitals, be designed for a lateral force equal to 2 percent of the weight. The Field Act, which applied specifically to schools, specified the horizontal force as a minimum weight percentage that, for low-rise unreinforced masonry structures, was slightly higher than that found in the UBC. It also required rigorous design review and construction inspection.

The 1943 edition of the Los Angeles code contained a formula for C that included the number of stories in the building. This was a step forward, because for the first time the code recognized that a taller building with a longer natural period of vibration was subjected to smaller seismic accelerations and forces than a squat, rigid building with a short period of vibration. This new expression for C better approximated the seismic effects on tall buildings, requiring the use of larger accelerations for upper stories:

$$C = \frac{60}{(N + 4.5)\,(100)}$$

where N was the number of stories above the one being considered. For a single-story building, N was zero.

In other areas of the country, a variety of earthquake provisions were finding their way into the building codes; however, some state codes did not require seismic analysis until the last decades of the twentieth century.

11.3.3 The SEAOC codes

A committee of California structural engineers was formed in 1957 to develop a new, more realistic seismic code to replace the many existing codes dealing with seismic design. The 1959 Structural Engineers Association of California (SEAOC) Code was pivotal in many respects, and its basic notions were incorporated in most seismic codes developed later. Among other firsts, it introduced the notion of ductility, discussed previously, by means of the ductility coefficient K, which depended on the type of lateral-load-resisting system in the structure. In the SEAOC code, the total earthquake force on the building—the base shear V—was computed as

$$V = KCW$$

The factor K reflected the fact that buildings with ductile moment frames possess a high degree of redundancy, as multiple plastic hinges can form in them under overload, allowing substantial deformation to take place and dissipate the applied seismic energy. In contrast, brittle shear walls have little redundancy, and failure in them can occur abruptly. Therefore, the factor K ranged from 0.67 for buildings with a ductile moment-resisting space frame system capable of withstanding the total seismic force to 1.33 for buildings with "box type" systems where the forces were resisted by shear walls or braced frames. To put it another way, buildings that relied on shear walls or braced frames had to be designed to withstand twice the seismic forces as otherwise identical buildings with moment frames.

The base shear seismic coefficient C was not a fixed value but was instead a function of the building's fundamental (longest) period of vibration T:

$$C = \frac{0.05}{\sqrt[3]{T}}$$

The period of vibration T could be computed as a function of the building's height H and width D:

$$T = \frac{0.005H}{\sqrt{D}}$$

For buildings with a moment-resisting space frame system capable of withstanding the total seismic force, and in which the frames were not enclosed or adjoined by rigid elements, the fundamental period of vibration T could be found as one-tenth the number of stories N:

$$T = 0.1N$$

The weight W included only the dead load, except for storage buildings and warehouses. The SEAOC code required all concrete and masonry walls to have a minimum reinforcing to provide some measure of ductility.

The SEAOC code was periodically revised to reflect the accumulated knowledge about the seismic performance of buildings. In the early 1970s, it became painfully evident that many concrete structures designed in accordance with the 1959 SEAOC code had sustained major damage in the 1971 San Fernando earthquake and in other seismic events. Some buildings had to be demolished. The SEAOC ad hoc committee, reflecting the experience gathered during those damage investigations, recommended some changes to the SEAOC code. The fourth edition of the *Recommended Lateral Force Requirements and Commentary* was prepared in 1974 and issued in 1975. The revised edition contained a new equation for base shear,

$$V = ZIKCSW$$

where a zoning factor Z was added to reflect the building's location and the expected severity of the design earthquake. This factor ranged from 1.0 for the areas of highest seismicity to zero for areas with an extremely small probability of earthquake occurrence. The importance factor I accounted for the higher reliability demanded of "essential facilities" that had to remain in operation after the disaster; it was set as 1.5 for essential facilities and 1.0 for others. The site factor S, ranging from 1.0 to 1.5, reflected the soil conditions at the site. The revised equation for C was given as

$$C = \frac{1}{15\sqrt{T}}$$

with a maximum value of 0.12.

The modifications even affected the applicability of the main method of analysis, in which complex wavelike seismic forces were approximated by the equivalent static forces. While the simplified method was retained for most structures, dynamic analysis was now required for buildings with complex configurations. This requirement has since become a staple of seismic codes. Unfortunately, the practical value of dynamic analysis methods is questionable. As Green[13] points out, "highly irregular structures…are exactly the structures for which it is difficult or impossible to make a good realistic dynamic analysis."

Those using the 1974 SEAOC code would generally arrive at higher design seismic forces than those using the previous editions of the code, largely because of the new values of *C*. The provisions of that code were included in a seismic section of the 1976 UBC.

In 1987, SEAOC adopted a revised set of seismic provisions, which were incorporated into the 1988 edition of the UBC; the next revision followed in 1990. The latter contained the following equation for base shear:

$$V = \frac{ZIC}{R_w} W$$

Here, a new numerical coefficient R_w reflected the type of lateral-load-resisting system of the existing building, in lieu of the old factor *K*. The more ductile systems were assigned higher values of R_w and thus could be designed for smaller base shear levels. The highest value, 12, was reserved for special moment-resisting frames, while the lowest, 4, was to be used for loadbearing concrete or heavy-timber braced frames.

11.3.4 The ATC and NEHRP codes

As the volume of seismic-related and damage assessment information increased, it became clear that SEAOC, consisting of volunteer practicing structural engineers, could not keep up with code development affecting the whole country. To continue SEAOC's efforts, a new organization called the Applied Technology Council (ATC) was formed in 1971. In contrast with SEAOC, the ATC was staffed largely with scientists rather than with practicing engineers, a fact that affected the scope and practicality of its main document, ATC 3-06, *Tentative Provisions for the Development of Seismic Regulations for Buildings,* published in 1978.

In ATC 3-06, the equivalent static load design approach was retained, but many of its provisions were modified. Introduced was the need to consider the "orthogonal effects" of seismic loads—simultaneously applying 100 percent of the lateral loads acting in one direction and 30 percent of the loads acting in the perpendicular direction. This design approach has since been mandated by many model codes. Among other, and perhaps more important, changes were explicit recognition of the inelastic action of buildings under seismic loading, an emphasis on strength-design procedures, and extensive requirements for detailing.

As already mentioned, the 1988 edition of the UBC essentially followed the recommendations of the 1987 SEAOC provisions. The latter, in turn, were largely based on ATC 3-06. This moved the UBC ahead of both the *BOCA National Building Code* and the *Standard Building Code*; both of those reflected the seismic provisions of ANSI A58.1, based on an older approach.

In the 1970s, it became obvious that the many code-writing bodies needed to arrive at a broad consensus concerning the development of future seismic code provisions. In 1977, the U.S. Congress authorized funding for the National Earthquake Hazards Reduction Program (NEHRP). Under this program, a countrywide extensive discussion of the available documents took place, and consensus-based recommended provisions were published in 1985. The *NEHRP Recommended Provisions for the Development of Seismic Regulations for New Buildings*[17] essentially endorsed ATC 3-06 with some modifications. The provisions were revised in 1988, 1991, and 1994.

The federal government has also become a player in this field. Executive Order 12699, which took effect in 1990, required that all new buildings owned, leased, assisted, or regulated by the federal government be designed and constructed using "appropriate seismic design and construction standards" by 1993. Affected was a considerable share of the U.S. building stock, from a storage building backed by the Small Business Administration to a home financed by the Veterans Administration. The "appropriate" seismic standards were defined as those found in the 1991–1992 editions of the three model codes.[18]

11.3.5   Too many codes?

It should be quite evident by now that, as we noted in the beginning of this chapter, the state of seismic design is a rather chaotic field, with many rapidly changing competing standards. An undocumented existing building constructed between 1970 and 2000 could have been designed following any number of guidelines, from the state code valid at the time to the NEHRP or SEAOC recommended provisions. A project of military importance could have been designed under the so-called tri-service (Army, Navy, and Air Force) design manuals.[19] The task of comparing various code provisions is beyond the scope of this book; it has been done by others.[20,21]

Perhaps the main difference among the various codes is that the 1991 and 1994 editions of the *NEHRP Recommended Provisions* and ASCE 7-95 emphasize the strength-design approach, whereas SEAOC and the UBC 1994 use working-stress design methods. The codes also contain different methods of seismic analysis, procedures for computing displacement amplification, and importance factors. Each of the codes incorporates extensive detailing requirements.

The new *International Building Code* uses its own set of seismic provisions, which place great emphasis on the location and type of the building. Some U.S. government agencies have their own seismic codes. One of the most stringent is the Veterans' Administration's *Earthquake Resistant Requirements for VA Facilities* (VA standard H-08-8).

11.4   Code Provisions for Seismic Upgrading

11.4.1   Is a lateral-load upgrade needed?

The old codes and documents mentioned in the previous section deal mostly with the design of new buildings. Are they relevant to the design of renovations? They are, because it may be necessary to evaluate the building's ability to resist lateral loads on the basis of the governing code in force at the time of the original construction. Also, knowledge of the preexisting design standards helps pinpoint the probable weak links in the original design approach. The primary importance of these documents lies in assessment of the existing building capacity. But which documents will tell us how to go about evaluating and upgrading the wind and seismic resistance of an existing building?

The most obvious source is again the governing building code. Most state building codes take as a reference one of the three model building codes or the new *International Building Code*. The renovation provisions of all these codes are discussed in Chap. 1 and need not be repeated here. In general, most of the codes will allow existing buildings to retain their existing lateral-load resisting systems if no (or only trivial) changes to the structure are proposed *and* the building use remains unchanged. The codes require upgrading of buildings when major changes or tied-in additions are planned, and when the proposed alterations reduce the existing lateral-load-resisting capacity. A lateral-load upgrade may also be required if the proposed changes move the building into the category of "essential" or "hazardous" facilities, or if the proposed renovation work exceeds 50 percent of the building's value.

For additions, the typical code sentiment is reflected by the tri-service design manual[19]:

> When the addition is not separated and a significant change occurs in the total weight, in the weight distribution, or in the building lateral force resisting system's rotational or translational stiffness, an upgrade will be done.

The seismic provisions of the *International Building Code* and of BOCA 1999 attempt to be more specific by quantifying the meaning of "significant change." Both require that the addition itself be compliant with the code for new construction, and require a seismic upgrade of the existing building if the addition increases the seismic forces in any existing structural member by more than 5 percent (unless, of course, that member is already strong enough to comply with the code). Similarly, the addition is not allowed to weaken the seismic capacity of any existing structural member to a level below that specified for new construction.

There are still plenty of questions as to how to interpret these provisions. For example, should the code for new construction or the original design code be used in computing the 5 percent maximum overstress? Some code authorities suggest that the current code be used, but this means that every old building to which additions are planned must be reanalyzed under the present-day code. What if, as is likely, some provision of the code is not met even prior to making any additions—must the whole building be strengthened if, for example, an exterior ramp is added to comply with the Americans with Disabilities Act (ADA)? And what about the vast majority of existing buildings that were never explicitly designed for lateral loads in the first place? Or, what if an opening is cut in an existing masonry wall during renovations and another opening of the same size is filled—according to the code, is the wall considered weakened, so that the whole structure must be upgraded? Similarly, many engineers might simply replace a removed wall with a new wall of the same rigidity but placed at a different location that is more appropriate from a planning standpoint. Assuming that the new wall is in line with the removed wall (so as not to get entangled in the issues of added torsion and changed load distribution among walls), has the building been weakened or not?

When building codes prescribe full compliance with their current seismic provisions, they are rarely explicit in telling the users what measures to take to seismically upgrade the building. There are exceptions, of course. On the West Coast, the San Francisco Building Code requires upgrading of existing structures to 75 percent of the strength required by the code for new construction. On the East Coast, the *Commonwealth of Massachusetts Building Code*[22] offers an elaborate path for determination of the required remedial measures; in some cases it allows the use of lower values of seismic forces than those used for new construction.

In some regions of high seismic activity, state and local codes and ordinances may require a seismic upgrade even for buildings that are *not* undergoing any renovations. Perhaps the best known of these is a seismic retrofit ordinance passed by the Los Angeles City Council on Feb. 24, 1994, in the wake of the Northridge earthquake. This sweeping ordinance required seismic rehabilitation of about 20,000 multistory wood-framed apartments and hotels located in the city—nearly one-half of the total such stock of 47,000. Among other steps, it required strengthening of ground-floor parking garages with steel moment-resisting frames and overlaying shear walls of stucco or drywall with plywood. The average cost per building was estimated at $200,000. After a public uproar and protests by apartment owners, the city proclaimed the ordinance voluntary and offered low-interest loans from the city and tax credits from the state as inducements to comply.[23]

California has taken some steps in this direction on the state level as well. For example, it required more than 450 acute-care facilities in the state to submit seismic evaluation and compliance plans showing how the facilities will withstand a code-level earthquake. The latter is defined as a seismic event with a 10 percent probability of being exceeded in 100 years. By 2008, all general acute-care facilities found to be vulnerable to collapse must be removed from service.[24] Given the toughness of these standards, some hospitals in California are being demolished rather than upgraded to meet them. One example is the UCLA Medical Center, the nation's second largest building in terms of area (about 1 million ft²). The building was damaged in the 1994 Northridge earthquake, and it was more economical to replace it at a cost of $1.3 billion than to reinforce it, which would have cost $2 billion.[25]

Beyond California, the situation is even less orderly, and the code provisions tend to use the 5 percent rule mentioned previously or to be even more vague. It is not uncommon to have one engineer declare that a building must undergo a complete seismic upgrade, while another states that nothing of the sort is needed. Then there is a question of money. Should any minor addition to a building, or even just a rearrangement of its plan, automatically bring with it the onerous duty to completely upgrade the lateral-load-resisting system to comply with the current building code? If the codes that happen to be enforced today say yes, the owner might judge the renovation uneconomical and abandon it, depriving both the owner and the public at large of the benefits the expansion could bring. Or, more likely, the owner will shop for an engineer in whose opinion a total upgrade is not needed and who is willing to justify this reading of the code to the building officials.

These real-life observations are not meant to imply that a seismic upgrade is not needed when minor renovations are involved. On the contrary: Many renovation plans that allegedly do not weaken the structure call for removal of existing partitions that are acting, intentionally or not, as shear walls. Strengthening these buildings for lateral loads is quite sensible. But what if the owner informs the engineer that the funds for such an upgrade simply do not exist? Should a partial upgrade be made in order to bring at least some improvement? Or should the engineer, weary of the liability involved, decline to do anything at all? Issues like these generate heated discussions among design professionals involved in building renovations.

And, of course, there is a raging debate on the question how realistic is the risk of major earthquakes in the eastern two-thirds of the United States.[26] Are code provisions that make sense in California equally applicable in New England? As of June 2000, there was no documented evidence that a single person had ever been killed by an earthquake in New England, or that a single building had collapsed because of one. Los Angeles–style regulations that spell out the extent of the required seismic upgrade for existing unreinforced-masonry buildings and other vulnerable structures may be prudent for high-seismicity areas but are wastefully expensive in others.

One conclusion is clear: Guidance on this issue from an authoritative source is sorely needed. One source has attempted to fill the void.

11.4.2   FEMA-273, *NEHRP Guidelines for Seismic Rehabilitation of Buildings:* The background

The job of developing a set of guidelines equally applicable to all areas of the country fell to the Federal Emergency Management Agency (FEMA). In 1989, the NEHRP was given another major assignment, a two-phase project dealing with the seismic upgrade of existing buildings. Phase I was handled by the ATC and culminated with the publication of FEMA Report 237, *Seismic Rehabilitation of Buildings—Phase I, Issues Identification and Resolution.* As the title implies, the report focused on the development of basic issues to be addressed in the next phase. This pioneering document included many important concepts, presented in a systematic manner.

Among the report's recommendations was one endorsing the use of performance-based design solutions to problems of life safety and damage containment. Performance-based design procedures allow designers to choose their own approaches to achieve the desired results, rather than blindly following the prescriptive requirements of the codes. Instead of dictating *how* to achieve a given design goal, performance-based design emphasizes the goals that must be met and sets the criteria for acceptance. This way, experienced engineers are free to innovate without running afoul of some overly specific code provision tied to yesterday's practice.

Phase II started immediately after the Phase I report was released. Under an agreement between FEMA and the National Institute of Building Sciences (NIBS), the latter, through its Building Seismic Safety Council (BSSC), coordinated the development of Phase II by two main subcontractors: ATC and ASCE. The result of this monumental effort was the publication in October 1997 of FEMA-273, *NEHRP Guidelines for the Seismic Rehabilitation of Buildings.* This document was intended to be not only a reference tool for designers and building officials, but also a foundation for the development of future building codes dealing with seismic issues. A companion report, FEMA-274, *NEHRP Commentary on the Guidelines for the Seismic Rehabilitation of Buildings,* was published to explain some provisions of FEMA-273. Another related report, FEMA-276, *Example Applications of the NEHRP Guidelines for the Seismic Rehabilitation of Buildings,* helps illustrate its use. The reports can be used for all types of buildings.[27]

The adoption of the NEHRP guidelines involved extensive consultations and user workshops for the interested parties around the country.

11.4.3   FEMA-273: The general philosophy and main provisions

The FEMA documents outline criteria and methods for ensuring the desired performance of buildings at various "performance levels" selected by the owners and their design professionals. The reports provide some guidance for selecting a performance level but do not impose a solution—a far cry from yesterday's one-size-fit-all mandates. For example, the guidelines allow owners to select a level of seismic upgrade that not only protects lives—a goal of all building codes—but also protects their investment in the bricks and mortar. The following three performance levels can be chosen:

- *Basic Safety Objective.* At this most common performance level, life safety must be assured for earthquakes with a 10 percent return probability in 50 years, and building collapse must be prevented for earthquakes with a 2 percent return probability in 50 years.

- *Limited Objectives.* This performance level is less stringent than the basic level. The goal may be to assure life safety for earthquakes with return probabilities higher than 10 percent in 50 years or to assure limited life safety for earthquakes with a 10 percent return probability in 50 years.

- *Enhanced Objectives.* This performance level is more stringent than the basic level; the goals may include assuring life safety and avoiding building collapse for earthquakes with smaller return probabilities.

The guidelines describe the systematic rehabilitation process, but they also allow a simplified rehabilitation method for meeting the limited objectives, should those be chosen. The simplified method can be used for small basic buildings located in areas of low and moderate seismic risk. In addition, the guidelines include specific steps that can be taken to reach a chosen level of performance, discuss some new methods of seismic analysis, and outline procedures for checking the acceptability of existing structural members made of wood, concrete, steel, and masonry. A section on seismic upgrade of nonstructural elements is also included.

It is important to keep in mind that all this information is advisory in nature and is not intended to have the force of law. The most important question—whether or not seismic strengthening must be undertaken for a given building—is beyond the scope of these documents. The task of selecting an appropriate performance level (objective) is also left to the owner's and the designer's judgment.

A final point about this and other seismic-design codes and standards: They operate with statistical probabilities, not absolute certainty. It is clearly wasteful of resources and just plain impractical to construct new or upgrade existing buildings to be able to withstand any possible earthquake without a scratch. Instead, new and upgraded buildings conforming to the codes are expected to sustain *some* damage during a strong level of ground shaking. Worse, there is a real though very small probability of collapse of these buildings when a "design" seismic event occurs.

11.5 Typical Elements of Lateral-Load Upgrading

11.5.1 Evaluation of lateral-load resistance in existing buildings

The process of building evaluation is described in detail in Chap. 2. Additional information can be found in ASCE 11-90, *Guideline for Structural Condition Assessment of Existing Buildings,*[28] also discussed in Chap. 2. ASCE 11-90 suggests a multilevel approach to structural evaluation of buildings. In the first level, preliminary assessment, the existing construction documents are reviewed, field inspection takes place, and a preliminary structural analysis is made. Depending on the conclusions and recommendations yielded by this effort, a second, more detailed level of evaluation might be needed. This two-step approach saves unnecessary efforts where they are not needed but allows for detailed assessment where it is justified.

Another relevant document was recently renamed FEMA-310, *Handbook for Seismic Evaluation of Buildings: A Prestandard.* (The earlier version of this popular publication was known as FEMA-178, *NEHRP Handbook for the Seismic Evaluation of Existing Buildings.*) It contains a set of checklists for various types of building structures, helping the user identify their weak points. Different checklists are provided for areas of high, medium, and low seismic risk. The seismic evaluation procedure in this document has been modified to incorporate the latest knowledge in this field and to be compatible with FEMA-273. The handbook has also been expanded to be useful for the multiple performance objectives of FEMA-273, and now contains six examples of seismic evaluation.

11.5.2 Addressing the common building deficiencies

Wind or seismic upgrade of buildings of moderate size and height typically means strengthening their horizontal and vertical lateral-load-resisting elements. These can be reinforced in place, or new elements can be added to them; if the existing lateral-load-resisting structure is grossly deficient, it can be replaced. Whenever shear walls, braced frames, or moment frames are designed to resist a larger seismic load, their foundations must be checked for the new level of loading and reinforced if needed.

Experienced engineers know that certain types of building structures—and a few specific components of those—have repeatedly failed in earthquakes and are prime candidates for renovation and strengthening. According to Forell,[29] some of these are

- Buildings with irregular configurations, such as those with abrupt changes in stiffness, large floor openings, very large floor heights, reentrant corners in plan, and soft stories (levels with significantly lower lateral stiffnesses than the other floors, such as the first elevated level in Fig. 11.5)

- Buildings on sites prone to liquefaction

- Buildings with walls of unreinforced masonry, which tend to crack and crumble under severe ground motions

- Buildings with a lack of ties between walls and floors or roofs

- Buildings with nonductile concrete frames, where shear failures at beam-column joints and column failures are common

Figure 11.5 A building with a first-floor soft story. Soft-story design was rather common in the past because it allowed for parking or walkways under the building.

- Concrete buildings in which insufficient lengths of bar anchorage and splices were used

- Concrete buildings with flat-slab framing, which can be severely affected by large story drifts

We would add some other elements that tend to fail during ground shaking: unreinforced masonry parapets and chimneys, and also nonstructural building elements, which can fall down or topple, blocking exits and injuring people.

The rest of this chapter is devoted to practical methods of upgrading various building elements to resist lateral loads. An excellent source of additional information on this topic is FEMA-172, *NEHRP Handbook for Seismic Rehabilitation of Existing Buildings.*[30]

11.5.3 Upgrading unreinforced masonry buildings

Perhaps the largest class of buildings in need of seismic upgrade is unreinforced-masonry (URM) buildings. These structures account for the majority of nonresidential buildings and have certain problems in common. URM buildings are typically loadbearing structures supporting wood or steel floor and roof framing. It is these buildings that are most commonly marred with scars after a string of powerful ground excitations.

Their biggest problem is a lack of proper connections between the floors and the walls, especially in buildings erected before the 1930s. If they are present at all, these connections are likely to consist of iron rods hooked into joists about every 6 ft and embedded into the brick or through it with large exterior washers. At walls parallel to the joists, even these anchors were typically omitted or, if used, were simply hooked into solid blocking between the joists. The anchors frequently failed by pulling out of the walls or joists, or because of corrosion.[31] The result was a wall falling away from the building, with all the predictable consequences, including partial or total building collapse if the wall was loadbearing.

The most common seismic strengthening steps for URM buildings involve providing new anchorages between the new or existing floor diaphragms and the walls. Some of the attachment details are discussed and illustrated in the following section and in Chap. 12. It has been observed that accelerations are higher near the middle of flexible diaphragms. Therefore, some authorities, including the UCBC (see Chap. 1), require the design anchor forces to be increased by 50 percent for the middle half of the diaphragm.

Another common strengthening step is to anchor or remove unreinforced parapets, as discussed and illustrated in Chap. 13. Existing parapets can frequently be observed to be bowed or displaced because of differential thermal and moisture movements between the parapets and the walls below. The typical result is horizontal or diagonal cracks near the roof level. When coupled with the weakening effects of through-the-wall flashing at the base of the parapet, a typical detail of the past, these cracks create the conditions for a serious seismic hazard. A detail for parapet bracing is shown in Fig. 11.6. Obviously, the roof members receiving the bracing must be strong and rigid enough for the task; it must be realized that typical roof joists are rather flexible and cannot prevent the parapet from cracking—they can only prevent it from falling down or suffering a major displacement. Some engineers consider parapets with height-to-thickness ratios of less than 2.5 to be stable and in no need of strengthening.

Sometimes URM walls are not strong enough to resist wind or seismic forces acting normal to their plane. This can be remedied by shortening

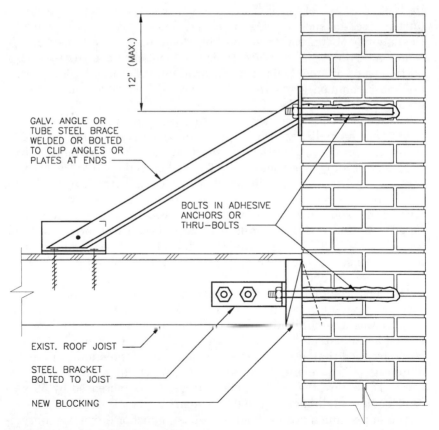

Figure 11.6 A detail for parapet bracing.

the span of the walls (by providing diagonal kickers at some distance below the floor, for example) or by reinforcing the walls with drilled-in bars, shotcrete overlay, or vertical supports, as discussed in Sec. 11.7.4 and in Chap. 9. In any case, as Avvakumovits[31] demonstrates, minor wall buckling does not necessarily equate to failure, because the dead load on the wall tends to stabilize it and arrest further damage.

11.6 Reinforcing Diaphragms

We begin our discussion with the evaluation and upgrade of diaphragms, rather than with that of the vertical rigid elements, because, as we shall see, quite often the characteristics of the existing diaphragms will control the course of action chosen.

11.6.1 Diaphragms: The basics

What makes floor decking a diaphragm—a deep horizontal beam capable of lateral-load transfer among the vertical rigid elements? Two things:

1. The ability of the decking to resist horizontal shear forces, meaning that it must possess a certain degree of strength and rigidity in its plane. This also means that various decking elements must be attached to each other and to the supporting floor structure with fasteners capable of transmitting these shear forces. To put it another way, the decking must be able to function as a web of the beam that does not break and does not deflect excessively under load.

2. The presence of flanges at the opposite ends of the diaphragm perpendicular to the applied forces. These flanges, called *chords,* must be attached to the diaphragm's web with fasteners capable of transmitting the desired forces (Fig. 11.7).

The horizontal distribution of load among the various walls or frames depends on the type of floor and roof diaphragms in the building. So-called flexible diaphragms (plywood or thin-gage metal deck) are assumed to distribute lateral loads to walls or frames in proportion to their tributary areas. So, for a building with three shear walls of the same size, two of which are located at the opposite exterior walls and one in the center, the center wall will take one-half of the total lateral load.

In contrast, rigid diaphragms (those made of concrete, for example) distribute lateral loads to walls or frames in proportion to their relative rigidities. In our example of the building with three shear walls, the middle wall will take one-third of the load. Rigid diaphragms can distribute horizontal forces by developing torsional stresses; this is

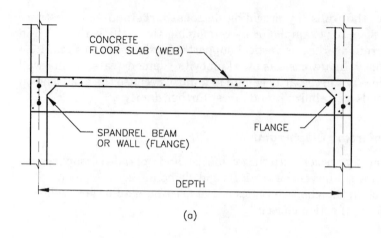

(a)

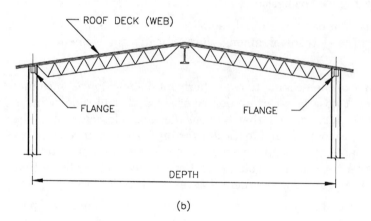

(b)

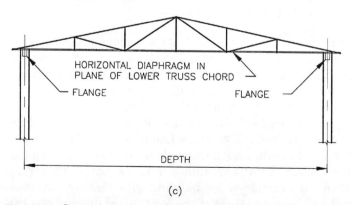

(c)

Figure 11.7 Components of a diaphragm. (*a*) Concrete floor diaphragm; (*b*) steel-deck or plywood roof diaphragm; (*c*) horizontal-truss diaphragm. (*After Ref. 19.*)

helpful in a building with a C-shaped or other irregular wall layout. Flexible diaphragms are considered too weak to work in torsion. The majority of real-life floor structures fall in between the two categories, and good engineering judgment is required to predict the behavior of these semirigid and semiflexible diaphragms. Some guidance on diaphragm classification can be found in Ref. 19.

The type and function of the existing diaphragms must be evaluated prior to making a decision about how to strengthen the vertical rigid elements of the building. For example, it is unwise to add shear walls or braced frames in an asymmetric manner if this introduces torsional stresses into the existing diaphragm and leads to its possible distress. If shear walls or braced frames are placed in the interior of the building, *collector* elements must be present in the diaphragm to carry the inertial forces to them.

If the existing diaphragm is inadequate for the design loads, it can be strengthened or replaced. Replacing a diaphragm means taking out the building's floors—major surgery that is reserved for the most critical of conditions. The methods of strengthening diaphragms depend on their composition and the nature of their weaknesses. The deficiencies of existing diaphragms typically fall into two categories: insufficient strength or stiffness and the absence of chords or proper connections to them.

11.6.2 Reinforcing wood diaphragms

Most old wood-framed buildings have weak diaphragms of wood boards nailed to the joists. Apart from their weakness in shear and their excessive flexibility, these diaphragms are usually not tied into the walls. The only positive connection may be so-called government anchors or similar devices that do not provide a reliable means of load transfer. Reinforcing wood-board diaphragms usually involves adding a layer of properly nailed plywood. A common approach is to disregard any diaphragm action of the existing flooring and to design the new overlay as the only diaphragm, at least for strength calculation. Any contribution of the existing flooring or roofing boards to diaphragm stiffness may also be conservatively ignored.

The renovation documents should show the typical nailing details, the schedule of nail spacing at various points, and the minimum nail penetration into existing framing. Typically, nail spacing is specified at the diaphragm boundaries—the edges where the web is attached to the chords—as well as at continuous and discontinuous edges of plywood panels and at intermediate joists (Fig. 11.8*a* and *c*). Where necessary, wood blocking should be added at the edges of the plywood panels (Fig. 11.8*b*). The procedure for designing plywood diaphragms is well established; it can be found in publications of the American

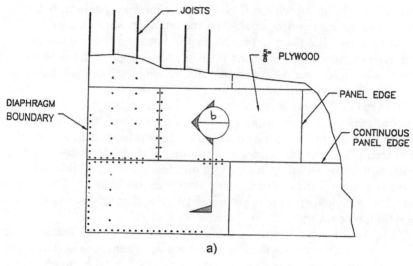

a)

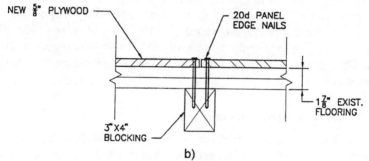

b)

NAILING SCHEDULE	
NAILING LOCATION	NAIL SPACING
DIAPHRAGM BOUNDARY	4" O.C.
CONTINUOUS PANEL EDGE	4" O.C.
OTHER PANEL EDGES	6" O.C.
INTERMEDIATE JOISTS	10" O.C.

NOTES:

1. TYPICAL MINIMUM PENETRATION OF NAILS SHALL BE $1\frac{1}{2}$" INTO FRAMING.

2. TYPICAL NAIL SIZE AT EXISTING FRAMING SHALL BE 20d.

c)

Figure 11.8 An example of strengthening an existing wood floor diaphragm with a new plywood overlay. (*a*) Partial plan of overlay; (*b*) detail of added blocking under plywood edges; (*c*) nailing schedule. (*Maguire Group Inc.*)

Plywood Association (APA). The shear values for plywood diaphragms and shear walls are included in APA literature, the *Uniform Building Code,* ASCE 7, and some other codes.

The details of the chord connections are important and should be shown. The chords of wood diaphragms can consist of wood blocking or ledger pieces, properly spliced to transmit the chord forces, or of wall elements that contain horizontal reinforcement that is able to develop these forces. Figure 11.9 represents a typical connection detail. Here, the chord is a bond beam in a masonry shear wall, to which a continuous wood ledger is bolted. The plywood attachment to the chord is made by nailing to the ledger and by bearing of the bolts against the wood and masonry. Today, many codes do not allow this ledger connection for exterior walls, the concern being the possibility of the ledger's failing in cross-grain bending under tensile load in the plywood. Unfortunately, a lot of these ledger connections did in fact fail in the 1971 San Fernando earthquake, allowing the walls to fall out and the roof to fall inside the building.[30] So, since the mid-1970s, the ledger detail has gradually lost favor with engineers.

A better connection utilizes a steel angle in lieu of the ledger (see Fig. 12.28 in Chap. 12) or some other detail that avoids the problem of cross-grain bending and provides adequate out-of-plane anchorage for the walls. A few of such details are shown in FEMA-172, *NEHRP Handbook for Seismic Rehabilitation of Existing Buildings,*[30] an excellent source of practical information on seismic strengthening of various building elements. For example, the detail shown in Fig. 11.10 illustrates a method in which a new steel strap is bolted to the underside of new solid blocking placed between wood joists running parallel

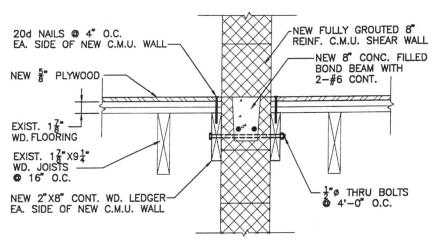

Figure 11.9 A detail of plywood attachment to the chords at interior walls. (*Maguire Group Inc.*)

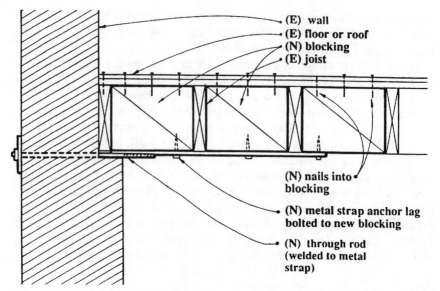

Figure 11.10 Providing out-of-plane anchorage for the wall and the new or existing diaphragm by means of a new bolted steel strap. (*From Ref. 30.*)

to the wall. The new blocking should receive nails from the existing floor sheathing, or from the new plywood overlay, if that is installed. The strap is anchored into the wall with a steel rod.

If the joists run perpendicular to the wall, the straps can be attached directly to them, or some other details can be used. Figure 11.11 shows two alternative methods of attaching the new rod to the sides of the existing joists. In the upper detail, it is simply welded to a new steel strap attached to the joist by nails, screws, or bolts. The obvious disadvantages of this connection are the need for field welding and the necessity of aligning the edge of the rod with the strap. In the lower detail, field welding is avoided; instead, the rod is attached to a bolted bracket, with a sleeve and torque nuts in between. A horizontal slotted hole could be provided in the bracket to compensate for misaligned rods.

The new connections can be placed on top of the existing floor or roof boards, with or without a new plywood diaphragm. Figure 11.12 illustrates a method of providing out-of-plane anchorage for the wall and the new diaphragm by means of a new bolted steel strap and continuous steel angle at the top of the floor.

In Figs. 11.10 through 11.12, rods are shown anchored through the wall. This is definitely the most reliable connection available, but it may not be architecturally acceptable in all cases. Another possibility is to grout the anchor inside the wall (an alternative detail shown in Fig. 11.12). The minimum angle is shown as 22° not because some exhaustive research required it, but because the committee members

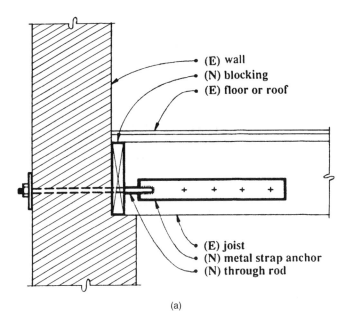

(a)

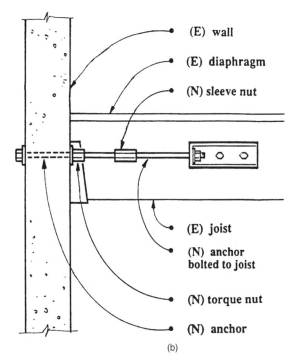

(b)

Figure 11.11 Two alternative methods of providing out-of-plane wall anchorage for joists supported on the wall. (*From Ref. 30.*)

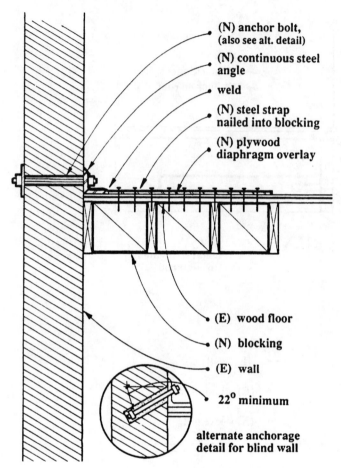

(N) anchor bolt, (also see alt. detail)

(N) continuous steel angle

weld

(N) steel strap nailed into blocking

(N) plywood diaphragm overlay

(E) wood floor

(N) blocking

(E) wall

22° minimum

alternate anchorage detail for blind wall

Figure 11.12 Providing out-of-plane anchorage for the wall and the new diaphragm by means of a new bolted steel strap and a continuous steel angle at the top of the floor. (*From Ref. 30.*)

who reviewed this detail were split between those who wanted a minimum angle of 45° and those who wanted a 0° angle; a compromise resulted. The author's own preference is to use the reliable adhesive anchors available today, installed perpendicular to the wall. If the holding capacity of the wall material is questionable—if the wall is made of hollow tile or weak brick, for example,—anchor tests can be commissioned, as illustrated in Case 2 of Chap. 12.

A rather common problem arises in URM buildings with heavy timber framing and masonry walls. Here, overlaying the existing wood-board diaphragm with plywood may not provide the desired level of strength and rigidity in situations where torsion must be considered, and therefore a semirigid or rigid diaphragm is required. One solution in this case is to add diagonal pieces of heavy timber between the existing ones on the perimeter of the building to make a horizontal truss. The new and existing timber pieces can be tied together with steel straps to make them into continuous chords.[31] A similar design can be used to provide collector elements in the floor.

An alternative step that can be taken when a plywood overlay will not do is to remove the whole floor structure and replace it with another diaphragm type. For example, in the same Case 2 of Chap. 12, replacement of some areas of the wood floor with concrete-filled steel deck was specified. Or, a more economical solution might be to add a lightweight concrete overlay only around the perimeter of the building, where diaphragm shear stresses are most critical.

Another approach is to supplement the existing wood diaphragms with horizontal steel truss bracing. The outer chord of the new truss bracing diaphragm could be a continuously spliced angle braced against buckling by ties to the wall. The inner chord could be a line of the existing wood beams, properly spliced to transmit the diaphragm-chord forces, or new laterally braced steel members.

11.6.3 Reinforcing steel-deck and concrete diaphragms

Steel-deck floors are usually found in structural steel buildings. The deck can be attached to the supports by plug welding (the most common case) or by pneumatically applied shot fasteners (a relatively recent innovation). In some cases, self-drilling screws have also been used in these connections. Diaphragms made of steel deck typically fall into the semirigid and semiflexible categories. The flexibility characteristics and shear resistance of a steel-deck diaphragm depend on the depth and gage of the deck, the length of the span between supports, and the method of attachment to structural beams or joists. The design procedure from either the tri-service manual[19] or the Steel Deck Institute manual[32] can be used.

When the existing steel-deck diaphragm lacks proper attachments to chords or to intermediate beams, the attachments can be upgraded. Additional plug welding requires removal of the floor or roof finishes, and a better course of action might be to use overhead fillet welds from below. The attachments to the chords are usually made in the same manner, since the deck is normally supported at the perimeter on steel angles bolted to the walls.

In many cases, the existing steel-deck floor diaphragm will contain concrete topping. What is the role of concrete here, a material with a much higher rigidity than steel deck? The answer depends on the method of diaphragm connection to the structure. If the connections are made by welding the deck to steel beams and perimeter angles or with shot fasteners, the diaphragm can be considered a steel-deck diaphragm with concrete fill. But if the connections are made with reinforcing dowels extending directly from the shear walls into the concrete topping, it is a concrete diaphragm.

As with wood floors, there are cases in which replacement is the most expedient method of upgrading steel-deck diaphragms, with or without concrete topping. Figure 11.13a shows a detail of a new steel-

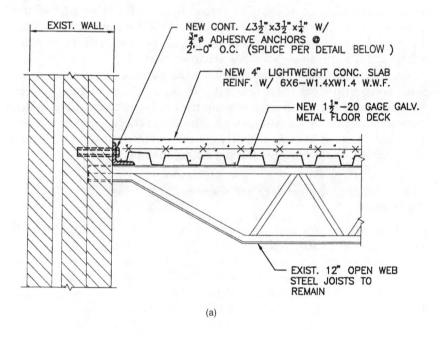

(a)

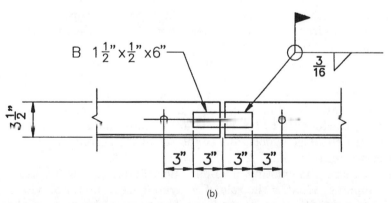

(b)

Figure 11.13 New steel-deck diaphragm with concrete topping installed on top of existing open-web steel joists. (a) Section at wall; (b) field splice of the diaphragm-chord angle. (*Maguire Group Inc.*)

deck diaphragm with concrete topping installed on top of existing open-web steel joists. Figure 11.13*b* illustrates a field splice of the diaphragm-chord angle designed to resist the chord forces.

Concrete diaphragms are sturdy elements that rarely require major upgrade except at their connections to the chords, where the just-mentioned continuous angles or similar sections can be used.

11.7   Reinforcing Wood, Steel, and Masonry Buildings

11.7.1   Adding shear walls and braced frames

Adding new shear walls or braced frames conforming to the current code detailing provisions is among the most common steps taken to strengthen the lateral-load-resisting systems of buildings. The new walls and frames can either complement the existing elements or be designed as the sole means of providing vertical rigidity in the building. In the former case, analysis of comparable rigidities must be done to determine what percentage of the total lateral loading the new construction will carry. If the latter course is chosen, the existing rigid elements that are now considered to be nonstructural must be isolated from the floors, lest they unintentionally get engaged—and damaged—during lateral movements. In any case, new foundations must be provided under the new elements and dowels placed around them for proper transfer of loads.

A common complication of adding shear walls and braced frames is that they tend to interfere with the building layout, circulation, or fenestration—issues that are important to the owners and zealously guarded by the architects. Engineers proposing this strengthening method must have a lot of patience, good interpersonal skills, and flexibility of thinking. Quite often, shear walls with openings or braced frames of unusual configurations may be needed to accommodate window or door openings. In some rare cases where a compromise cannot be reached, exterior buttresses or counterforts (Fig. 11.14) might be considered.

Adding braced frames, usually of structural steel, can be economical in buildings where the existing steel framing will not require strengthening to accommodate the bracing and where the existing framing is readily accessible. In other cases, this upgrade method might require a lot of framing reinforcement and demolition—and cease to be economical.

A design example involving the addition of shear walls and braced frames can be found in Chap. 12, Case 1.

But what about reinforcing the so-called light buildings—those framed with wood or metal studs? Here, shear walls of plywood or oth-

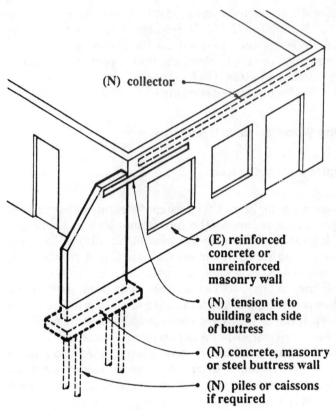

(N) collector

(E) reinforced concrete or unreinforced masonry wall

(N) tension tie to building each side of buttress

(N) concrete, masonry or steel buttress wall

(N) piles or caissons if required

Figure 11.14   Adding external buttresses to improve the lateral rigidity of the building. (*From Ref. 30.*)

er sheathing-type materials can be added. The shear capacity of plywood-sheathed walls is well established, but that of walls sheathed with other materials, such as oriented-strand board (OSB) and gypsum sheathing, may be subject to interpretation. Serrette et al.[33] studied the racking behavior of those materials under static loading and concluded that the shear values of plywood and OSB panels installed over metal studs were comparable, but that the strength of walls sheathed in gypsum board was relatively low. Research has also shown that connections between plywood and studs were about 20 percent stronger than connections between OSB and studs.

Occasionally, light buildings are reinforced with steel-plate shear walls, especially where circumstances require toughness or bullet resistance. The analysis of shear walls made of thin steel plate is discussed by Elgaaly and Liu.[34]

More often than not, the problem with light-frame buildings is not the load capacity of their shear walls sheathed with plywood or other materials but lack of connection between the walls in different tiers.

Strengthening these buildings for lateral loads should include proper load transfer between the chords of the shear walls. Two methods of doing this are shown in Fig. 11.15.

11.7.2 Using concrete frames with infill masonry

In many cases, the existing concrete or steel skeleton is stiffened by filling the space between the beams and columns with masonry (or, occasionally, with cast-in-place concrete). These infill walls can be a cost-effective method of increasing the lateral strength and rigidity of the building—if they are spaced uniformly throughout the building. Figure 11.16 illustrates the typical cracking pattern between a reinforced-concrete frame and the infill after a moderate earthquake. Perimeter cracks (in the picture, they have been repaired, and are thus exaggerated) opened up at the interface of the infill and the concrete frame, but no other failure is evident. Indeed, Youssef[35] states that concrete frames infilled with unreinforced masonry performed fairly well in the 1994 Northridge earthquake. Still, he cautions that they should be considered with care during seismic evaluation and renovations, because some of the infilled areas were damaged.

Designers should avoid counting on some of the infill walls in structural analysis but not on others, because the stiffness of the frames filled with this "nonstructural" masonry will increase, whether the designers realize this fact or not. In an earthquake, these panels attract unexpectedly (to the designers) large lateral forces and are damaged, or the perimeter columns, beams, and their connections break.

Some examples of this type of column failure are given by Green.[13] After the 1967 Caracas earthquake, he observes, many multistory concrete buildings in which concrete frames were designed as the sole means of resisting lateral loads had their corner columns crushed. The failures occurred in the bays filled with clay tile. The drastic overstress was explained by the fact that the stiffened bays took more seismic loading that they could handle, producing huge overturning moments on the columns. The corollary: When a frame, however well designed, is filled with rigid material, however brittle and weak, the fundamental behavior of this structural element is changed from that of a frame to that of a shear wall. (Those interested in reviewing the results of experiments with infilled frames are referred to Mosalam et al.[36])

11.7.3 Seismic reinforcing of existing steel framing and old moment connections

Structural steel framing often can be economically strengthened in the field, as explained in Chap. 3. Reinforcing the existing braced frames is relatively straightforward and is often preferable to adding new

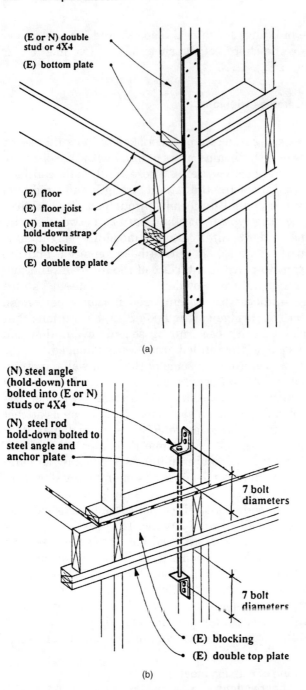

(E or N) double
stud or 4X4

(E) bottom plate

(E) floor

(E) floor joist

(N) metal
hold-down strap

(E) blocking

(E) double top plate

(a)

(N) steel angle
(hold-down) thru
bolted into (E or N)
studs or 4X4

(N) steel rod
hold-down bolted to
steel angle and
anchor plate

7 bolt
diameters

7 bolt
diameters

(E) blocking

(E) double top plate

(b)

Figure 11.15 Two methods of load transfer between shear-
wall chords located above each other. (*a*) Using a steel strap;
(*b*) using commercially available or custom-made hold-down
anchors. (*From Ref. 30.*)

Figure 11.16   Perimeter cracking around a panel of unreinforced masonry placed within a reinforced-concrete frame. (The cracks' appearance is exaggerated by the repair.)

ones. The work includes adding new cover plates, angles, or similar shapes and making new welded or bolted connections. For existing bolted connections of the bearing type, new welds can be designed to take the whole load, or the existing fasteners can be removed and replaced with new, stronger ones. When welded reinforcement is contemplated, it is wise to check the existing steel for weldability, unless some other welding to that steel is already in place.

As mentioned previously, many old steel buildings are framed with semirigid "wind" connections made with a pair of hot-rolled angles or tee sections attached to the beam and the column by rivets. A common variation on this scheme involves a stiffened seat angle in lieu of the web angles of Fig. 11.2. Both these types of connections can be upgraded to make them moment-resistant.

Some options for seismic retrofit of riveted structural steel connections made with stiffened seat angles were studied by Bruneau and Sarraf.[37] According to them, these connections, while capable of developing substantial moment resistance even in their original condition, cannot easily dissipate energy and thus suffer from a lack of ductility, a crucial property of seismic-resistant construction. The most intuitively appealing option, simply welding all the beam-to-column interfaces to approximate the design of Fig. 11.4, is not easily achievable. Such a design would involve overcoming interference with the existing framing, difficulties in placing backup bars and in cleanup, and the need for a major amount of the weld material to fill the (usually) large gaps between the beam and the column. Even if a full moment connection could be made, plastic hinges would probably develop in the relatively weak columns that are typically found in old buildings, rather than in the beams. That would be a highly undesirable situation, defeating the purpose of the seismic retrofit.

Instead, Bruneau and Sarraf proposed two main methods of reinforcement. The first method involves selective welding at carefully chosen locations and replacement of some rivets by high-strength bolts 19 mm ($^3/_4$ in) in diameter. The only rivets replaced are those attaching the top angles to the column flanges, because they are critical for improving ductility. The welding is done at the top and bottom beam flanges at their interfaces with the angles; the welds are designed for the full tensile capacity of the angle legs. Another area of welding is between the stiffener angles and the seat angles, to prevent a gap from developing in this area under load.

In the second method, knee braces are added both at the bottom *and* at the top of the beam-to-column connections. The knee braces in the study consisted of 63 × 63 × 6.3-mm (2.5 × 2.5 × 0.25-in) double angles welded to 100 × 50 × 12.7-mm (4 × 2 × $^1/_2$-in) tab plates and extended 542 mm (21 in) from the faces of the column. These knee braces were found to be sufficiently ductile and capable of good energy dissipation. Indeed, the relatively light angle sizes were chosen to make certain that they would become the energy-dissipating weak links; overly strong knee braces could result in plastic hinges being formed in the columns. (Unfortunately, the obvious problem with this method is that adding knee braces at the top of the beam will not be acceptable in many circumstances.)

And what of the welded moment connections found in numerous buildings around the world (see Fig. 11.3)? In regions of high seismicity, they may need strengthening. The upgraded connection must not only be strong enough to resist the stresses resulting from gravity and seismic loading, but also be flexible enough to have adequate plastic rotation capacity (at least 0.025 to 0.03 rad is required by FEMA documents). The usual reinforcing involves welding to make an old connection that might look like that in Fig. 11.3 look more like that in Fig. 11.4. Another approach common after the Northridge earthquake involved reinforcing the welds with cover plates welded to the top and bottom beam flanges and to the column. Yet another possible solution under consideration by researchers is to repair steel connections damaged by earthquakes with additional welding, typically an overlay of higher-grade weld material over the damaged area.[38]

Repairing connections usually involves, in addition to structural work, removal of wall and ceiling finishes and some disruption of the user's operations, even when the repair is done after working hours. It is obviously expensive. According to the published estimates quoted by Mosallam,[12] these repair costs can exceed $20,000 per connection (the inspection alone might cost $1500). Another estimate put the number at about $125 per square foot of floor space! Not surprisingly, researchers are busy developing and testing alternative approaches to strengthening of welded connections. According to Mosallam, some of these involve steel or polymer composite stiffeners that are adhesively bonded or bolted to the steel beams and columns. These fillet-type stiffeners involve no welding and are easy to apply—two major pluses for work in occupied buildings.

Those interested in obtaining further information on seismic upgrade of pre-Northridge welded moment connections are referred to AISC Design Guide 12, *Modification of Existing Welded Steel Moment Frames for Seismic Resistance.*[39] The guide provides information on three designs for reinforcement: reduced beam section, welded haunch, and bolted bracket. In addition to the technical discussion, it also covers several practical implementation issues, such as reducing tenant disruption in occupied buildings and dealing with safety issues.

11.7.4   Reinforcing existing shear walls

There are several possibilities for reinforcing existing shear walls:

- Increasing the wall's thickness by applying reinforced shotcrete to the wall surface. Shotcrete—a mixture of aggregate, cement, and water sprayed by a pneumatic gun at high velocity—is widely used for strengthening walls because it bonds well with both masonry and concrete. Some prefer application by the dry-mix method (sometimes called gunite) because the slump and stiffness can be better controlled by the nozzle operator and because gunite is applied at higher nozzle velocities, promoting superior bonding. The common procedures for surface preparation described in Chap. 5 apply in this case as well.

- Placing reinforcing bars into the wall by drilling and grouting them from the top or by cutting slots into the wall at regular intervals. An illustration of the use of the former in the renovation of masonry can be found in Chap. 9; a design example of the latter is described in Chap. 12, Case 1.

- Post-tensioning walls made of unreinforced masonry, also described in Chap. 9.

- Adding bolted steel plates to unreinforced-masonry shear walls. The plates act as external steel reinforcement.

- Increasing the shear strength of the walls by using carbon- and glass-fiber wrapping attached with epoxy. (A general description of this technique is given in Chap. 5.) The wrapping and the epoxy combined are only about 0.25 in thick, so that additional foundations are not needed. According to published reports,[40] this procedure has already been used to stabilize a brick building in Redwood City, California.

Of these methods, reinforcing the walls with shotcrete (gunite) is especially popular. The following approach is described by Green.[13] A typical application includes a continuous 3- or 4-in-thick layer of shotcrete on the face of the masonry (concrete walls can be reinforced as well). The shotcrete is reinforced at mid-depth with a mat of small-size (#3 or #4) rebars and is cut into the wall at regular intervals, usually 6 to 8 ft. The resulting vertical shotcrete ribs are reinforced as columns, with vertical bars and closely spaced ties. For practicality of placement, the width of the ribs should be at least 1.5 times their depth, as shown in Fig. 11.17.

The vertical ribs span between the floors and are anchored into their diaphragms as illustrated in Fig. 11.17. At the bottom, vertical dowels are drilled into the existing foundations at each vertical bar in the shotcrete layer and each stiffening rib. The shotcrete layer (slab) spans horizontally between the ribs, resisting wind and seismic forces perpendicular to the wall. The spacing of the ribs is determined either by the spanning capacity of the shotcrete layer in bending or by its buckling resistance to shear forces in the plane of the wall.

In this design, the existing masonry is assumed not to be effective, a sensible assumption for badly deteriorated walls. For walls in good condition, the gunite overlay may not be necessary, and the reinforcement may consist only of gunite ribs supporting the existing brick wall, which spans horizontally between them. According to Green, this approach works best for walls without openings.

In Fig. 11.17, shotcrete is applied to the interior face of the wall. Figure 11.18 illustrates the case in which it is placed on the exterior of the building. The interior application is more difficult, because of interference with the wall-supported framing members, conduits, etc., but it may be necessary when the appearance of the building must be preserved. One problem with this approach is that the shotcrete must extend above the bottom of the joists, unless the shear resistance of the existing wall and the shotcrete ribs alone is insufficient. If the shotcrete must stop at the bottom of the joists, the space between the joists can be filled with cast-in-place concrete.

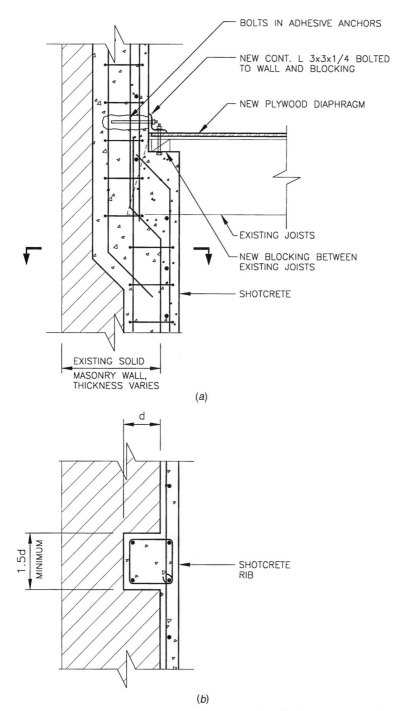

Figure 11.17 Strengthening existing masonry walls with shotcrete applied to the interior face of the wall. Shown are (*b*) a plan of the shotcrete rib and a cross section taken through the shotcrete rib (*a*). Also note the connection of the new plywood diaphragm to the wall. (*After Ref. 13.*)

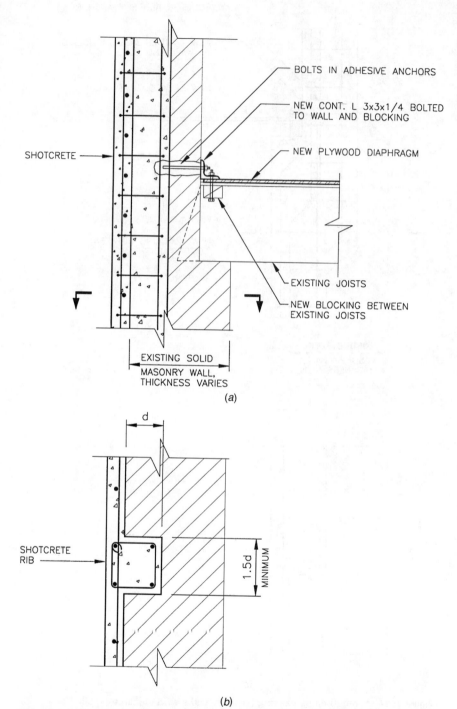

Figure 11.18 Strengthening existing masonry walls with shotcrete applied to the exterior face of the wall. Shown are (b) a plan of the shotcrete rib and (a) a cross section taken through the shotcrete rib. (*After Ref 13.*)

When interior shotcreting is used, attention must be directed toward stabilizing the exterior walls and any exterior ornamental elements of the structure. These may have to be tied back into the new shotcrete course by drilled-in dowels set at regular intervals or by some other means, including adhesive foam (mentioned in Chap. 13). Dowels placed in exterior elements that are exposed to moisture should be given a measure of corrosion protection, such as galvanizing.

In cases where it is desirable not to increase the wall size, the outer course of bricks can be removed and replaced with shotcrete. The same can be done with interior shotcreting, except that any members framing into the wall may have to be shored during this operation. The added bonus of this approach is that the vertical load on the existing wall foundations changes very little, and they may not need the otherwise-necessary enlargement.

A design example involving strengthening existing shear walls of unreinforced masonry can be found in Chap. 12, Case 2.

11.8 Reinforcing Concrete Buildings

11.8.1 Common areas of deficiency

A properly designed reinforced-concrete building that complies in all respects with today's codes for new construction has an excellent chance of surviving a strong earthquake. But what about the majority of solid-looking concrete buildings that do not meet the detailing requirements of these codes? In the 1970s, after the 1971 San Fernando earthquake and other seismic events, the codes shifted emphasis from simply requiring a specified level of strength to prescribing detailing provisions to assure ductile behavior of concrete. The code changes have made many previous detailing standards obsolete—and the buildings that relied on them theoretically deficient. Among the affected structures were nonductile concrete frames, beams and columns with insufficient lengths of bar anchorage and splices, and flat-slab framing. It was suggested that in all these cases, a failure might occur at the poorly reinforced beam-column joints and in the areas of inadequate confinement or lap splicing. Yet it is far from certain that a failure *will* occur simply because the concrete detailing does not comply with today's standards.

As Sabol and Islam[41] correctly observe, a code that requires, for example, that column hoop ties be spaced 4 in on centers is silent about what will happen if the ties are spaced 12 or 18 in apart. They also note that the lack of reliable information about the actual performance of concrete buildings leads to inconsistencies in the proposed renovation solutions: The planned level of strengthening is often either excessive or inadequate. To continue our list, they mention some common "apparent" seismic deficiencies in concrete buildings, all of which reflect previous detailing practices that are now considered poor. These include beams that lack the minimum amount of shear reinforcement, reinforcing bars that are not confined at the ends, and bottom reinforcing bars that lack the proper development length. Also, among the noncomplying elements are columns and piers with inadequate lap splices or too-sparse transverse and shear reinforcement, including beam-column joints that lack confinement, as well as discontinuous shear walls and lightly reinforced coupling beams between them.

According to the results of this research, in some cases the actual behavior under load of these presumably seismically deficient areas would probably prove satisfactory. For example, the lightly reinforced coupling beams between shear walls could be expected to undergo serious damage but still provide enough shear capacity to make the adjacent shear walls act together. Some other deficiencies can be economically remedied by reinforcing the most critical areas or by providing added confinement to the columns and boundary elements in shear walls that lack it.

El-Attar et al.[42] studied the earthquake behavior of reinforced-concrete buildings designed only for gravity loads and reached a similar conclusion. Despite very large deformations that resulted in significant P-Δ effects and damage, the researchers concluded that the old design details "will not necessarily lead to collapse or to a complete failure mechanism."

The problems that are most difficult to fix are those caused by the irregular configuration of the building (e.g., abrupt changes in stiffness, soft stories, large floor openings, and reentrant floor corners). These cases may require the addition of vertical or horizontal rigid structural elements, as well as strengthening of existing foundations or addition of new ones.

A good source of information on this topic is ATC-40, *Seismic Evaluation and Retrofit of Concrete Buildings*.[43] The two-volume report describes recommended methods for seismic evaluation and strengthening of concrete buildings, especially those constructed from the early 1900s to the 1970s, and includes four case studies.

11.8.2 Reinforcing concrete frames

Many pictures of earthquake damage show sheared-off columns that used to be parts of a frame. Sometimes, the concrete cover is spalled, the column bars are buckled, and the concrete inside is broken up. In reviewing the damage to reinforced-concrete structures in the Northridge earthquake, Youssef[35] states that, consistent with the research mentioned previously, most problems in concrete frames involved bar splices and failures of beam-column joints that lacked confinement and in which reinforcement was stopped prematurely. The failed frames with flat slabs included a seven-story building in which most of the columns in the five lower stories were damaged, and a building whose whole second story collapsed. Parking garages, even those constructed after 1973 (and presumably reflecting the post-San Fernando code changes), fared much worse than other concrete buildings.

Observing the problems of flat-slab systems, Youssef describes a characteristic case of seismic damage to lightweight concrete waffle slabs supported by round columns. The spirally reinforced columns were essentially undamaged, but many slabs sustained punching-shear or joint failures, and the building collapsed.

Without a doubt, many old buildings with flat-slab and flat-plate floor systems (the latter do not have any dropped beams at the column lines) are vulnerable to earthquakes. However, many such buildings have successfully survived seismic events. The question is: Must all of them be strengthened to meet today's detailing requirements? A good number probably should be, but some might be proved by sophisticated analysis to be able to resist the desired level of lateral loads without strengthening.

A case in point: The massive 800,000-ft^2 former Montgomery Ward regional distribution center in Portland, Oregon, was converted into a mix of trade-show and other commercial uses. This flat-plate structure dating from the 1920s was originally designed for a 250-psf live load, with 9-in-thick slabs spanning between columns spaced 20 ft on centers. Because the building was obviously not designed for lateral loads and lacked a delineated lateral-load-resisting system, the preliminary analysis indicated that a large amount of shear walls would be needed, a very expensive proposition given the size of the building. However, extensive finite-element and dynamic analysis concluded that the building could resist the design earthquake forces without strengthening. It was also determined that the floor slabs had a sufficient level of ductility and did not require modifications. The engineers observed that the structural analysis, which took 9 months to complete, probably required more engineering than most other projects, even though no changes to the building were ultimately needed.[44]

If an upgrade is required, the available methods of strengthening include encasing the beam-column joints in steel or high-strength-fiber jackets (Fig. 11.19). One such design uses jackets consisting of four U-shaped corrugated-metal parts, two around the beam and two around the column, as described in a paper by Biddah et al.[45] The column jackets are bolted to the end of the beam, the pieces are welded together, and the space between the jackets and the frame is filled with grout. The paper describes the methodology for computing the jacket thickness. It recommends that the jackets extend above and below the beam reinforcing bars for a minimum distance equal to the depth of the joint, but not less than one-sixth of the clear column height or 450 mm (18 in).

Frame joints damaged during earthquakes can be repaired with epoxy injection, and badly fractured concrete can be removed and replaced. To minimize shrinkage, the replacement concrete should be made with preplaced aggregate or shrinkage-compensating (type K) cement, or should utilize shrinkage-reducing admixtures, as described in Chaps. 4 and 5. Frame members that have been pushed out of alignment during an earthquake should be jacked back into proper position before reinforcement. Damaged columns can also be strengthened with fiber-reinforced plastic (FRP) wraps or other methods of exterior concrete confinement; this is common for seismic strengthening of bridge columns in California.

One seismic-upgrade task that should not be forgotten in renovation of concrete frames—or any other concrete buildings, for that matter— is transfer of load from the floor diaphragms to the vertical frames (or

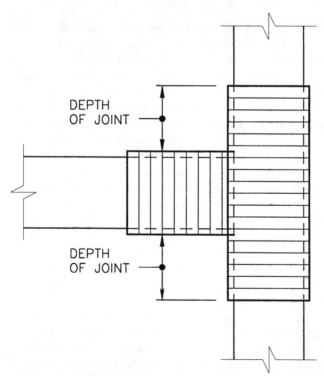

Figure 11.19 Reinforcing jacket around a beam-column joint. (*After Ref. 45.*)

walls). For this, new drag struts (collectors) may be required. These elements can be added by attaching new concrete or structural steel sections to the underside of the existing floors. They are typically placed against cleaned and roughened concrete surfaces and anchored to the floors and to the frames by drilled-in dowels or through-bolts.

11.8.3   Adding new concrete shear walls

As discussed in Sec. 11.7.1, adding new shear walls at a few strategic locations can be a very cost-effective approach to seismic retrofit. The usual challenge is to be able to convince the architect and the owner of the benefits of this upgrade solution. The schematic method of adding concrete shear walls within a concrete frame is shown in Fig. 11.20. The new wall is connected to the adjoining frame by drilled-in dowels in a relatively straightforward fashion. Its foundations are similarly doweled into the existing column footings. To accommodate wall shrinkage, the wall can stop short some distance (2 in, for example) from the existing concrete at the top; the space can be later filled with nonshrink grout.

One example of a successful use of this method is presented by Miller and Reaveley,[46] who describe the process of adding shear walls in the historic Hotel Utah in Salt Lake City, constructed in the early 1900s. After comparing various methods of seismic upgrade, the engineers concluded that adding shear walls to the steel-framed 10-story hotel was the most economical course of action—and they were able to do it without major changes in the interior layout. In the process, a dangerous soft-story condition at the lower levels of the building was corrected. Negotiations with the architects, which made the shear-wall solution possible, also required the walls to be relatively thin. To improve their shear capacity, concrete with a 28-day compressive strength of 5000 psi was used. Because of congested reinforcement, concrete mix with $^3/_8$-in coarse aggregate was specified and careful placement and vibration were required. The new shear walls contained many openings that had to be reinforced. The new walls were connected to the existing diaphragms by attaching shear studs to the adjacent steel beams and partially encasing them in concrete corbels that formed part of the walls. One other complication involved the shear-wall foundations. Since the walls were often placed between the existing columns, the column footings were in the way of the wall foundations. In some areas, the existing columns were shored and their footings were removed and replaced with footings of different configurations to make space for the shear-wall footings.

Another example of reinforcing a concrete building by adding shear walls is a seismic upgrade of several Naval Station buildings on the island of Guam (Fig. 11.21). The two-story buildings were made of cast-

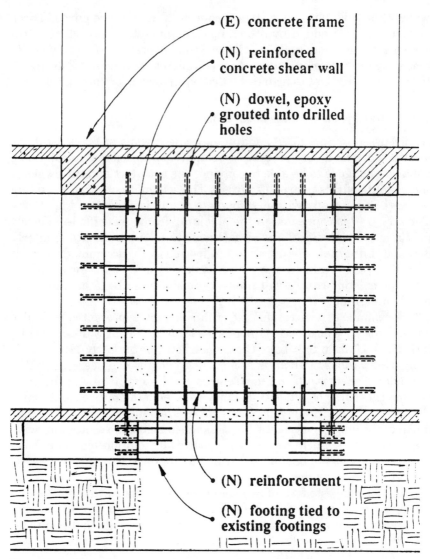

(E) concrete frame

(N) reinforced concrete shear wall

(N) dowel, epoxy grouted into drilled holes

(N) reinforcement

(N) footing tied to existing footings

Figure 11.20 Strengthening an existing concrete-frame building with added concrete shear walls. (*From Ref. 30.*)

in-place reinforced-concrete frames supporting precast floor and roof panels with a 2-in topping at the second floor. The first floor was framed with slabs on grade. The buildings were enclosed by 8-in-thick cast-in-place walls with large openings, some of them with masonry infill. There were full-height concrete shear walls at each end of the building.

A seismic analysis of the buildings indicated that the existing shear walls were adequate, but the capacity of the precast floor diaphragm was insufficient. To correct this deficiency, in each building the diaphragm span was shortened by adding two interior shear walls in the transverse (short) direction. Precast panels were also bolted together to improve shear transfer. Each of the two shear walls consisted of two halves separated by a corridor opening and connected by a coupling beam. At the base, the new shear walls were supported by new foundation walls connected by dowels into the exterior walls of the same depth. The new shear walls were placed in lieu of existing masonry partitions, and the architectural layout of the building did not suffer. The construction was completed without incident.

Case 1 in Chap. 12 further illustrates the challenges of adding concrete walls inside the concrete framing.

11.8.4 Reinforcing existing concrete shear walls

There are relatively few cases of failure of buildings with shear walls; the exceptions typically involve major overload or grossly deficient design. Overload can occur, for example, when, because of an irregular layout, one or a few walls take most of the lateral load. Concrete shear

Figure 11.21 One of the buildings on Guam that was seismically upgraded by adding shear walls.

walls that lack ductility may fail by crushing of their boundary elements, horizontal ("sliding") shear cracks along construction joints, or diagonal cracking caused by combined flexure and shear.[35] Among the most common areas of damage are the coupling beams. These can be repaired by through-bolted side plates extending onto the faces of the walls. Short and rigid piers between wall openings also tend to attract an inordinate amount of seismic loading and fail.

Shear walls and short piers that were damaged during an earthquake or that require strengthening because of inadequate capacity can be upgraded by adding a shotcrete overlay, as discussed in Sec. 11.7.4. The existing rigid infill panels within the frame can be strengthened in the manner described in that section, with removal of brick layers and addition of thickened ribs around the openings where necessary.

The major differences between shotcreting of masonry and concrete walls lie in the surface preparation and in the fact that existing concrete may be counted as part of the strengthened wall. All loose and cracked concrete must be removed from the existing wall, and its surface cleaned and roughened by sandblasting or by similar means. To assure composite action, the overlay is mechanically connected to the wall by closely spaced shear dowels. In addition, steel reinforcement placed in shotcrete is developed at the ends by grouted-in dowels or by continuation into an adjacent overlay space. This involves drilling through the perimeter beams or columns, filling the drilled openings with epoxy, and splicing the bars with those in the adjoining overlay areas. If the existing wall openings must be filled, the infill should be connected to the roughened edges of the opening with perimeter dowels set in epoxy.[47]

Cracked coupling beams can be repaired—and the shear capacities of the undamaged beams improved—by adding side plates extending on the faces of the walls, as described by Harries et al.[48] and illustrated in Fig. 11.22. In this procedure, envisioned for moderate seismic zones, the plates are attached with both epoxy adhesive and anchor bolts. The epoxy provides adequate connection strength under a low level of load; the bolts become effective under major cycling loading. The plates may be attached to only one face of the wall or can be placed at both faces for extra strength, with the opposite plates through-bolted together. The required plate areas and bolt shear capacities can be computed by formulas included in the paper. Another possibility for strengthening coupling beams is carbon-fiber wrapping, described in Chap. 4.

11.9   Energy-Dissipating Devices

So far we have discussed methods of strengthening the lateral-load-resisting systems of existing buildings in a belief that a stronger system

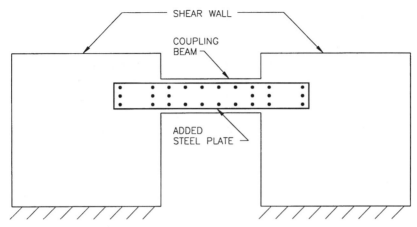

Figure 11.22 Repairing a cracked coupling beam by adding side plates. (*After Ref. 48.*)

should be able to resist a larger lateral force. However, the actual behavior of these "strong" or "stiff and stout" buildings in major earthquakes has not been outstanding. A strong ground shaking tends to find and break the weak links in the structure and, like a skillful military leader, to bypass the strongly fortified defenses. As Arnold[49] puts it, "This approach of arm wrestling with nature is neither clever nor subtle, and it involves considerable compromise." And, it may be simply impractical to eliminate all the weak points in an existing structure without taking it apart.

A more direct and economical approach is to provide some means of dissipating seismic energy within the structure. It is now considered *de rigueur* to provide built-in weak links at preselected locations in the lateral-load-resisting structures of new buildings in areas of high seismicity. These weak links are intended to absorb seismic forces by yielding or by controlled deformation and to extend the building's ductility into inelastic regions. A building that can achieve this type of ductile behavior is said to possess *hysteretic damping.*

One example of a lateral-load-resisting system with hysteretic damping is the eccentric braced frame system, in which the middle part of the beam is allowed to deform and absorb the forces of ground shaking. Another example is a so-called dog-bone moment connection, in which a section of the steel beam near the column is deliberately weakened to make it yield at that point. (In contrast, the pre-Northridge earthquake steel moment connections of Fig. 11.3 did not prove ductile enough.) Unfortunately, it may be difficult to use these strategies during seismic retrofit. Also, a building in which hysteretic damping takes place is likely to deform more than a "strong" building designed for an elastic response. Large deformations can be unacceptable in some older buildings with valuable brittle finishes and ornamentation. In addition, a permanently deformed steel beam may jeopardize the building's function.

But what if, instead of meeting lateral loads head on, the building could cushion them in some way? After all, most of us ride in cars with shock absorbers and have storm doors with pneumatic closing devices that prevent the doors from slamming into the house. This train of thought has led engineers to the idea of using energy-dissipating devices to resist lateral loads. The most familiar of these are *fluid inertial dampers*.

Dampers filled with viscous liquids have been used since the late 1800s in artillery, where they help to control the recoil of large cannons. In one design, a piston rod carrying the recoil force is connected to several moving plates that radiated like a fork from the end of the rod. The moving plates are surrounded by parallel fixed plates, with a viscous fluid in between; the whole assembly is contained in a sealed cylinder. The rod's movement is attenuated by shearing stresses generated in the fluid. Similar fluid inertial dampers have been used in the military ever since their development. With the end of the Cold War, advanced damper technology was made available for general construction applications, and one of the most popular damper designs is based on that used for the B-2 "Stealth" bomber.[50]

A typical fluid inertial damper is illustrated in Fig. 11.23. As the manufacturer explains, when the piston rod that carries the load moves to the right, it compresses the silicone fluid in chamber 2, and some of it flows into chamber 1. The damping force generated by the fluid is proportional to the pressure differential between the two chambers. The accumulator at the right helps prevent the otherwise inevitable reduction in the liquid's volume and the accompanying development of the springlike restoring force. The device can be made to exhibit no measurable stiffness for piston motions below a certain frequency, or it can be fabricated without any such cutoff frequency and made effective for frequencies up to 2000 Hz. Having a device with a cutoff frequency is often desirable: A fluid damper of this kind can provide additional damping and stiffness for higher modes of vibration that occur above the cutoff frequency, and help suppress them.

Fluid inertial dampers help reduce the wind and seismic forces transmitted to the building frame. This, in turn, reduces stresses and deflections in the frame and leads to smaller reinforcement requirements, an especially desirable outcome when strengthening existing buildings. The manufacturer states that these dampers can be sized to resist forces from 10 to 2000 kips and that the devices require no maintenance. The technology is relatively new, but it has already been used in dozens of buildings. In typical applications, the dampers are used as

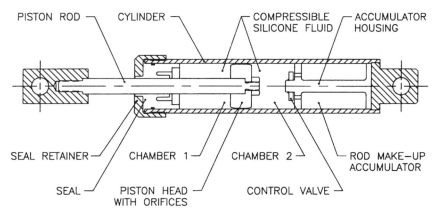

PISTON ROD ⌐   CYLINDER ⌐   ⌐ COMPRESSIBLE ⌐ ACCUMULATOR
                                 SILICONE FLUID / HOUSING

SEAL RETAINER ⌐/   CHAMBER 1 ⌐/   CHAMBER 2 ⌐\   ⌐ ROD MAKE−UP
                                                  ACCUMULATOR

SEAL ⌐/   PISTON HEAD ⌐/   CONTROL VALVE ⌐\
          WITH ORIFICES

Figure 11.23   A typical fluid inertial damper. (*Taylor Devices, Inc.*)

a part of diagonal vertical bracing or chevron bracing; they can also be combined with a base isolation design (Fig. 11.24). An example of a large-capacity fluid inertial damper installed as part of a diagonal brace is shown in Fig. 11.25.

One advantage of these fluid viscous dampers, and of similar energy-dissipating systems, is that they can be installed in the bracing, away from the regular gravity-resisting floor framing, so that any damage to them does not make the building unsafe while they are repaired (unless, of course, another major earthquake occurs during that time). Another advantage has already been mentioned: These devices substantially reduce building deformations. According to one reported efficiency study, dampers may cut the estimated building damage by 90 percent and, although they required an up-front cost of some $2 million, produce a life-cycle saving of $10 to $20 million.[51] A specific advantage of fluid viscous dampers is the fact that, because the force in them varies with stroking velocity, the damping force is applied out of phase with the forces on the frame. Translation: These dampers do not increase column forces as a result of their action, as other damper systems might do.

And, yes, there are several alternative damper designs. For example, *viscoelastic dampers* are made of stacked plates separated by an inert polymer; their shear deformations dissipate the seismic energy. Unlike the fluid devices, viscoelastic dampers add stiffness, sometimes an undesirable property for tall buildings. There are also *hysteretic dampers* that use curved steel plates intended to yield in an earthquake. Yet another type of damper relies on friction generated by steel plates sliding against each other. These *friction dampers* produce a constant damping force that can be tailored to specific requirements,

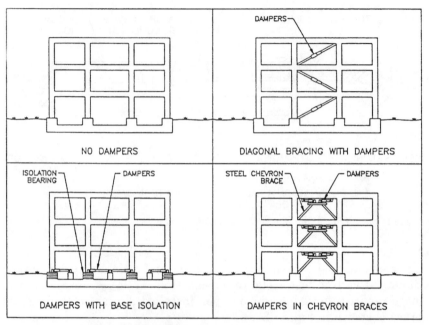

Figure 11.24   Typical applications of fluid inertial dampers. (*Taylor Devices, Inc.*)

Figure 11.25   This large-capacity fluid inertial damper was installed as part of a diagonal brace during seismic upgrade of a high-rise building. (*Taylor Devices, Inc.*)

but these requirements may be difficult to define. Fluid and viscoelastic dampers are temperature-sensitive; friction and hysteretic dampers are not.

These damper varieties are collectively called *passive dampers* because they need no outside source of energy. There are also *active dampers,* which require electricity to operate. Obviously, electricity may not be always available after the first few seismic shocks.[51]

The foregoing discussion focused on seismic resistance, but these energy-dissipating devices can also be used to reduce lateral drift of tall buildings caused by wind. Installing these devices can help address the concerns of office workers occupying the top floors of high-rise buildings, who sometimes complain of motion sickness.

11.10 Seismic Isolation

Energy-dissipating devices acting as shock absorbers present one alternative to the "stiff and stout" design approach. But why not go one step further and use shock absorbers to reduce the inertial forces transmitted to the building at the base? The idea of seismically isolating buildings makes intuitive sense. The best place to apply the isolators would be at the interface of the building's columns and foundations. The columns are part of the building's superstructure, while the foundations essentially move with the soil.

It is practically impossible, of course, to totally decouple a building from its supports, but today's seismic isolation methods promise to reduce the magnitude of seismic forces transmitted to the building by a factor of 5 to 10. If achieved, a reduction of this kind would not only help protect the building from collapse caused by unanticipated catastrophic seismic forces, but also help reduce building drift and contain property damage from ground shaking of a moderate magnitude.

The principle of seismic isolation has been used for a long enough time for some of the buildings reinforced in this fashion to have been tested in actual earthquakes. This real-world validation has allowed seismic isolation to overcome its experimental status and take its place in the arsenals of structural engineers involved with strengthening of existing buildings.

A good seismic isolation system should provide lateral flexibility under major seismic loading but remain rigid under normal circumstances, including lateral loading from strong winds. At the same time, the lateral movements (sliding) at the now-flexible supports should remain within manageable limits. The difficulty of achieving a functional seismic isolation system lies in integrating the two criteria. After all, it is relatively easy, at least conceptually, to place a building on rollers or ball bearings that transmit virtually no seismic forces to the structure. The challenge is not to let the building roll away—it must return to its original position! Many actual isolator designs use elastomeric materials that are capable of bringing the structure back, but still allow for large horizontal movements that are sufficient to reduce the transmission of seismic forces.

Most seismic isolation systems aim to increase the natural periods of vibration of the buildings they protect. As the period increases ("shifts"), the transmitted seismic force decreases (see the equations in Sec. 11.3). Also, making the building move much more slowly than the waves propagating in the ground typically helps eliminate the possibility of its resonance with the ground excitations. Resonance, a condition in which the structure moves in unison with the applied cyclical force (think of a parent pushing on the back of the swing holding a child), can lead to huge building movements at the top and probable destruction.

But there is a downside to increasing the natural periods of vibration of buildings: The magnitude of lateral drift from seismic loads also tends to increase. To contain these potentially large building motions, an increase in damping (energy dissipation within the building) is usually needed. Also, buildings must remain stable under normal circumstances and keep their rigidity under wind loading. The building structure must be relatively stiff to preclude wholesale yielding and energy dissipation at the upper levels, rather than at the base isolation system. Both these goals—increasing damping and assuring rigidity under wind loading—can be achieved with the use of the energy-dissipating dampers discussed previously.

The basic components of a typical seismic isolation system are shown in Fig. 11.26. The illustrated system utilizes an elastomeric bearing laminated with steel plates; the bearing is sandwiched between the steel collar and bearing-plate assemblies. This laminated design has been shown to be able to produce large enough lateral movements under seismic loading while carrying the weight of the building. A flat-jack (see Chap. 9 for a description of this device) on top of the bearing is intended to be used to prestress the elastomeric material of the bearing. Seismic isolation systems of this or similar construction—and there are other competing systems on the market—have been installed in many new buildings around the world. In the United States, new construction utilizing these systems most commonly takes place on the West Coast.

Such a system has also been used for seismic strengthening of some of the most prestigious buildings in the West, including the first major retrofit use of seismic isolation in the historic Salt Lake City and County Building in 1989. Here, base isolation was chosen because the alternative, adding full-height shear walls, would have largely destroyed the

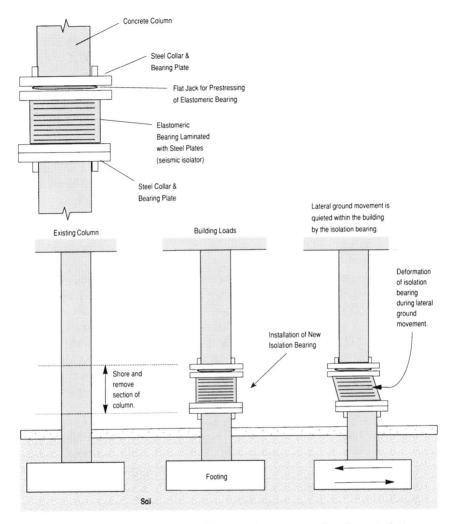

Concrete Column

Steel Collar &
Bearing Plate

Flat Jack for Prestressing
of Elastomeric Bearing

Elastomeric
Bearing Laminated
with Steel Plates
(seismic isolator)

Steel Collar &
Bearing Plate

Existing Column

Building Loads

Lateral ground movement is
quieted within the building
by the isolation bearing.

Deformation
of isolation
bearing
during lateral
ground
movement.

Installation of New
Isolation Bearing

Shore and
remove
section of
column.

Footing

Soil

Figure 11.26   The basic components and installation sequence of a seismic isolation system. [*From* Concrete Repair and Maintenance Illustrated *by Peter H. Emmons, copyright R.S. Means Co., Inc., Kingston, Mass., 1993, (800) 334-3509, all rights reserved.*]

rich interior of the building. The base isolation work was combined with some traditional elements of seismic strengthening, such as adding plywood diaphragms and shear walls at the attic level, anchoring the walls to the floors, and adding a lightweight concrete overlay around the perimeter of the floors to strengthen the existing floor diaphragms. The base isolation system consisted of 504 bearings. It reportedly increased the fundamental period of the building's vibration from 0.5 to 2.5 s; this had the effect of reducing the seismic forces by a factor of about 6.[49]

The Salt Lake City building utilized isolators of bonded metal and rubber. The isolators under the exterior walls had added lead cores that were designed to remain rigid under wind loading but yield in a major earthquake. These "fuses" also provide inelastic viscous damping that helps absorb seismic energy and control lateral drift. In addition, a concrete retaining wall was placed around the building with a 12-in gap to arrest the building's lateral movement should it become excessive.[52]

Figure 11.26 shows the typical installation sequence for a seismic isolation system in an existing building. Here, the column scheduled to receive the seismic isolator is shored, a section of the column near the bottom is removed, and the isolator assembly is installed. If necessary, the foundation is strengthened or rebuilt before the bearings are placed. Flatjacks are typically used to install the bearings, because they are small in height and can be inflated to the specified axial load on the isolator.

The Los Angeles City Hall also received a seismic isolation system during renovations in the late 1990s. The structure, built in 1927, was damaged in the 1994 Northridge earthquake. Two false starts took place in 1995 and 1996, but the installation work eventually commenced in 1997 (Fig. 11.27). As with the Salt Lake City building, a 12-in maximum lateral movement was anticipated and provided for.[53] When the work is completed, this 460-ft-tall building will be the tallest base-isolated building in the world. The work involved the installation of 416 rubber bearings and 90 sliding bearings.[54]

Will seismic isolation become the predominant method of strengthening the lateral-load-resisting systems of existing buildings? Forecasting is a hazardous business, but it is likely that this system will become commonplace in renovation of buildings with large historic value that fall within the limiting parameters listed here. It is also probable that the use of seismic isolation systems for other buildings will increase, but perhaps not to the degree hoped for by the system's proponents. Seismic isolation is certainly not a panacea. Weissberg et al.[55] list some conditions that must exist if a seismic isolation system is to be practical:

1. The underlying soil does not produce ground motions of predominantly long periods.

2. The building is relatively squat.

3. Design wind or other nonseismic lateral loads amount to no more than 10 percent of the building's weight.

4. The natural period of vibration for a fixed-base structure should be less than 1.5 to 2 s.

Figure 11.27 The Los Angeles City Hall received a seismic isolation system during renovations in the late 1990s.

In addition, as should be clear by now, using this system requires sophisticated analysis and expertise that is not often available in engineering offices.

11.11   Reinforcing Nonstructural Elements

Nonstructural building elements include exterior and interior ornamentation, wall finishes, mechanical units, ceilings, suspended fixtures, furniture, and equipment. What do these components have to do with seismic strengthening of buildings? Actually, quite a lot. While by definition excluded from the scope of work of structural engineers, failures of nonstructural building elements during strong ground shaking tend to cause most of the bodily injury. Think of people getting out of an earthquake-damaged building being bloodied by falling glass and pieces of exterior masonry. Think of free-standing full-height shelving, partitions, and file cabinets toppling onto office workers or trapping them inside a partly damaged building by blocking the exits. Visualize suspended light fixtures swinging wildly and, again, spraying the people below with glass slivers. Does this sound far-fetched? Unfortunately, all of these things have happened—many times. In one tragic example, a student fleeing a parking garage was struck and killed by a precast panel that fell on him.[56]

We outline here some typical areas of concern that should be addressed during renovations of buildings located in seismically active regions. For new construction, the governing building code will contain specific design prescriptions for nonstructural building elements. Those interested in the theoretical basis for these provisions are referred to Villaverde,[57] who provides a good overview of the code requirements and research efforts in this area. For renovations, the code's guidance may be less certain, but some sophisticated owners may provide their own criteria. For example, the U.S. General Services Administration, the agency that oversees the operation of most federal buildings, requires bracing of nonstructural components during renovations and provides guidance for doing so. Some of the most vulnerable nonstructural elements are

- *Unreinforced masonry chimneys.* Since Colonial times, chimneys have been known to topple during ground shaking. Accordingly, in the past some builders braced particularly tall chimneys with steel rods that extended through the chimneys and were anchored on the outside with decorative star-shaped plates (Fig. 11.28). The obvious concern here is preventing the chimney from falling on the pedestrians below, as the rod does little to prevent a fall inward or to brace the chimney in its strong direction. In today's renovations, it may be better to have this detail than nothing at all, but, if the aesthetics permit, a more substantial design similar to that of Fig. 11.6 provides better protection.

- *Exterior ornamentation, parapets, large signs, and rooftop antennas.* Like chimneys, these elements can fall down and injure pedestrians below or occupants fleeing the building. The design of seismic bracing for these elements is usually straightforward, except in cases of historic ornamentation (think of gargoyles), which can be rather challenging. Strengthening of parapets is addressed in Sec. 11.5.3.

- *Partitions.* Possessing little ductility, unreinforced masonry partitions can crack and break under seismic loading. For critical appli-

Figure 11.28 The typical method of bracing tall masonry chimneys employed by the builders of yesterday.

cations, such partitions can be removed and replaced with partitions of reinforced masonry or steel studs. Of particular concern are partitions (made of any material) that are laterally unrestrained at the top—these can topple and block emergency exits or injure the occupants. A simple remedial solution involves a pair of steel angles attached to the structure above but not to the partition. This design allows for lateral restraint without transferring vertical loading from the floor above to the partition. It also avoids transmitting to the partition racking forces from large interstory drifts, a common cause of damage to brittle wall finishes. A connection at the bottom may also be needed. Partial-height masonry partitions can be stabilized by V-shaped bracing made of steel angles and bolted at the top of the partition above the ceiling line. (The bolted connection should be made in a vertically slotted hole.) Partial-height metal partitions, common in toilet rooms, may be fixed at their bases. In any case, the partition structure should be capable of taking some minimal lateral load, such as the 5 psf typically mandated by codes for new interior partitions.

- *Wall veneers.* Many old public buildings are adorned with interior marble or thin-stone wall veneers or with exterior thick-stone panels. While some exterior installations have a better chance of survival than others, interior veneers often lack proper attachment to the structure and can easily fall down in a earthquake. Reanchoring the veneers is not easy, and the operation may leave noticeable traces. Occasionally, the panels are damaged during this operation and require replacement.

- *Tall bookcases, shelves, and furniture.* As with partitions, the main concern with these components is their toppling over. Shelves and furniture can be secured at the top with clip angles or similar devices that are unobtrusively attached to the back or sides of the element and anchored into the adjacent braced partitions or floors. In addition, open bookcases and shelving can be fitted with small lips at the front to prevent the contents from sliding off during an earthquake.

- *Large glass areas.* In many modern buildings, large expanses of glass are unavoidable. The potential damage to people both inside and outside the building can be lessened if tempered or laminated glass is used. In critical cases, plastics can be considered instead.

- *Suspended light fixtures, pipes, and ceilings.* These can swing during ground shaking and break or fall, injuring the occupants below or blocking their path. Fluorescent light fixtures, for example, are often simply laid on top of the suspended ceiling. To prevent their becoming dislodged in an earthquake, these light fixtures should be attached directly to the underside of the slab above with at least two wires per unit. Suspended light fixtures, pipes, and ceilings can be laterally braced by horizontal wires (or steel angles in the case of heavy elements such as pipes). Laid-in heavy ceiling panels can be equipped with retaining clips. In active seismic zones, brace-suspended light fixtures made of plastic rather than glass might be appropriate.[58]

- *Mechanical and electrical equipment.* HVAC and electrical equipment—heaters, boilers, storage tanks, air conditioners, transformers—should be secured to the roof or the floor structure to prevent their overturning during ground shaking. Proper anchorage and functioning of this equipment not only removes the problem of its falling down and causing damage, but also prevents the pipes carrying oil, gas, and water to the equipment from rupturing and considerably worsening the situation. The working equipment can also help in the postearthquake recovery efforts. Vibration isolators deserve special attention in this respect. If used, they should have retaining rings to prevent the equipment from sliding off. Or, rigid stops designed to prevent excessive lateral movements can be installed at the base of the equipment (Fig. 11.29).

- *Raised floors.* Raised floors have been used in computer rooms for a long time. Recently, they have become popular as energy management devices in "green" buildings, where the entire underfloor

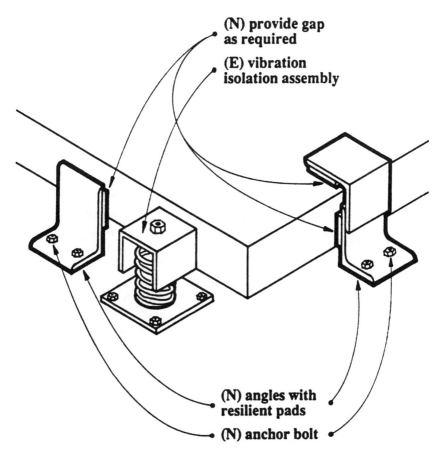

(N) provide gap as required

(E) vibration isolation assembly

(N) angles with resilient pads

(N) anchor bolt

Figure 11.29 Rigid stops designed to prevent excessive lateral movements are installed at the base of the mechanical equipment. (*From Ref. 30.*)

space is used as a mechanical plenum. Raised floors are supported on pedestals that are typically spaced 2 ft apart in both directions, with square floor panels placed on them. The pedestals can
be attached to the structural slab only with an adhesive, and the
resulting assembly is very flexible. This flexibility can lead to
damaging raised-floor motions or even connection failures and collapse. FEMA-172 recommends either securing the pedestals to the
slab with expansion anchors or providing new diagonal angle
braces at regular intervals (Fig. 11.30).

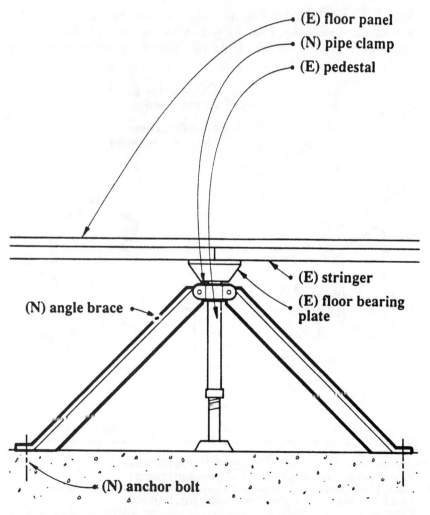

(E) floor panel

(N) pipe clamp

(E) pedestal

(E) stringer

(E) floor bearing plate

(N) angle brace

(N) anchor bolt

Figure 11.30 Strengthening raised floors with diagonal angle braces. (*From Ref. 30.*)

References

1. Data provided by A.M. Best, as reported in *ENR*, April 26, 1999, p. I-22.
2. Samuel J. Brown and Victor Perez, "Curing What's Curable," a sidebar to "Hurricane Retrofit," *Civil Engineering*, May 1991, p. 60.
3. ANSI/ASCE 7, *Minimum Design Loads for Buildings and Other Structures*, American Society of Civil Engineers, New York, 1995.
4. Alexander Newman, *Metal Building Systems: Design and Specifications*, McGraw-Hill, New York, 1997.
5. Paul A. Devlin, "Resisting Wind Losses," a sidebar to "Wind Endurance," *Civil Engineering*, December 1997, p. 62.
6. Ahmed M. Tarabia and Rafik Y. Itani, "Seismic Response of Light-Frame Wood Buildings," *Journal of Structural Engineering*, vol. 123, No. 11, November 1997, pp. 1470—1477.
7. *Manual of Steel Construction*, vol. II, *Connections*, American Institute of Steel Construction, Chicago, 1992.
8. Roberto T. Leon and Kyung-Jae Shin, "Performance of Semi-Rigid Frames," *Proceedings of Structures Congress XIII*, Boston, 1995, p. 1020.
9. FEMA-274, *NEHRP Commentary on the Guidelines for the Seismic Rehabilitation of Buildings*, FEMA, Washington, D.C., 1997.
10. Charles W. Roeder, "Seismic Performance of Older Steel Frames with PR Connections," *Proceedings of Structures Congress XIII*, Boston, 1995, pp. 1016–1019.
11. James O. Malley, "Performance of Steel Framed Buildings in the January 17, 1994 Northridge Earthquake," *Proceedings of Structures Congress XIII*, Boston, 1995, pp. 320–332.
12. Ayman S. Mosallam, "Making the Connection," *Civil Engineering*, April 1999, p. 57.
13. Norman B. Green, *Earthquake Resistant Building Design and Construction*, Elsevier, New York, 1987.
14. James R. Harris, "Overview of Seismic Codes," BSCES/ASCE Structural Group Lecture Series, Boston, 1991.
15. William J. LeMessurier, "40 Years of Wind Engineering: A Personal Memoir," *Proceedings of Structures Congress XIII*, Boston, 1995, pp. 1243–1252.
16. Jim Rossberg, "Fundamental Changes in the Basic Wind Speed Map in ASCE 7-98," *Structure*, Spring 1999, pp. 22, 23.
17. *NEHRP Recommended Provisions for the Development of Seismic Regulations for New Buildings*, Building Seismic Safety Council, Washington, D.C., 1994.
18. "New Seismic Code Has Widespread Implications," *Civil Engineering*, November 1992, p. 22.
19. *Seismic Design for Buildings*, Joint Departments of the Army, Navy, Air Force, USA, TM 5-809-10/NAVFAC P-355/AFM 88-3, Chapter 13, Oct. 20, 1992.
20. R. W. Luft, "Comparison among Earthquake Codes," *Earthquake Spectra*, vol. 4, no. 5, 1989.
21. R. D. McIntosh and S. Pezeshk, "Comparison of Recent U.S. Seismic Codes," *Journal of Structural Engineering*, vol. 123, no. 8, August 1997, pp. 993–1000.
22. *The Commonwealth of Massachusetts Building Code*, 6th ed., Boston, 1997.
23. "Seismic Retrofit Ordinance Is Only an Advisory," *ENR*, March 9, 1998, p. 9.
24. "California Law Shakes Up Hospitals," *Building Design and Construction*, April 1999, p. 14.
25. "UCLA Hospital Will Meet Strict Seismic Standards," *Civil Engineering*, April 1999, p. 30.
26. Robert M. Dillon, "Development of Seismic Safety Codes," in *Societal Implications: Selected Readings*, Earthquake Hazard Reduction Series 14, Federal Emergency Management Agency, Washington, D.C., 1985.
27. Christopher Rojahn and A. Gerald Brady, "FEMA 273 Guidelines for the Seismic Rehabilitation of Buildings," *Structure*, Spring 1999, pp. 24–27.
28. ASCE 11-90, *Guideline for Structural Condition Assessment of Existing Buildings*, American Society of Civil Engineers, Reston, Va., 1990.
29. Nicholas F. Forell, "Seismic Strengthening of Existing Buildings," BSCES/ASCE Structural Group Lecture Series, Boston, 1993.

30. FEMA-172, *NEHRP Handbook for Seismic Rehabilitation of Existing Buildings,* FEMA, Washington, D.C., June 1992.

31. Otto Avvakumovits, "Seismic Evaluation and Retrofitting of Existing Unreinforced Masonry Buildings," *Proceedings of Structures Congress XIII,* Boston, 1995, pp. 1821–1824.

32. *Diaphragm Design Manual,* 2d ed., Steel Deck Institute, Canton, Ohio, 1987.

33. Reynaud L. Serrette et al., "Static Racking Behavior of Plywood, OSB, Gypsum, and FiberBond Walls with Metal Framing," *Journal of Structural Engineering,* vol. 123, no. 8, August 1997, pp. 1079–1086.

34. Mohammed Elgaaly and Yinbo Liu, "Analysis of Thin-Steel-Plate Shear Walls," *Journal of Structural Engineering,* vol. 123, no. 11, November 1997, pp. 1487–1496.

35. Nabih Youssef, "Behavior Characteristics of Reinforced Concrete Structures in the Northridge Earthquake, California 1994," *Proceedings of Structures Congress XIII,* Boston, 1995, pp. 1419–1422.

36. Khalid M. Mosalam et al., "Static Response of Infilled Frames Using Quasi-Static Experimentation," *Journal of Structural Engineering,* vol. 123, no. 11, November 1997, pp. 1462–1469.

37. Michel Bruneau and Majid H. K. M. Sarraf, "Seismic Retrofit of Riveted Stiffened Seat Angle Connections," *Proceedings of Structures Congress XIII,* Boston, 1995, pp. 791–806.

38. "Dynamic Load Welds for Repair of Existing Steel Moment Frame Buildings Damaged from Earthquakes," DLW Task Group, c/o Sitech Ltd., Winnetka, Calif., 1997.

39. AISC Design Guide 12, *Modification of Existing Welded Steel Moment Frames for Seismic Resistance,* American Institute of Steel Construction, Inc., Chicago, 1999.

40. "Glass Fiber Wrap Increases Wall Strength," *Civil Engineering,* December 1998, p. 70.

41. Thomas A. Sabol and M. Saiful Islam, "Applications of Current Research to Seismic Evaluations of Reinforced Concrete Structures," *Proceedings of Structures Congress XIII,* Boston, 1995, pp. 738–741.

42. Adel G. El-Attar et al., "Behavior of Gravity Load Designed Reinforced Concrete Buildings Subjected to Earthquakes," *ACI Structural Journal,* vol. 94, no. 2, March–April 1997, p. 133.

43. ATC-40, *Seismic Evaluation and Retrofit of Concrete Buildings,* Applied Technology Council, Redwood City, Calif., 1997.

44. Christopher Olson, "Renovation Updates the Goods of a Landmark Warehouse," *Building Design and Construction,* May 1988, pp. 76–81.

45. Ashraf Biddah et al., "Upgrading of Nonductile Reinforced Concrete Frame Connections," *Journal of Structural Engineering,* vol. 123, no. 8, August 1997, pp. 1001–1010.

46. Jeffery T. Miller and Lawrence D. Reaveley, "Historic Hotel Utah Remodel and Seismic Upgrade," SP 160-5, in ACI Special Publication No. 160, *Seismic Rehabilitation of Concrete Structures,* American Concrete Institute, Farmington Hills, Mich., 1996.

47. James Warner, "Methods for Repairing and Retrofitting (Strengthening) Existing Buildings," Workshop on Earthquake-Resistant Reinforced Concrete Building Construction, University of California, Berkeley, July 11–15, 1997.

48. Kent A. Harries et al., "Seismic Retrofit of Reinforced Concrete Coupling Beams Using Steel Plates," SP 160-4 in ACI Special Publication No. 160, *Seismic Rehabilitation of Concrete Structures,* American Concrete Institute, Farmington Hills, Mich., 1996.

49. Christopher Arnold, "Seismic Design: Now Comes Base Isolation," *Architecture,* March 1987, pp. 64–67.

50. Douglas P. Taylor and Michael C. Constantinou, *Fluid Dampers for Applications of Seismic Energy Dissipation and Seismic Isolation,* Taylor Devices Inc., North Tonawanda, N.Y., 1996.

51. Eric Rasmussen, "Dampers Hold Sway," *Civil Engineering,* March 1997, pp. 40–43.

52. James S. Bailey and Edmund W. Allen, "Massive Resistance," *Civil Engineering,* September 1988, pp. 52–55.

53. "Shaky Past, Stable Future," *Building Design and Construction,* September 1998, p. 16.
54. Nabih Youssef, "Tall Order," *Civil Engineering,* October 1999, pp. 58–61.
55. Stephen M. Weissberg et al., "An Alternative for Earthquake Design," *Modern Steel Construction,* January 1991, pp. 41–44.
56. N. Taly, "The Whittier Narrows, California, Earthquake of October 1, 1987—Performance of Buildings at California State University, Los Angeles," *Earthquake Spectra,* vol. 4., no. 2., pp. 277–317 (quoted in Ref. 57.)
57. Roberto Villaverde, "Seismic Design of Secondary Structures: State of the Art," *Journal of Structural Engineering,* vol. 123, no. 8, August 1997, pp. 1011–1019.
58. Pat Guthrie, *The Architect's Portable Handbook,* McGraw-Hill, New York, 1995, p. 90.

12

Case Studies in Seismic Upgrading

These case studies illustrate the process of upgrading a building's lateral-load-resisting system. The emphasis here is on explaining how and why specific decisions were made by the designers and on illustrating the design process. Some step-by-step calculations are included for relatively uncommon tasks, but the reader is spared routine computations that are familiar to most structural engineers.

12.1 Case 1: Seismic Upgrading of a Former Industrial Building*

12.1.1 Project background

The Massachusetts Bay Transportation Authority (MBTA) needed a new building for its police headquarters. The MBTA police force is a fully accredited and highly professional law-enforcement agency serving the mass-transit system of Boston, Massachusetts, and vicinity. Having outgrown their old building, the police were looking for a new home for their expanded personnel and state-of-the-art equipment. After considering several other locations, the Authority purchased a building in South Boston (Fig. 12.1).

The building, originally designed in 1967 as a meat-packing plant, contained more than an acre of single-story space plus a partial second story. The building had last been used as a commercial printing facility. Maguire Group Inc. (MGI), the design firm selected to develop the

---

*This case study is based on a project by Maguire Group Inc., which supplied the accompanying illustrations, except that Fig. 12.2 is by Briggs Associates, Inc.

Figure 12.1   The new MBTA Police Headquarters, Boston, after renovations.

renovation program, was able to obtain copies of the original drawings, structural calculations, and foundation data. (This took some detective efforts, mentioned in the beginning of Chap. 2.) In addition to the drawing review and visual inspection, MGI hired a testing agency to conduct an extensive field investigation program. The results of the field investigation helped to provide much-needed information about the actual composition and structural condition of the building.

12.1.2   Existing conditions

The roofs over the main single-story area and over the second floor were framed with steel beams, bar joists, and narrow-rib steel roof deck. The columns supporting the structural steel roof framing were 6-in-diameter steel pipes or 6-in-square tubes. The roof structure was partly discontinuous in some areas, with parallel roof girders separated by a gap, apparently for reasons of thermal isolation. At columns, however, there was no structural separation, and both girders were eventually supported by the same columns.

The partial second floor was framed with three different structural systems: steel beams carrying a 3.5-in concrete slab on corrugated steel deck reinforced with 6×6–4×4 welded wire fabric; an 8-in-thick two-way concrete slab supported on concrete beams; and one-way concrete joists with a 3-in slab. This concrete framing was supported by cast-in-place square and round concrete columns.

The first floor structure, a two-way flat-plate slab supported by pile caps, was about 4 ft above the finish grade. It had a ventilated crawl space underneath. At the perimeter, grade beams 12 in wide and 6 to 8 ft deep spanned between pile caps. According to the original drawings, the 7.5-in-thick slab under the former freezer was designed for a 250-psf live load. This slab supported a 6-in layer of rigid insulation and a 4-in finished slab, both of which could be removed if needed. The rest of the first floor consisted of a 7.75-in flat slab designed for a 150-psf live load and formed directly against the soil. All the slabs utilized straight and bent reinforcing bars specified to conform to ASTM A15 and A305 specifications for hot-rolled intermediate-grade steel. Concrete strength at 28 days (f'_c) was specified to be 3000 psi.

The foundation system consisted of composite concrete and untreated wood piles with a design capacity of 22 tons. The bottom portions of the piles were wood; the tops were 12-in-diameter concrete-filled corrugated steel shells. There appeared to be no positive attachment capable of tension transfer between the two parts of the piles. The piles supporting columns were grouped and joined by cast-in-place concrete pile caps; there were also single piles supporting the slab.

The exterior skin of the building included 4-in brick veneer and 8-in concrete masonry units (CMU). The walls terminated in low parapets and were covered with insulation and wall finishes on the inside. There were no wall control joints, and the original drawings did not show any wall reinforcement. Field investigation uncovered ladder-type joint reinforcement interconnecting the layers but no vertical reinforcing bars (Fig. 12.2), so the walls could be considered unreinforced masonry (URM). The walls were tied to the building frame by a few strap anchors, a poor solution for transfer of either in-plane or out-of-plane loading.

The building did not appear to have any clearly identified lateral-load-resisting system. Some cross-bracing made of 4-in by $\frac{5}{16}$-in flat bars provided the only means of resisting wind and seismic loads, and this cross-bracing was not positioned in a logical manner. It was also unclear how well the roof functioned as a diaphragm. The concrete beams and columns were not detailed to act as frames, and apparently there were no dowels between them.

12.1.3 Dealing with code requirements

The *Commonwealth of Massachusetts Building Code,* 5th edition, governed the design and construction of this facility. This code recognized the need to preserve the existing building stock that did not

Figure 12.2 A field investigation uncovered ladder-type joint reinforcement interconnecting the wall layers, but no vertical reinforcing bars.

conform to the requirements of the latest building codes. Article 32 offered guidance about existing buildings undergoing renovations or changes in occupancy and the required degree of compliance with the code for new construction.

From Table F-1 of the code's Appendix F, "Reference Data for Repair, Alteration, Addition and Change of Use of Existing Buildings," a hazard index and a use group classification could be determined as a function of the building's use. The previous use, meat packing, was assigned use group F and a hazard index of 3. The proposed use, police station, could be classified as use group B and a hazard index of 2. Since the proposed use had a lower hazard index, the renovated building did not need to fully comply with the code for new construction.

However, another section of the code contained a provision requiring replacement or strengthening of any portion of the building "which will not safely support the loads of the proposed use group" as specified in the code for new construction.

Based on this code provision—even temporarily putting aside the issue of seismic loading—the building's lateral-load-resisting system needed to be upgraded, because, as analysis indicated, the existing building could not safely support wind loads of any appreciable magnitude. In other words, however the engineers felt about the chances of a devastating earthquake occurring in Boston, the possibility of a strong hurricane could not be ignored. Given the fact that the use of the building required it to be in operation following such natural disasters, a complete program of strengthening its lateral-load-resisting system was warranted, even if this was not specifically mandated by the code. The designers were also aware that the upcoming edition of the state code would specifically require buildings undergoing renovations to be analyzed for earthquake loads and strengthened if needed.

All parties eventually agreed that from the standpoint of function and safety, a police headquarters needed to be capable of withstanding major earthquake and storm events. The decision was to proceed with a structural upgrade for wind and seismic loads.

12.1.4 Design for lateral-load upgrade

One of the main questions facing the designers was what to do with the existing URM walls. The walls were obviously not strong enough to resist the wind loads: In many areas, the 8-in unreinforced CMU walls were vertically spanning about 18 ft. Wall removal and replacement was not considered economical, as that would require tearing down the whole exterior envelope. Even if it were strong enough, in its present condition the masonry could not function as shear walls because it was not attached to the roof diaphragm. Making proper attachments at the level of the roof deck would require removal and rebuilding of all the wall parapets and of the roofing and decking along the whole perimeter of the building—a major task.

The designers opted instead for another approach: reinforcing the walls only vertically, so that they could resist wind loads, and introducing other lateral-force-resisting elements to the building. The latter included concrete or CMU shear walls and steel-braced frames. In addition, there were already some X-braced frames, although these were in need of strengthening.

To provide support for out-of-plane wind loads on the existing exterior walls, but to keep these walls from functioning as shear walls, the existing exterior masonry was tied to the structure with wall braces. The wall braces were designed to resist wind loads acting normal to the walls but to be flexible enough not to transfer forces parallel to the walls. The braces were spaced about 4 ft apart, close enough to avoid overstressing the unreinforced CMU spanning horizontally between them, but far enough apart to keep their number to a minimum.

Here is a summary of the steps taken to upgrade the building's lateral-load-resisting system:

1. Attachment of existing exterior walls to the building frame by fabricated steel braces anchored into the walls and field-welded to the underside of the frame beams, as discussed previously. The resulting torsion in the beams was relieved by diagonal braces or by perpendicular beams (Fig. 12.3).

2. Division of the existing building into three separate units, plus a fourth unit for a new sallyport, for the purpose of lateral-load analysis (Fig. 12.4). Each unit had a different roof elevation and was provided with its own shear walls and braced frames; their spacing and location was selected to allow the existing steel deck to function as a diaphragm without further modifications. The only deck upgrade involved connections between discontinuous deck areas to transfer the diaphragm shear forces (Fig. 12.5).

3. Conversion of the existing unreinforced-masonry walls into vertically reinforced walls. This was accomplished by cutting vertical slots at 4 ft on centers or less and grouting reinforcing bars into the

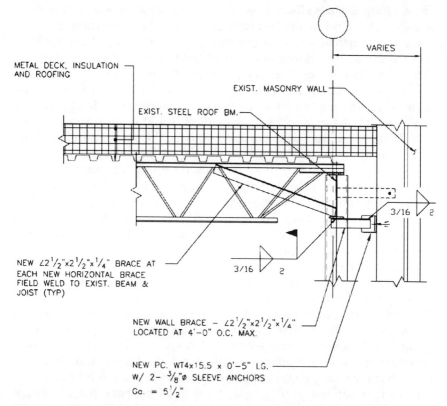

Figure 12.3 New wall braces tie the existing masonry walls to the building frame.

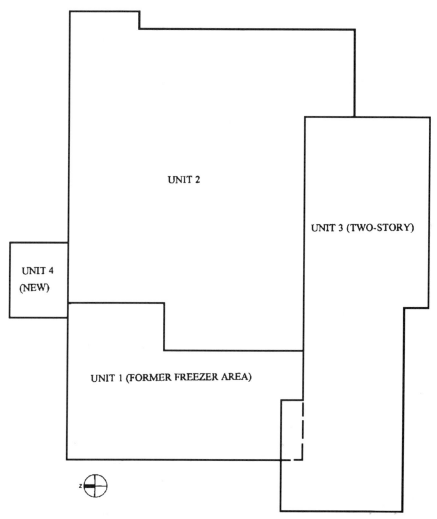

Figure 12.4 The existing building divided into four separate units.

middle of the CMU blocks (Fig. 12.6). In some places, where new CMU walls were built behind the existing ones for architectural reasons, these new walls were used as structural backup for the existing walls. In that case, the existing walls were left unreinforced and in effect treated as veneer; they were connected to the new walls by anchors (Fig. 12.7).

4. Removal or strengthening of some existing cross-bracing (Fig. 12.8), supplemented by new steel X- or K-bracing and shear walls. Many custom details were developed for various existing conditions; some

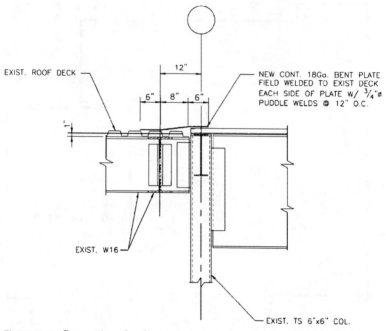

Figure 12.5 Connecting the discontinuous deck areas to transfer diaphragm shear forces.

MASONRY WALL REINFORCEMENT SCHEDULE

WALL HEIGHT FROM FLOOR TO SUPPORT POINTS	VERT. BARS GROUTED INTO EXIST. CMU		NEW VERT. WALL BARS (f'=2000 PSI, TYPE S MORTAR)	
	LOCATED MORE THAN 10' FROM CORNERS	LOCATED WITHIN 10' FROM CORNERS	LOCATED BEYOND CORNERS	LOCATED WITHIN 10' FROM CORNERS
13.1' TO 17.7'	#8 @ 48"	#8 @ 24"	#7 @ 48" OR #4 @ 16"	#7 @ 24" OR #4 @ 8"
UP TO 13.1'	#7 @ 48"	#7 @ 24"	#7 @ 48"	#7 @ 24"

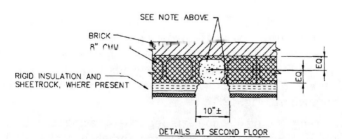

DETAILS AT SECOND FLOOR

NOTE: THE REBARS MAY ALSO BE INSERTED INTO THE EXISTING CMU BY DRILLING FROM THE TOP AND HIGH–LIFT GROUTING.

Figure 12.6 Existing CMU walls strengthened by grouted reinforcing bars.

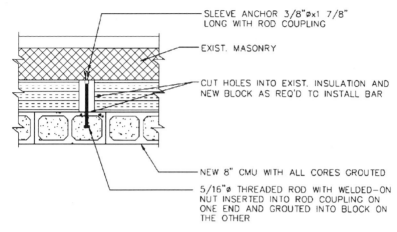

Figure 12.7   Connection of the new reinforced CMU walls to the existing exterior walls.

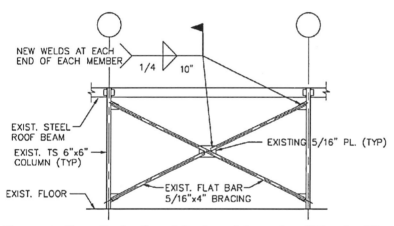

Figure 12.8   Strengthening the existing braced frames by additional welding at connections.

of these are shown in Figs. 12.9 to 12.11. (Forces "A" and "B" were summarized in a separate table not included here.) Details of connections for Fig. 12.9 are shown in Fig. 12.11. The new concrete shear walls located within the existing concrete framing are shown in Fig. 12.12; the new CMU infill shear walls within the existing steel framing are shown in Fig. 12.13.

5. Addition of steel cross-bracing to the free-standing concrete and steel columns along the south elevation. Previously, the area under the second floor had been a hazardous "soft story" susceptible to collapse during an earthquake (Fig. 12.14).

This strengthening of the building for lateral loads was part of the comprehensive renovation work, which was successfully completed in 1994.

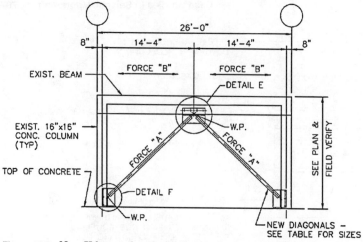

Figure 12.9 New K-brace placed within existing concrete framing.

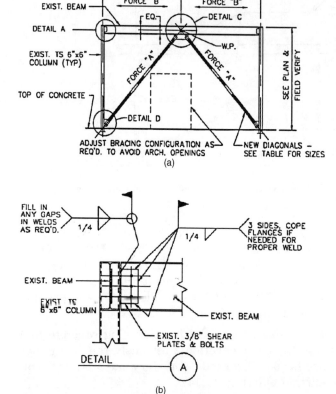

(a)

(b)

Figure 12.10 New K-brace placed within existing steel framing.
(a) Overall view. (b) Detail A.

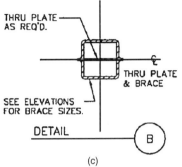

THRU PLATE
AS REQ'D.

THRU PLATE
& BRACE

SEE ELEVATIONS
FOR BRACE SIZES.

DETAIL   B

(c)

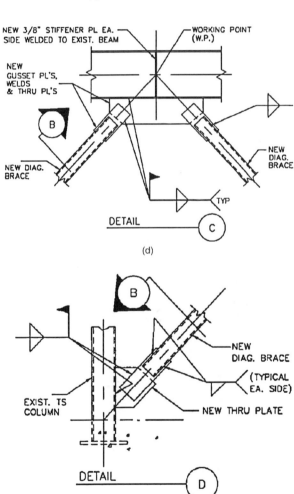

NEW 3/8" STIFFENER PL EA.
SIDE WELDED TO EXIST. BEAM

WORKING POINT
(W.P.)

NEW
GUSSET PL'S,
WELDS
& THRU PL'S

B

NEW
DIAG.
BRACE

NEW DIAG.
BRACE

TYP

DETAIL   C

(d)

B

NEW
DIAG. BRACE

(TYPICAL
EA. SIDE)

EXIST. TS
COLUMN

NEW THRU PLATE

DETAIL   D

(e)

Figure 12.10   (*Continued*) (c) Detail B. (*d*) Detail C. (*e*) Detail D.

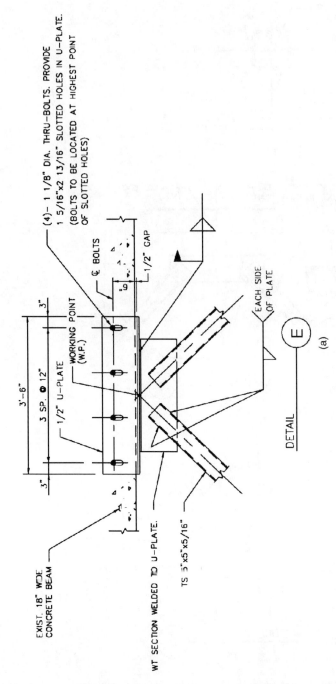

(4)– 1 1/8" DIA. THRU-BOLTS. PROVIDE
1 5/16"x2 13/16" SLOTTED HOLES IN U-PLATE.
(BOLTS TO BE LOCATED AT HIGHEST POINT
OF SLOTTED HOLES)

℄ BOLTS

1/2" GAP

6"

WORKING POINT
(W.P.)

3"

3'-6"

3 SP. @ 12"

1/2" U-PLATE

3"

EXIST. 18" WIDE
CONCRETE BEAM

WT SECTION WELDED TO U-PLATE.

TS 5"x5"x5/16"

EACH SIDE
OF PLATE

DETAIL

E

(a)

Figure 12.11 Details of new K-braces. (a) Detail E.

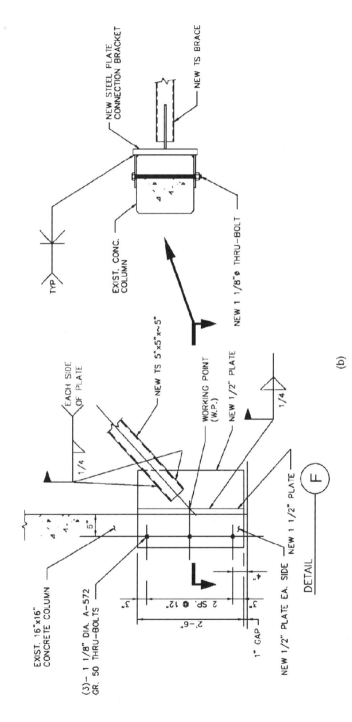

Figure 12.11 (*Continued*) (b) Detail F.

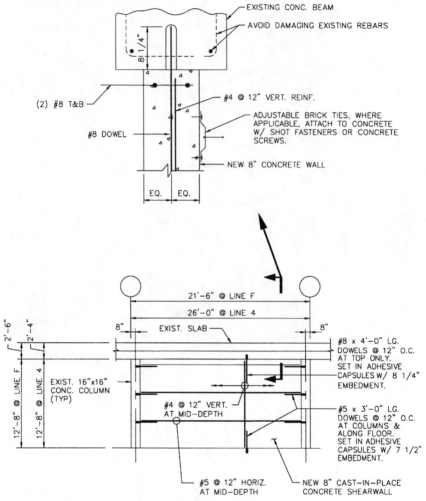

Figure 12.12 New concrete shear walls within existing concrete framing.

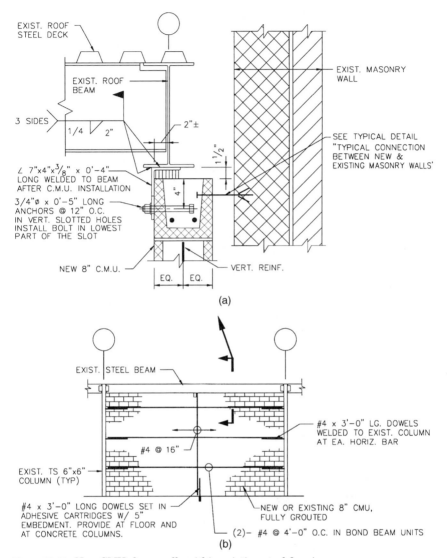

EXIST. ROOF
STEEL DECK

EXIST. ROOF
BEAM

3 SIDES
1/4 ⎺⎺ 2"

2"±

EXIST. MASONRY
WALL

SEE TYPICAL DETAIL
"TYPICAL CONNECTION
BETWEEN NEW &
EXISTING MASONRY WALLS'

∠ 7"x4"x³⁄₈" x 0'-4"
LONG WELDED TO BEAM
AFTER C.M.U. INSTALLATION

3/4"⌀ x 0'-5" LONG
ANCHORS @ 12" O.C.
IN VERT. SLOTTED HOLES
INSTALL BOLT IN LOWEST
PART OF THE SLOT

1¹⁄₂"

4"

NEW 8" C.M.U.

EQ. EQ.

VERT. REINF.

(a)

EXIST. STEEL BEAM

#4 x 3'-0" LG. DOWELS
WELDED TO EXIST. COLUMN
AT EA. HORIZ. BAR

#4 @ 16"

EXIST. TS 6"x6"
COLUMN (TYP)

#4 x 3'-0" LONG DOWELS SET IN
ADHESIVE CARTRIDGES W/ 5"
EMBEDMENT. PROVIDE AT FLOOR AND
AT CONCRETE COLUMNS.

NEW OR EXISTING 8" CMU,
FULLY GROUTED

(2)- #4 @ 4'-0" O.C. IN BOND BEAM UNITS

(b)

Figure 12.13 New CMU shear walls within existing steel framing.

Figure 12.14 New exterior steel bracing eliminated a dangerous "soft-story" condition.

12.2 Case 2: Proposed Renovation of an Unreinforced-Masonry Building*

12.2.1 Project background

In the early 1990s, an agency of the federal government was considering renovating an old unreinforced-masonry (URM) building that it owned in a large New England city. The building, surrounded by plenty of open space and parking, was situated outside the downtown area. It had been constructed in three stages and consisted of three semi-independent parts. The oldest part, called the Center Building (Fig. 12.15), was built around 1894, the East Wing was added in the 1920s, and the West Wing (Fig. 12.16) was added in the 1940s. A schematic plan of the whole structure is shown in Fig. 12.17.

---

*This case study is based on a project by Maguire Group Inc., which supplied the accompanying illustrations, except that Figs. 12.18, 12.19, and 12.21 are by Briggs Associates, Inc.

Figure 12.15 Center Building (south elevation).

Figure 12.16 West Wing.

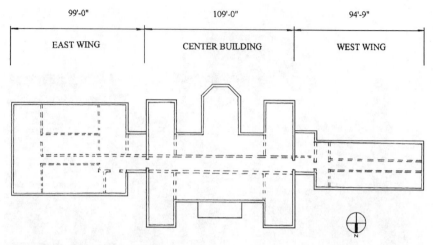

Figure 12.17 Schematic building plan.

The building had once been a religious seminary, but at the time of the proposed renovation it was essentially vacant, except for some areas used for storage. The building had working heating and sanitary systems. The structure consisted of three floors, a flat roof, and a partial basement, with a total building area of over 75,300 ft². The floors of the three parts were at slightly different elevations, and the transitions were made by corridor ramps. The agency wanted to convert the building into office space for its personnel; the renovated building was to conform to the *Uniform Building Code* (UBC). The agency required a frequent schedule of design submittals, holding design reviews after each submittal.

While they shared an unreinforced-masonry exterior—with wall thicknesses that varied at different locations—the three parts were built with different materials and structural systems. The Center Building had mostly wood joist framing, except for one area above the boiler room that was framed with steel beams supporting clay-tile and concrete slabs. The framing of the East Wing consisted of plain-concrete slabs carried on expanded metal and supported on open-web steel joists. These joists were of a proprietary Kalmantruss design (see Fig 3 4 in Chap. 0) and were supported on steel beams or masonry walls. The West Wing was framed with wood joists bearing on masonry walls.

The only available existing drawings were architectural, and so the first order of business was to undertake an extensive field investigation program. The investigation was conducted by the design firm's own personnel and by an independent testing laboratory. Guided by the project structural engineers, the laboratory made a number of floor and wall openings to determine the sizes and spacing of the framing. Several wall cores were also taken to measure the wall thicknesses at various locations. A representative core is shown in Fig. 12.18. Among the other items investigated by the laboratory were the building foundations, which were uncovered by digging several test pits. The test pits brought a rather unpleasant surprise: the presence of loosely laid rubble foundations at the Center Building (Fig. 12.19). The laboratory prepared a report of its findings, which was used to develop the preliminary framing drawings reflecting existing conditions.

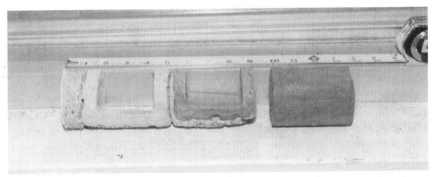

Figure 12.18 Typical core taken through an exterior wall to determine its composition.

Figure 12.19 The test pit uncovered the presence of rubble foundations at the Center Building.

12.2.2   Meeting the code requirements

After the field investigation was largely completed, the design team formulated a program of proposed code compliance. The structural engineers evaluated the load-carrying capacities of the existing framing and proposed a program of modifications and upgrades to make the building conform to the current code for live, wind, and seismic loads.

For maximum planning flexibility, the agency wanted the live-load capacity of the renovated building to be 100 psf. The preliminary analysis of the existing framing had made it clear that this capacity could not be practically obtained. Instead, the engineers proposed using typical live loads used for office buildings: a minimum live load of 50 psf plus a partition allowance of 20 psf. The floor framing was also to be able to support a 2000-lb concentrated load.

Meanwhile, the design architects determined that the existing building layout could not meet the new requirements and proposed that the existing corridor walls be moved in order to widen the corridors and to rearrange the available space for maximum benefit. Even with these changes, the available space was deemed insufficient for the new use, and some minor expansion by adding to the building's floors on each level was proposed. The proposed architectural modifications also included installation of new stairs and elevators in the East and West Wings and replacing the first floor over the mechanical room in the Center Building.

This plan presented some challenges. Most of the corridor walls were loadbearing, and their removal would require a large amount of reframing. There were also many proposed floor openings, filled areas, and extensions. The load-carrying capacity of the framing in the East Wing was unknown, as the framing consisted of plain-concrete slabs on expanded metal supported by proprietary Kalmantruss joists. To determine what loads the floors could carry, the engineers proposed load testing for uniform and concentrated loads.

A preliminary analysis of the building's lateral-load system concluded that there was no clear load path for lateral loading and that major remedial work was needed in order to meet the requirements of the UBC. As in many such buildings, the existing masonry walls were unreinforced, the exterior walls were not tied to the floors, and the diaphragm capacities of the floors were deemed insufficient. It was clear that the building did not comply with the current building codes for seismic loads and, for that matter, for wind.

The extent of the upgrade for lateral loads depended on close reading of the building codes and on the code to be followed.

12.2.3 Design for lateral-load upgrade in accordance with UBC-88

All three parts of the building had been constructed well before the current seismic design methodology was developed. As discussed in Chap. 11, this methodology relies on building floors acting as diaphragms distributing seismic loads to shear walls or other vertical bracing elements. To act as a diaphragm, a floor needs to be constructed in a specific fashion, with properly designed attachments between the floor and the walls and between the floor elements. The existing building floors did not qualify as proper horizontal diaphragms, and new plywood overlays, wall anchors at each floor joist, and additional bracing were proposed.

The unreinforced-masonry and terra cotta walls presented another major problem, as the use of these walls in seismic-resistant design was a subject of controversy. For example, while the 1990 *BOCA National Building Code* allowed their use with severe penalties in terms of additional seismic loads that needed to be resisted, the 1988 UBC did not even list unreinforced masonry among the accepted seismic-resisting elements. The UBC defined reinforced masonry as masonry with reinforcement in both directions spaced not over 48 in on centers. As a point of interest, the state building code did not require seismic design for buildings under 75 ft high with an office occupancy. The designers therefore suggested that the agency waive the seismic-design provisions of the UBC for this building, but the request was denied.

To substantially meet the seismic requirements of the *Uniform Building Code,* 1988 edition, the following upgrade items were required:

1. Place 6- to 8-in-thick reinforced-concrete walls on both sides of the existing loose-rubble foundation walls that supported the brick bearing walls. To accomplish this work, selective demolition of the basement slab in the affected areas was necessary.

2. Construct additional shear walls of reinforced CMU.

3. Add wood blocking between the existing wood joists.

4. Provide joist anchors at all masonry walls.

5. Place new plywood sheathing over the existing flooring at all floor levels, with proper nailing into the floor joists.

6. Install reinforcing bars in the existing brick walls by drilling holes in the walls vertically (at 48 in o.c.) and horizontally (at each floor level and above and below openings) and grouting the holes after installing the reinforcing bars.

7. In the East Wing, in addition to these items, provide new horizontal diagonal bracing at all levels as well as new concrete floor diaphragms at the west end.

8. Separate the superstructures of the three wings with 2-in-wide expansion joints to assure their independence in resisting lateral loads. Continuing the joints through the foundation level was considered unnecessary, because the foundations of old masonry buildings are rarely damaged during actual earthquakes.

It was clear that, unlike in Case 1 of this chapter, a program of seismic upgrade that called for field drilling and grouting of rebars into the existing walls in order to make them conform to the letter of the UBC would be extremely difficult and expensive. Furthermore, the brittle terra cotta could be damaged by the procedure, in which case replacement of the damaged walls would be needed. If both the walls and the floors ended up being substantially replaced, it might be more cost-effective to demolish and rebuild the whole structure.

As expected, this program was perceived by the client as unnecessarily conservative and expensive. The designers were directed to explore other options that would retain as much of the existing structure as possible.

12.2.4 Design for lateral-load upgrade in accordance with UCBC

In an attempt to find a more practical way to strengthen the building for lateral loads, the engineers looked to the 1991 *Uniform Code for Building Conservation* (UCBC) to provide the answers. As discussed in Chap. 1, this code is intended to offer guidelines for seismic upgrade of existing URM buildings in order to preserve the existing building stock—exactly the task at hand.

The building was in UBC seismic zone 2A. The UCBC requirements for that zone were concerned mainly with anchorage of walls and material strengths. The code assumed that some damage would be sustained by the structure during an earthquake. The specific provisions applicable to this building were

- Connecting all wythes of the masonry walls with ties (see the following discussion)
- In-place shear testing of mortar at various locations to verify masonry strength
- Fully grouting walls made of hollow clay tile
- Providing wall anchorages to resist out-of-plane wall stresses

Since the UCBC provisions did not require reinforcement of existing masonry walls, drilling and grouting vertical and horizontal bars into the walls—the most expensive part of the first program—could be avoided.

Now, the most expensive of the four efforts just listed was tying the masonry wall wythes together with special anchors and grouting all the voids. Why was this needed? The UCBC provisions required that the height of walls made of hollow concrete block and clay tile be limited to two stories. Since the four-story building exceeded that limitation, the walls had to be of solid masonry—meaning grouted hollow walls and bonded wythes. For multiwythe walls, the bonding requirement could be satisfied if all the wythes were bonded by closely spaced headers (which was not possible in this case) or if the wythes were anchored and made composite with the backup masonry with $^3/_{16}$-in diameter anchors for each 2 ft^2 of wall area. Ties of different sizes and spacing could be used if they provided an equivalent strength, and drilled-in adhesive anchors $^5/_8$ in in diameter spaced at 4 ft on centers in each direction were selected in order to reduce field labor. The anchors shown in Fig. 12.20 were specified to be drilled into all existing masonry walls.

The UCBC further required that all existing walls be tested and that they meet the material standards established in the code. Repointing of deteriorated mortar joints was also required (which could be done before testing).

12.2.5   Testing masonry anchors

To obtain reliable shear and pullout data on the capacities of anchors drilled into the existing brick, the designers specified field testing of the anchors. The testing was conducted in accordance with ASTM E488 and included gradual loading of the anchors until they failed.

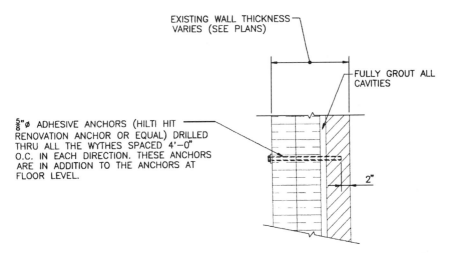

Figure 12.20   Existing masonry walls strengthened by grouting the cavities between the wythes and by drilled-in anchors.

The anchor manufacturer's representative was present during the tests to ensure proper installation techniques. The magnitudes of ultimate loading and displacements at failure were tabulated, and load-displacement curves were constructed (Fig. 12.21). The results of the testing revealed that, because of the poor condition of the masonry, the anchors had lower capacity in tension and shear than was typically recommended by the anchor manufacturer. The allowable anchor capacities were adjusted accordingly.

12.2.6 Testing masonry mortar

The requirements for testing the existing masonry were contained in Appendix Chapter 1 of the UCBC, Sec. A106(a). That section required in-place testing of the mortar to verify that it met the minimum acceptable strength level and to establish the actual mortar strength that could be used in design. The in-place shear tests of mortar, also called in-place Coulomb shear tests, were to be done in accordance with UBC

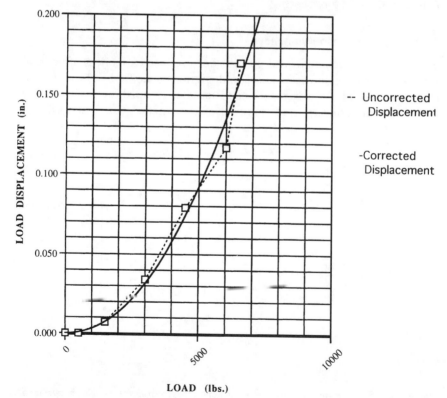

LOAD (lbs.)

Figure 12.21 A load-displacement curve for tested drilled-in anchors.

Standard No. 24-7. (Alternative tests could be used for materials other than brick.) The test locations were to be determined at the building site by the design engineer and were to be based on quality of construction and the state of repair, deterioration, or weathering.

The minimum number of tests per class of masonry was at least two at the top and bottom stories and at least one at each of the other floors, counted per wall or per line of wall elements. The minimum number of tests was eight, or at least one test per 1500 ft^2 of wall. Based on these requirements, a total of 72 in-place shear tests was specified for the project. Since the total number of tests seemed quite high, the engineers decided to reduce the number of tests if the wall conditions were consistent enough to warrant it.

The in-place shear test of mortar involved identifying a brick to be tested, removing the mortar at its head (vertical) joints, removing an adjacent brick in the row by sawing (or by drilling and excavating), and inserting a small hydraulic ram with steel loading plates in its place. The mortar in the collar joints behind the brick was to remain. The ram was then used to apply a horizontal load to the brick until the mortar joints failed, and the loading information was recorded. After the test, the bricks were replaced. The test setup is illustrated in Fig. 12.22. (The illustration is based on the information contained in *Proceedings of an International Seminar on Evaluating, Strengthening, and Retrofitting Masonry Buildings,* The Masonry Society, October 1989.)

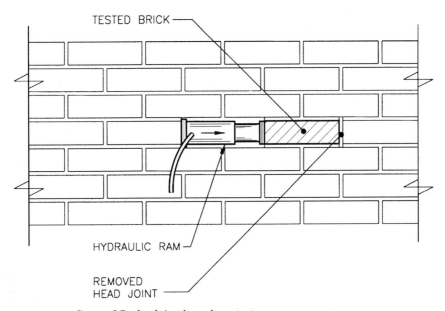

Figure 12.22 Setup of Coulomb in-place shear test.

The test results indicated that the shear strength of the mortar was generally quite good. The mortar joints were well filled and of uniform quality, with little observed deterioration. The condition of the collar joints, which were inspected during the test as required by UCBC Sec. 106, was also found to be satisfactory, with an estimated mortar coverage of 80 to 90 percent.

For purposes of testing and design, all existing brick masonry was considered to be of one class, so that the test values from various places could be averaged. The design mortar test value v_{to} was determined from UCBC equation A1-1:

$$v_{to} = (V_{test} - P) / A_b$$

where v_{to} = mortar shear test value

V_{test} = failure load of the in-place shear test

P = vertical load at the time of the test (estimated)

A_b = area of bed joints above and below the test specimen; for a typical brick (8.125 × 3.375 × 2.125 in) , $A_b = 2 \times 8.125 \times 3.375 = 55$ in².

The following test values were obtained or calculated for this test:

	V_{test}, lb	v_{to}, psi
Maximum	6000	94
Median	4000	57
Average	3800	54
20th percentile	2600	32

The 20th percentile in the last line represents the mortar shear strength v_t that was computed as the number exceeded by 80 percent of the test values. This value of v_t was used for determination of the allowable masonry shear value. In this case, v_t was 32 psi, which exceeded the minimum code value of 30 psi. The mortar quality, therefore, was acceptable.

The allowable shear stress in unreinforced masonry v_a could now be computed following UCBC Sec. A108(b):

$$v_a = 0.1 \, v_t + 0.15 \, P_u / A$$

For $P_d = 840$ lb (the wall weight per brick), A (the brick area) = 55 in², and $v_t = 32$ psi,

$$v_a = 0.1 \, (32) + 0.15 \, (840) / 55 = 8 \text{ psi}$$

12.2.7   Load testing in East Wing

As mentioned already, a load test for uniform and concentrated loads was needed to determine the load-carrying capacity of the existing floor assembly framed with proprietary open-web joists and plain-concrete slabs. To establish the magnitude of the test load, the engineers followed Chap. 20 of ACI 318. The code stated that the total test load should not be less than

$$0.85 \ (1.4D + 1.7L)$$

where D is the dead load and L is the live load on the structure, including the dead load already in place.

Using 25 psf for the dead load of the existing slab, 20 psf for the partition allowance, 10 psf for mechanical and miscellaneous loading, and 50 psf for the live load, the total load was computed:

$$0.85 \ [1.4 \ (25 + 20 + 10) + 1.7 \ (50) \ ] = 137.7 \text{ psf}$$

Subtracting the dead loads already in place, the minimum test load was

$$137.7 - 25 = 112.7 \text{ psf}$$

Rounding off, the design uniformly distributed test load was specified to be 115 psf. Since the testing lab had indicated its desire to use water as a testing medium, the equivalent height of water was $115/62.4 = 1.84$ ft or 22 in.

The actual load test for uniform loading was conducted by constructing a 25- by 8-ft pool made of wood framing with a plastic liner. The dimensions allowed loading of several steel joists for their full length.

The test for a 2000-lb concentrated load was conducted using piled-up sandbags. The magnitude of the test load was determined by calculating

$$0.85 \ (2000) \ (1.7) = 2890 \text{ lb}$$

For both tests, the ceiling below the test area was partly removed, and a combination of dial gages and steel rulers was attached to the underside of the structure. The gages were hung from threaded rods clamped to the joists or nailed into the concrete. The gages were also connected to a system of base plates and steel rods mounted on the floor below. The deflections of the steel joists and the concrete slab between them were measured.

After the base readings were recorded, the load was applied in four equal increments at half-hour intervals. The initial deflections were measured 24 h after the test, and the load was removed. The residual deflections were measured 24 h after that.

ACI 318 assumed that the test was satisfactory if the structure did not show any evidence of failure and if the measured deflection of the member did not exceed the value

$$l_t^2/20,000h$$

where l_t is the member's span and h is its overall thickness. For the 21-ft span of steel joists in this test, this value was

$$(21 \times 12)^2 / (20{,}000 \times 12) = 0.26 \text{ in}$$

which was less than the observed deflection of 0.32 in from the uniform load (it resulted in larger deflections than the concentrated loading). However, Chap. 20 of ACI 318 also allowed the test to be considered satisfactory if the deflection recovery within 24 h after the load removal was at least 75 percent. Since the actual deflection recovery was at least 86 percent, the test of the joists was deemed successful.

The deflections of the slabs were similarly measured and evaluated. Although the unreinforced slab exhibited relatively large deflections under load—from 0.27 to 0.33 in—the deflection recovery was at least 80 percent, and so the test of slabs was also considered successful.

12.2.8 Summary of the proposed work

Center Building. Because some existing interior bearing walls were scheduled to be removed, an extensive reframing with wood joists was proposed. At the central area of the building, the wood framing was to be replaced with lightweight concrete topping on steel deck and steel beams (Fig. 12.23). The replacement was specified to accommodate several proposed large floor openings for mechanical ductwork and other plan changes that would seriously decrease the effectiveness of the wood-floor diaphragms. Another reason was to provide the required fire rating above the mechanical room on the ground floor.

The new masonry walls and a few new steel columns were to be supported on spread footings placed as high as possible to minimize the depth of excavations close to the existing foundations. The existing loose-rubble foundations were to be injected with grout to make them act as a unit. In one area, underpinning of the stone-rubble foundation was required in order to accommodate a large mechanical louver.

The proposed lateral-load-resisting system consisted of masonry shear walls and plywood or concrete diaphragms. The cavities in the existing brick walls were to be fully grouted and wall ties provided between the wythes, as discussed previously. The existing brick walls were to be supplemented by new reinforced CMU shear walls. The new plywood diaphragms were specified to be installed over existing and new joists and flooring (Fig. 12.24). The wood or concrete diaphragms were to be anchored to the floors and walls by bolted steel angles. The detail of attachment to a typical interior wall is shown in Fig. 12.25.

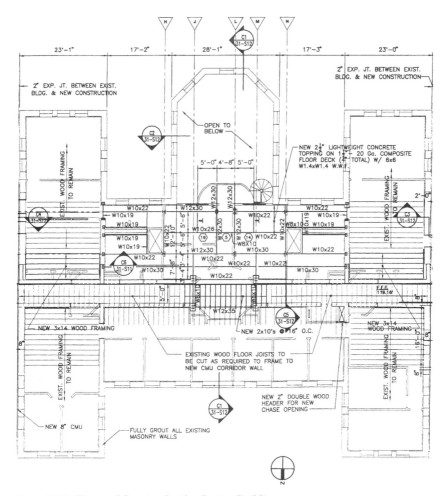

Figure 12.23 Proposed framing for the Center Building.

The design for lateral loads followed the provisions of the UCBC, except where extensive changes were proposed to the interior framing, such as the replacement concrete diaphragm, which was designed according to the UBC requirements for new construction.

West Wing. The existing wood flooring, roofing, and joists were to be retained as much as possible after the removal of all existing interior bearing and nonbearing walls. The new steel framing and concrete floor diaphragm were to be installed in a fashion similar to that for the Center Building. The new and remaining existing floor joists were to be supported by 10-in-deep steel beams bearing on new steel columns or new reinforced-CMU walls (Fig. 12.26).

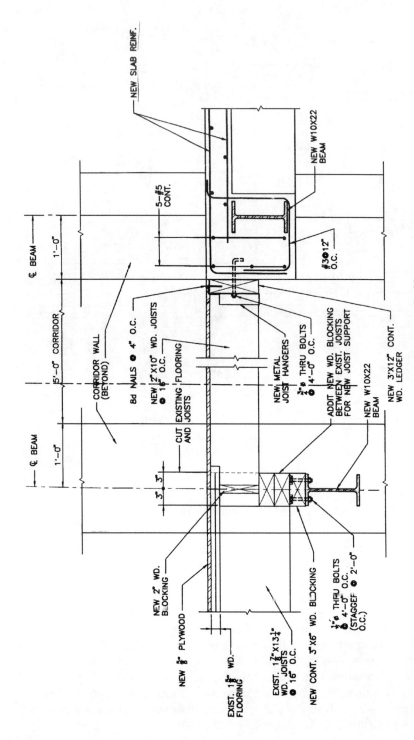

NEW SLAB REINF.

€ BEAM

5'-0" CORRIDOR

€ BEAM

1'-0"

1'-0"

5-#5 CONT.

NEW W10X22 BEAM

#3@12" O.C.

CORRIDOR WALL (BEYOND)

8d NAILS @ 4" O.C.

NEW 2"X10" WD. JOISTS @ 16" O.C.

CUT EXISTING FLOORING AND JOISTS

NEW METAL JOIST HANGERS

$\frac{3}{4}$"⌀ THRU BOLTS @ 4'-0" O.C.

ADDIT NEW WD. BLOCKING BETWEEN EXIST. JOISTS FOR NEW JOIST SUPPORT

NEW W10X22 BEAM

NEW 3"X12" CONT. WD. LEDGER

3" 3"

3" 3"

NEW 2" WD. BLOCKING

NEW $\frac{5}{8}$" PLYWOOD

EXIST. 1$\frac{5}{8}$" WD. FLOORING

EXIST. 1$\frac{7}{8}$"x13$\frac{1}{4}$" WD. JOISTS @ 16" O.C.

NEW CONT. 3"X6" WD. BLOCKING

1$\frac{1}{2}$"⌀ THRU BOLTS @ 4'-0" O.C. (STAGGEF @ 2'-0" O.C.)

Figure 12.24 Proposed framing at removed loadbearing corridor walls in the Center Building.

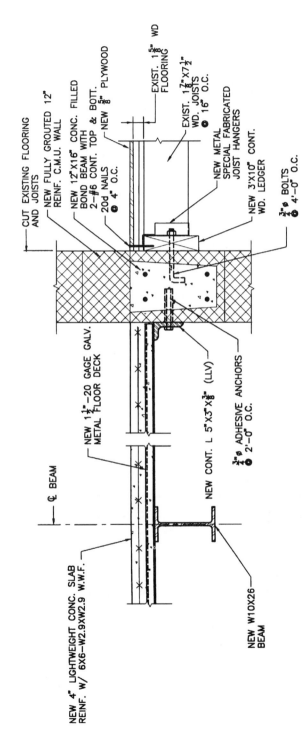

CUT EXISTING FLOORING AND JOISTS

NEW FULLY GROUTED 12" REINF. C.M.U. WALL

NEW 12"X16" CONC. FILLED BOND BEAM WITH 2-#6 CONT. TOP & BOTT.

NEW $\frac{5}{8}$" PLYWOOD

20d NAILS @ 4" O.C.

EXIST. 1$\frac{5}{8}$" WD FLOORING

EXIST. 1$\frac{7}{8}$"X7$\frac{1}{2}$" WD. JOISTS @ 16" O.C.

NEW METAL SPECIAL FABRICATED JOIST HANGERS

NEW 3"X10" CONT. WD. LEDGER

$\frac{3}{4}$"ϕ BOLTS @ 4'-0" O.C.

℄ BEAM

NEW 1$\frac{1}{2}$"-20 GAGE GALV. METAL FLOOR DECK

NEW CONT. L 5"X3"X$\frac{3}{8}$" (LLV)

$\frac{3}{4}$"ϕ ADHESIVE ANCHORS @ 2'-0" O.C.

NEW 4" LIGHTWEIGHT CONC. SLAB REINF. W/ 6X6—W2.9XW2.9 W.W.F.

NEW W10X26 BEAM

Figure 12.25 Attachment of new concrete and plywood diaphragms to a proposed interior shear wall.

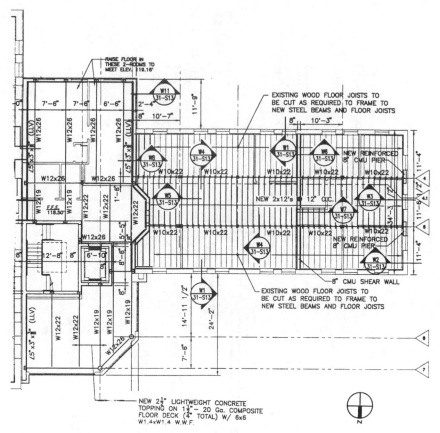

Figure 12.26 Proposed framing for the West Wing.

New columns and masonry walls were to be supported on spread footings placed flush with the existing ground floor to minimize the depth of excavations close to existing foundations. The new CMU shear wall adjacent to the Center Building was specified to be supported by the existing stone-rubble foundation of that building, which was to be grouted and widened to accommodate the new wall (Fig. 12.27).

The lateral-load-resisting system was detailed similarly to the masonry shear walls and plywood diaphragms of the Center Building. The new plywood diaphragm installed over the existing and new joists and flooring was to be anchored to the exterior walls by bolted steel angles (Fig. 12.28) to prevent cross-grain bending from the out-of-plane wall loads. (This was not a problem with attachments to the interior walls shown in Fig. 12.25, where plywood was connected to the wall by a wood ledger that had to resist only in-plane loading.)

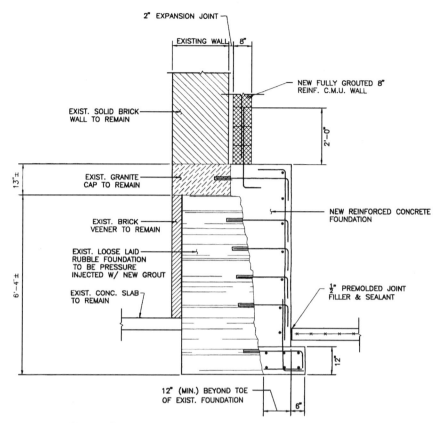

Figure 12.27 Proposed strengthening and widening of the existing foundation to support a new CMU shear wall.

East Wing. Modern building codes do not allow the use of plain concrete in elevated floors without a continuous support. Still, the results of the load tests of the first and second floors indicated that the existing floor structure, including steel bar joists and plain-concrete slabs, was capable of resisting the proposed live load and partition allowance. Therefore, this archaic structure could be largely retained.

In the western part of the East Wing, where extensive reframing and floor expansion were necessary, all the elevated floor and roof slabs were to be replaced with 2.5-in-thick lightweight concrete topping on 1.5-in composite steel deck supported on steel beams and columns (Fig. 12.29). This area could function as a floor diaphragm, obviating the need for a separate horizontal bracing structure. The perimeter anchorage of that diaphragm to shear walls would be accomplished by bolted steel angles (Fig. 12.30).

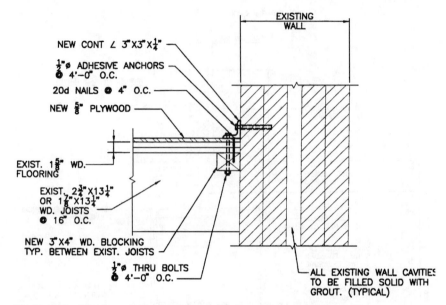

NEW CONT ∠ 3"x3"x¼"

½"∅ ADHESIVE ANCHORS
@ 4'-0" O.C.

20d NAILS @ 4" O.C.

NEW ⅝" PLYWOOD

EXIST. 1⅝" WD.
FLOORING

EXIST. 2¾"x13¼"
OR 1⅞"x13¼"
WD. JOISTS
@ 16" O.C.

NEW 3"x4" WD. BLOCKING
TYP. BETWEEN EXIST. JOISTS

½"∅ THRU BOLTS
@ 4'-0" O.C.

EXISTING
WALL

ALL EXISTING WALL CAVITIES
TO BE FILLED SOLID WITH
GROUT. (TYPICAL)

Figure 12.28 Attachment of new plywood diaphragm to an exterior wall.

The existing steel columns, which seemed to have had no anchorage to the foundations at all, were to be positively connected to the foundations by drilled-in adhesive anchor bolts. The existing Kalmantruss joists were specified to receive new top- and bottom-chord bridging. At the interface between the joists and the new steel beams, the bridging was to be cross-braced to ensure the stability of the joists (Fig. 12.31).

12.2.9 Renovate or demolish and rebuild?

This dilemma is debated in general terms in Chap. 1. Here, the answer was not evident, although the informal opinion of everyone involved seemed to be that it would be too difficult and expensive to upgrade the building, and that a new building could be built at a comparable or lower cost. Unfortunately, the only way to prove this to the client was to prepare a reasonably complete renovation design and a realistic cost estimate.

The cost estimates were performed by both the design team and an outside construction manager, who estimated the cost of the proposed renovations at $8.7 million (of which the seismic upgrade accounted for only about $0.6 million). The construction manager also estimated that for about $8.4 million, including demolition, a new three-story office building of comparable size could be built at the rear of the existing structure.

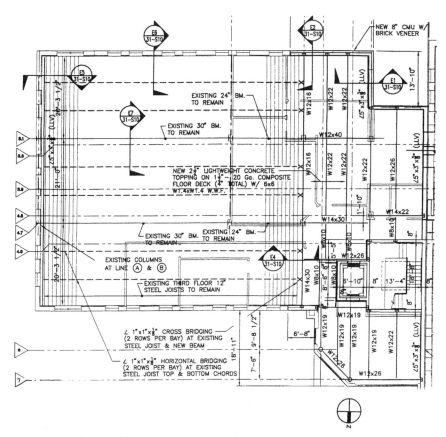

Figure 12.29 Proposed framing for the East Wing.

Armed with this information, the client selected the demolish-and-rebuild option. The building was finally demolished in 1999. It is yet to be rebuilt.

While this decision affirmed the designer's "gut feeling," there remained some unease about the whole process. Indeed, there remained a lot of questions to be answered. Why would a fairly good-looking building without any signs of distress, not located in an area of exorbitant land prices, and with functioning heat, electricity, and plumbing be demolished only because complying with all the current code requirements and with the architect's proposed design scheme was too expensive? Why did the architect insist on using a scheme that required so much reframing? Why did the federal agency insist on complying with the UBC, which was not generally followed in New England, rather than with the more liberal state code? And why couldn't there be some mechanism for dispensing

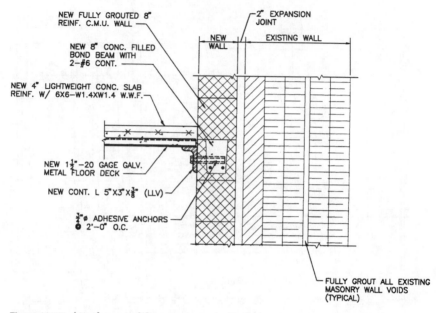

Figure 12.30 Attachment of the new concrete diaphragm to the new shear wall to be constructed next to an existing exterior wall (the 2-in expansion joint separates two parts of the building).

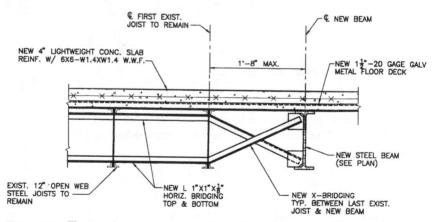

Figure 12.31 The interface between the existing bar joists and the new steel framing.

with the federal mandates for ADA, seismic upgrade, and thermal efficiency in cases like this? Certainly, there are legitimate explanations for each of those actions. But the lesson of this project is unmistakable: The more changes are proposed and the more codes, mandates, and constraints there are to satisfy, the dimmer the outlook for a successful and economical building adaptive reuse becomes.

12.3   Case 3:   Seismic Upgrade of Terminal 1, Oakland International Airport*

12.3.1   Project background

In 1994, the Port of Oakland, owner and operator of the Oakland International Airport, commissioned Hratch Kouyoumdjian & Associates (HK&A), Consulting Engineers, to evaluate the seismic condition of the airport. Following the evaluation, the Port commissioned HK&A in 1995 to develop a detailed design for seismic retrofit of Terminal 1.

This terminal had been the first one constructed at the airport in 1960 and was still extremely busy, with approximately one-half of the nearly 10 million airport passengers arriving and departing through it. Terminal 1 housed various ticketing areas, airline offices, car rental counters, baggage claim carousels, and airport support facilities (Fig. 12.32). The terminal was largely contained in the Main Terminal Building (M101), a unique structure made of precast-concrete barrel arches arranged on a radial plan, which became the focus of the upgrade.

The analysis concluded that the Main Terminal Building could not resist the minimum seismic force levels deemed appropriate for reasonable protection of passengers, workers, and visitors. The existing structural system was likely to fail during a strong earthquake occurring on one of the major active faults in the area. Figure 12.33 illustrates the seismic response spectra for structures on this site, including the spectrum for an earthquake with a 10 percent probability of being

Figure 12.32   A view of Terminal 1.

---

*This case study is based on a project completed by Hratch Kouyoumdjian & Associates, Consulting Engineers, San Francisco and Oakland, who supplied the accompanying illustrations.

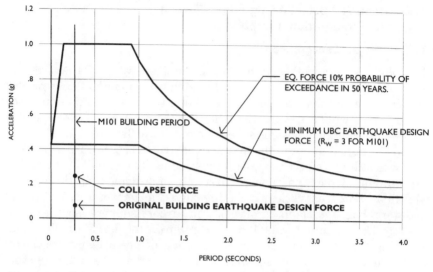

Figure 12.33   Seismic response spectra for structures on the site.

exceeded in 50 years and the spectrum constructed in accordance with the UBC that was current at the time of the renovations. The design force level stipulated by the original code at the time of construction and the load level causing probable collapse are also shown for the building period indicated.

The probable mode of structural failure was deemed to be of a sudden, brittle, and potentially catastrophic nature, because of the very heavy concrete roof elements, the lack of ductility in the original details of construction, and the magnification of ground accelerations caused by the loose sandy soils beneath the structure. A study of the existing design documents confirmed that the original elastic design was based on about 0.075 g. It was determined that the building could collapse at an acceleration level of about 0.25 g acting either in the short or in the long direction, a level that was quite probable in that area. For reference, several instruments at nearby sites had recorded ground accelerations of about 0.2 g at the time of the Loma Prieta earthquake, which had had an epicenter more than 60 mi away from this site.

The Port of Oakland was not seeking a permanent solution, because its long-range plan was to expand the airport by the year 2005 and replace Terminal 1 with a larger facility. Therefore, the renovation efforts for Building M101 were limited to the structural work required in order to reduce the risk of death or injury over the anticipated 10 years of the building's use. Such objectives as reduction of property damage and limiting of postearthquake downtime were not pursued in this program. The Port challenged the design team with an aggressively modest budget to implement the retrofit.

After a detailed analysis, several retrofitting schemes were explored and evaluated. The selected scheme was developed and, after reviews and discussions, used for preparation of contract documents.

12.3.2 Existing conditions

Building M101 was a 70,000-ft^2 single-story structure, with a footprint of about 100 ft by 550 ft. Its overhanging roof (approximately 130 ft by 550 ft in size) was framed with large segmented barrel-arch units. The arches, laid out in a unique curved fashion, were supported on individual cast-in-place concrete columns cantilevered from the foundations. One column supported the corners of four different precast barrel-arch units. The entire front and sides of the buildings were open, and the back wall had numerous openings for access to adjacent support spaces. A partial utility basement extended along the entire length of the terminal.

The lateral system of the original structure relied only on the bending capacities of the flagpole-type columns supported on pile caps, two small concrete X-brace elements at the front (provided more for visual than for structural effect), and some masonry infill walls on the back. There was no effective roof diaphragm, because the barrel-arch units were not connected to each other, except for small welded-plate connections at the columns. The system had very low redundancy: All the columns could fail at the same time if they were subjected to a relatively moderate horizontal deformation. If only one barrel arch slipped off a column during an earthquake, progressive collapse would be quite possible.

Each precast-concrete barrel arch was approximately 19 ft wide by 60 ft long and weighed about 30 tons. The existing drawings indicated that the units contained adequate steel reinforcing to resist dead loads and the light live loads generally used for arches (20 psf or less). Figure 12.34 shows a typical cross section of the Terminal 1 structure.

The interior columns were typically 16 by 20 in, with #11 bars at the corners and a #10 vertical bar at each long side. Typical ties were two legs of #3 bars at 16 in on center, making the columns susceptible to brittle shear failure well before the full bending capacity of the section could be developed. The exterior columns were 16 in square with four #11 and eight #6 vertical bars and with four legs of #3 ties at 16 in on center. The exterior columns were judged adequate for shear but insufficient for bending demands. The columns were supported by piles extending into the so-called bay mud.

12.3.3 Seismic evaluation

Lessons learned from recent earthquakes have led to changes in many established structural design and construction practices, and public

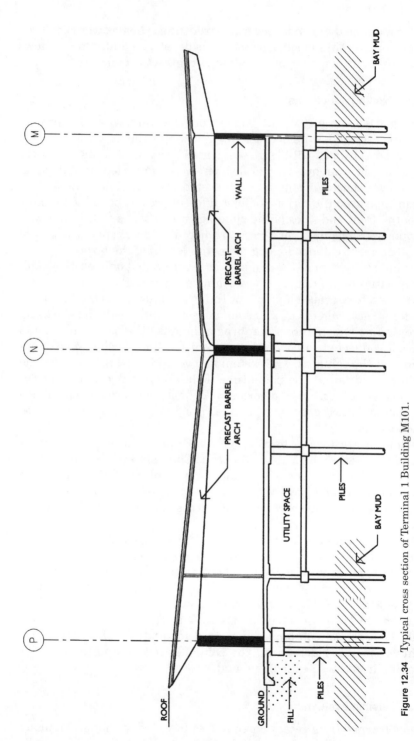

Figure 12.34 Typical cross section of Terminal 1 Building M101.

awareness of the hazards associated with many older structures has increased. Various methods of evaluating the earthquake resistance of structures have been developed by researchers, government agencies, and engineering groups; some of these methods are described in Chap.11.

The seismic analysis of Building M101 for a moderately strong earthquake concluded that the top-heavy structure would undergo excessive lateral drift under load. This lateral movement would induce simultaneous shear and bending failures in most of the columns, and the heavy roof would probably collapse.

The magnitude of the gravity and seismic loads at a typical column line is shown in Fig. 12.35. A simplified analysis indicated that a code-level static earthquake force would induce a 3.8-in horizontal roof movement. The permissible limit set by the code was 0.72 in. Figure 12.36 illustrates the deflected shape of the structure at the expected drift movements caused by the design forces; it also shows the permissible drift limits and the probable points of failure.

To substantially reduce seismic risk, lateral drift at the roof had to be limited to a level at which the shear and bending capacities of the existing columns would not be exceeded. Figure 12.37 defines the demands on and capacities of interior columns along the strong axis, and Fig. 12.38 defines the demands on and capacities of exterior columns as oriented in their actual positions.

Because of the nonductile nature of the column details, the inelastic capacities of the columns could not be counted on. Also, the capacity of the connection between the group of four barrel arches and the column was insufficient to deliver the lateral forces generated by each barrel arch.

12.3.4 A summary of seismic deficiencies and design constraints

The seismic deficiencies of the building included

1. Lack of an effective roof diaphragm

2. Lack of column ties to prevent early column shear failure

3. Limited stiffness of the structure—only about 25 percent of the minimum required for stability

4. Lack of ductility in the joints between the arches and the columns

5. Lack of continuity across the joints

6. Inability to undergo deformations in excess of $H/200$ without failure (H is the height of the structure)

7. Lack of resistance to load reversals

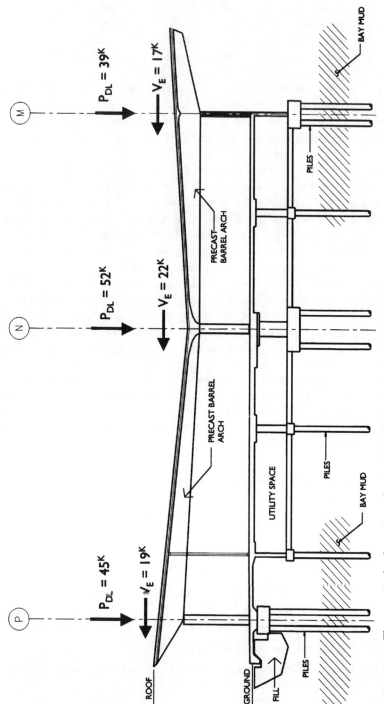

Figure 12.35 The magnitude of gravity and seismic loads at a typical column line.

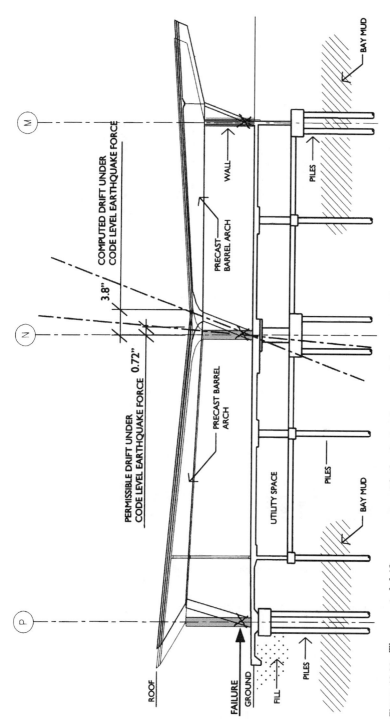

Figure 12.36 The expected drift movements with the design forces and the permissible drift limits.

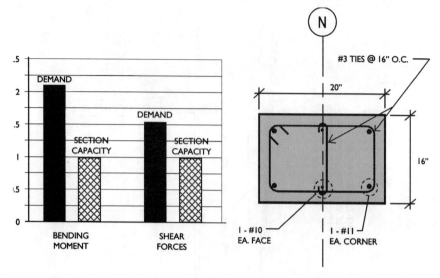

Figure 12.37   The demands on and capacities of a typical interior column.

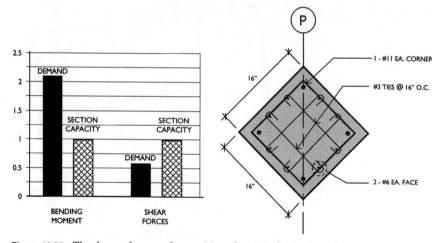

Figure 12.38   The demands on and capacities of a typical exterior column.

8. Impossibility of utilizing inelastic reserves in members because of poor connection details

Some of the design constraints were

1. The airport and all terminal functions were to remain fully operational at all times during construction.

2. Construction work was to be completed within the Port's tight schedule and very modest budget.

12.3.5 Options for seismic retrofit

A number of seismic strengthening schemes were developed and studied. As a basis for establishing the required strengthening, the guidelines of the *NEHRP Handbook for Seismic Rehabilitation of Existing Buildings* (see Chap. 11) were utilized. The loads derived from the guidelines were very close to the minimum loads prescribed by the UBC.

The various strengthening schemes were examined to determine their

- Compatibility with existing construction
- Reliability
- Redundancy
- Potential interference with the Port's operations
- Impact on architectural, mechanical, and electrical systems
- Suitability for phasing of construction work
- Costs
- Detailed implementation requirements

Each retrofit option had to fulfill the following requirements:

1. Form an effective diaphragm.
2. Develop a mechanism for transferring diaphragm forces from the barrel arches to the load-resisting elements below.
3. Prevent overstress and failure in the existing concrete columns.
4. Limit drift to the deformational capacities of the elements.
5. Develop an overall resistance to the seismic forces greater than that required by the code, because of poor ductility in the existing members and connections.

12.3.6 Determination of design seismic forces

As discussed in Chap. 11, the design seismic forces prescribed by modern codes depend on a number of factors, including seismic zone, type of soil, level of stiffness and ductility of the structure, and type of facility ("ordinary" vs. "essential").

The existing building structure fell outside the range of common structural systems described by modern codes. It could not be classified as either a frame or a wall structure, it did not possess an effective diaphragm, and it resisted lateral forces through bending of cantilever columns with partial base fixities. In addition, those columns lacked sufficient ties for ductile performance.

To ensure the survival of the structure and satisfaction of some minimum life-safety requirements in an earthquake, a dual-level analysis was made:

1. An elastic analysis based on an assumed R_w of 4, most closely representing the existing structural system

2. A stability analysis for $V = 1.0g$, selected as the life-safety threshold

These criteria would be considered to be satisfactorily met if the roof structure was prevented from collapsing and the existing columns continued to support vertical loads at a seismic force on the structure equal to 1.0g. As already noted, such considerations as the extent of damage to structural systems and nonstructural elements and potential economic or functional loss of the terminal were all outside the scope of the life-safety objective established for this effort.

The selected design seismic force equal to 1.0g represented the "probable maximum force" that would be expected at this site subject to the frequency of occurrence and probability of being exceeded associated with the standard design spectra noted on Fig. 12.33. This unusually high force level was selected to compensate for the lack of redundancy in the existing system, the susceptibility of the existing columns to sudden brittle failure, and the columns' lack of capacity to resist load reversals. A retrofit structure that could survive this level of seismic force would also be able to limit lateral drift and damage, as well as protect the occupants, in a moderate earthquake.

12.3.7 Solution for seismic strengthening

The final design solution included the strengthening measures described in this section.

The precast roof barrel arches were interconnected by a series of bolted plates designed to transfer the lateral forces applied to the roof (based on accelerations equal to 1.0g). In the middle of the building, the barrel arches were connected to a new continuous collector, a steel tube running in an arc over the entire length of the roof. The location of the collector permitted engaging every barrel arch. The collector transferred seismic forces to four steel diagonal braces located under it. Finally, the braces transferred lateral forces for the long direction of the structure to the ground-level concrete slab.

At the front of the building, a new continuous steel frame with special moment connections, described later on, was used to receive seismic forces from the barrel arches and to connect the individual arches to one another. At the back side of the terminal, a series of bracing tubes was inserted to connect with a separate concrete structure, added at a later date, to prevent pounding.

The lateral-load-resisting system in the long direction of the building, therefore, consisted of three lines of lateral resistance: a 27-bay steel moment frame in the front, four diagonal steel braces under the new roof collector in the middle, and some existing walls plus new

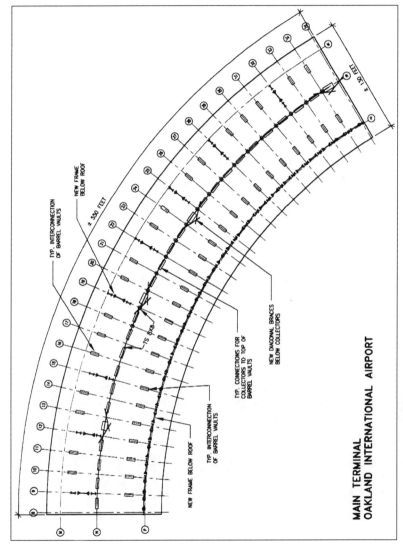

Figure 12.39 Schematic building plan showing the points of barrel arch connections and the locations of various braces and frames.

745

Figure 12.40 Typical new steel frame with bolted moment-resisting connections.

Figure 12.41 Typical new exterior steel moment-resisting frame with bolted connections. The connections at the front columns were enlarged for aesthetic reasons.

braces on the back. The connections between the existing structure and these rigid elements completed the load path.

In the short direction of the building, eight new steel moment-resisting frames with bolted connections were used to stabilize the structure and transfer all diaphragm forces down to the foundations.

Figure 12.39 shows a schematic building plan. It identifies the points of the barrel-arch connections one another and to the collector, and the locations of the various braces and frames that form the entire retrofit solution for Terminal 1.

A new bolted moment connection that is capable of undergoing high rotations when girders are fully plastic was utilized in this building. HK&A originally developed this connection with locally weakened flanges for a girder-to-column moment-connection used in a special moment resisting frame structure. Figure 12.40 shows a typical steel frame with bolted moment-resisting connections. The bolted connection was modified at the front columns for aesthetic reasons to form an oversized strong supporting base for the barrel arches (Fig. 12.41).

The new lateral-load-resisting systems were proportioned to maintain compatible deformations with the existing systems in order to reduce torsion and limit uplift forces and base movements to the capacities of the existing foundations.

Finally, to complete the transfer of shear forces to the foundations through the partial utility basement, it was necessary to introduce new transfer elements between the grade and the basement levels. To keep the design shear forces in the various elements from exceeding the member capacities—and the uplift capacities of the existing foundations—concrete shear walls were shotcreted under five interior frames.

The design work was finished in 1996. As a public-works project, the seismic retrofit work was competitively bid the same year. Construction was successfully completed in 1998, within the original budget, on schedule, and without any appreciable disruptions to airport operations.

12.4 Case 4: Seismic Retrofit of the Administration Building, San Francisco State University*

12.4.1 Project background

After the Loma Prieta earthquake, a campuswide seismic survey and evaluation of the buildings in the California State University (CSU) system was commissioned. The Administration Building at San

---

*This case study is based on a project completed by Hratch Kouyoumdjian & Associates, Consulting Engineers, San Francisco and Oakland, which supplied the accompanying illustrations. The case study was previously reported in *Modern Steel Construction* (March 1999), and the illustrations are reproduced with permission of both *Modern Steel Construction* and the case author.

Francisco State University was found to be one of the most seismically hazardous. Hratch Kouyoumdjian & Associates (HK&A) was retained to implement a seismic retrofit program.

The seismic requirements adopted by the CSU stipulated that buildings at the San Francisco campus must be able to resist ground accelerations of at least 0.4g without structural collapse. An earlier analysis had concluded that if the Administration Building were to be subjected to ground accelerations of this magnitude, the resulting drifts would probably be so large that they would cause exterior column failures, subjecting the occupants to unacceptably high risks of injury or death.

12.4.2 Building description

The Administration Building was designed in 1973 and constructed in 1974. It had a cast-in-place conventionally reinforced concrete structure with exposed concrete exterior frames, walls, and spandrels. Figure 12.42 shows a general view of the south elevation prior to retrofitting.

The building was rectangular in shape and had five floors approximately 115 by 190 ft, with a single parking level below ground. There was a partial fifth floor in the western half of the building, of approximately 10,000 ft². The overall area of the building, including the underground parking, was approximately 130,000 ft². Typical floor-to-floor heights were 13 ft.

Figure 12.42 South elevation of the Administration Building prior to retrofitting.

There were three separate core structures, containing elevators, stairs, and service shafts, located at the east, west, and south exterior walls. These core structures formed prominent architectural features of the building but were not detailed to participate as lateral-load-resisting elements.

A typical floor structure included a 16 ½-in-thick waffle slab system formed by using 30 × 30 × 12-in pans spaced at 36 in on center. There were wide interior beam strips running in two directions between the columns. The beam strips varied in width between 4 ft 10 in and 6 ft 6 in and were of the same thickness as the overall waffle structure. They were transversely reinforced with #3 ties at 7 in on center.

Up to level 4, the exterior columns at the north and south walls were 24 × 36 in, with twelve #18 vertical bars and #4 ties at 10 in on center. The columns were also restrained by 48-in-deep concrete spandrels and cast-in-place windowsill sections. Since these columns formed part of the building's lateral resisting system in the east/west direction, their shortened clear height made them overly stiff and could lead to early shear failure under moderate earthquakes. Alternating between the main columns were 24 × 24-in, lightly reinforced, U-shaped non-load-carrying vertical decorative sections. The east and west exterior walls were similar to the north and south walls except that at each of the corner bays there was a short solid shear wall.

The building was founded on square spread footings under each of the interior columns and continuous strip footings under the exterior columns and walls. A typical floor plan is reproduced in Fig. 12.43.

12.4.3 Existing lateral-force-resisting system

The building was most likely designed in accordance with the minimum static force provisions of the 1973 *Uniform Building Code*. At the time of the study, the governing building code was the *California Administrative Code*—the latest edition of the UBC along with the State of California amendments. The code at the time of the study stipulated significantly higher minimum seismic design forces for buildings than those required by the 1973 UBC; it also included provisions for limiting lateral drifts, factoring of certain loads, and extensive detailing requirements. The goal was to ensure that minor and moderate earthquakes that might occur during the normal life of a building would not adversely affect the structure, and that the building would retain sufficient structural integrity to support all loads and not become a hazard to its occupants during a catastrophic earthquake. Additionally, the code required that all building components and contents be properly anchored and braced.

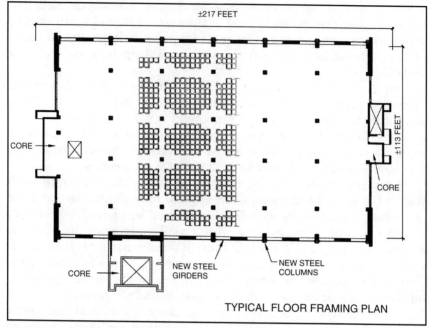

Figure 12.43   A typical floor plan.

Under the probable original design code (1973 UBC), the normal design base-shear load for the moment frames in the east/west (long) direction would be equal to 0.04*g*. For the shear walls along the north/south (short) direction of the building, the design base shear would be 0.07*g*.

A detailed evaluation of the existing structure indicated that it was capable of resisting those loads. Figure 12.44 represents the seismic design spectra at this site, together with the anticipated performance of the building as originally designed.

The three cores on the east, south, and west sides of the building projected outward. It did not appear that their concrete walls were designed or detailed to provide the main building structure with lateral resistance. The performance of the walls during the Loma Prieta earthquake indicated that they not only did not offer any structural resistance to the building, but were in fact damaged themselves.

12.4.4   Analysis and evaluation

A three-dimensional computer model was developed to study the seismic response of this building and evaluate retrofitting options. Both gross and cracked sections of the existing frame elements were modeled

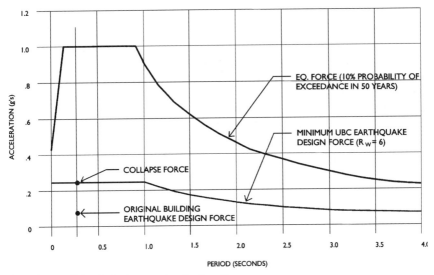

Figure 12.44 Seismic design spectra for the site.

in order to better understand their available capacities and expected future performance.

The magnitudes of horizontal drifts that could be safely accommodated by the existing columns were then calculated, based on their available moment capacities, modulus of elasticity E, and moment of inertia I, and using end conditions that approximated the existing framing. Each column, girder, and joint at each level and on every frame line was analyzed and evaluated for earthquake moments acting concurrently with gravity loads and moments.

After allowing for gravity loads and moments and the code-stipulated magnifier $3R_w/8$ of the computed Δ (lateral drift), representing conditions of overstress, the so-called *safe floor Δ* (delta) was established. The term safe floor Δ related to the controlling capacities, in shear and bending, of the existing columns to ensure stability. The safe floor Δ was also used to check the columns and girders to ensure that their load-carrying capacity would be safely maintained and that punching-shear stresses at the girder/column joints would remain within permitted limits.

The evaluation continued separately for the two main building directions.

North/south (short) direction. The original structural system in the north/south direction of the building was most probably designed using a building period, computed in accordance with the 1973 UBC, of about 0.30 to 0.50 s, which dictated a design base shear of about 2,125,000 lb.

An analysis model of the building's dynamic modes and periods indicated that the period in the north/south direction was between 0.50 and 0.60 s, depending on whether cracked or gross member sections were assumed. The *California Administrative Code* limited the period computed for this building to 0.46 s and stipulated R_w equal to 8 for wall and frame structures. Using these values and a site factor S equal to 1.2 resulted in a minimum static seismic design force of 3,275,000 lb, more than 1.5 times the 1973 UBC design force. It was clear that building movements (drifts) associated with such large forces would cause shear failures of existing exterior frame columns and could cause partial collapse.

The short shear walls would also be loaded to approximately 1-½ times their 1973 UBC loads; accordingly, their boundary elements, stress levels, and overturning stability would all be well past the allowable member and stability limits. The resulting drifts would be excessive, and columns could fail and contribute to the overall building instability.

East/west (long) direction.   If the perimeter columns were assumed to be 13 ft tall, the building period would be 0.50 s if computed using the 1973 UBC, and between 1.15 and 0.90 s using computer analysis based on the current code. However, the actual clear height of the columns was less than 6 ft. Although the columns had been heavily reinforced vertically with twelve #18 bars, or about 50 in^2 of vertical steel in a 24 × 36-in section, the lateral ties were only #4 at 10 in on center. Therefore, the column shear capacity would be very quickly exceeded by even a small floor drift, and a failure mechanism would be formed at relatively low seismic loads—well below a code-level earthquake. The movements associated with the current-code design forces, factored by $3R_w/8$, would greatly exceed the shear capacity of the column, threatening floor collapse.

In contrast, the interior columns had a clear height of 11 ft 7½ in and were able to resist induced movements of 0.75 in per floor, factored by $3R_w/8$. Their stability in cases of severe seismic overload was ensured.

Requirements for upgrade.   To correct these major deficiencies, any proposed seismic upgrading scheme in both directions needed to

1. Provide lateral resistance to the structure so that 100 percent of the lateral forces stipulated by the current code were resisted by the new structural system in a predominantly elastic mode.

2. Provide sufficient reserve strength to ensure that, in the event that code-level design forces were exceeded, the structure could still maintain overall stability through several cycles of seismic overload, with yielding of new and existing members but without collapse.

3. Limit the movements of the upgraded structure under code-level forces, so that the existing elements could continue to carry design loads.

12.4.5 Upgrading schemes

Numerous options for upgrading the building were explored. The main approaches included the following:

A. Strengthen the existing flexural members, shear walls, foundations, and joints to the levels of strength and stiffness required by the current code.

B. Provide a new lateral-load-resisting system that would work in conjunction with or independent of the existing system to meet 100 percent of the current strength and stiffness requirements. A verification would be made that, with the new system in place, all existing members could perform satisfactorily under loads or deformations induced by earthquakes of various levels of intensity.

C. Isolate the existing structure from the ground at or near the base plane and install mechanical isolators, so that, under the current-code earthquake level, the magnitudes of forces and deformations transferred to the structure would be within the capacities of the existing members.

Fifteen schemes were explored and brought to a sufficient level of detail to ensure that the performance objectives were fully met, design details developed, and meaningful cost estimates made. All schemes were assumed to offer adequate and comparable performance. The following schemes are the most representative of the 15 variants.

Concept A, interior strengthening. This seismic strengthening scheme was furnished by CSU as the basis for soliciting engineering proposals. It included providing the following:

- Four bays of new internal concrete shear walls in both directions. Each wall was approximately 30 ft long and was bearing on new foundations.

- Steel jacketing of all ten exterior columns at all five levels, for a total of 50 jackets required.

- Steel collars at all (28) column-girder joints at six levels.

This system would require removal of interior partitions, relocation of utilities at the various work locations, altering circulation paths to install shear walls, losing some parking space, and evacuating the building for at least one year.

Figure 12.45 illustrates this concept.

Concept B, exterior shear wall strengthening system.   This system would utilize concrete shear walls located at the exterior corners of the building. The walls would extend from the foundations to the roof level and would be designed for the total seismic forces. The walls would be proportioned to limit the building drifts so that they were within the capacities of the existing elements. To transfer the floor forces into the walls, a series of new collectors would be introduced at each floor. The foundations under the new shear walls would consist of a number of interconnected deep augured caissons, approximately 60 ft deep.

Figure 12.46 illustrates this concept.

Concept C, exterior steel braces.   This system would utilize diagonally braced bays to provide strength and rigidity to the structure. The individual floor forces would be collected by the slab through new collectors connected to the exterior braced bays. The brace columns in this

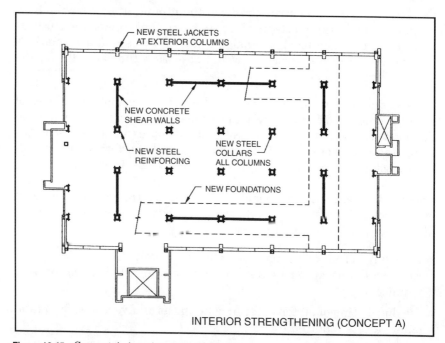

Figure 12.45   Concept A, interior strengthening.

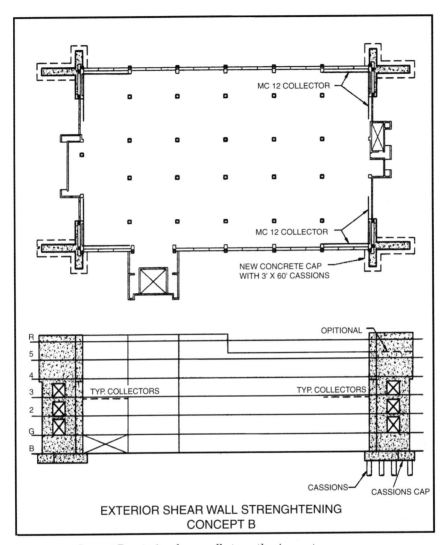

Figure 12.46 Concept B, exterior shear wall strengthening system.

system would be subjected to high overturning forces and would need substantial foundations for anchorage. A new series of walls would be required on four sides of the building to distribute the loads. To resist uplift, the foundations for this system would require several interconnected large augured piers 60 ft deep.

Figure 12.47 illustrates this concept.

Concept D, exterior steel moment frame. In another approach, an exterior moment-resisting ductile steel frame, also referred to as a "special

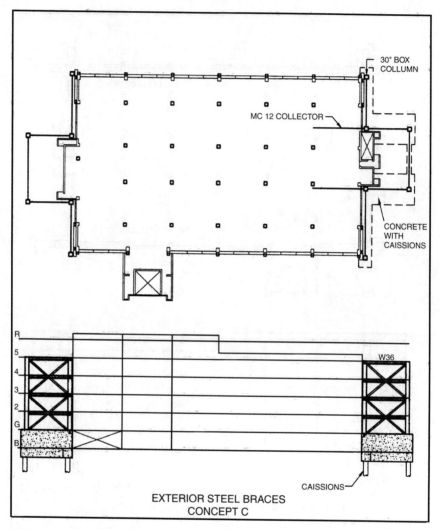

Figure 12.47   Concept C, exterior steel braces.

moment-resisting frame" (SMRF), would be designed and detailed to resist 100 percent of the required lateral load. The frame would be proportioned to limit drifts in the building to within the capacities of the existing columns and beams. Its members would be rolled sections connected to the exterior walls by dowels and bolts. Key advantages of this system include the wrap-around design over the entire building perimeter, which eliminates the need for new foundations at interior bays. Also, to minimize the visibility of the repairs, the new columns could be located in front of the existing main columns, and the new girders could be placed in front of the existing concrete spandrels. All spaces between the new frame and the existing concrete would be completely grouted, and the system would be covered with an architectural cladding of glass-fiber reinforced concrete (GFRC) to replicate the original elevations and details of the exposed concrete building.

Concept E, base isolation.   This scheme would introduce a plane of seismic isolation between the building structure and the ground. Building loads would be supported on base isolators, discussed in Chap. 11.

For this building, an isolation plane could be introduced in the garage immediately below the first floor slab. The garage walls and columns would need to be cut to install the isolators, and horizontal steel trusses would be installed to stabilize the tops of the cut walls and columns. Elevator pits and stairs protruding through the plane of isolation could be suspended from the level above. Utilities would be cut and reattached with flexible connections to facilitate movement.

12.4.6   The selected scheme

After careful consideration of all 15 schemes, concept D, exterior steel moment frame, was selected. The new exterior steel frame was designed to provide 100 percent of the required lateral strength and to support the dead and live loads at the building's exterior should the existing concrete columns fail.

Typical frames included W14 × 426 column sections and W36 × 170 girders, to provide the necessary strength, stiffness, and ductility. The material specified for columns was ASTM A572, Gr. 50, and that for girders was ASTM A36, Gr. 36. Connection bolts were specified as $1\frac{1}{2}$-in-diameter A490-X, and all full-penetration welds were to be made with E7018 rods. Several "push-over" analyses verified that the retrofitted structure could resist seismic forces well in the range of 1.0g and could undergo large drifts while remaining stable and suffering limited damage to the exterior concrete columns. Therefore, it fully satisfied the CSU's seismic safety program.

In the course of this work, it was realized that a new moment connection, reflecting the lessons learned in the Northridge and Kobe earthquakes, was necessary. The new connection, designed and developed by HK&A following the SAC guidelines, is shop-welded to the columns to form a tree-shaped assembly. Girders are field-bolted to the flanges and the web to make standard connections, thereby completely eliminating the concerns associated with field welding of the flanges of heavy sections, a detail common in earlier moment-resistant designs.

A sample full-scale assembly of the new moment connection was tested at the ATLSS Laboratory at Lehigh University in 1996, prior to construction. The test verified the joint stability and satisfactory inelastic performance of the assembly. This connection delivered the necessary strength and inelastic rotations of more than 7 rad without any joint overstress, degradation, or yielding; every bolt and weld remained intact and within its elastic range (Fig. 12.48).

Figure 12.49 illustrates the installation of a new frame against the south wall. Figure 12.50 shows a close-up of a typical prefabricated column and connection being installed.

To evaluate and verify the performance of the retrofitted structure, several elastic and inelastic analyses were performed. Figure 12.51 graphically depicts the relationship between increasing loads, deformations, and the stability of the overall structure.

Through a number of push-over analysis cycles, the initial sequence of shear failure in the exterior columns was developed. The associated "softening" of the structure was then modeled to determine the incremental deformations with each subsequent hinge formation, until the entire lateral load was transferred to the new frames. The structure was further analyzed to ensure complete stability at extremely large drifts, well into the inelastic range of members. Figure 12.52 graphically identifies the seismic load and the associated deformations during the sequence of hinge formation in the retrofitted structure.

Figure 12.48 A test of the new moment connection developed by HK&A. Note the full plastic hinge, including web yielding, outside the connection zone.

Figure 12.49 Installation of a new frame against the south wall.

Figure 12.50 A close-up of a prefabricated column-joint assembly at the south wall.

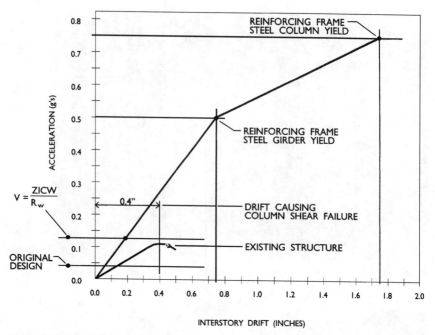

INTERSTORY DRIFT (INCHES)

Figure 12.51 The relationship between increasing loads, deformations, and the stability of the overall reinforced structure.

12.4.7 Conclusion

The selected scheme of seismic strengthening for the Administration Building at San Francisco State University evolved from many studies and concepts. HK&A evaluated more than 15 different concepts. After this, a unique retrofitting scheme utilizing a structural steel moment-resisting frame placed on the exterior of the building was formulated. Exterior steel moment frames do not obstruct windows, spread out the lateral resistance, increase redundancy, and reduce overturning effects. Rolled steel sections were selected to improve the capacities of boundary elements at a few existing concrete walls along the short direction of the building. By limiting drifts to small amounts, well within the capacities of the existing concrete columns and joints, costly reinforcing of existing columns and joints was avoided.

The chosen scheme provided a very high level of strengthening, but it was also the most cost-effective solution. In addition, it permitted continued use of the building during construction and required no changes to the interior of the building. Moreover, none of the existing windows was affected. When GFRC cladding was applied over the exterior steel frames, the building's appearance was close to the original look (Fig. 12.53).

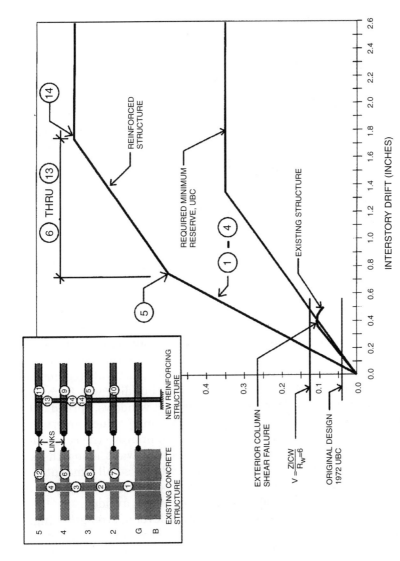

Figure 12.52 The sequence of hinge formation in the retrofitted structure.

Figure 12.53 The strengthened building looks almost the same as prior to retrofitting.

The project was publicly bid in 1997, and construction was success-fully completed below the original budget and on schedule in 1998. The Administration Building is now retrofitted and capable of resisting the seismic loads prescribed by the California State University Seismic Safety Program. Consistent with the university's program, the building is expected to remain structurally unaffected by strong earthquakes and stable under very strong earthquakes.

13

Renovating Building Façades

13.1 General Issues

13.1.1 Introduction

Exterior building walls have three roles. The first role is structural—resisting wind, seismic, and gravity loads, and then transferring them to primary framing. The second is environmental—protecting the building from the elements, moisture, and temperature extremes. The third is architectural—the façade is the face of the building, as the name implies. Failures can occur in all three areas.

In this chapter, we discuss the main causes of distress in exterior walls and examine the available repair solutions. The first façade problem, the most dramatic and difficult to ignore, is basic disintegration of material, caused by aging, deterioration of supports, and freeze-thaw cycles. This problem can be painfully obvious and exposed for everyone to see. In extreme cases, it can lead to the too-familiar story of a brick falling down and injuring or killing somebody. Less dramatically, façade disintegration may take the form of crumbling old mortar.

The second cause of façade distress is movement of the building structure or differential shortening of the building frame. The typical result is cracked, bulging, or out-of-plumb walls. Foundation settlement can cause not only cracked walls, but also out-of-square doors and windows that don't operate properly.

The third type of façade problem is a lack of weathertightness, resulting in water penetration and excessive air infiltration from the outside. The opposite process can also take place: moisture cannot escape from inside the building and produces condensation. Either way, the building insulation and interior finishes may be ruined and energy wasted.

There are two main classes of exterior walls: loadbearing and curtain (nonbearing) walls. Both of them can perform all three roles—structural, environmental, and architectural—but a loadbearing wall can also support the structural framing of the floor and the roof, while a curtain wall is not intended for this purpose.

This chapter deals specifically with exterior walls, with an emphasis on curtain-wall systems, which constitute the bulk of today's façades. Accordingly, we first discuss the basic issues of façade deterioration, then move to the general problems of curtain walls, and then examine the specifics of renovation of various wall systems. The issues related to rehabilitation of loadbearing walls are examined in other chapters: Chap. 9 for renovation of masonry, Chaps. 4 and 5 for renovation of concrete, and Chap. 10 for renovation of metal panels.

13.1.2 Façade inspection programs

From the standpoint of harm to the general public, façade disintegration is probably the most serious problem. Grimm[1] estimates that a piece of masonry falls off a building façade in the United States at least once every 3 weeks, on average. In the last few years, over 100 pedestrians have been hurt or killed by falling pieces of exterior walls. Tragically, all this occurs under normal circumstances and not because of natural disasters.

It is these front-page news stories that typically prompt cities to institute façade-inspection ordinances. Chicago started its first façade-inspection program after a pedestrian was killed in October 1974 by two pieces of glazed tile that fell from the cornice of a 17-story building. In short order, about 2400 downtown buildings were visually inspected. The result? Nearly half were found to have "loose and potentially unsafe building materials." Chicago's latest Exterior Wall Maintenance and Repair Program requires owners of tall buildings to provide either an annual visual inspection of cladding or a detailed examination every 4 years. The affected buildings are defined as those being at least 80 ft, or six stories, in height.[2] Most of the problems are with terra cotta tiles in buildings constructed during the first two decades of the twentieth century. Terra cotta problems seem to be particularly widespread in New York and Chicago, where the material was used the most.

Similar façade-inspection programs exist in other cities, including Detroit, Michigan, and Columbus, Ohio. New York's program was recently made more stringent after several well-publicized collapses of walls and parapets. The revised inspection requirements call for owners of buildings more than six stories high to have the façades inspected every 5 years by licensed architects or engineers. The inspectors can rate the exterior walls as either safe, unsafe, or "safe with a repair and maintenance program" and must immediately notify the Department of

Buildings if they discover an unsafe condition.[3] Boston enacted its façade-inspection ordinance, CBC9-9.12, in 1996 and began notifying the owners of buildings over 70 ft high the next year. The inspection is targeted toward identification of areas of major façade deterioration, any unsafe conditions, movement of façade components, and problems with watertightness. Such inspections must be made every 5 years.

How are these inspections performed? The easiest method is a ground-based binocular survey, but in many cases this approach fails to detect cracks and other serious problems. Schwartz[4] advocates using a combination of a binocular survey and hands-on close-up inspection that covers at least 25 percent of each building elevation. In many cases, only a close-up observation can find damaged elements. Schwartz recommends, among other specific techniques, spot sounding of cladding, which is prone to delaminations and separation from the supporting framing, attempting to move suspicious-looking stone pieces to check if they are firmly attached, and pushing on glass to check for rattling. Finally, he suggests making selective checks of the interior wall surfaces for signs of leakage.

A close-up inspection can be made from swing staging (Fig. 13.1), from ground-based lifts, from cherry pickers, and even by means of rope descent. The newest inspection procedure employs a video camera that is moved up and down the wall and records its condition. Reportedly, this equipment can inspect 25,000 to 35,000 ft^2 per day, compared with 5000 to 7000 ft^2 per day using a swing stage—and is less expensive. Still, this inspection method does not permit physical probing and spot sounding; it is most effective for quantifying repetitive deficiencies that have already been diagnosed.[5] The general procedure for conducting exterior wall inspections is described in Chap. 2.

The primary focus of the city-mandated inspections seems to be older buildings with massive masonry and tile exteriors, but new wall materials are not immune from problems. As the exterior walls of urban buildings have gradually evolved from solid brick and stone to thinner and lighter components, architects and engineers have had to learn how to deal with performance and durability issues relating to the new cladding. And we are still learning, as more and more of these curtain walls deteriorate and need attention. In the words of one engineer quoted by *ENR*,[6] "Thirty to forty years ago, we built these time bombs that are now just starting to show up." Another calls the façade problems "epidemic." Even more worrisome, there is a feeling among some engineers that a few architects are now designing walls with even less foresight, creating time bombs with fuses far shorter than 40 years.

13.1.3 Curtain walls: The neglected child?

In most cases, poor performance of exterior walls is a direct result of poor design and execution. The details of proper workmanship are beyond the

Figure 13.1   Façade inspection from swing staging. (*Maguire Group Inc.*)

scope of this book, but some recurring design problems are worth examining. As we shall see, the two main sources of curtain-wall problems, apart from construction deficiencies, are improper design and incompatibility with the building frame. Both stem from poor coordination between architects and structural engineers. Why does this happen?

Curtain walls are essentially cladding, and structural engineers tend to regard them as mere dresses that can be removed and replaced without affecting structural performance. Many engineers do not consider the design of curtain walls to be their responsibility and do not get involved in this architectural—as they see it—issue. (Other engineers would like to get involved in curtain-wall design, but are rebuffed by architects protecting their turf.)

Nicastro[7] points out that structural engineers are typically called to investigate curtain-wall failures, but rarely design the walls in the first place. "Engineers seem to have abdicated responsibility for curtain-wall design," he concludes. He believes, as the author does, that the frequency of wall failures will be reduced only when the situation changes and engineers get involved.

13.2   Curtain-Wall Problems Caused by Structural Forces and Movements

13.2.1   Problems caused by lateral loads

Curtain walls *do* play a structural role—and a critical one at that—in resisting wind loads acting perpendicular to their plane. If they fail in this role, the result is an unusable building, whether the structural frame is damaged or not. A typical occurrence is described in Chap. 11: In a hurricane, excessive deflection of window frames allows rain to be blown inside the building and destroy its interior and contents. The cost of the resulting damage to finishes and of lost time vastly exceeds the cost of fixing any structural problems, and the consequences are certainly disastrous. But how often are window frames designed by the structural engineer of record? Similarly, improperly selected overhead doors may deflect so much in a hurricane that they lose their grip on their tracks. The doors may end up being blown inside the building, and once that happens, the building is open to the wind-driven rain and its contents and finishes will be ruined. So, the first general problem with curtain walls is their improper design for wind loading.

The second common problem occurs at the interface of curtain walls with the structural frame. Even if the curtain wall is properly designed for wind by a specialized supplier, it is the design professional of record who is responsible for its proper integration into the building. A case in point: A flexible building frame that deflects too much under wind loading in the direction parallel to the curtain wall may unintentionally impose racking forces on the cladding. A brittle curtain wall made of masonry that was not designed for this shear-wall action could crack or break. An excessive story drift under lateral loading (and even under gravity loading in buildings of gable rigid frame construction) acting perpendicular to the wall surface can lead to similar results. This often occurs in metal building systems that were originally clad in metal siding and later "improved" by being reclad in brick veneer, as discussed in Chap. 10 and in Newman.[8]

The signs of curtain-wall failure from heavy winds include cracks in and breakage of the wall, fracture or pullout of fasteners connecting the wall to the structure or to the backup wall, and tearing of sealants and flashing.

Wind causes most curtain-wall problems, but cladding can also be damaged during earthquakes. Like wind, the violent seismic motions can lead to huge racking forces that literally tear the walls apart, destroy the integrity of flashing and sealants, and break the connecting fasteners. Also, when adjacent structures collide, their wall materials may be destroyed by being smashed together with tremendous force. For these reasons, ductility of connections and isolation of rigid curtain walls from the primary building structure is paramount to their survival in an earthquake. The specific requirements for accomplishing this goal are contained in the model building codes.

13.2.2 Problems caused by structural movement

According to Bell,[9] lack of provision for movement between the curtain wall and the building frame is the most common source of structural distress in cladding. He lists the following types of *short-term* building frame movement:

- Deflection or twist of spandrel beams under live load
- Frame shortening under live load
- Lateral drift under wind and seismic loads and under transient gravity load
- Thermal expansion and contraction of the frame

The main types of *long-term* building frame movement are

- Deflection or twist of spandrel beams caused by creep and shrinkage (for concrete and wood framing)
- Frame shortening from creep and shrinkage (for concrete and wood framing)
- Frame shortening and camber from prestressing or post-tensioning forces (for concrete)
- Settlement of foundations

In turn, curtain walls also undergo many types of movement. Among them are

1. *Expansion and contraction.* Short-term thermal expansion and contraction affects all materials. Long-term moisture expansion is characteristic of brick and wood. Lateral thermal ratcheting affects masonry and stone over the long term. (It occurs when mortar in vertical panel joints cracks and wedges in the joint during stone panel expansion. The panels are then prevented from returning to their original position during the contraction phase, and thus the façade progressively grows. This phenomenon can often be seen at masonry parapet corners.)

2. *Bowing.* Short-term bowing caused by a thermal gradient affects all materials. Any two materials with different thermal and moisture expansion characteristics that are joined side by side can be affected. Long-term bowing caused by a moisture gradient can occur in stone, masonry, and wood. Long-term bowing and expansion from thermal hysteresis primarily affects marble. Long-term bowing and shortening is a typical design issue for prestressed concrete.

3. *Shrinkage and shortening.* Long-term shrinkage occurs in concrete, concrete masonry units, and wood.

4. *Creep deflection.* This is a long-term phenomenon in wood and concrete walls and columns.

It may be difficult to accommodate the various permutations of movement types in curtain walls, in the building frame, and between these two. While there is no single solution, Bell addresses a few common situations. To deal with frame shortening, horizontal soft joints or sliding connections between the curtain wall and the frame can be used. The excessive deflection of spandrels can be accommodated by two-point support of cladding panels and by horizontal and vertical control joints in them. The effects of panel expansion can be minimized with horizontal and vertical control joints and with sliding connections. Panel bowing will be less of a problem if statically determinate lateral support of panels is provided (i.e., one pin and one roller attachment for a two-point support). Settlement of foundations will cause little damage if closely spaced vertical control joints are provided in the walls. Isolation joints and sliding connections between cladding and frame and between adjacent panels can help deal with excessive story drift (frame sway).

Whenever the existing building cladding shows signs of distress attributed to movement of the curtain wall or building frame, it is wise to check whether any of the steps just listed were taken by the original designers. If, after examination of the existing wall and its connections to the frame, the answer is no, cosmetic repairs will not last long. More likely than not, a complete replacement of the building skin, or at least major surgery, will be required. One example of such surgery is the rehabilitation of Chicago's Reliance Building.

The building was reportedly the first skyscraper to use a curtain wall of terra cotta stone panels hung from a steel frame. Being clay masonry, the terra cotta had gradually expanded over the years to the point where its pieces started to rest on each other. At the same time, the steel frame had shortened under load. As a result of these two factors, terra cotta panels that were never intended to be load-bearing became just that and began to crack and spall. One of the first remedial measures was to cut horizontal expansion joints in the terra cotta panels at the floor levels, using a plunge-cut chain saw. Separating the adjacent panels helped to transfer their weight back to the frame. This operation was done before the cracked units were removed and replaced, in order to prevent redistribution of weight to the remaining units, which could have cracked under the added load.[10]

13.3   Water Leakage

13.3.1   What water leakage does

Water leakage is the most common and easily noticed problem with exterior walls. Masonry of all kinds is permeable to water, especially to the driving rains that are common in many areas of the United States. According to Grimm,[1] there are at least 52 cities in the contiguous states that receive rain on 15 or more days during one or more months of the year.

The signs of water penetration in walls without interior finishes are unmistakable (Fig. 13.2). Fortunately for the pictured wall, the water that left the stains eventually evaporated and caused little other damage. The situation is more serious when there are interior finishes, as water evaporation is slowed down, and the moisture lingers within the wall, causing deterioration of the finishes, corrosion of metal fasteners, and degradation of insulation. Prolonged exposure to moisture can lead to growth of mold and mildew and to decay of wood walls. Air-quality problems are sure to follow, with the occupants' absenteeism climbing and morale plunging.

The Wall Street Journal[11] states in alarm: "A decade after a huge boom in shiny new glass-and-metal-cased buildings, many of those 10- to 15-year-old edifices are leaking." The paper mentions a few cases of people having to deal with excessive water penetration. In an office building in San Antonio, the workers covered their desks with plastic each day before going home. In Chevron Corp.'s laboratory and office complex in California, people rigged tubes to the leaking walls to water their plants. In a new casino in Nevada, workers put out dish tubs to catch wall leaks whenever rain was forecast. In all three cases, repairs were eventually made.

On the exterior, a continuous or frequent flow of water down the walls can ruin wood siding, absorptive stone veneer, and even masonry. Figure 13.3 shows the exterior surface of a masonry wall where mold and lichen are growing because of frequent wetting owing to a missing downspout.

13.3.2   The barrier wall

The problems of water penetration are as old as buildings. To deal with them, the designers of yesterday typically relied on heavy masonry walls coated with plaster. This *barrier wall* approach was reasonably effective. (In fact, some of the medieval buildings that survive to this day and are still occupied were constructed with their walls directly in the water.) If continuously exposed to water, barrier walls absorb it and stay saturated for a long time, keeping the building interiors constantly damp. Thus, the best way to keep the interiors dry is to pre-

Figure 13.2 Water stains on the inside of single-wythe masonry are unmistakable signs of leakage.

vent frequent wetting of the walls and to shield the walls, their joints, and their windows from rain. This was traditionally accomplished by roof cornices, heavy reliefs, and precast "water-table" courses with protruding drip edges.

Barrier walls are still being built. However, they have lost some of their bulk and must rely on other means of preventing water intrusion; the most common are impermeable materials and properly sealed joints, even though these measures often have only limited success. A typical contemporary example is the single-wythe wall made of concrete masonry units (CMU). Walls of this type are popular because of their low cost, good fire resistance, and toughness. Yet while these walls are common, so are problems with leakage, as already shown. Some types of CMU are more prone to leakage than others: Lightweight blocks, for example, are more porous and generally more susceptible to water penetration than normal-weight blocks.

Figure 13.3 Mold and lichen growth resulting from frequent wetting caused by a missing downspout.

To reduce water penetration through single-wythe CMU walls, masonry blocks with integral water-repelling admixtures have been developed. The Achilles' heel of these is masonry joints, which must also receive some sort of water-repellent in the form of an admixture or coating. Another possible solution that could be used to reduce moisture penetration *from outside* through a barrier wall is to cover its entire surface with an impermeable vapor-seal coating or a penetrating sealer. However, according to Schwartz,[12] vapor-seal coatings are rarely effective for more than a few months, because they tend to crack or wear out. Even worse, they prevent migration of water vapor *from inside* the building and trap moisture within the wall, leading to its premature degradation. Penetrating sealers such as silanes, and siloxanes, discussed in Chap. 6, penetrate the surface of masonry and concrete and fill pores that would otherwise be open to water entry. These sealers are said not to create vapor barriers, with all their problems. But before they are widely accepted for waterproofing walls in the United States, penetrating sealers need to build a sufficient track record to prove their long-term effectiveness.

Coatings cannot negate the fact that construction of a single perfect line of defense is a practically impossible task. Every crack or construction defect provides a conduit for moisture entry, and there are plenty of reasons for masonry cracking. Among them are deflection under wind load, restrained shrinkage (at wall openings, for example), and temperature movement. Figure 13.4 illustrates cracking of a solid wall made of concrete masonry units caused by temperature expansion and contraction.

This type of cracking can be reduced by providing expansion and contraction wall joints, but at the cost of introducing another problem: The joints must receive proper sealants, and these have shorter service lives than masonry and require frequent maintenance. While masonry units, mortar, and even stucco have estimated lives in excess of 100 years, sealants last only 5 to 15 years. Even during their short lives, joint sealants frequently fail to provide a watertight bond owing to poor surface preparation or improper sizing. Sometimes the width of expansion joints is arbitrarily set by architects at $^3/_8$ in to correspond to the size of a masonry mortar joint, even though this width may be inadequate to accommodate the actual wall movements.

13.3.3 The cavity wall

The obvious limitations of the barrier wall have led to a better concept for resisting moisture penetration—the *cavity wall* design. The design of cavity walls, otherwise known as rainscreen walls, recognizes the futility of making a single perfect seal against water. Instead, in a cavity wall, there are two barriers to moisture, separated by an air space.

The concept is not new: Primitive cavity walls can be found in some surviving ancient Greek and Roman structures. The idea of a cavity wall was resurrected in the nineteenth century, probably by the British, and made its way into the United States in the middle of that century. The cavity walls of that era consisted of two 4-in-thick masonry layers connected by iron ties. Cavity walls started to gain popularity in the 1930s and 1940s, first in low-rise and later in high-rise buildings.[13,14]

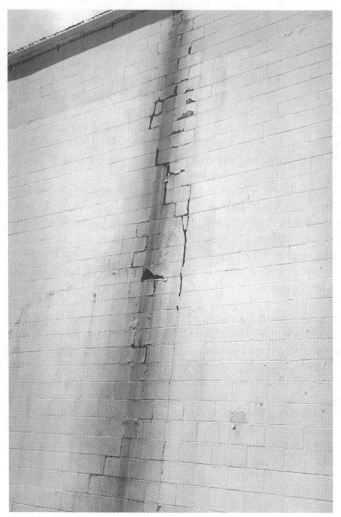

Figure 13.4 Cracking of a CMU barrier wall caused by temperature movement.

In the contemporary cavity wall, the exterior veneer prevents most water from entering the wall and protects the interior air and moisture barrier from degradation. The interior barrier is typically a layer of waterproofing material, such as building paper or elastomeric membrane, that cannot last without this exterior protection.

It is assumed that some water will eventually penetrate the first line of defense and get into the cavity. The third part of the rainscreen wall, the air space, allows this water to drain away with the help of the fourth and fifth cavity wall components—flashing and weep holes.

To be effective, flashing must be placed at the proper locations—at each shelf angle, at the head and sill of each window and other wall penetrations, at the wall base, under copings, and at other spots shown in Fig. 13.5. To minimize the water flow inside the cavity and to prevent the collected water from staying within it, flashing should be sloped toward the exterior. Flashing must also project beyond the exterior surface of the veneer, to prevent water from flowing back under it, and be dammed at the sides to keep water from escaping sideways. A wall made of brick veneer over wood or steel studs represents a good example of a cavity wall, and we continue this discussion in Sec. 13.6.

Cavity walls generally function very well, but they are not without drawbacks. For one thing, they require wall ties to attach the exterior veneer to the backup wall, and these can corrode and fail in the moist environment within the cavity. To prevent leaks, the flashing, weep holes, and waterproofing layer must be in flawless condition, and every wall joint must be filled with a well-maintained sealant.

13.3.4 The pressure-equalized rainscreen wall

The concept of a *pressure-equalized rainscreen wall* is an attempt to improve on the good idea of the cavity-wall system. In theory, pressure-equalized walls have no exposed sealants and rely instead on some specific design steps taken to block the various pathways of water intrusion into the cavity.

There are five possible mechanisms of entry into a cavity wall through deliberately open joints[15]:

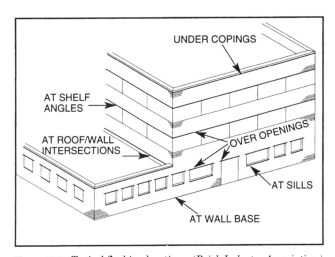

Figure 13.5 Typical flashing locations. (*Brick Industry Association.*)

1. *Kinetic energy:* Water is carried into the cavity by wind-driven rain.

2. *Surface tension:* Water clings to and flows along the wall until it finds an opening.

3. *Gravity force:* Water travels downward along the exterior surface, looking for entry.

4. *Capillary action:* Water seeps into a thin joint or wall crack as through a wick.

5. *Air-pressure drop and air currents:* Water is sucked into the cavity by a pressure differential between the cavity and the outside air.

The cavity can be defended against the first four entry mechanisms by making the wall joints conform to some specific designs. To guard against water carried by kinetic energy, a vertical ridge can protect the joint opening. Penetration by means of surface tension can be remedied by a drip edge. Water carried by gravity can be deflected by sloping the joint surfaces upward. Capillary action essentially disappears when the joint is at least $1/2$ in wide. However, no joint design can prevent water intrusion via the last entry mechanism, air-pressure drop; for that reason, we must still rely on the integrity of the interior waterproofing to prevent leakage.

Here is where the principle of pressure equalization comes in. According to this principle, to stop water from entering the cavity, the air-pressure differential between the cavity and the exterior must be eliminated; that is, the pressures must be equalized. Pressure equalization is accomplished by leaving openings of a certain size in the exterior veneer, which allow the cavity to become a part of the outside world. In order to restrict air movement within the cavity and allow for rapid changes in air pressure, the cavity is divided into small compartments. Then, the air pressure within the cavity can almost instantaneously follow that of the outside air, and the pressure drop is eliminated. To prevent air movement through the interior waterproofing, the latter must also be an effective air barrier. Indeed, in pressure-equalized walls, the need for an air barrier theoretically exceeds the need for waterproofing.

Is it possible to achieve in practice a nearly perfect pressure equalization? Can all exposed sealants really be eliminated? The answer to both questions is, It remains to be seen, as our experience with this type of curtain wall is still relatively limited. It seems quite difficult to ensure the continuity of the air barrier within the cavity and across all its compartments and different materials. And what about windows and other wall penetrations? Most likely, sealants will still be required around these openings, bringing with them the problems of frequent maintenance. For these reasons, some engineers feel that the concept of the pressure-equalized rainscreen wall is an unnecessary complication of the sound idea of a cavity wall.

13.3.5 Water leakage vs. condensation: Telling the difference

The unmistakable signs of water penetration can be readily detected in single-wythe masonry walls without finishes (see Fig. 13.2). In walls of cavity construction, in those with interior finishes, or in massive masonry structures, identifying the source of moisture may not be as easy. Deterioration of finishes and insulation, corrosion of fasteners, growth of mold and mildew, and rotting of wood are typical symptoms of moisture damage, but these can also be caused by condensation. Krogstad and Weber[16] provide some guidance on this issue, as discussed here.

It is obvious that water leakage is occurring when water puddles on the floor or drips from window heads during a heavy rain. In cavity walls, however, moisture can easily travel some distance from the point of leakage to the point where the symptoms appear. Consider a typical case of a brick and block cavity wall with gypsum board on furring channels and a vinyl wall covering. Moisture that gets absorbed into the masonry finds its way into the space between the furring channels and lingers there because evaporation into the interior is prevented by the wall covering, which acts as a vapor barrier. When the exterior temperature falls below the dew point of this air, moisture can condense behind the wall covering. If this continues, mold and mildew will grow behind the wall covering and around wall penetrations, and the gypsum board will deteriorate. A problem originally caused by leakage in a remote location now affects the whole wall.

There are also some obvious cases in which *condensation* is most likely to blame for moisture problems, typically when warm and humid interior air meets colder exterior walls, or when moisture appears only in negatively pressurized rooms but not in others.

Sometimes finding the cause of water damage is not easy. The symptoms of moisture damage can be caused by infiltration of moist air into the cavity by means of the air-pressure differential and its eventual condensation. Moisture can get into the wall both from the inside (migration through the interior barrier) and from the outside (penetration through the exterior veneer). Once within the wall, moisture that came from the exterior can seep into negatively pressurized areas inside the building and can be easily mistaken for condensation. In some cases, determining whether leakage or air infiltration and condensation is to blame may be nearly impossible; in some cases, both may be at work.

Lotz[17] provides a vivid example of this. Black and green mold was growing in every room of a new 18-story hotel located on the South Carolina oceanfront. The walls were constructed of an exterior insulation and finish system (EIFS), described in Sec. 13.9, and had an interior drywall finish with vinyl wall covering. Lotz's investigation revealed that the drywall finish—which, incredibly, had been installed *before* the exterior cladding—had been soaked during a summer squall. The damp drywall was then covered by the EIFS on the outside and by the vinyl wall covering on the inside, sealing the moisture within the wall. As if this one-time soaking were not enough, the wall continued to absorb moisture after construction. Water was entering through faulty joint sealants in the EIFS wall and condensing behind the wall covering. The latter acted as a vapor barrier installed on the wrong side of the wall, since vapor migration was mostly from the outside. The recommended solution included removing the wall covering, replacing caulking, and refinishing the interior with a vapor-permeable finish.

13.3.6 Investigating water leakage in curtain walls

What procedure should be used in order to determine the source of moisture? Turning again to Krogstad and Weber,[16] the following approach to investigation of water problems in curtain walls is suggested.

Document review. The original drawings can contain a wealth of information, including the wall type and construction and the layout of the mechanical system. The source of problems is evident if the design lacks flashing at critical points or contains similar obvious errors. The documents can also show where sealants, vapor retarders, and air barriers are. If these are not properly and clearly detailed, chances are that the installation did not make the situation any better. To be effective, vapor retarders must be placed on the side of the insulation facing the more humid environment. In hot and humid climates, they should be on the outside; in cold and dry ones, on the inside. In many locales, the theoretically desirable position changes from season to season; in this case, the location of the vapor retarder is controlled by the most critical environmental effect.

The mechanical drawings and reports, if available, can show whether the building is intended to be pressurized, that is, to contain more exterior intake (supply) air than exhaust air. Pressurization is particularly common in hot and humid environments. Another point of interest is whether the fresh air can effectively reach the perimeter rooms without losing its pressure. To answer that question, the locations of intake and exhaust outlets should be examined.

Why is any of that important to a structural engineer or an architect? The objective of this review is to find out whether a consistent negative pressure differential exists at the perimeter rooms. If it does, the air pressure outside the building exceeds that on the inside wall surfaces, allowing constant air and moisture infiltration. According to Krogstad and Weber, most air infiltration problems are caused by long-term pressure differentials in the mechanical system, not short-term fluctuations of wind speed and direction. As discussed in the previous section, the air-pressure differential is one of five major pathways for water entry into walls, and the one most difficult to guard against.

Interior survey and air-pressure measurements. A survey of the inside surfaces of exterior walls can include measuring the moisture content of the gypsum board with a moisture meter, looking for signs of mold and mildew near the electrical outlets, and checking for water stains. If walls protected by wide overhangs do not show signs of water problems but other walls do, the culprit may be either wind-driven rain or moisture from the cavity. If the signs of damage appear only in rooms with a negative pressure differential, the moisture was most likely brought by air infiltration.

Another investigation task involves taking actual air-pressure measurements at various points in the building, using sensitive micromanometers. The goal is to verify the information contained in the mechanical drawings and to find the areas where the interior pressures are lower than elsewhere—the areas prone to leakage.

Conducting water leakage investigation. If the two previous steps yield no obvious conclusion, detailed investigation is required. This might include some local in-depth condition surveys, making exploratory wall openings, and spray testing. For the testing, Krogstad and Weber prefer using a modified version of ASTM E1105, *Test Method for Field Determination of Water Penetration of Installed Exterior Windows, Curtain Walls, and Doors by Uniform or Cyclic Static Air Pressure Difference,* without any additional air pressure. Water is applied to all wall surfaces by means of a spray rack. In some cases, the source of leakage becomes fairly evident; in others, further exploratory wall openings may be needed.

According to Rutila,[18] a typical well-constructed brick-veneer wall will start to leak into the cavity after about half an hour of this testing. The typical avenues for water entry include cracks in the veneer, unsealed ends of flashing, fastener penetrations at windows and through waterproofing, and window jambs without proper flashing.

13.3.7 Alternative methods of investigation

Some firms prefer full-scale testing in accordance with the procedure of ASTM E331.[19] This test is performed by spraying water at a specific rate and pressure (the minimum flow is 5 gal/h per square foot of wall) while maintaining an air-pressure differential between the inside and outside wall surfaces. The spraying is done by a calibrated spray rack. The testing procedure begins with the isolation and spraying of perimeter or transition elements. Then, other wall assemblies and elements are isolated one at a time with a waterproof barrier seal made of polyethylene masking and tested.

But what if there is no budget for these rather costly investigations? Should we decline to advise the client at all? Perhaps not before checking the obvious avenues of water entry first. Says consultant William A. Lotz: "Regardless of façade material, the most frequent cause of leakage is failed caulk. The second most frequent source of leaks are holes in the roof."[17] Roof holes can be the result of poor workmanship, lack of maintenance, or storm damage. The obvious leaks can be detected by some very basic testing, such as soaking the wall with a garden hose. Lotz also points out that the location of a building affects its leakage potential, as waterfront structures or those built on hilltops may be subject to horizontal or even upward rain during windstorms. When rain attacks from below, water can enter into places that are not usually associated with leakage—under flashing and at roof eave vents, for example. In one case, Lotz reports, water was blown through eave vents and soaked the drywall ceiling for as far as 20 ft away from the wall!

There are other tests that can be useful for investigating leakage through exterior walls made of brick veneer. If only the joint sealants need to be tested, a calibrated hose nozzle test, assembled in accordance with AAMA 501.2, *Field Testing of Metal Curtainwalls for Water Leakage*, can be used. Wall flashing and weep holes can be checked by a wall drainage test. This test involves making a small opening in the veneer and spraying some water directly into the cavity at regular intervals under pressure of 30 psi. Either the water will be effectively removed through the weep holes or signs of leakage will appear at the backup wall. If leakage is observed, the wall can be opened up in a few locations to determine the cause. What may be found is mortar bridging, flashing terminating within the brick (this is especially bad if the brick is hollow), discontinuous or poorly spliced flashing, or plugged weep holes.[20]

Other methods of investigating masonry problems are described in Chap. 2.

13.3.8 Repairing water leakage

Naturally, repairs should address the most probable source of water intrusion. If water is getting inside because of wind-driven rain, attention should be directed toward closing off the points of easy entry. Among the possible repairs are repointing of deteriorated mortar for masonry; installing roof overhangs, if practical, for small wood buildings; replacing broken siding; removing and replacing old joint sealants; and stripping siding that cannot be effectively protected from moisture infiltration. Of these, replacement of sealants is the most common and cost-effective. Figure 13.6 shows a failed brick expansion joint that does not provide adequate moisture protection.

Whenever barrier-type curtain walls are used, proper functioning of joint sealants is critical, because in these walls there is no other line

Figure 13.6 This deteriorated brick expansion joint provides no moisture protection at all.

of defense. Even in cavity wall construction, sealants play an important role by bearing the brunt of the weather attacks. Unfortunately, the service life of sealants is much shorter than that of the rest of the wall, and sealants must be periodically inspected and replaced when needed. The owner who does not pay attention to sealant maintenance will soon be faced with leaks.

Selection and installation of sealants requires some knowledge, and the proverbial "caulking warriors" can do more harm than good by indiscriminate sealant application that plugs weep holes and traps moisture produced by condensation (or other causes) inside the curtain wall. The trapped moisture can corrode the wall components and encourage the growth of mold and mildew.

During the 1980s, many curtain wall buildings were sealed with polyurethane caulks that have since deteriorated under exposure to heat and sunlight. Now, most of them must be replaced with other products, usually silicone sealants. Partly for this reason, the sales of silicone sealants for retrofit work outstrip those of sealants for new construction by a 2-to-1 margin. These projects can become quite expensive and take a long time to complete. For example, in one project involving a 50-story tower in downtown Dallas, the buyer expected to spend about $1 million to replace the sealants. The task of removing thousands of linear feet of old caulk, grinding down the surfaces, and applying new sealants was expected to take nearly a year, with the work being done after hours and on weekends to avoid distracting the tenants.[14]

Moisture repairs in hot and humid climates should include removal of vinyl wall covering, which tends to act as a vapor barrier on the wrong side of the wall. In severe cases, it might even be desirable to remove the gypsum board and replace it with a more vapor-permeable finish, such as wood paneling. (Of course, the paneling may warp if the underlying problem is left unsolved.)

Total wall replacement should be used as a remedy of last resort, after all other "fixes" have failed. Lotz describes a case in which the owner became convinced that the only solution to leakage through his new brick exterior walls was to replace them and ordered the contractor to do so. The investigation by Lotz concluded that all that had to be done was to replace a few areas of failed caulking! Lotz muses that the expensive water-spray test undertaken by the owner prior to condemning the wall was completely unnecessary, as the locations of the failed caulk were evident.

Once the source of water infiltration has been found and the proposed repair program implemented, the effectiveness of this program can be verified by additional spray testing of a representative area.

13.4   Rehabilitating Solid Masonry Walls

13.4.1   Solid masonry vs. brick-veneer walls

Masonry exteriors suggest beauty, stability, security, and fire resistance. The thick loadbearing barrier walls of yesterday provided all four, but today such walls are uneconomical to build. Since the mid-twentieth century, those monumental walls have been replaced by single-wythe walls of concrete masonry units (CMU) and by brick-veneer walls, in which the thickness of brick is less than 4 in (10 cm). The old massive brick walls and today's single-wythe CMU walls are barrier walls, whereas brick veneer, discussed in the next section, is an example of cavity wall construction.

Solid masonry walls made of brick or single-wythe CMU can suffer from the three main causes of distress in exterior walls discussed in Sec. 13.1.1: façade disintegration, problems stemming from thermal and moisture movements or movement of building structure, and water penetration or condensation.

13.4.2   Repair of façade disintegration

The most common signs of damage in masonry walls are deterioration of joint mortar and delamination and cracking of exterior brick or masonry units. Crumbling and cracked mortar is the result of freeze-thaw cycles and the attack of harmful chemicals. The aging process accelerates if the mortar is not sufficiently compacted, or if the joint profile encourages water to linger there. When this water freezes, it damages both mortar and brick.

Disintegration of brick most often involves delamination and cracking of the exterior brick surfaces as a result of weathering and freeze-thaw cycles. The speed of this process depends greatly on the type of brick. Excessively soft brick absorbs a lot of water, and therefore invites more freeze-thaw damage, but very hard-burned brick also presents a problem—its sometime poor bond with mortar tends to invite water entry at the joints. Brick with a water-boiling absorption rate of 6 to 9 percent (measured in accordance with ASTM C216) usually bonds well.[21] Softer brick is usually lighter in color; the higher the temperature at which the brick was fired, the darker the resulting color and the harder the brick. In fact, in the past, the softest under-burned brick was called salmon brick because of its pink color; its use was limited to interior fill in solid walls.[22]

The natural aging process is hastened when brick is covered with nonbreathing paint or is sandblasted (often to remove that paint). Such painting leads to a buildup of moisture in the brick under the paint barrier. Eventually, brick that is subjected to water pressure sheds the unwanted coating—and a thin surface layer spalls with it. The process is essentially the same as that described in Chap. 5 for concrete coated with impermeable finishes. Sandblasting speeds up deterioration because it removes the dense exterior surface that was imparted during the oven firing of brick and opens the porous interior to the elements. Figure 13.7 shows the resulting damage.

A somewhat similar process occurs in glazed masonry: The glaze tends to trap moisture behind it, and in some cases the masonry spalls or cracks. The deterioration is usually most severe at parapets, where the full effects of freezing and thawing cycles can be felt, while the walls below are kept warmer by the escaping building heat. In the worst-case scenario, freezing of water trapped behind the parapets can lead to loosening of the bricks and to pieces of masonry falling to the ground. The replacement parapets should incorporate a drainable cavity with flashing and weep holes and should be anchored to the structure, as discussed in a later section.

One of the most common causes of brick deterioration is rusting steel lintels. The lintels can rust on the underside, as a result of expo-

Figure 13.7 Brick deterioration caused by moisture buildup under an impermeable coating and delamination from freeze-thaw damage.

sure to moisture in the air and poor corrosion protection, or at their top surfaces under the brick. The latter is usually caused by defective flashing and weep holes that allow for moisture buildup at the top of the lintel.

The repair of deteriorated masonry façades is relatively straightforward in principle, but tedious and expensive in practice. Loose and cracked bricks can be resecured by the injection of epoxy grout that can penetrate behind the bricks. Broken and spalled bricks or blocks can be removed and replaced by cutting out the mortar around them with a chisel, removing the units, and cleaning the opening to receive new masonry pieces. If deterioration is caused by freeze-thaw cycles or excessive weathering, replacement facing brick conforming to ASTM C216 Grade SW should be used. The degree of expected weathering can be found in the standard. It includes a map of the United States showing areas subjected to negligible, moderate, and severe weathering. When badly rusted steel lintels are encountered, they should be replaced, an operation that involves wall shoring.

The most critical issue to keep in mind is that the repairs must be compatible with the original construction. If they are not, more harm than good can result. Snodgrass and Boornazian[23] note that "it is not uncommon to see previous restoration work that is aesthetically incorrect and/or has caused further damage due to incompatibility of materials." Color matching, while perhaps not the first priority for structural engineers, should receive careful attention, as nothing gives away a cheap repair more than color differences. In some cases, exact matching is impractical, and the designers have no choice but to settle for an approximation. Then, it is usually better to select a shade that is darker than the original rather than one that is lighter. Another important concept in historic masonry restoration is to use the least possible amount of restoration and cleaning and to increase the effort only when the gentler approach does not work. Masonry restoration experts are unanimous: Do not overclean existing masonry!

What about the cracked and loose terra cotta tiles described at the beginning of the chapter? Repairs to them are approached in the same general manner. First, the avenues of water infiltration, if that is a factor, are closed by repairing the roofing and flashing. The cracked terra cotta units are glued together with epoxy adhesives or other materials; their support brackets are evaluated and replaced if necessary. Units that have been damaged beyond repair are replaced with like materials if practical or with fiberglass-reinforced polyester (FRP) or cast-stone look-alikes.

13.4.3 Repairing mortar joints

Deteriorated mortar joints are among the main causes of deterioration and loosening of bricks. The joints can be repaired by repointing (also called tuckpointing). In this procedure the old mortar is typically removed to a depth of at least 0.75 in, and deeper if the mortar is badly deteriorated. Grinders with circular abrasive blades are often used to remove mortar from horizontal joints. For short vertical mortar joints, these tools are too risky, as they tend to cut the adjacent bricks. Also, there are special raking tools for mortar removal, and there is the old standby, the mason's chisel.

In the next operation, the joint surfaces are cleaned by compressed air, a brush, or a stream of water, and the new mortar is placed. To minimize shrinkage, the mortar should be of as dry a consistency as is practical. Which joints should be repointed? Typically, those where the mortar is crumbling or cracked, has fallen out of the joint, or has eroded by a depth of more than $1/4$ in. It is often better to repoint the whole wall than to repoint only the visibly deteriorated joints, because some mortar damage may not be evident to the naked eye.

Which mortar should one specify for repointing? Ideally, the new mortar should match the existing mortar in strength and appearance. The strength of the existing mortar can be found by testing. If the old mortar is weak, the new one must be weak also. A strong repointing mortar in combination with a soft and crumbly one in the rest of the joint is not desirable: It will introduce a hard point, on which, over time, the weight of the wall will become concentrated. A small area of the brick near the exterior edge may then crack and spall. Also, strong mortar will move differently from the weak existing mortar with changes in temperature and moisture and may even shrink away from the surfaces it is intended to protect. Usually, mortar for repointing is specified to conform to ASTM C270 Type N or O.

The BIA[24] recommends that the repointing mortar be prehydrated to reduce shrinkage. Prehydration involves mixing the ingredients with only enough water to produce a damp consistency—so that the mortar can retain its shape when formed into a ball—and leaving the mortar to stand for 1 to 1.5 h. After this, some more water can be added prior to placement, still keeping the same mortar consistency. (Allowing mortar to partly set and then reworking it may weaken the material. This may not be a problem if a rather weak mortar is desired in the first place.)

The success of repointing greatly depends on good workmanship. Without it, the work will do more harm than good. The joint surfaces of brick should be dampened prior to placing the mortar, but, to avoid shrinkage, concrete masonry need not be dampened. In hot climates, the whole brick wall can be periodically moistened to prevent premature drying. Mortar should be placed in thin ($\frac{1}{4}$-in) layers rammed into the joint; each layer should be allowed to stiffen "thumbprint" hard before installation of the next one. Shortly after placement, the excess mortar is scraped and swept from the masonry, and the mortar is tooled to compact it and push it into close contact with the masonry. The most weather-resistant joints are those tooled to a concave or V configuration. Vertical joints should be tooled first. An aged look can be given to new mortar by imprinting it with a stiff brush.

A few trial installations should be attempted in order to match the color of the new mortar to that of the rest of the building. To check for color match, Beall[25] suggests soaking a portion of the wall and checking the color of the wet old mortar against that of the new. If need be, the proportions of the new mortar can be slightly modified, provided that they still fall within the confines of ASTM C270. The issues of color and texture matching of mortar, which are especially acute in the repointing of historic masonry buildings, are addressed in further detail in Ref. 26.

13.4.4 Distress caused by thermal movement or movement of building structure

When expansion or contraction of masonry from temperature or moisture changes is restrained, substantial stresses can accumulate in the walls. In old monumental brick walls, the relatively weak mortar provided some cushioning, but in contemporary single-wythe hollow CMU walls, especially those of unreinforced or lightly reinforced construction, these stresses must be relieved by properly sized and spaced expansion joints. If the joints are too narrow or are spaced too far apart, they will not provide the desired movement relief or protection, and the wall may become cracked, bulging, or out-of-plumb.

Figure 13.8 shows what can happen to a CMU wall that has insufficiently wide expansion joints: The joint filler has been literally squeezed out, and the wall itself is cracked and bulging right near the joint because of a buildup of compressive stresses. Cracking of brittle masonry as a result of tensile forces, as in Fig. 13.4, would be understandable, but the fact that the wall cracked from *compression* is sobering.

Structural movement, such as differential frame shortening, lateral drift, or foundation settlement, can also be to blame for cracked and broken exterior walls. A common example of this can be found in concrete-frame buildings with brick infill walls. Here, the combined effects of concrete-frame shortening and brick expansion from moisture changes can result in especially severe distress, usually on the part of the brick. Interconnected composite walls of brick and CMU can be broken apart by differential movements caused by temperature

Figure 13.8 Joint filler in this single-wythe CMU wall is squeezed out and the wall itself is cracked and bulging because the width of the joint is too small.

or moisture changes in these two different materials, as will be elaborated upon in Sec. 13.5.

Masonry walls can crack at the bearing points of supported beams and slabs, either because of movement or because of corrosion of those members. Cracking caused by rusting of wall-bearing steel beams and lintels occurs because the expanding rust products exert tremendous stress on the surrounding masonry. Concrete slabs placed directly on masonry can shrink, and their ends can curl and separate from the masonry, causing horizontal cracks to appear below the slab, especially at the corners.[27]

Repairs of distress caused by temperature or moisture movements typically involve replacement of the broken masonry units and enlargement of the existing expansion joints. The latter can be done by saw-cutting one side of the existing joint, removing the existing filler and brick slivers, and refilling the joint with new material. More often than not, new expansion joints will also be required at some critical locations, such as at the corners, changes in wall height, and wall offsets and junctures. Where necessary, the rigid joints between concrete frame and brick or between CMU and brick may have to be changed to "soft" joints filled with compressible materials. Needless to say, such forced separation and rejoining is a painful and disruptive process.

The methods of repair of cracked walls damaged by foundation settlement depend on whether the settlement process has essentially stopped or is ongoing. Repairs should never be undertaken until this determination has been made and the underlying problem addressed. If the settlement is determined to have stopped, the walls can be repaired by simply filling the cracks and replacing the cracked units. If the settlement is continuing and no foundation work is planned, new vertical expansion and contraction joints cut into the wall will probably be needed to reduce the panel size and minimize further cracking.

13.4.5 Repair of moisture-related problems

The effects of water penetration and excessive air infiltration from the outside have been addressed in previous sections. However, one additional point needs to be made: Water penetration can initiate a process of wall deterioration that can be continued by other destructive mechanisms. In Fig. 13.9, persistent water leakage from a missing downspout has led to the eventual disintegration of the single-wythe CMU wall as freeze-thaw cycles came into the picture. The constant flow of water provided the needed "fuel" for the freezing and thawing process. In this case, it took about 20 years to destroy the blocks, but the process can happen much faster.

Water leaking into the wall and freezing there can exert great force on the surrounding bricks or concrete blocks—confined water can expand with pressures of nearly 15,000 psi. If these forces become large enough, the brick or block with the weakest bond may pop out. Figure 13.10 shows a case in which persistent leakage has led to bricks falling out of the wall. A similar situation may occur when leakage leads to rusting of steel columns placed inside exterior brick walls: The expanding rust can push one or more bricks outward.[28] Steel columns often rust where roof drains are placed between their flanges, an unfortunate design that has been popular for a long time.

The effects of water penetration can be highlighted by the appearance of efflorescence—white deposits of leached salts, mostly water-soluble calcium compounds. In massive brick walls, there is plenty of material to leach from, and the effects of efflorescence can be dramatic. Figure. 13.11 shows the condition of an interior brick wall in the nineteenth-century Ft. Knox, Maine, where heavy roof leakage has contributed to a cavelike appearance. (The wall is illuminated from below.)

Figure 13.9 Water leakage from a missing downspout and freeze-thaw cycles resulted in disintegration of this single-wythe CMU wall.

Some brick is more prone to efflorescence than others. The efflorescence test contained in ASTM C67, *Standard Methods of Sampling and Testing Brick and Structural Clay Tile,* can be used to determine the brick's degree of susceptibility. The test involves placing the bricks in distilled water to a depth of about 1 in for a week. Units rated "not effloresced" would show no efflorescence visible from a distance of 10 ft at a certain level of illumination.

To repair moisture-related problems, attention should first be directed to the repair of roofing and flashing, as well as to gutters and

Figure 13.10 Persistent leakage has led to bricks falling out of the wall.

downspouts, to stop further water penetration inside the wall. Provisions for proper drainage of any water that does get inside the wall cavity should also be included. Only then should the deteriorated mortar and masonry be replaced as described previously. If the wall cracking is caused by structural steel corrosion, the masonry around the corroded member must be removed, and the member cleaned and reinforced if the rusting has substantially weakened it. To prevent the problem's recurrence, the steel may need a coating to protect it from corrosion. Finally, the masonry is reinstalled, maintaining the desired relationship between the column and the wall—they can be either tied or totally separated.

What can be done if the wall itself is leaking badly? Stockbridge[29] has investigated the various methods of joint repair, including repointing, parging the whole wall with grout, grouting only the joints, and covering the wall with a clear waterproof coating. His conclusion: Repointing was the most effective method of decreasing leakage through masonry walls.

There is a limit to how much a barrier wall can be weatherproofed without substantially altering its appearance or composition. A solid wall made of porous brick will still absorb a lot of moisture during a heavy driving rain, and some interior finishes may get moist despite the best efforts of the restorers. Attempts to prevent this by painting the interior wall surfaces with vapor-retardant paints

Figure 13.11 Heavy efflorescence on the interior of brick walls in the nineteenth century Ft. Knox, Maine, caused by roof leakage.

should be discouraged because they help trap moisture inside the walls and lead to further damage. Instead, it is best to remove from barrier walls any interior finishes that are sensitive to moisture.

Another dubious remedy commonly proposed for porous walls is to coat their exterior surfaces with sealers. As already discussed (and to be mentioned again), this solution is short-term at best—the sealer would have to be frequently reapplied—and harmful at worst, if it impedes evaporation. If a coating is desired, a more substantial solution for CMU walls is to use a breathing heavy-body paint-type product that can actually fill the masonry pores. Unfortunately, this heavy coating is usually proposed only after multiple coats of masonry sealers have been applied, making its proper adhesion extremely difficult.

In those cases, it might be better to cover the wall with stucco or with a carefully made EIFS system.

A question often asked is: Should ivy growing on old masonry façades be removed? Opinions vary. On the one hand, the ivy roots damage the mortar and contribute to its disintegration. On the other hand, densely growing ivy may act as a rather effective rainscreen that prevents most water from being absorbed into the wall. If the ivy grows very dense, it probably does more good than harm. Otherwise, it can be removed by cutting it away from the wall, leaving the roots in place. The roots will die in a couple of weeks and can be scrubbed off at that time.[24] Ivy should never be pulled from the brick, as it will pull pieces of mortar with it.

Another common question is: Do barrier walls need flashing at the base or over the windows? Again, opinions vary. Since there is no cavity, flashing is certainly not as critical as in cavity walls. In any case, no intermediate flashing (at the floor levels in a multistory building, for example) is typically required, and such flashing can actually cause harm if it becomes an avenue for water entry through the wall. The main function of the base flashing is to prevent the moisture in the foundation from traveling up the wall, but the same result can be achieved more easily by a dampproofing coating under the masonry. Also, the flashing over the windows protects the top surfaces of the steel lintels from moisture and covers the otherwise exposed collar joints, but many existing barrier walls do not have any flashing and have performed well. The flashing becomes more important when the brick is cracked—it helps prevent moisture from attacking the top of the lintel and thus protects it from corrosion. But even in that case, it is difficult to justify major surgery to install new flashing where none has existed before.

13.5 Brick-Veneer Walls with CMU Backup

13.5.1 Two kinds of brick-veneer walls

There are two main kinds of brick-veneer walls: those with CMU backing, shown in Fig. 13.12, and those with a backing of steel or wood studs, discussed in Sec. 13.6. Many brick-veneer walls have performed satisfactorily over the years, but others exhibit a multitude of characteristic rainscreen-wall problems ranging from efflorescence to corrosion of the metal ties attaching the brick to the structural backup. Some of these problems are shared by both kinds of brick-veneer walls; others are unique to one or the other.

13.5.2 How the system works

Brick-veneer walls with CMU backing benefit from the durability and fire resistance of CMU walls as well as from the good water resistance

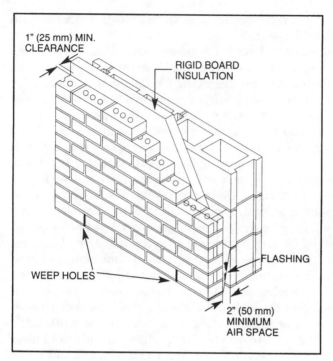

Figure 13.12 Brick veneer and concrete masonry cavity wall. (*Brick Industry Association.*)

of rainscreen cavity walls. The brick veneer and the blocks are typically connected by horizontal joint reinforcement with integral or adjustable ties; the ties transfer lateral loads between the two wall layers (wythes). The term *veneer* accurately suggests that the brick's role is that of pure cladding. From a structural standpoint, other materials, such as split-face CMU or stone, can be used as veneer in these walls instead of brick without affecting the design of the CMU backup. However, some designers assume a composite action between the two materials.

The Brick Industry Association (formerly Brick Institute of America),[13] differentiates between cavity walls and brick-veneer walls. According to the BIA, cavity walls are those in which the brick resists some fraction of the applied vertical and lateral load; the amount depends on the relative rigidities of the brick and CMU layers. Brick-veneer walls, in contrast, are those in which the brick's function is nonstructural and all the loads are resisted by the CMU. Our discussion is equally applicable to both of these wall types, but in this section we will focus solely on brick-veneer walls, even when we call them cavity walls.

As shown in Fig. 13.12, rigid insulation is often placed in the cavity to improve the insulating properties of a wall with otherwise poor insulating characteristics. (Another possible place for the insulation is on the inside surface of the CMU, where it is typically covered with gypsum board placed on furring channels. This design introduces its own set of problems, such as inability of moisture in the CMU to evaporate in the interior direction.) The minimum air space is 1 in for walls with insulation and 2 in without it; anything less is difficult to keep free of mortar droppings, which can carry water across the cavity or trap water in it and short-circuit its function.

Keeping true to the classic cavity wall design, we recommend that the exterior face of the CMU be covered with a waterproofing membrane. In real life, this is not always done; in fact, the BIA's Note No. 21[13] states that this is not necessary, although it also states that it might be prudent in cases where mortar could bridge the cavity or accumulate at its bottom owing to poor workmanship. Like any cavity wall, this one should also include flashing and weep holes.

Flashing should extend completely through the wall and protrude at least $\frac{1}{4}$ in beyond the walls's exterior face. The protruding point should form a drip by being bent down at 45°. Some architects try to avoid exposed flashing, but flashing that terminates within the brick allows water to travel under it back into the cavity and to freeze there, breaking the wall from the inside. Flashing should be extended at least 8 in above the penetration point and be embedded into the interior CMU layer. It is important to turn up and seal the discontinuous ends and edges of flashing for at least 1 in to stop water from escaping sideways. Continuous flashing should be overlapped a minimum of 6 in.[13] The most durable flashing materials are stainless steel and lead-coated copper. Weep holes should be located directly above the flashing at 16 to 24 in on centers.

13.5.3 The fundamental problem

The main problem of walls made of brick veneer and concrete block is that these two materials, even though both are called masonry, are radically different in nature. Brick is made of clay; it is fired in the oven and contains practically no moisture immediately after fabrication. As brick ages, it gains moisture and steadily expands in the process. A concrete masonry unit begins its life as a wet mix and is later cured saturated with water. As a CMU wall ages, it gradually loses moisture and shrinks.[30]

The rates of moisture expansion for brick and shrinkage for CMU are much more difficult to quantify than the thermal expansion coefficients for these materials. BIA's Technical Notes on Brick Construction

No. 18[27] suggests that a 100-ft-long wall of a heavyweight CMU will expand 0.62 in when subjected to a 100°F temperature increase, and CMU with cinder aggregate will expand 0.37 in. The rate of moisture expansion for brick is listed as 0.02 percent, and the shrinkage coefficient for CMU is listed as 0.05 percent. The latter is much greater in arid regions and almost nonexistent in structures that are continually wet. (An even larger rate of moisture expansion for brick is used in Sec. 13.6.)

Ironically, the incompatibility of the two materials is reduced by the delays in getting them to the site. By the time brick and CMU are combined in the same wall, some (one hopes most) of the initial volume changes will have already taken place. But brick and CMU still move differently with moisture changes. They react to temperature changes differently, too: Concrete masonry expands and contracts up to 15 percent more than brick. However, the brick in the exterior wythe is directly subjected to changes in temperature and exposed to the sun's rays, whereas the CMU (separated from it by an air space) is not. Thus the brick undergoes larger swings in temperature than the block and may move more. (Incidentally, light-colored brick and other materials absorb less solar radiation and move less.)

Several factors allow brick and block walls to avoid self-destruction from differential thermal and moisture movements. First, many of these walls are interconnected by adjustable ties that permit some independent movement of the two wythes, although many are joined by horizontal joint reinforcement with integral tabs that are more restrictive. Second, most of the time, the height of these walls is relatively modest—e.g., when the buildings are single-story. In taller buildings, the problems with differential vertical movements become more pronounced. Third, the horizontal differential movements are reduced by the expansion and control joints that are usually installed in these walls.

One particular problem with brick and block walls is that they typically require two types of mortar. Why? For exterior brick veneer, the most important qualities of the mortar include the ability to absorb slight movements and deformations, because, as was noted, brick may move more than block. Since the brick veneer transfers all the applied lateral loads to the block backup, it does not need a very strong mortar, but resistance to freeze-thaw cycles might be beneficial. Mortar conforming to ASTM C270 Type N is usually appropriate, being relatively weak and more pliable than the stronger mortars of the other types. Incidentally, masonry cements are too rigid and are not recommended for brick veneer. Conversely, CMU backup walls are structural in nature, often contain steel reinforcement, and tend to move less. They typically require stronger mortars of ASTM C270 Type

S or M. Obviously, with two different types of mortar, proper supervision of masonry installation is vital.

The different nature of bricks and blocks is highlighted in Fig. 13.13. In this unfortunate design, a rowlock course of bricks was placed to cap the parapet of a wall made of brick veneer and concrete block. Differential movement between the two materials (expansion of the brick and shortening of the CMU) broke the mortar bond and lifted the rowlock course above the CMU. As a result, the rowlock course separated from the CMU along the whole perimeter of the building, and many of its bricks cracked. After water intrusion and freeze-thaw cycles took their toll, the mortar below the brick deteriorated, and the whole rowlock course became vulnerable to displacement and falling down. This large building was less than 14 years old at the time the picture was taken.

13.5.4 Wall joints

Expansion and control (contraction) joints are often confused, but they are installed with opposite goals in mind. Expansion joints in brick are filled with compressible materials and, logically, are intended to provide space for brick expansion. (As already noted, the price that must

Figure 13.13 This rowlock course of bricks placed to cap the parapet of a wall made of brick veneer and concrete block was lifted up by differential movement.

be paid for these sealant-filled joints is a lifetime of maintenance.) Control joints in interior CMU are included to provide for shrinkage; these joints contain rigid inserts to transfer loading among the adjacent blocks. Expansion and control joints are normally offset in relation to each other in order to minimize moisture entry (Fig. 13.14).

To be effective in accommodating both these movements, expansion joints should be spaced at relatively close intervals, typically not more than 18 to 25 ft apart. These joints should also be placed near building corners, at wall offsets and intersections, and at the points of changes in the foundation system.

The BIA's Notes 28B[31] gives the rate of moisture expansion for brick as 0.0005 in/in, and the design coefficient for its thermal movement as 0.000004 in/°F. (These numbers are averages, as moisture and temperature differentials vary widely across the country.) According to the Notes, the total formula to be used in joint design is

$$w = [0.0005 + 0.000004(T_{max} - T_{min})] \, L$$

where w is the total expected movement of the brick (in), T_{max} and T_{min} are the maximum and minimum mean temperatures of the brick (°F), and L is the length of the wall (in). So, a 100-ft-long brick wall will expand about 1.08 in when subjected to a 100° temperature increase. Despite this seemingly huge value, brick movements can be even larger

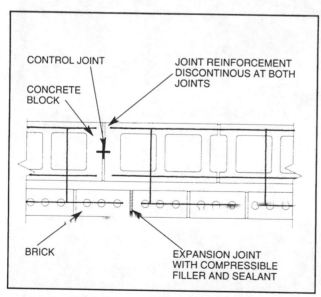

Figure 13.14 Expansion joints in brick and control joints in CMU are normally offset. (*Brick Industry Association.*)

in some areas with continental climate, where the annual extreme temperature range is near 200°F.

This formula can help establish the width of an expansion joint as a function of the geographic location and joint spacing. Using the numbers just given, it would seem that for an "average" location, a 25-ft-long brick wall will undergo a likely movement of about $1/4$ in. So is a $1/4$-in-wide joint sufficient? No, because the compressibility of the expansion-joint filler and sealant has not yet been considered. When the wall undergoes temperature and moisture expansion, the filler and sealant are compressed; when the wall contracts, they expand. No joint material of today can expand from zero thickness to $1/4$ in, and none will expand from $1/4$ to $1/2$ in, a 100 percent rate. Instead, joint filler that is placed in an overly narrow joint will tend to be squeezed out of the joint by wall expansion in the summer and will separate from the joint edges when the wall contracts in the winter. Either way, the weathertightness of the wall will be compromised.

As Szoke and Carrier[32] point out, many joint fillers have a compressibility of only 25 to 50 percent. A $3/8$-in expansion joint with a filler compressibility of 35 percent can accommodate about $1/8$-in movement. Therefore, when the brick expansion joints are spaced 25 ft apart and $1/4$ in of wall movement must be accommodated, the required width of a joint with a filler having a 35 percent compressibility would be $3/4$ in. If a filler with 50 percent compressibility was used, the minimum joint width would be smaller, $1/2$ in. Despite these requirements, many architects specify $3/8$-in expansion joints simply because they correspond to the width of the mortar joints, not because of any rational analysis.

Successful functioning of joint fillers and sealants depends on proper installation techniques. The joints should be kept free of mortar droppings that could bridge them and compromise their function. Some recommend taping or priming the joint filler at its interface with the sealant to improve sealant performance by preventing a bond between the two and allowing the sealant to move freely with the wall movement. If this is not done, the sealant would be restrained by the filler and could fail by debonding at one of the edges, letting water inside the joint. Another suggestion is that the popular closed-cell foam joint fillers not be punctured during installation, because this could slowly release minor amounts of chemicals from the foam, causing the uncured sealants to bubble and fail.[32]

The correct place for expansion joints in walls with isolated ("punched") windows is midway between the windows (Fig. 13.15). This location is preferable to making the joints at one of the window sides (Fig. 13.16). The latter is a common but improper joint detail that cuts across the lintel and allows it to bridge the joint and transmit at least some of the brick movement across it. (This statement does not apply to continuous ribbon windows, because the brick above those is typically supported by hangers rather than by loose lintels.) If this detail must be used for some reason, at least the end of the lintel near the joint should be coated with oil or a piece of flashing placed there to reduce friction.

Note that in Fig. 13.16, the joint filler is squeezed out by wall movements, suggesting an insufficient joint width. Figure 13.17 illustrates an even more serious case—here, the filler is too rigid to be squeezed out of a narrow joint. As a result, the brick has cracked and bulged out from the restrained compression.

13.6   Brick-Veneer Walls with Steel Studs

13.6.1   Wall Composition

The essential components of this system, developed in the mid-1960s, are brick veneer and steel studs covered with exterior-grade sheathing and building paper. Brick and studs are separated by an air space. The interior surface of the studs is finished with interior gypsum wallboard. The steel studs are spaced 16 or 24 in on centers and are attached to the brick veneer with adjustable metal ties (Fig. 13.18). The space between the studs is typically filled with fiberglass batt insulation.

This wall type offers some advantages over brick-and-CMU construction: light weight, which allows for smaller supporting framing and foundations, ease of insulation, and cost-efficiency.

A wall made of brick veneer and steel studs is the most common type of cavity wall in contemporary construction. Like any rainscreen wall, it is expected to admit some water into the cavity, on the assumption that properly functioning waterproofing, flashing, and weep holes will promptly escort this water out. Therefore, to assure leak-free performance, brick-veneer-and-steel-stud walls must be meticulously designed, detailed, and built. There is little margin for error.

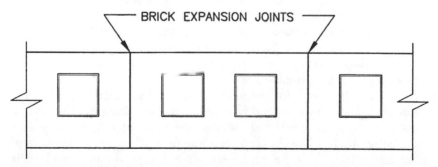

Figure 13.15   Location of expansion joints in walls with punched windows. (*After Ref. 32.*)

Figure 13.16 This brick expansion joint not only cuts across the steel lintel, but also is too narrow. Note that the joint filler has been squeezed out by wall movements.

However, this system is now used so widely that in many cases essential details are improperly executed or simply left out. Some knowledgeable architects and engineers argue that this rather complex wall system has become popular too fast, before all the fine points of its design and construction, including the issues of long-term durability, have been researched. There have been a number of well-publicized failures. In one case, the defects in a 19-year-old 150,000-ft^2 federal office building in Norfolk, Va., were so pervasive that the building had

Figure 13.17   A combination of overly rigid joint filler and a narrow joint has led to cracked and bulged brick.

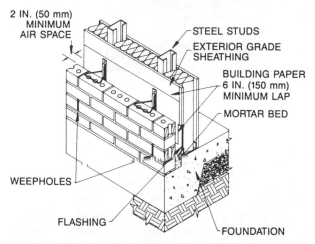

Figure 13.18   Brick veneer and steel stud wall. (*Brick Industry Association.*)

to be totally stripped of its exterior walls and reclad with precast concrete panels.[33]

According to Rutila,[18] the first version of this system, shown in a 1968 manufacturer's catalog, was quite primitive by today's standards of good practice. The recommended details did not show any weather barrier on top of the gypsum sheathing at all; the flashing did not extend through the wall (it was stopped at the toe of the shelf angles and was tucked flat under the steel studs); there were no soft joints under the relieving angles; and corrugated 24-gage brick ties were specified. The standards slowly improved over the years, until in the early 1980s they were in substantial agreement with present-day practice. This means that most pre-1980s walls, even if properly constructed in accordance with then-current recommendations, can suffer from durability problems and may require expensive repairs or even replacement.

There are still relatively few sources of information on the proper design and construction details of this wall system. One of the best is BIA's Technical Notes 28B.[31] This document contains structural design criteria, as well as recommendations on avoiding water penetration, the minimum size of the air space, maximum tie spacing, and other important aspects of design. As we shall see, ignoring these rules leads to failures that are very expensive to fix.

13.6.2 The issues of horizontal deflections and adjustable ties

The most fundamental problem with brick veneer and steel stud walls lies in the concept itself—the assumption that brittle brick veneer, supposedly a purely cosmetic cladding, can be effectively provided with lateral support by flexible steel studs. In reality, the brick can crack before the studs assume their deflected shape. The only way to avoid this cracking is to make the studs more rigid than the brick. The needed rigidity can be gained by using deeper sections or thicker metal than would be required for strength alone; this can reduce the cost-effectiveness of the system. Not surprisingly, lateral stiffness requirements for studs have elicited heated controversy.

The magnitude of lateral (horizontal) stud deflections is measured in relation to their length L. The BIA's Notes 28B recommend that the maximum deflection of "steel stud backup, when considered alone at full lateral design load, be $L/600$ to $L/720$." There are experts who consider this limit not stringent enough; they say that the brick will crack at deflections less than one-third of that, or $L/2000$.[34] The steel-stud industry, on the other hand, for many years insisted that a liberal limit of $L/360$ was adequate.[35]

So which lateral deflection criterion should be used? One way to find out is to undertake a comprehensive full-scale testing program covering a variety of wall heights, attachment methods, and boundary conditions. A less expensive alternative is to conduct a rigorous finite-element analysis with proper allowances for the stiffnesses of steel studs, brick veneer, and brick ties. The published results of an analysis of this type performed by Gumpertz and Bell[21] conclude that the BIA's limit of $L/600$ is reasonable. We recommend it as well.

Other structural issues are how closely to space the brick ties connecting the brick to the steel studs, and for what wind loading to design them. A typical adjustable tie is attached to an anchor connected to a steel stud by screws or other fasteners. A good anchor and tie assembly should permit vertical adjustment equal to at least one-half the brick height. Some common kinds of adjustable ties are shown in Fig. 13.19.

Adjustable ties act as mini-columns that transfer wind load from the brick to the studs. They should be made of thick wire, the design thickness of which depends on the width of the cavity but should be a

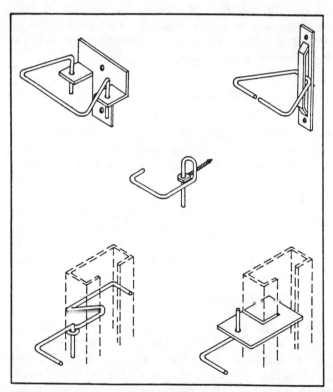

Figure 13.19 Common types of adjustable brick ties. (*Brick Industry Association.*)

minimum of $^3/_{16}$ in. The wider the cavity, the longer the length of the "column." The BIA's Notes 21[13] recommend limiting the cavity width to 4.5 in, to allow the use of commonly available ties. The thin corrugated-metal ties that were formerly common in housing construction should never be used in new buildings; unfortunately, they have been installed in numerous existing structures.

The BIA's Notes 28B propose using one tie for each 2 ft² (0.18 m²) of wall area and a maximum tie spacing of 24 in (600 mm). ACI 530/ASCE 5[36] is more stringent—it limits the area to 1.77 ft² per tie and tie spacing to 16 in o.c. in every direction. The ACI 530 code is typically used as a reference by the model building codes and is therefore considered more authoritative. In either case, it is assumed that each brick tie receives the wind load in accordance with its tributary area of spacing, so if the wind load on the wall is 21 psf, the force on a tie with a tributary area of 1.77 ft² would be 21 × 1.77 = 37 lb.

The actual distribution of the load may not be as straightforward because of stud and brick deflections. It is logical to assume that the ties in the middle of the wall, where stud deflections relative to brick are the largest, carry less load than the ties at the top and bottom of the wall, where little differential displacement takes place. An investigation by Gumpertz and Bell[21] has discovered that wind-force distribution among various brick ties is indeed nonuniform and that two or three ties located at the top and bottom of a stud carry a significant fraction of the total wind force on the stud. Thus, a simplified method of arriving at the design load on the ties would be to take the total wind load tributary to a stud of moderate height and divide it by 4 or 6.

13.6.3 Preventing moisture penetration

The problems of moisture penetration through brick-veneer-and-steel-stud walls have received a lot of attention. In addition to the usual moisture problems of barrier walls and veneer walls with CMU backup, in this type of construction there is another serious issue: corrosion of the brick-tie connections to the studs and of the studs themselves as a result of excessive moisture penetration into the cavity. Corrosion of steel studs can be devastating and can take place surprisingly fast. Gumpertz and Bell[21] tell of a building less than 10 years old in which the painted (not galvanized) steel studs were found to be totally rusted in some places. Figure 13.20 illustrates adjustable brick ties that are corroded (and also too short). However, steel studs and brick ties made of galvanized steel are not always the weakest link in this system—most commonly the weakest link is the *fasteners* attaching the brick ties to the steel studs.

Figure 13.20 Corroded (and also too short) adjustable brick ties. (*Photo: Simpson Gumpertz & Heger Inc.*)

In the vast majority of installations, these fasteners are self-drilling screws driven through the sheathing into the studs. The screws are usually protected only by a corrosion-resistant coating that can be easily damaged during fastener installation, making the threads of the screws vulnerable to corrosion. The edges of the holes made by the screws are also unprotected, even if the studs are galvanized. A moisture attack can be most harmful here, especially if the moisture contains salt. Salt does not have to come from some sea mist; Grimm[34] points out that all masonry contains some of it. He warns that the brick "is literally hanging on the building by a thread, and the fine thin arris of the unprotected thread of a steel screw may be periodically bathed in a salt solution." It might be argued that in most cases the situation is not so dire, but the threat of corrosion-induced failure is clearly present.

What steps can be taken to improve the situation? Stainless steel screws, suggested by some, have a superior corrosion resistance on their own, but they can initiate galvanic action when they are in contact with plain or galvanized steel. As they are very expensive, use of these screws is probably not worth the effort (and, of course, they don't help with corrosion of the hole edges). A more practical approach is to specify a special baked-on copolymer coating on the screws. Such coatings have been found more effective in corrosion protection and abrasion resistance than standard cadmium or zinc plating.

In addition to coating the screws, there are other steps that can reduce the likelihood of fastener corrosion. One is to require some minimum thickness of the steel studs in order to increase the grip of the screws and prolong the life of the metal around the holes. The author's practice is to specify galvanized studs of at least 18 gage, regardless of strength requirements, simply to provide a larger thickness of metal at the connection. Rutila[18] recommends even thicker studs, at least 16 gage. Another step is to restrict the entry of water into the connection to the maximum extent possible. Self-drilling screws normally have neoprene compressible washers under their heads, but some additional protection will help. Gumpertz and Bell[21] suggest installing a piece of compressible gasket made of a synthetic rubber membrane (EPDM) or similar material behind the base anchor.

A radical solution to the problem of thread corrosion would be to use pop rivets, small bolts, or similar nonscrew fasteners that are not yet commonly used in this application. One downside of these is the special waterproofing required around the large resulting holes. Another approach would be to eliminate the fasteners from the anchors altogether, as some of the products shown in Fig. 13.19 are intended to do. It is not clear how these anchor designs affect the long-term integrity of the waterproofing, a critical component of any cavity wall.

Incidentally, the recommended waterproofing is a No. 30 asphalt-saturated felt conforming to ASTM D226, Type II. In some cases, this waterproofing is unwisely omitted when moisture-resistant gypsum board is used, in the hope that the latter will provide sufficient weather protection. As Rutila[18] demonstrates, this improved gypsum board alone is not sufficient to protect the interior from leakage. He also suggests the ultimate step in preventing water penetration through a brick-veneer–steel-stud system: switching to a CMU backup, which might not be any more expensive than a properly designed wall of steel studs constructed as discussed here.

13.7 Repairing Brick-Veneer Walls

Like all exterior wall materials, brick-veneer walls with either CMU or steel-stud backup can suffer from excessive thermal and moisture movements, veneer disintegration, and water penetration. They can also be damaged by movement of the building structure and settlement of foundations. Of these, repair of brick disintegration is addressed in Sec. 13.4. Repairs of the other problems are discussed in this section.

13.7.1 Repairing problems caused by thermal movement or by movement of building structure

When expansion (relief) joints are not provided or are of insufficient width, brick-veneer walls with either CMU or steel-stud backup can crack. Vertical cracks can occur at such typical locations as wall offsets, setbacks, and corners—anywhere the wall movement is restrained. Diagonal cracking may occur at the piers between "punched" windows, where the long expanses of brick above move more than the short piers, and shearing stresses develop at the interface, as shown in Fig. 13.21. (The brick below the windows is restrained by friction against the foundation and tends to move less.) Some other cases of damage attributed to differential movement are

- *Masonry on exposed concrete foundation walls.* When a masonry-clad wall attempts to contract after a temperature drop, the concrete resists this contraction, and the masonry can crack near the corner. Conversely, when the masonry expands more than the concrete wall,

Figure 13.21 Diagonal cracking in a brick wall caused by temperature movement.

which stays at a cooler temperature, a crack at the corner of unreinforced concrete can appear.

- *Parapet walls.* These are exposed on two sides and thus tend to move more than the masonry below. The differential movement can result in horizontal or diagonal cracks near the base of the parapet and in parapets that are bowed or displaced, especially at the corners. This condition, together with the weakening effects of through flashing at the parapet base, a common detail of the past, makes brick parapets potentially hazardous during earthquakes.

- *Encased columns.* When steel or concrete columns are encased in masonry, any column movement under load will tend to be transferred to the enclosure and cause vertical cracks in it.

- *Frame shortening.* When the frame of a high-rise building shortens under load (this usually occurs more in concrete buildings because of the added creep, although steel framing will also compress) and when the brick veneer is not properly isolated from the frame with soft joints, the frame will transfer some of the load to the masonry. Not being able to take this load, the masonry veneer can bow, crack, or break. Signs of this problem include horizontal cracks in the piers above the punched windows and vertical cracks at the corners or window heads. A similar situation can occur even when frame shortening affects only the backup wythe, because any bowing in the latter will be transferred to the veneer by the ties.

Fixing problems caused by movement or foundation settlement involves the same steps as for barrier walls, including replacing the broken bricks, enlarging the existing expansion joints by saw cutting, and making new expansion joints at the proper locations, such as at corners and at changes in wall height. (A point to remember: Whenever new expansion joints are introduced, new wall ties must be provided on each side of the joint to support the veneer at both edges.) Other remedial steps include introducing or enlarging horizontal "soft" joints filled with compressible materials under shelf angles and frame beams. Leaning or cracked parapets are best removed, if possible, because masonry parapets are troublesome in many respects. If they cannot be removed, the parapets should be completely rebuilt with proper flashing and anchorage into the supporting structure (Fig. 13.22).

To prevent steel studs or CMU from becoming loadbearing elements as a result of frame shortening, creep, or deflection, the backup walls should not be rigidly attached to the underside of the structural beams. Instead, there should be soft joints filled with compressible materials. To assure the lateral stability of the backup walls, slip connections of an appropriate design should be made. For CMU, these

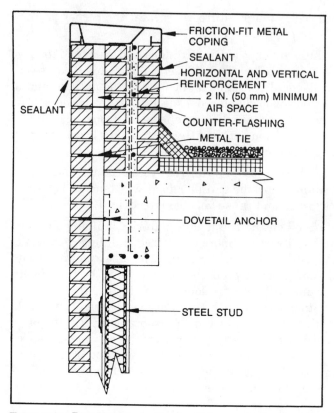

Figure 13.22 Distressed parapets should be removed, or at least totally rebuilt with proper anchorage into the supporting structure. (A slip connection should be provided at the top of steel studs, as discussed in the text.) (*Brick Industry Association.*)

connections might include a pair of angles attached to the frame but not to the wall; steel studs can be placed in an extra-deep-leg track attached to the beam but not to the studs. Unfortunately, it is all too easy to simply shot-fasten the track at the tops of steel studs to the beams or to have CMU walls solidly mortared into them. (This poor detail is so common that it seems to be present even in Fig. 13.22!)

What can be done if the existing steel-stud backup does not have enough rigidity to provide the required lateral bracing for the veneer? A wall like that can be strengthened by adding new studs in between the existing ones. This seemingly simple solution requires removal of the interior wall finishes and rerouting of any conduits running within the wall. Also, it is practical only if the existing brick ties are of proper design and spacing. More often than not, however, the improperly designed studs are accompanied by cheap ties that are spaced too far apart. In that case, it is best to replace the wall, or to leave it alone if it at least provides good resistance to moisture penetration. Unfortunately, poor-quality builders tend to be consistent, and these walls probably have ineffective flashing and weep holes as well, making wall replacement the most desirable option.

13.7.2 Repair of problems caused by moisture

Repair of problems caused by moisture penetration through brick-veneer walls requires that the source of the water entry be found first. In some cases the source is fairly obvious, as in Fig. 13.23. Here, localized efflorescence at the top of the brick, a sign of water penetration, is easily traced to the loose flashing on top, which allows wind-driven rain to enter the wall. The sources of leakage—roofing, flashing, gutters, downspouts, or negative air-pressure differential—should be fixed before the wall itself is repaired. Once the water penetration has been stopped, the deteriorated mortar, bricks, ties, flashing, sealants, and the elements of interior backup wall can be repaired or replaced. Replacement of bricks and repointing have already been discussed. Replacement of the other materials may be quite difficult.

Field investigation often finds that through-wall flashing or flashing under the masonry copings in parapet walls is deteriorated, missing, or improperly placed. Sometimes, the wall flashing is intact but sags within the cavity; this may happen with soft materials such as PVC. The sagging flashing will collect water, which can attack the metal elements, freeze, and damage finishes. If the flashing cannot be readily fixed, it should be replaced.

Replacement of flashing requires removal and rebuilding of some of the brickwork above it. For continuous flashing, contractors typically choose to remove brick in short alternating sections (5 to 10 ft wide), replace the flashing there, and rebuild the masonry on top. After waiting 5 to 7 days to allow the mortar to cure, the work can move to the remaining sections. The pieces of flashing should be spliced by lapping them a minimum of 6 in.[24] A similar procedure can be used to replace the flashing on top of shelf angles, except that it may be possible to refill the whole opening at once if the brick removed in alternating sections is temporarily shored—supported by 2-by-4 wood pieces placed on top of the shelf angles, for example. The case study in Sec. 13.7.5 illustrates a method of shoring the brick with adjustable jacks.

Flashing made of fully adhered rubberized asphalt is easier to place and patch with mastic than most other materials. It can be simply adhered to the existing backup wall, instead of cutting a recess in the wall for the otherwise required flashing termination. For these reasons,

Figure 13.23 Water penetration into this wall, evidenced by the efflorescence at the top of the brick, was traced to the loose flashing on top, which allowed wind-driven rain to enter.

replacement flashing of rubberized asphalt is often favored by renovation contractors. However, this material deteriorates when it is exposed to ultraviolet radiation and thus cannot be left protruding through the wall, as recommended in Sec. 13.5.2. The problem can be solved with the help of a stainless-steel drip attached to the edge of the flashing with mastic; this drip edge can extend through the wall.[37]

Whenever continuous flashing or lintels are replaced, difficulty in keeping the joint at the top of the replacement brick adequately filled with mortar often arises. Some mason's ingenuity is required here in order to make certain that the joint is wide enough for mortar placement and that the mortar is well packed for the whole depth.

Proper functioning of the existing weep holes is no less critical than that of the flashing. If the weep holes are plugged, they can be opened up by careful drilling, but a more reliable way is to remove some bricks and replace them with others that have unfilled head (vertical) joints on one side.

Joint sealants are equally as important as, if not more important than, flashing and weep holes in keeping the building dry. Replacement of deteriorated sealants involves cutting out the existing material, cleaning and priming the joint edges, installing a backer rod for joints wider than $^3/_8$ in or deeper than $^3/_4$ in, and placing a new sealant.

In many old brick-veneer walls, the width of the cavity is insufficient. Anything less than 2 in, the minimum required today, may not assure adequate drainage. In some cases the cavity width may be nearly zero, with one wall layer touching the other. If the exterior brick is replaced for some reason, the drainage of the cavity can be improved (this is impractical to do otherwise). Although admittedly little can be done with a zero-width cavity, a cavity between $^1/_2$ and 1 in wide can benefit from a drainage board, a product that was originally intended to promote the flow of water behind foundation walls. A drainage board can be made of dimpled plastic with a fabric filter attached on one side or of woven fabric. When placed in a cavity as narrow as $^1/_2$ in, a drainage board prevents mortar bridging and allows water drainage.[37]

What can be done about efflorescence? The remedial steps depend on its causes. Efflorescence on a brand new building is most likely caused by the salts in the mortar; when the white stains appear on an old structure, moisture penetration is probably to blame. Dealing with efflorescence may not require surgical steps, but it can be maddening. The white powder of the leached salts is relatively simple to remove, but more of it reappears.

In new construction, steps to reduce efflorescence include using low-alkali cement, and limiting admixtures in the mortar, testing brick for efflorescence potential (see Sec. 13.4.5), and adhering to the proper design and construction techniques for cavity walls discussed previously. In existing construction, these steps cannot be taken unless the veneer is replaced. The worst thing to do in this situation is to coat the wall with a sealer: The coating may indeed prevent evaporation from the face of the brick and temporarily halt its effects, but at the cost of potential brick spalling caused by vapor buildup. Also, the coating will make removal of any salt deposits that do appear extremely difficult. A better way to curtail evaporation is to remove the source of water penetration into the wall. If that is done but the problem continues, the brick itself or the mortar is to blame, and the most effective, if frustrating, solution is to continue to remove the efflorescence—or to live with it. Fresh stains can be simply washed off with water because they contain water-soluble calcium hydroxide, but when this calcium hydroxide is exposed to carbon dioxide in the air, it turns to difficult-to-dissolve calcium carbonate. Thus old efflorescence stains require diluted acids for removal.[38]

13.7.3   Reattaching brick veneer to backup walls

The most dangerous effect of moisture penetration is the corrosion of the steel ties attaching the brick to the backup wall. Corroded ties cannot transfer the wind loading from the brick veneer to the structure behind and thus make the veneer vulnerable to collapse. It would seem that repair or replacement of these ties would require taking down the wall and rebuilding it, but, fortunately, there are other solutions. Among them is installation of retrofit fasteners drilled through the face of the brick into the backup wall. To minimize the visibility of these fasteners, they are installed through the mortar joints, if those are in good condition. The fasteners can also be placed from the inside.

There are several available types of remedial anchors; three of them are shown in Fig. 13.24. The recommended installation procedures and details of construction vary among different manufacturers, but each of these anchors requires special setting tools. The helical-type anchors (Fig. 13.24a) are drilled into the supports (and for this reason are best used with concrete or solid masonry backup walls) and are secured into the brick veneer with polyester-based resin. The chemical anchors in Fig. 13.24b can be used with backup walls of hollow block, because their holding action is provided by adhesive cartridges placed into predrilled and cleaned holes. The anchor rods are inserted into the cartridges and recessed into the veneer.

The anchors in Fig. 13.24c utilize expansion-bolt principles; slightly different versions are used for hollow and solid backup walls. These unusual but very effective anchors resemble double-sided expansion bolts (Fig. 13.25). Installation involves placing the anchor with an expansion shield on one end into a predrilled and cleaned hole and tightening it with a locking key and a setting wrench. Then, the outer expansion shield is threaded onto the body of the anchor and expanded by special attachments to the torque wrench. Filling of the hole completes the repair.

The anchors in Figs. 13.24 and 13.25 are intended for masonry or concrete backup walls, but similar anchors can be anchored into steel-stud walls using self-tapping screws. The screws should receive one of

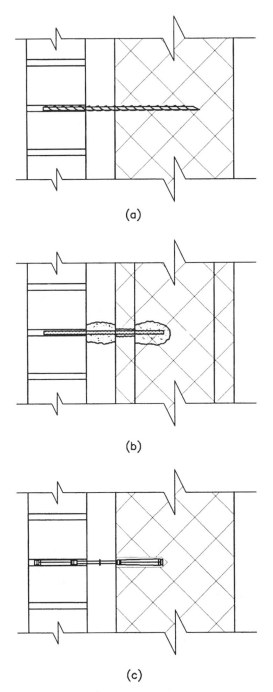

Figure 13.24 Retrofit fasteners for reanchoring
brick veneer to backup walls. (*a*) Helical-type; (*b*)
adhesive; (*c*) expansion-bolt type anchors.

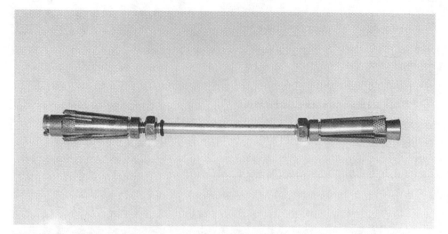

Figure 13.25 These distinctive-looking façade-stabilization anchors by Dur-O-Wall, Inc., resemble double-sided expansion bolts.

the methods of corrosion protection discussed in Sec. 13.6.3. In fact, all remedial anchors should be made of corrosion-resistant materials, such as brass or stainless steel, or at least receive a durable coating.

What can be done with the old corroded ties? If the rusted anchors cause cracking or spalling of masonry, they can be cut out with a grinder during repointing (but not before the new ties are installed). Otherwise, the old ties can be left in place.[20]

13.7.4 Wall replacement

In some cases, the problems of brick-veneer walls are so severe that a complete wall replacement is more economical than repairs. One example of this scenario involving walls of brick veneer and steel studs was mentioned in Sec. 13.6.1. But brick-veneer walls backed up by CMU are not immune from fatal problems.

One example described by Fisher[39] involves the Commodore Perry Apartments in Buffalo, N.Y. The six high-rise buildings were built in 1956. By the mid-1970s, the brick veneer was cracking and spalling, and in the late 1980s, emergency repairs were made. Still, the deterioration continued. The detailed investigation revealed an array of problems that could be traced to two major causes: The buildings had no brick expansion joints, and there were too few brick anchors. As a result of movement, the walls had bulged outward and at the corners, shearing off or weakening the few brick ties that were installed there. That allowed even more movement. The resulting cracks allowed water to enter, leading to corrosion of the ties and relieving angles and to further deterioration of the brick.

In addition, there were no soft joints under the relieving angles, and so the moisture-induced vertical expansion of the brick displaced the angles and loosened their fasteners. It also broke some of the wall ties. At the roof, this expansion pushed the coping upward and opened the cavity to the intrusion of rainwater. Numerous other design and construction deficiencies were found, among them mortar droppings that filled the cavities, and flashing that was simply adhered to the masonry backup instead of being inserted into the joints.

Prior repairs had not improved the situation and may have worsened it. Some expansion joints had been cut into the walls, but the requisite wall ties were not inserted on each side of the joints and the new weep tubes did not actually extend into the cavity. This time, because of the multiplicity of problems, the designers and the owner made the wise decision to completely replace the walls with a properly designed brick-veneer system rather than undertake another futile round of incremental repairs.

In another project, the author investigated a cracked and slightly bowed brick-veneer wall backed up by CMU. The wall, built in the mid-1970s, was covered with efflorescence and was reportedly leaking. Some mortar joints at the top of the wall had deteriorated. A test cut revealed that the brick ties were too short and were not embedded into the brick, meaning that the veneer was laterally unbraced. The wall above the veneer was made of wood siding, and the metal flashing between the two walls had pulled away, allowing water into the brick veneer (this wall was shown in Fig. 13.23). The other areas of the building did not show any signs of deterioration or leakage. The proposed solution was to remove both the brick and the wood siding above it and rebuild the whole affected area with properly anchored brick veneer.

13.7.5   Repair of a brick-veneer façade: A case study*

The Calvert Building in Baltimore, Maryland, suffered from many of the typical problems of early brick-veneer/steel-stud wall systems. This five-story, steel-framed building with a 360- by 200-ft footprint was built in 1981. The exterior wall backup system consisted of prefabricated panels made of steel studs, gypsum sheathing, and polyethylene waterproofing sheets. The panels were welded to the building frame, followed by installation of the brick veneer and windows. The windows were of both horizontal strip and punched-opening type with aluminum

---

*The author is grateful to Mr. James C. Myers of Simpson Gumpertz & Heger Inc., Arlington, Mass., for supplying this case study and the accompanying illustrations (Fig. 13.26a through d).

frames. The building corners at the top floor were truncated to form triangular patios with perimeter planters. Limestone strips accented the brick veneer at the corner patio planters, below the roof coping, and at the lower spandrels on an elevated portion of the building over a parking area. Here are some of the problems that the investigation uncovered and the solutions implemented to correct them.

Field investigation and its findings.   The investigation included surveying the exterior and interior conditions along the walls, making over 100 openings to verify the concealed wall construction conditions, and spray-testing 50 areas to determine the paths and severity of water leakage (Fig. 13.26a). Water leakage and structural distress were detected in several places.

(a)

Figure 13.26   (a) Testing building façade for leakage using a spray rack.

(b)

Figure 13.26 *(Continued)*
(*b*) Bowing of brick veneer at this corner has distorted brick ties and fractured anchor screws.

(c)

Figure 13.26 *(Continued)*
(*c*) Elevation of the building after repairs with new limestone bands above and below windows and at corners.

(d)

Figure 13.26   (*Continued*)
(*d*) Installation of new window head flashing and steel support plates between studs while
supporting the brick wall with screw jacks.

Water leakage at window heads.   Water leakage into the building from
above the strip windows was a large source of the widespread interior
staining. Water leaked inside swiftly and voluminously during rain-
storms, as well as during the water tests, often dripping from the window heads within a minute or two after simple wetting of the wall. The
typical leakage path was as follows: Water penetrated the brick veneer
and drained down to the copper-paper through-wall flashing on top of
the shelf angle above the strip window. The sealant placed over the
shelf angle acted as a dam and prevented the escape of the water,
which then bypassed the concealed front edge of the flashing and
drained inward on top of the window head.

Windowsill anchorage.   The bottom of a typical strip window was simply
screwed into the top of a brick soldier course. The soldier course was
placed on flashing, which interrupted the mortar bond and separated
the soldier bricks from the rest of the veneer. Since the soldier bricks
were not separately tied to the stud wall, they—and the windowsills
they were intended to support—were not adequately anchored.

Corner distress.   The brick veneer was found to be cracked and bowed
near the building corners. The distress was attributed to the permanent

moisture-induced expansion of the veneer walls owing to most of the control joints in them being clogged by mortar accumulation. These large veneer movements had pushed out the opposing walls at the building corners and in some areas nearly eliminated the brick bearing on the shelf angles. This, in turn, had cracked the brick veneer and distorted and damaged the brick wall anchors, bending some of them sideways, buckling others, and fracturing anchor screws (Fig. 13.26b). The movement had also twisted the steel studs in the backup wall and broken the welds connecting the backs of the studs to the steel building frame.

The proposed façade repairs. The remedial plan included the following repairs:

- Installing through-wall flashing at the heads and sills of strip windows
- Reanchoring strip windowsills
- Reconstructing building corners
- Installing new strip windows on the fourth floor in an area of existing solid wall

Some repair challenges. The proposed repairs had to resolve the following challenges, among others:

- The repairs had to be done with minimal disruption to building occupants. This included not only the noise associated with demolition, but also such things as not removing the windows to continue to allow the occupants to use the space, and coordinating site access to minimize loss of parking. Unlike the situation with new construction, these nontechnical occupancy constraints can often dictate the manner of repairs.
- Since the entire wall was not removed, the details had to allow the wall to be supported temporarily while the repairs were being made.
- The replacement masonry had to match the existing masonry as closely as possible.

The solutions

Selecting material for façade repair. Since large areas of brick masonry had to be removed to install the new wall flashing and rebuild the corners, replacement brick was needed to patch the demolished areas. Unfortunately, the original manufacturer no longer produced the brick used in the building, and other manufacturers were unwilling to produce a small volume of special brick. Instead of settling for an imperfect match, the decision was to highlight the repair areas by introducing contrasting limestone. The limestone added interesting variety to the façade and echoed the existing limestone at the roof coping, fifth-floor patio walls, and second-floor spandrels (Fig. 13.26c).

Temporarily supporting the walls. A common method of temporarily supporting a wall during remedial work is to remove the brick in alternating 5-ft-wide sections ("leg-and-leg"), so that the remaining sections support the wall. After installing the flashing in the open sections and rebuilding the masonry, the contractor repeats the process in the remaining sections to provide a continuous flashing. This leg-and-leg approach is time-consuming, because all operations must be repeated twice in order to complete an area and because waiting for the mortar to cure delays the start of demolition in the alternate areas.

To improve production rates on this project, another approach was taken. After 5-ft-wide alternating pockets of brick were removed, adjustable screw jacks from ordinary pick-up trucks were placed between the existing shelf angle and the masonry (Fig. 13.26d). Once the jacks were supporting the wall, the masons removed all of the remaining brick, creating a clear work area in which materials could be installed in a continuous manner. Then, other jacks were placed under the completed sections while the flashing at the original jack locations was installed. This approach avoided delays while waiting for the mortar in the "legs" to cure. It also allowed each trade to complete its work during a single visit to the area.

Installing flashing at the window heads. At the window heads, the workers removed the existing brick masonry and flashing and installed more than a mile of new stainless steel through-wall flashing. In contrast to the existing flashing, the new flashing extended through the wall and terminated beyond its outer plane with a drip edge to prevent water from returning into the wall. The drip edge was hidden from view by the limestone above, which slightly protruded beyond the face of the brick veneer.

Anchoring the limestone. Unlike brick anchorage from studs into the bed joints, the limestone anchors had to be located at each vertical stone joint and required discrete points of attachment. (The usual method of anchorage into the bottom of the stone would reduce the reliability of the flashing and increase costs; this method was not selected here.) Since the existing studs generally did not coincide with the joints in the stone, a continuous line of supports was needed. The remedy included attaching a steel plate to the shelf angle between the studs (Fig. 13.26d). The anchor brackets were fastened through a rubber gasket to this steel plate placed behind the stone at the vertical joint locations.

Repairing the window sills. The same screw jacks that were used at the window head supported the windows temporarily at each window mullion, allowing the windows to remain in place and reducing occupant disruption. At the strip windowsills, a stainless steel flashing extended down from behind the sill and through the wall beneath the limestone. A steel bent plate supported the windowsill and connected it to the existing stud wall. During installation, the steel plates were placed between the jacks, and then the jacks were removed.

Reconstructing the corners. The corner repairs included removing the existing brick in 6-ft-wide vertical strips on each side of the corner, rebuilding the structural supports (anchoring the cantilevered shelf angles and replacing the studs), and installing limestone panels with new control joints (Fig. 13.26c). To prevent future veneer movement from accumulating at the corners, new $3/_4$-in-wide joints were cut at all existing brick control joint locations.

Summary. The multimillion-dollar, two-year façade repair program implemented at the Calvert Building successfully solved the building envelope problems. The repairs stopped water leakage to the building interior and stabilized the walls at the corners. While, as is typical, the remedial work was tedious and difficult, supporting the wall with jacks allowed for good production rates and allowed the owner to maintain operations. This project handled the common problem of blending the patched areas into the existing façade by using limestone to treat the repairs architecturally and improve the overall façade appearance.

13.8   Repairing Stone and Stone-Panel Walls

In this section, we conclude our discussion of renovating masonry walls. Masonry includes brick, stone, and CMU. Having already addressed the problems of brick (both solid and veneer) and CMU walls, we will take a closer look at stone-clad buildings. Like brick, stone can be found in two configurations: solid masonry and stone veneer, the ultimate version of which is thin-stone panels.

13.8.1   Solid stone walls

Buildings with stone loadbearing walls were among the very first structures made by humans. Almost all ancient buildings that remain standing today have loadbearing stone masonry exteriors, and many of those have survived in very demanding environments. Figure 13.27 shows an 1836 stone building on the Cape Cod, Massachusetts, waterfront. Despite being exposed to frequent high winds and salt spray, the building is still in use, although it has evidently required many mortar repairs.

Figure 13.27 Candle House, Woods Hole, Massachusetts, is an example of solid stone construction. Built in 1836 and located close to the ocean, it has undergone many repairs to mortar and wood elements but is still in use.

The basic processes of stone aging have been known at least since Vitruvius. The causes of stone decay are both natural (weathering and freeze-thaw cycles) and human-made (staining, pollution, application of coatings). With the passage of time, the sharp edges become dull and then are totally washed away as the stone is eroded. The rate of deterioration greatly depends on the type of material, its age, and the environmental exposure. The range of stone condition could run from good to failing. At one extreme is stone masonry that is structurally sound but simply dirty. (Some general methods of cleaning masonry are addressed later.) At another extreme, the material loses its integrity and crumbles, a common fate of brownstone, limestone, dolomite, and sandstone. All those are more vulnerable to water-caused degradation than igneous (granite) and metamorphic (marble) stones.

Brownstone, of which many monumental buildings and townhouses were built in the Northeast between 1860 and the 1920s, is a particularly soft material that is prone to absorption of moisture and deterioration. Its main advantage was its ease of carving, but it has proven too soft to be durable. Brownstone is a brown variety of sandstone, a sedimentary material, and is essentially a cemented sand. Like other sedimentary stones, brownstone was deposited in horizontal layers on the bottom of the ocean. It is strongest when it is placed in a building the same way, with bedding planes horizontal, and is weakest when it is loaded parallel to the bedding planes. Unfortunately, many builders in the nineteenth century did not realize this fact and placed brownstone pieces with bedding planes vertical to expose their larger sides. As a result, this porous material was eroded easily by freeze-thaw cycles and tended to delaminate along its bedding planes and flake off.

Biggs and Keating[40] tell about the rehabilitation of a historic brownstone house covered with intricate carvings. Degradation of brownstone had led to pieces falling out, and the owner considered stripping the stone completely. Instead, the house was repaired, although a large percentage of the façade had to be removed and replicated with a synthetic patching material that was carved in place. The surface patching material was a two-component, latex-modified cementitious compound that closely matched the brownstone in color and texture. Behind it was placed a less expensive base material made of polymer-modified cementitious mortar.

The famous French Library in Boston, where as much as 70 percent of the brownstone ornamentation had been lost as a result of weathering and ineffective prior renovation efforts, was restored with custom-formulated repair mortar. The mortar was applied with a trowel, allowed to set for 1 to 1.25 h, and formed with a straight edge or raked back with metal templates that matched the original moldings. It was applied after the surface had been cleaned of old plaster and coatings, although some of those were left in place because they were nearly impossible to remove. Fortunately, percolation tests of the coating have shown it to be water- and vapor-permeable, and, since the repair mortar was also "breathing," delamination of the patch material owing to vapor buildup was not considered likely.[41]

There are several repair options for patching stone. Typically, they start with removal of all deteriorated stone areas and preparation of the substrate as for concrete patching, described in Chap. 5. One repair option is to make a wedgelike cut in the remaining stone and build up the original profile with a repair mortar specially formulated for the task, such as an epoxy mortar mix. To complement the wedge action, additional mechanical anchorage can be provided by stainless steel pins. For example, pins of $\frac{1}{8}$-in diameter can be set into $\frac{3}{16}$-in drilled-in holes filled with epoxy mortar. A large patch can be reinforced with stainless steel wire mesh (Fig. 13.28). The pin spacing depends on the application; in the repairs of the French Library in Boston, the spacing was 4 in.[41]

To repair spalled corners and edges, a slightly different approach can be used. For large spalls, the corner can be formed and anchored

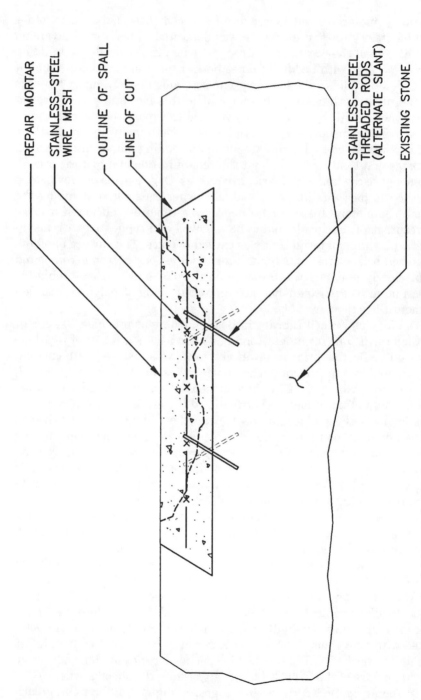

Figure 13.28 Repair of deteriorated stone by patching. (*After Michael J. Scheffler and Deborah Slaton, "Stone Claddings," Building Renovation, March–April 1993, pp. 53–56.*)

REPAIR MORTAR

STAINLESS–STEEL WIRE MESH

OUTLINE OF SPALL

LINE OF CUT

STAINLESS–STEEL THREADED RODS (ALTERNATE SLANT)

EXISTING STONE

with the same stainless steel pins as in the preceding case. For smaller spalls, where pins cannot be installed, small keys can be drilled into the stone for anchorage. Gere[42] suggests drilled-in keys of $3/_{16}$ in diameter and $3/_8$ in deep spaced 1.5 in apart. Once the repair mortar sets, its surface can be given a texture matching that of the surrounding stone. To repair stone with a polished surface, a polyethylene liner can be used under the formwork. Cracks in stone can be saw cut to a depth of $1/_2$ in and filled with the patching material.[43]

A relatively new repair approach is to use so-called consolidants, a class of chemicals that penetrate into the stone and bind its grains together. The use of consolidants is considered controversial by some experts, who worry that if the chemicals are allowed to penetrate too deeply, they could make the problem worse.[44] However, consolidants have been successfully used in a number of brownstone repair projects.

To repair stones that are loose, but otherwise in good condition, an adhesive foam or epoxy can be used to resecure them in place. The foam, a relatively recent invention, can be expected to last for at least 20 years.[2] A piece of stone that has cracked to the point of nearly total separation can be reattached by methods similar to those used to reattach brick veneer. For example, small holes can be drilled through the cracked area into solid stone, and, after cleaning, stainless-steel pins can be placed into the holes, which are then filled with epoxy. The pins should engage both pieces of stone. The exposed holes are then filled with a repair mortar. Additional information about this procedure can be found in Gere.[42]

A serious situation arises when stone deterioration has gone so far that replacement of some or all of the masonry is the only option. In some cases, the appearance of stone can be replicated with other materials. For example, Chicago's Reliance Building, mentioned in Sec. 13.2.2, had its deteriorated terra cotta cornice replaced with a cast aluminum one—presumably, a much safer solution for the pedestrians.

13.8.2   Thick-stone panel walls

The foregoing discussion applies to both solid stone and thick-stone panels. Many multistory buildings constructed in the first half of the twentieth century are clad in thick stone panels of limestone or granite. The panels, 4 to 6 in thick, are typically attached to solid masonry backup walls with metal anchors embedded into the panel joints.

The most common mode of panel distress is not erosion but cracking caused by poor anchorage details or inability of the panels to expand and contract with temperature and moisture changes. The latter is exemplified by problems in the 500-ft-tall Metro Dade Center in Miami, where limestone panels cracked and a piece fell down only two years after construction. The investigating engineers from Simpson Gumpertz & Heger Inc. determined that most of the problems were caused by poorly designed and constructed attachments of panels to the concrete frame. The attachments did not permit differential movement between the frame, which tended to shrink and creep, and the panels, which were undergoing thermal expansion and contraction. The thermal ratcheting phenomenon described in Sec. 13.2.2 was also at work, as the joints between the panels were solidly mortared and expansion "soft" joints were lacking. The repairs included cutting in horizontal and vertical soft joints and replacing 44 badly damaged panels.[45]

Another common problem with thick-stone panels is spalling caused by corrosion of the metal anchors securing the panels to the masonry backup walls. Figure 13.29 illustrates a typical spall in a limestone panel and the rusted anchor that caused it. As discussed in Chap. 5, rusted steel drastically expands in volume, applying bursting forces to the surrounding material, in this case the stone. The monumental building in which this occurred was constructed in the 1930s and had several such spalls. The author was able to remove easily the cracked limestone pieces, some weighing more than a pound, by hand. Needless to say, this constituted a serious safety hazard. Some previously spalled areas had been patched. During field investigation, the patches were checked by hammer sounding. Some sounded hollow, some didn't.

Figure 13.29 Spalled limestone panel at a rusted anchor.

Why were this particular anchor and several others corroding, but not nearby anchors? One clue was the previously repaired mortar above the spalls, suggesting a leak at those locations. Another avenue of water penetration into the wall was traced to missing cap flashing at the top of the parapet, which left the collar joint unprotected.

The recommended repairs for this condition include drilling out the rusted anchors, to prevent their further destructive action, and making new attachments to the backup walls. At parapets and other places where the backup wall is accessible, it can be rebuilt in place, with new stone anchors inserted in the stone joints. Most often, the repairs must include embedding face-drilled anchors into the backup wall (some of these are shown in Fig. 13.24) and fixing the spalled areas with repair mortar as described previously. Mortar manufacturers will frequently work with designers to prepare a mix to match the existing stone.

13.8.3 Thin-stone panels

Beginning in the 1950s, façades of solid stone construction were being increasingly challenged by curtain-wall stone panels. These panels typically consisted of a thin layer of stone attached to the backup system with stone anchors. The backup could be a steel framework (such as a steel truss system or panels made of steel studs), precast panels, and CMU or concrete backup walls. These thin-stone panels eventually became the dominant method of stone cladding. As the technology of stone-cutting improved, stone panels as thin as 1 in could be produced. In some extreme cases, stone cut as thin as a slice of bread was put on buildings. (The author was involved in the design and production of panels with $1/_2$-in-thick stone sheets, although that was a *synthetic* stone.)

As with brick veneer replacing solid brick, the economy of the new system was obvious, but some of its pitfalls were not. As a result, there were several much-publicized failures of thin-stone panels, particularly those made of marble. When in 1985 three marble panels fell from a 23-year-old high-rise building in Kansas City, the owner replaced all 4400 panels with a glass curtain wall. The 27-story Lincoln Tower in Rochester, New York, in 1973 among the first high-rise buildings clad in thin marble, was eventually reclad in aluminum. Amoco Corp. decided to replace the 1.25- and 1.5-in-thick white marble panels of its 80-story Amoco Tower in Chicago with 2-in granite panels.[46,47]

All these and other cases involved the same problem: The thin marble panels bowed and warped. Some panels disintegrated. Some loosened from the supports. The researchers studying these failures concluded that under repeated temperature and moisture cycles, marble gradually weakens and warps. One explanation of this phenomenon: The exterior layers in marble expand during temperature increases but do not fully contract when cooled. Since the interior surface is heated or cooled less, it moves less, and eventually the panel bows. The thinner the panel, the more pronounced the warping. Marble also suffers from some other problems that make it unsuitable for façade applications, such as flaking and disintegration under acid-rain conditions. The properties of marble are extremely variable, and some specimens are much more susceptible to problems than others.[47]

Of all the stone materials used for cladding, marble is near the bottom of the weathering durability scale, granite is near the top, and sandstone and limestone are near the middle. Polishing can somewhat increase the durability of any stone, because it reduces the surface porosity. In addition to the already mentioned bowing and weathering of marble panels, thin-stone panels in general can suffer from several other common problems. Some of them are the same as in other curtain walls[48]:

- Distress from large deflections or rotations (twists) of the support structure, or from its deterioration
- Cracking and other distress caused by inability of the stone to move with temperature and moisture changes (such as when stone is rigidly attached to a precast backup)
- Fracture of stone at points of stress concentration
- Moisture penetration when panels are made as barrier systems without cavity or flashing

Other problem areas are specific to stone. Because the stone is so thin, it is susceptible to damage during handling and to mechanical damage when installed. Figure 13.30 shows a broken stone panel over an overhead door; the damage was probably caused by a backing vehicle. Also, by their nature, some stone materials absorb moisture, sometimes with drastic consequences. In 1996, one building in Stamford, Connecticut, only 9 years old at the time, had to be stripped of its limestone cladding because the cladding had absorbed so much water that it was certain to fall off. The building was reclad in granite at a cost of $5 million.[11]

Thin-stone panels are usually secured to the backup with stone anchors inserted into sawn kerfs in the edges of the stone. For several reasons, this is the most troublesome location in the panel. First, the thickness of the stone at the kerf is less than one-half that of the already thin stone slab, and both mechanical damage and environmental degradation are felt most acutely here. Second, the anchor, being a discrete point of support, introduces stress concentrations into this area. Third, if they are not made of stainless steel, the anchors will

Figure 13.30 Breakage of a thin-stone panel.

eventually corrode, and the resulting rust will break the edges of the kerf (and, of course, the anchor will lose its holding capacity). And finally, water can enter into the kerf, freeze there, and cause its fracture.

Many experts realize that the edges of the stone panels are the most critical areas of the wall. Larkin[49] observes that panel designers tend to focus on flexural design of thin stone, even though midspan flexural breakage is rarely a problem. The far greater threat is localized edge damage caused by poor anchor design or poor fabrication and installation. The most dangerous condition is when the stone is spalled on the back side: The damage is hidden from view but can result in the stone falling out. Field-cut kerfs and drilled holes are especially troublesome, because the margin of error is so small. Larkin also notes that damage to thin-stone panels often occurs during delivery and erection, when panels are typically subjected to stresses exceeding those encountered in service. One of the examples he mentions is a 26-story university building in Chicago, where 15 percent of its 2-in-thick limestone panels suffered irreparable damage during shipping and handling.

Occasionally, cracking and spalling of field-applied stone panels is caused by erection shims left in the joints by the installers. The shims act as hard points of restraint, preventing the stone from expanding with temperature and moisture changes.[50] If left in horizontal joints, shims act as unintended points of concentrated support, loading the stone in the wrong places. If the panel joints are caulked, these left-in-place shims are very difficult to find.

Unlike solid stone and thick-stone panels, broken thin-stone panels are nearly impossible to repair. Most often, the damaged panels are replaced, although in some cases the cracked or broken edges are fixed with epoxy, leaving visible patches. Of course, if the problems are widespread, the whole cladding system needs to be replaced, as was done in the cases mentioned previously.

When the problems involve panel attachments, repairs may have a better chance of success. In one widely publicized case, the spandrel marble-cladding panels in the 31-story FirstCity office tower in Houston were resecured with a waterproofing foam. The procedure started with removal of a capstone covering the top of the spandrels and installation of a backer rod into the 1-in cavity between the concrete spandrel beam and the back of the panel. Then, a mix of a patented polyurethane, water, and a catalyst was poured into the cavity. The resulting foam filled up the cavity within minutes. The engineers who pioneered this solution are experimenting with other methods of application.[51]

13.8.4 Cleaning stone and masonry walls

Stone and masonry can be cleaned with various degrees of effectiveness and abrasiveness. As a rule, the gentlest method capable of doing the job should be tried first. The effectiveness of various methods can vary, and the best way of verifying the effectiveness of a particular method is by testing in an inconspicuous area. Crude mechanical methods such as sandblasting or grinding are certainly effective in removing dirt, but they can cause irreparable damage to stone masonry and should be avoided if possible. The "kinder, gentler" methods include water cleaning, which typically involves soaking the stone, scrubbing it, and rinsing. Experienced contractors know which water pressure, temperature, and method of spraying are appropriate for various conditions and can make the necessary adjustments if required.

Stone façades suffering from the attack of pollution may be difficult to clean with water alone; in this case, chemicals are used. Field testing of various compounds and concentrations can determine the weakest weapon that will work. Chemicals applied to a dry stone wall can burn it, and the wall in the area being worked on and below must be kept continuously wet during chemical application. The cleaning solution can be applied by spraying and brushing. Typically, the chemicals are allowed to "dwell" on the surface for 3 to 5 min; then the additional materials are applied, and the whole application is rinsed off. Improper application of chemicals can result in brown, yellow, and green streaks. Brown stains contain manganese; green and yellow streaks, vanadium salts. All of these may be leached onto the surface of the wall by overly aggressive acid cleaning.[44]

To remove encrusted soot and grime from urban masonry buildings, several steps may be needed. This was the case, for instance, with the historic terra cotta façade of Chicago's Reliance Building. The terra cotta was first coated with a weak alkaline gel, which was allowed to dwell, and then was scrubbed and rinsed. Then, a weak acid wash was applied to neutralize any remainders of the previous chemical and again rinsed. Another water rinse followed. Sometimes the process had to be repeated two or three times.[10]

The removal of paint from masonry may be the most difficult of all the cleaning chores. Old paint is likely to contain lead, and its removal typically requires involving specialized contractors and sealing the area. In many cases, this circumstance alone may prompt the decision to leave the building as is, but if the paint still must be removed, proper procedures are a must. Above all, sandblasting of brick should almost never be done, because it tends to remove the hardened brick "skin" and lead to rapid deterioration, as already stated. Chemical paint removers can work effectively while sparing the brick's integrity. Finding the right paint stripper takes the expertise of a knowledgeable contractor and some trial and error.

There are two main types of chemical paint removers for masonry: organic solvents and alkaline strippers. Perhaps the most effective organic solvent is methylene chloride, although it has come under attack for being a suspected carcinogen, being harmful to the ozone layer, and producing hazardous waste that requires proper disposal. Still, methylene chloride works fast and can remove most types of paint by dissolving their organic components. It is typically applied with a brush or roller, allowed to dwell as required (usually for 15 to 30 min), and rinsed off. Other organic paint removers are not as effective or versatile as methylene chloride but are safer and thus are likely to overtake it in popularity in the future. Alkaline (caustic) strippers are based on lye (potassium hydroxide) or on sodium hydroxide. They work best for latex and linseed oil–based paints and can be applied as sprays, gels, or paste (poultice). To prevent efflorescence, alkaline strippers must be neutralized by an acid-based rinse. These chemicals, too, are highly toxic.[52]

13.9 Rehabilitating Exterior Insulation and Finish Systems

13.9.1 The available systems

Invented in Europe, exterior insulation and finish systems (EIFS) were introduced into this country in the late 1960s. For a while, the product was called by the name of the company that pioneered its application in the United States: Dryvit Systems, Inc. Within 20 to 30 years, similar systems were available from many other companies. Today, these companies are represented by a trade organization called EIFS Industry Members Association (EIMA), which publishes guideline specifications, technical notes, and other useful information about the product.[53]

Exterior insulation and finish systems can be field-applied to masonry walls or panelized with steel-stud and sheathing backup. The panelized system typically includes steel studs, exterior sheathing, rigid insulation attached to the sheathing with adhesive, base coat, reinforcing mesh, and finish coat (Fig. 13.31). The last three components are sometimes called lamina.

There are two generic classes of these systems, differentiated by the type of base and finish coating and methods of attachment. Class PB (polymer-based) systems are made with thin, flexible coating materials; the insulation board is attached to the substrate with adhesives. By contrast, Class PM (polymer-modified), or "hard-coat," uses cementitious products and requires mechanical attachment of mesh and insulation. Today, polymer-based Class PB systems are far more popular than the thicker Class PM products, accounting for over 80 percent of all EIFS. Some typical design details of Class PB systems can be found in Ref. 54.

The rigid insulation commonly used for Class PB systems is molded expanded polystyrene (MEPS), an open-cell material also known as bead board. Class PM systems generally use extruded polystyrene (XPS), a closed-cell material found in, among other things, disposable coffee cups. Closed-cell extruded polystyrene has much better vapor-retarding properties than open-cell molded expanded polystyrene.

EIFS have enjoyed tremendous popularity because of several advantages, among them low cost and excellent insulation value. Architects celebrate the design flexibility of the systems, which allows a variety of shapes and surface textures to be used on projects with tight budgets. Contractors appreciate the fact that the systems can be applied over existing surfaces and manufactured in panels with relatively inexpensive light-gage steel framing backup. But, like the early versions of brick veneer, the systems became popular too fast, before their weak points were well understood. Indeed, even at the turn of the millennium, there was no comprehensive ASTM standard covering EIFS. There are, however, two good sources of information from the ASTM: Manual 16, *Exterior Insulation and Finish Systems: Current Practices and Future Considerations,*[55] and STP 1187.[56]

The high failure rate of this system has been well publicized, from case studies in the technical literature to investigative reports on TV. Many lawsuits have been filed alleging defective design and construction of EIFS; one of the most famous is in North Carolina, where water

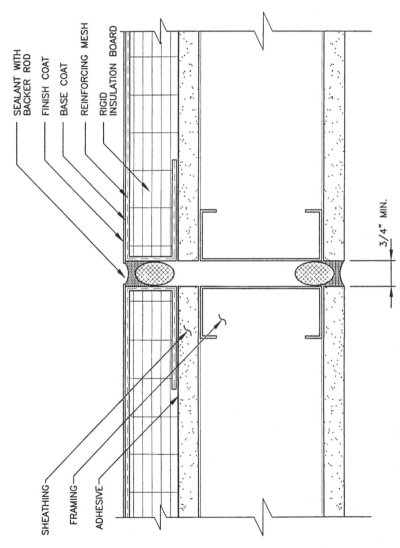

Figure 13.31 Typical components of EIFS and the expansion joint in type PB panels.

SEALANT WITH
BACKER ROD

FINISH COAT

BASE COAT

REINFORCING MESH

RIGID
INSULATION BOARD

SHEATHING

FRAMING

ADHESIVE

3/4" MIN.

damage has prompted removal of the skin from about 250 homes. Similar widespread water damage to structure behind EIFS was observed in at least six states. As a result, some banks no longer insure home builders who use this product.[57,58] Property appraisers are trying to put a monetary value on the "stigma,"as they call it, associated with this type of construction.[59]

On a personal note, the author was intimately involved with production of EIFS panels in the early 1980s and vividly remembers salespeople stomping on the product samples to demonstrate how the plastic mesh made them "virtually indestructible." Many of the applications later failed, but not because of the strength: The finish peeled off like a snake's skin.

Why such a sorry fate? There are several causes. First, the early materials may have been less than satisfactory. There have been reports that the first EIFS introduced in the United States were not nearly as good as their European siblings.[60] Second, the brittle Class PM products simply were not suitable for the flexible backup (wood and steel studs) with which they were used. Figure 13.32 shows a failed EIFS wall with a cementitious finish that has separated from the substrate, cracked, and broken up. Third, the installation of this product is not as simple as some contractors believed. (Some fine points are mentioned in the next section.) Finally, and perhaps most importantly, the early systems functioned as barrier walls or, as some would call them, surface-sealed covering.

As discussed in Sec. 13.4, a 100 percent effective barrier wall is practically impossible to construct; some leakage will always occur through cracks, openings, and penetrations. Typically, water enters the EIFS wall around window and door openings and seeps through the wall into the stud cavity. Its escape is blocked by the vapor barrier on the inside wall surface and by the EIFS on the outside. Trapped moisture leads to decay of wood components. (In a widely reproduced picture, a huge mushroom is growing on the inside of an EIFS wall.) Sadly, the original European applications used masonry and concrete backup with a drainage cavity, while the American installations used wood-frame backup without any provisions for drainage. Also, the real stucco that EIFS was intended to supercede used building paper as waterproofing, but the imitation dispensed with it.

13.9.2 Building better EIFS walls

Contemporary EIFS products have been much improved over the first versions, with better materials and better technology. Most EIFS manufacturers have introduced versions of cavity wall design that claim to provide drainage within the cavity. How well these newer products will

Figure 13.32 This Class PM EIFS wall has separated from the substrate and failed.

do over the long term remains to be seen, but in the interim here are a few guidelines for assuring better performance of EIFS walls, as suggested by several sources[61-65]:

1. Regular exterior-grade gypsum sheathing, while allowed by the EIMA guidelines, can absorb moisture and lead to system disintegration. Instead, use cement fiberboard or proprietary products conforming to ASTM C1177, such as Georgia-Pacific's Dens Glass Gold.[*]

2. Even the flexible coatings of Class PB cannot bridge large gaps in the substrate without cracking. A common source of cracks is the joint between pieces of rigid insulation, which should be filled with slivers of insulation rather than being filled with base coat or adhesive or simply left unfilled. Figure 13.33 illustrates an attempted repair of this cracking that did not please the owner.

3. The cement content of the base coat should not exceed 33 percent, although there are many products on the market with a 50 percent cement content. The added cement makes the coating brittle. In

---

[*]Dens Glass Gold is a registered trademark of Georgia-Pacific.

Figure 13.33 Quite noticeable crack repairs in an EIFS wall.

addition, the extra alkalies may overcome the alkali-resistant coating on the reinforcing mesh and cause its corrosion.

4. EIFS should not be used as roofing, on surfaces with a slope shallower than 45° or as moldings, cornices, or belt courses, because the top coat tends to soften with prolonged exposure to water. EIFS work best on vertical walls.

5. To prevent delamination, the EIFS finish should be attached to the substrate with both mechanical fasteners and adhesives. This "belt-and-suspenders" approach can save the wall for a time from the dangers brought by water penetration and poor surface preparation.

6. The commonly used base-coat thicknesses of $^1/_{16}$ and $^1/_8$ in are inadequate for moisture protection. It is best to use a base coat at least $^3/_{32}$ in thick and apply the coat in two layers.

7. One of the most common areas of water penetration is caulked panel joints. To be effective, these should be constructed as explained in a later section.

8. Avoid using dark colors. Dark wall surfaces that receive direct sunlight or are subjected to reflections from adjacent windows can heat up to temperatures that will melt the molded expanded polystyrene rigid insulation.

9. Provide expansion joints in EIFS following the suggestions for size and spacing listed for brick veneer.

10. Avoid aligning joints in insulation board or in sheathing with the edges of doors and windows or with each other—this could cause cracking.

11. Some experts[64] recommend that EIFS systems use a complete flashing system, like any other curtain wall. This means parapet copings and through-wall flashing at wall openings and penetrations as well as at horizontal joints.

Not everybody agrees with all these suggestions, but one thing is clear: There are plenty of pitfalls awaiting the specifiers and installers of these deceptively simple structures.

13.9.3 Renovating EIFS walls

The first step in the renovation of EIFS walls is to make certain that the material is not real stucco. EIFS can be betrayed by the reinforcing mesh protruding through the wall surface in some places. Otherwise, striking the wall with a fist should help: A somewhat hollow sound identifies EIFS, while a solid sound means the real thing. Also, rigid insulation will deflect under pressure, while real stucco will not.

In most cases, rehabilitation of EIFS walls involves their removal and replacement. This simple solution is easy to justify when the structural backup has deteriorated. This occurred, for example, in the North Carolina case mentioned previously, in which moisture was trapped behind the EIFS. In cases where the damage is only local—caused by the impact of a golf ball, for example—and the backup structure is not affected, EIFS can be repaired by cutting out a small piece of the wall and replacing it with new material. The biggest problem here lies in assuring proper adhesion of the repair and making the waterproof joint between the repaired area and the rest of the panel.

Leaks in EIFS walls are investigated in the same manner as those in other curtain walls (Sec. 13.3). Water entering an EIFS might result in a number of problems, including degradation of the paper facing on the gypsum board and its subsequent delamination. Once inside the wall, water can travel between and perhaps even through the pieces of rigid insulation and collect near the sealed joints, leading to their failure. A prolonged moisture attack can corrode the steel-stud backup.

Nelson[64] recognizes that this system has not been around long enough for reliable techniques for its repair to be developed. Still, he offers the following suggestions for repairing EIFS, based on the best available practice:

- *Repairing lamina.* Damaged lamina (an assembly of the base coat, reinforcing mesh, and finish coat) can be repaired as follows: First, the EIFS is removed to the sheathing. The deteriorated sheathing is replaced as required, even beyond the originally intended area if necessary. Then, a 4-in-wide strip at the edges of the repair is ground off to remove the top coat and expose the base coat. The new pieces of polystyrene are tightly fitted next to the existing ones to fill the opening and covered with a base coat and mesh, with the mesh overlapped at least 4 in onto the adjacent base coat. The new top coat and finish are then applied to match the existing. It may be necessary to repaint the whole panel for proper color matching.

- *Panel recoating.* Faded, overly porous, or repaired EIFS walls can be recoated. There are many coating formulations on the market, ranging from acrylic-based paints to heavy finishes akin to texture paints. Recently, urethane and silicone coatings were introduced. The two primary concerns in specifying EIFS recoating are the material's compatibility with the given EIFS formulation and its ability to "breathe." As Nelson states, even products advertised as permeable may retard vapor flow through the system and cause damage to the wall.

- *Sealant replacement.* Replacement of sealants should follow the usual procedures, with the caveat that top coats are easily damaged during sealant removal and joint cleaning. When the size of a joint is insufficient or its surfaces are damaged, the joint can be reconstructed.

13.9.4   Repairing EIFS joints

Joints in EIFS material tend to be especially troublesome, perhaps because previous practice was not always conducive to their long-term durability. In the words of the noted architect and author Christine Beall, "Properly installed sealant joints may in fact be the single most important element in ensuring durability of the system."[65] Why is this so?

Since the sealants adhere poorly to rigid insulation, they must be placed against the coating. In previous practice, the whole lamina was carried around the joint corners, and sealants were placed against the finish coat. Being water-based, finish coats tended to soften with prolonged exposure to moisture, which was frequently trapped behind the sealants. Moisture was also retained by the open-cell backer rods, which acted as sponges. When the panels contracted with temperature changes, the tensile stresses between the sealant and the softened finish coat often resulted in delamination between the finish and base coats of EIFS. This, in turn, allowed moisture entry inside the wall and into the gypsum sheathing, and perhaps even beyond.

To repair deteriorated joints, some surgical action may be needed. If the existing joint is of the proper size and deterioration has not yet progressed into the sheathing, the repair may be as simple as removing the sealant and all traces of the finish coat from within the joint and installing a new sealant. The best method of removing the finish coat is by scraping it away with a scalloped utility knife or similar tool. Mechanical grinding should be avoided, because it can generate a lot of heat and melt both the finish and the insulation. Solvent cleaners should also be avoided, as they can penetrate the base coat and damage the polystyrene insulation.[65] Another common method of repairing EIFS joints is to cover them with a bridge seal made of silicone sheet.

If the joint is of an insufficient size or its edges are damaged, joint reconstruction is required. One method is to remove a strip of the EIFS on each side of the joint or around the opening, place new strips of rigid insulation (or special plastic J-molds) at the edges, and cover them with a base coat and fine fiberglass mesh back-wrapped around the edges for a minimum distance of 2.5 in. The finish coat is also removed from the adjacent areas in strips several inches wide. An additional strip of mesh is applied to bridge the interface between the repair and the adjacent base coat (an overlap of at least 4 in is suggested), with additional pieces of mesh placed at the corners of any openings. The new top coat and finish should terminate at the joints without turning the corner, so that the sealants are applied to the primed base coat, not to the finish coat. Low-modulus silicone sealants with 50 percent extension and compression capacity tend to work best with EIFS. For panelized applications, joints with dual sealants separated by an air space should be used (Fig. 13.31).

13.10 Rehabilitating Other Wall Types

13.10.1 Precast concrete panels

Precast concrete panels are often selected for applications requiring high fire resistance and protection against impact and sound. Precast panels can be of either loadbearing or curtain wall construction. The former is popular in metal building systems, the latter in parking garages, among other uses. In the past, even precast double-tee roof panels have been used for tall curtain walls. Precast panels can be reinforced with deformed bars and welded wire fabric or can be prestressed. The bars can be epoxy-coated or galvanized to protect against corrosion. To enliven the otherwise gray monotony, architects often specify deep reveals, three-dimensional panel configurations, and exposed-aggregate finishes. For guidance on the design of precast wall panels, the reader is referred to ACI 533R,[66] *Guide for Precast Concrete Wall Panels,* and ACI 318[67] by the American Concrete Institute, and to *PCI Design Handbook.*[68]

Precast concrete has a low thermal insulation value, and insulated panels consisting of a thick structural part, a thin exterior course, and rigid insulation sandwiched in between have been developed. A

12-in-thick insulated panel might consist of an 8-in structural part, 2.5 in of expanded polystyrene, and 1.5 in of exterior exposed-aggregate concrete. These panels are popular in food processing plants, where strict cleanliness requirements call for hard and smooth interior finishes.

One recurrent issue facing those who investigate distress of precast concrete panels is a lack of design information in the contract documents. It is a rare case in which every panel is designed and detailed by the architect or the engineer of record. More commonly, panel design is performed by the panel suppliers, and their shop drawings share the fate of other similar submittals—they are eventually lost. In any case, structural design of precast elements is frequently controlled by handling, transportation, and thermal stresses, and the actual strength of the panels may have little to do with their resistance to service loads.

Typically, precast concrete panels have a 28-day compressive strength of 5000 psi or higher, and their inherent durability is rarely of concern. The problems usually lie in incorrectly designed or installed supports, in damaging movement of the building structure, and in thermal bowing. Also, connections made by field welding clip angles to embedded plates are susceptible to rusting and degradation. Fortunately, these are betrayed by stains and are relatively easy to spot. Occasionally, problems arise when the panels are infilled with other materials, such as brick, and differential movement between the two materials is not accommodated.

Like any other curtain wall system, precast concrete requires periodic maintenance, and some products need it more than others. Nicastro[69] mentions a building in Chicago that was sheathed with precast concrete blocks simulating limestone. The porous material required periodic reapplication of a waterproof coating. However, the owner neglected the chore, and the precast blocks disintegrated as a result of freeze-thaw cycles. The entire façade had to be rebuilt.

13.10.2 Glass-fiber-reinforced concrete

Glass-fiber-reinforced concrete (GFRC) was introduced in the United States in the 1970s as a lightweight alternative to precast concrete and stone. GFRC is made of short glass fibers mixed into a slurry of sand and cement and sprayed on molds. The fibers provide an incredible difference in ductility, even though they constitute only about 5 percent of the weight. The main advantage of GFRC is the multitude of possible panel configurations, as molds of every conceivable shape can be made. Also, unlike precast concrete, which requires minimum corner radii dictated by the size of the coarse aggregate, GFRC panels are produced with very sharp details. The appearance of the panels is sometimes enhanced by a sandblasted finish or by a facing mix of decorative aggregate $\frac{1}{8}$ to $\frac{1}{2}$ in thick. GFRC is expensive, and at this time its use is confined to cornices, parapets, fascias, column covers, and similar high-visibility elements.

The thickness of the GFRC skin is only about $\frac{5}{8}$ in, and it requires structural backup, typically made of light-gage steel framing. The connections between the skin and the backup consist of rigid gravity anchors designed to support the weight of the skin, and flexible ("wind") anchors designed only for lateral loads. Both types of anchors are embedded in thickened panel pads attached to the back of the skin during its fabrication. Depending on the panel shape, the combined weight of GFRC and its structural backup typically ranges between 8 and 20 lb ft².

GFRC panels have had more than their share of problems. Their success or failure greatly depends on the qualifications of the panel fabricator and erector. A common problem involves delamination of the thickened panel pads from the back of the skin because of transportation damage. The pads can usually be reglued with epoxy adhesives. Another problem is panel cracking caused by mishandling during fabrication and installation. Unfortunately, this cracking may not become apparent until long after the panels have been erected, because GFRC tends to lose strength with age. According to PCI,[70] the ultimate flexural strength of GFRC can degrade as much as 50 percent over time. In one published case,[69] badly warped GFRC panels that were forcibly straightened in the field developed numerous cracks within a year after installation.

The best way to deal with these problems is, of course, to conduct a thorough inspection *before* the panels are erected, rejecting those that are discolored, cracked, or have delaminated anchor pads, although, as just noted, some latent defects will probably slip by. In existing buildings, the repairs are straightforward in principle, although difficult in practice. Cracks in panels may be filled with epoxy adhesives, if this is acceptable visually. Delaminated anchor pads can also be reglued.

If deterioration is widespread, or if the problems are traced to poor design and construction details, the best course of action is panel replacement. A classic example occurred in a 21-story office building in Texas, where many GFRC panels had cracked and the finish of other panels had become damaged. A field investigation found several serious underlying problems. In one area, cracking was attributed to the "wind" panel anchors being too rigid and restricting movement of the panels. In other panels, cracking was caused by the face mix of simple plain concrete. As it happens, the coefficients of thermal and moisture expansion of these two kindred materials are different. GFRC contains at least three times the amount of cement found in concrete and therefore tends to shrink much more than the face mix, which acts as a restraint. The resulting stresses from differential movements between the face mix and GFRC were blamed for numerous delaminations at their interface and cracking of the face mix. In addition to all these problems, there were a number of other deficiencies in design and workmanship, such as many missing anchors and most of the rest having less weld that was required. What was the remedy? The GFRC panels were replaced with a granite and glass cladding system.[45]

In another case, GFRC cladding of a 375,000-ft² building in Seattle was replaced with an aluminum system because of widespread cracking. The original GFRC panels were cast against ceramic tiles, which became the embedded facing material. Unfortunately, the aging and thermal and moisture expansion characteristics of the ceramic tile were incompatible with those of GFRC. As just noted, GFRC, being a cementitious material, shrinks with age; ceramic tile does not. The composite panel, which would naturally bow as a result but was prevented from doing so by the steel-stud backup, developed through-wall cracking.[71] The lesson from these two cases? GFRC panels should not contain finishes made of other incompatible materials.

GFRC has not been used long enough to reveal all its limitations, and procedures for its proper design, fabrication, and installation have been developed only recently. Among good sources of information is PCI's *Recommended Practice for Glass Fiber Reinforced Concrete Panels.*[70] To ensure proper controls during panel fabrication, the manufacturer's operations should comply with PCI's *Manual for Quality Control.*[72]

13.10.3 Wood siding

Buildings clad in wood siding or shingles are ubiquitous. Equally common is the sight of broken, missing, and decayed shingles (Fig. 13.34) and other wood components damaged by rot or insects. The most common areas of distress are the bottom courses of siding, where water is likely to be splashed and termites are likely to enter; exposed windowsills; gutter boards; and sheathing around penetrations. Damage to these nonstructural items may indicate deterioration of hidden structural members. These areas should not be missed during field investigation of wood cladding, described in Chap 2.

Some building exteriors are clad in wood materials that have poor performance records; some are even clad in materials intended for interior applications. One of the most troublesome is the so-called composition siding, also known as hardboard. This material, essentially an amalgam of wood particles and glue, was widely marketed in the mid-1970s as a substitute for wood siding. Since then, a multitude

Figure 13.34 Broken, missing, and rotted shingles—a familiar sight in wood buildings.

of failures have occurred. Most involved disintegration and rotting of the boards, in many cases even under properly applied paint. The only certain cure for this is to replace the boards.

As discussed in Chap. 8, wood rotting occurs in a damp environment brought about by water leakage and inability of the wood to dry out quickly. Water leakage through wood siding often occurs when water is allowed to run along the wall and enter under the flashing and through poorly sealed areas. Leaking walls are often betrayed by peeling paint. As Lotz[17] points out, the need to provide proper flashing in wood buildings is not universally understood, and many buildings with moisture problems do not have flashing. Another typical avenue for water entry is through eave and ridge vents.

However moisture gets into the wall, problems arise only when the wood siding cannot quickly dry out, especially when the air humidity is high or when the wood is enclosed by other materials. Despite the reassuring claims of aluminum- and vinyl-siding salespeople, rotting is widespread in wood shingles covered with these materials. This hidden deterioration presents two major problems. First, rotting can spread to the interior wood structure and compromise its integrity. Second, the new skin, which is typically attached only to the old siding, loses its anchorage and becomes vulnerable to being blown off in a hurricane.

13.10.4   Glass

Glass storefronts with problems are difficult to miss. Broken glass and windows that fall out tend to be noticed. The problems with cracking and spalling glass in the John Hancock Building, a Boston landmark, first arose in the early 1970s and became widespread by the late 1970s. An investigation by Simpson Gumpertz & Heger Inc. concluded that the cause of the distress was a product-related problem: The spacer separating the large tinted panes of glass was unsuitable for the application. Eventually, many of the glass panels were replaced. Bostonians still recall the sorry sight of the beautiful skyscraper covered with temporary plywood patches while investigation and repairs were progressing.

According to Nicastro,[69] most glass problems can be traced to improper installation. For example, in one case, the glass installer used shot-in nails in lieu of the specified expansion anchors to attach window frames to the surrounding concrete. The relatively weak nails were subsequently pulled out of the concrete by strong winds. Product defects can also be to blame. A patch of nickel sulfide inadvertently included in a tempered glass pane during glass manufacture will lead to a stress concentration and possible breakage. Schwartz[73] discusses the causes of glass fractures and examines the significance of various crack characteristics. Another example of material failure is a poor-quality sealant that fails to do its job.

As for window leaks, Nicastro observes that many of them are traced to such prosaic causes as gaskets that are too short or butt joints between frame extrusions that are loose. He is alarmed by an observation that many window components are simply left out during construction. Finding the cause of leakage may not be easy, although the most basic and common sources should be tried first. Many windows leak simply because their weep holes are plugged by paint or dirt, and water accumulating at the bottom of the window finds its way inside. Cleaning and opening up the weep holes may restore drainage. Window frames frequently leak at their corners.

Beyond these relatively easy fixes, the causes of leaks can be quite complex. Some of them are described by Louis,[74] who also discusses the appropriate methods for testing and investigation of glass. Those challenges are typically handled by specialized consultants rather than by generalist architects and structural engineers.

References

1. Clayford T. Grimm, "Preventing Fatal Façade Failures," *Civil Engineering,* March 1999, p. 96.
2. "Stemming Chicago Façade Crisis," *ENR,* March 10, 1997, p. 8.
3. "New York City Toughens Inspection Requirements," *CE News,* October 1998, p. 21.
4. Thomas A. Schwartz, "Constructive Advice: Boston's Façade Inspection Ordinance," *ArchitectureBoston,* vol. 2, 1998, pp. 41–43.
5. Gordon Wright, "Keeping Façades Safe," *Building Design and Construction,* March 1999, pp. 48–50.
6. "Not Just Back on Track—Forward with `Flash-Track,'" *ENR,* Jan. 27, 1997, p. 66.
7. David H. Nicastro, "Can Engineers Cut Curtain-Wall Failures?" *Civil Engineering,* November 1993, pp. 48, 49.
8. Alexander Newman, *Metal Building Systems: Design and Specifications,* McGraw-Hill, New York, 1997.
9. Glenn R. Bell, "Exterior Walls: Wind Effects, Structural Systems, and Durability," session 2 of *The Building Envelope: Solutions to Problems,* a national seminar series sponsored by Simpson Gumpertz & Heger Inc., 1996.
10. John Gregerson, "Old Reliable," *Building Design and Construction,* March 1997.
11. Neal Templin, "Leaks Cause Buckets of Woes," *The Wall Street Journal,* Nov. 12, 1997, p. B12.
12. Thomas A. Schwartz, Exterior Walls: Waterproofing," session 1 of *The Building Envelope: Solutions to Problems,* a national seminar series sponsored by Simpson Gumpertz & Heger Inc., 1996.
13. *Brick Masonry Cavity Walls: Introduction,* Technical Notes on Brick Construction No. 21, Rev., Brick Industry Association, Reston, Va., 1998.
14. *Cavity Walls: Design Guide for Taller Cavity Walls,* Masonry Advisory Council, Park Ridge, Ill., 1992.
15. Richard Keleher, "Rain Screen Principles in Practice," *Progressive Architecture,* August 1993, pp. 33–37..
16. Norbert V. Krogstad and Richard A. Weber, "Detecting Causes of Exterior-Wall Moisture Problems," *Masonry Construction,* December 1998, pp. 681–684.
17. William A. Lotz, "Creating Effective Barriers to Wind-Driven Rain," *Building Design and Construction,* September 1997, pp. 50–52.
18. Dean A. Rutila, "Innovations in Brick Veneer," *The Construction Specifier,* October 1998.
19. ASTM E 331, *Test Method for Water Penetration of Exterior Windows, Curtain Walls, and Doors by Uniform Static Air Pressure Difference,* American Society for Testing and Materials, Philadelphia.
20. Norbert V. Krogstad, "Tricky Brick," *Building Renovation,* Spring 1995, p. 46.
21. Werner H. Gumpertz and Glenn R. Bell, "Engineering Evaluation of Brick Veneer/Steel Stud Walls, Part 1—Flashing and Waterproofing and Part 2—Structural Design, Structural Behavior and Durability," *Proceedings, Third North American Masonry Conference,* Arlington, Tex., June 1985.

22. *Salvaged Brick,* Technical Notes on Brick Construction No. 15, Brick Institute of America, Reston, Va., 1983.
23. Joel C. Snodgrass and Glenn Boornazian, "Solutions to 7 Common Errors in Masonry Restoration," *Traditional Building,* July/August 1995, pp. 10–14.
24. *Moisture Resistance of Brick Masonry: Maintenance,* Technical Notes on Brick Construction No. 7F, Brick Institute of America, Reston, Va., 1987.
25. Christine Beall, "Tuckpointing Old Mortar Joints," *The Magazine of Masonry Construction,* June 1988, pp. 116–118.
26. Robert C. Mack and John P. Speweik, *Preservation Brief 2: Repointing Mortar Joints in Historic Masonry Buildings,* U.S. Department of the Interior, Washington, D.C., 1998.
27. *Differential Movement: Cause and Effect, Part I,* Technical Notes on Brick Construction No. 18, Brick Institute of America, Reston, Va., 1986.
28. Michael M. Monhait, "Masonry Wall Repair," *Plant Engineering,* May 28, 1981, pp. 229–231.
29. Jerry G. Stockbridge, "Finding the Best Way to Repair Mortar Joints," *The Magazine of Masonry Construction,* June 1988, pp. 112–115.
30. Stephen S. Szoke and Hugh C. MacDonald, "Combining Masonry and Brick," *Architecture,* January 1989, pp. 103–106.
31. *Brick Veneer/Steel Stud Panel Walls,* Technical Notes on Brick Construction No. 28B, Rev. II, Brick Institute of America, Reston, Va., 1987.
32. Stephen S. Szoke and Gerald J. Carrier, "Avoiding Cracks in Brickwork," *The Construction Specifier,* August 1986, pp. 44–56.
33. "No Longer Red in the Face," *ENR,* May 26, 1997, p. 33.
34. Clayford T. Grimm, "Brick Veneer: A Second Opinion," *The Construction Specifier,* April 1984, pp. 6, 7.
35. *Here Are the Facts about Steel Framing–Brick Veneer Systems Design,* Metal Lath/Steel Framing Association, Chicago, undated.
36. ACI 530/ASCE 5/TMS 402-98, *Building Code Requirements for Masonry Structures,* American Society of Civil Engineers, New York, 1999.
37. Michael Bordenaro, "New Solutions to Masonry Moisture Problems," *Building Design and Construction,* October 1992, pp. 62—66.
38. Bruce A. Suprenant, "Efflorescence: Minimizing Unsightly Staining," *Concrete Construction,* March 1992, p. 243.
39. Thomas Fisher, "Commodore Perry Apartments," *Building Renovation,* September–October 1993, pp. 30–35.
40. David Biggs and Elizabeth Keating, "Synthetic Patching Saves Historic Brownstone," *Masonry Construction,* November 1998.
41. Bill Schwartz, "Restoring Delaminated Brownstone," *Masonry Construction,* May 1995, pp. 226–230.
42. Alex S. Gere, "Repair of Exterior Stone on High Rise Buildings: Guidelines and Procedures," *Building Stone Magazine,* November/December 1989, pp. 37–40.
43. Thomas Fisher, "Center of Attention," *Building Renovation,* Summer 1995, p. 60.
44. Hugh Cook, "The Keys to Historic Masonry Restoration," *Building Design and Construction,* February 1997, pp. 42–46.
45. Gordon Wright, "Inappropriate Details Spawn Cladding Problems," *Building Design and Construction,* January 1996, pp. 56–60.
46. "Marble Panels Stripped from K.C. Building," *ENR,* March 31, 1988, p. 19.
47. "Amoco Tower Retrofit Underscores Problems with Marble Veneers," *Building Design and Construction,* June 1989, p. 18.
48. Glenn R. Bell, "Masonry: Brick, Stone, and Concrete Block," session 5 in *The Building Envelope: Solutions to Problems,* a national seminar series sponsored by Simpson Gumpertz & Heger Inc., 1996.
49. James H. Larkin, "Securing the Stone," *Civil Engineering,* January 1998.
50. James C. Myers, "Lessons Learned from Damaging Interactions between Masonry Façades and Building Structures," *Proceedings, 5th International Masonry Conference,* October 1998.
51. "Foam Eliminates 'Sticky' Problem," *Building Design and Construction,* May 1997, p. 9.

52. Carolyn Schierhorn, "Removing Paint from Masonry,"*Masonry Construction,* October, 1994, pp. 461–465.

53. *EIMA Guideline Specification for Exterior Insulation and Finish Systems, Class PB,* and other publications, EIFS Industry Members Association, Clearwater, Fla., 1994.

54. *Exterior Insulation and Finish System, Class PB Details,* EIFS Industry Members Association, Clearwater, Fla., 1994.

55. Mark F. Williams and Barbara Lamp Williams, *Exterior Insulation and Finish Systems: Current Practices and Future Considerations,* ASTM Manual 16, American Society for Testing and Materials, Philadelphia, 1994.

56. Mark F. Williams and Richard G. Lampo, eds, *Development, Use, and Performance of Exterior Insulation and Finish Systems (EIFS),* STP 1187, American Society for Testing and Materials, Philadelphia, 1995.

57. "EIFS Deemed Too Big a Financial Risk," *Masonry Construction,* December 1996, p. 565.

58. June Fletcher, "When Dream Products Turn into Nightmares," *The Wall Street Journal,* July 24, 1998, p. W8.

59. John A Kilpatrick et al., "The Performance of Exterior Insulation Finish Systems and Property Value," *The Appraisal Journal,* January 1999,

60. "Exterior Insulation and Finish Systems," in *Natural Hazard Mitigation Insights,* vol. 7, Insurance Institute for Property Loss Reduction, Boston, February 1997.

61. Richard Piper and Russell Kenney, "EIFS Performance Review," *Journal of Light Construction,* 1992.

62. Michael Bordenaro, "Avoiding EIFS Application Pitfalls," *Building Design and Construction,* April 1993, pp. 67–70.

63. Margaret Doyle, "Trends in Specifying EIFS," *Building Design and Construction,* August 1988, pp. 59–62.

64. Peter E. Nelson, "Exterior Insulation and Finish Systems," session 8 in *The Building Envelope: Solutions to Problems,* a national seminar series sponsored by Simpson Gumpertz & Heger Inc., 1996.

65. Christine Beall, "Repairing Synthetic Stucco Joints," *Architecture,* November 1993, pp. 117–119.

66. ACI 533R-93, *Guide for Precast Concrete Wall Panels,* American Concrete Institute, Detroit, Mich.

67. ACI 318, *Building Code Requirements for Reinforced Concrete,* American Concrete Institute, Detroit, Mich.

68. *PCI Design Handbook, Precast and Prestressed Concrete,* 4th ed., Precast Prestressed Concrete Institute, Chicago, 1985.

69. David Nicastro, "Uncovering the Reasons for Curtainwall Failure," *Exteriors,* Summer 1988.

70. MNL-128, *Recommended Practice for Glass Fiber Reinforced Concrete Panels,* Precast/Prestressed Concrete Institute, Chicago, 1994.

71. Tom Harrison, "Saving Face," *Civil Engineering,* May 1997, pp. 44–47.

72. MNL-130, *Manual for Quality Control for Plants and Production of Glass Fiber Reinforced Concrete Products,* Precast/Prestressed Concrete Institute, Chicago, 1991.

73. Thomas A. Schwartz, "A Pain in Your Pane?" *Building Renovation,* July–August 1993, pp. 53–55.

74. Michael J. Louis, "Weepy Windows," *Building Renovation,* Fall 1995, pp. 41–43.

Index

About the Author

Alexander Newman, P.E., is principal structural engineer with Maguire Group, Inc., a national architectural, engineering, and planning firm, in Foxborough, Massachusetts. With two decades of engineering and management experience, he has worked as a project engineer with a consulting engineering firm, design engineer with a light-gage framing panel manufacturer, and manager of fabrication for a steel fabricator. He has planned and supervised structural renovation of numerous buildings throughout the country, including a Boston Edison switching and conversion station that won the 1990 American Consulting Engineering Council of New England Award for Engineering Excellence. Mr. Newman holds an advanced degree in structural engineering from the Moscow Civil Engineering Institute in Russia, and a master's degree in business administration with high honors from Boston University. He is the author of the Bestselling *Metal Building Systems*, also from McGraw-Hill, and a number of award-winning articles that have appeared in leading engineering publications. In addition, he conducts continuing education seminars on metal building systems for design professionals sponsored by the American Society of Civil Engineers and other organizations, and teaches at Northeastern University.